D0764339

# FRANCIS CRICK

. . . . . . . . . . . . . . . . . . . . . . . . . . . . . .

## Hunter of Life's Secrets

# FRANCIS CRICK

## Hunter of Life's Secrets

1916—

### *Robert Olby*

*University of Pittsburgh*

COLD SPRING HARBOR LABORATORY PRESS
Cold Spring Harbor, New York • www.cshlpress.com

**FRANCIS CRICK** Hunter of Life's Secrets

© 2009 by Cold Spring Harbor Laboratory Press, Cold Spring Harbor, New York
All rights reserved
Printed in the United States of America

| | |
|---|---|
| **Publisher & Acquisitions Editor** | John Inglis |
| **Development Director** | Jan Argentine |
| **Developmental Editor** | Judy Cuddihy |
| **Project Coordinator** | Joan Ebert |
| **Permissions Administrator** | Carol Brown |
| **Production Manager** | Denise Weiss |
| **Production Editor** | Mala Mazzullo |
| **Desktop Editor** | Lauren Heller |
| **Interior Book Designer** | Denise Weiss |
| **Cover Designer** | Ed Atkeson |

*Front cover:* Crick as a physics student at University College London, c. 1937 (Courtesy of the Crick Family); *back cover:* Crick beside helical model (© Topham Picturepoint/The Image Works)

Library of Congress Cataloging-in-Publication Data

Olby, Robert C. (Robert Cecil)
   Francis Crick : hunter of life's secrets / by Robert Olby.
      p. cm.
   Includes bibliographical references and index.
   ISBN 978-0-87969-798-3 (hardcover : alk. paper)
   1. Crick, Francis, 1916-2004. 2. Geneticists--Biography. 3. Molecular biologists--Biography. I. Title.

   QH429.2.C75O43 2008
   572.8092--dc22
   [B]

                                                2009018695

10   9   8   7   6   5   4   3   2   1

All Cold Spring Harbor Laboratory Press publications may be ordered directly from Cold Spring Harbor Laboratory Press, 500 Sunnyside Boulevard, Woodbury, New York 11797-2924. Phone: 1-800-843-4388 in Continental U.S. and Canada. All other locations: (516) 422-4100. FAX: (516) 422-4097. E-mail: cshpress@cshl.edu. For a complete catalog of Cold Spring Harbor Laboratory Press publications, visit our World Wide Web Site http://www.cshlpress.com.

**Copyright Disclaimer**

# Contents

*See photo section between pages 220 and 221.*

# Time Line
## Francis Crick, 1916–2004

**Birth, 8 June 1916, Weston Favell, near Northampton, United Kingdom.**
Attends Northampton Grammar School, 1925–1930.
Scholar at Mill Hill School, 1930–1934.
Physics undergraduate, University College London, 1934.
Graduate student, 1937–1939.

**Civilian Scientist, Royal Navy, 1940–1947.**
Marries Ruth Doreen Dodd, 18 February 1940. Michael born,
25 November 1940.
Meets Georg Kreisel in Havant, 1943. Francis and Doreen divorced,
8 May 1947.
Resigns from the Admiralty, September 1947.

**Begins research in biology at the Strangeways Laboratory, Cambridge,
October 1947.**
His father dies, 28 January 1948.
Moves to the MRC Unit in the Cavendish Laboratory, Cambridge,
June 1949.
Marries Odile Speed and lives at The Green Door, 14 August 1949.
First research paper published, 1950.

**Becomes a member of Gonville and Caius College, Cambridge, and enters
the doctoral program, 1950.**
Gabrielle is born, 15 July 1951.
James Watson arrives at Cambridge, October 1951. First attempt with
Watson on the structure of DNA, November 1951.
First four theoretical papers on protein structure published, 1952.
Purchases 19 Portugal Place, Cambridge, 2 July 1952.

**Watson and Crick's second attempt on DNA, February 1953.**
Base-pairing discovered by Watson, 28 February 1953. Their paper
published in *Nature,* 25 April 1953.

Visit of Gerard Pomerat (Rockefeller Foundation) to the Cavendish, 1 April 1953. Visit of Linus Pauling, first week of April 1953.
Lectures to Zoology Department, 15 May 1953. First newspaper report on same day.
Takes oral exam for the PhD, 18 August 1953.

**Crick family leaves for New York on the Mauritania, 22 August 1953.**
Begins work at the Protein Structure Project, Brooklyn Polytechnic, directed by David Harker.
George Gamow visits Crick, December 1953.
Jacqueline born, 12 March 1954. Crick returns to the UK and the MRC Unit, Cambridge on seven-year appointment, 13 September 1954.
Alex Rich and Jim Watson come to Cambridge to work with Crick in 1955—Rich on polyglycine and collagen, Watson on virus structure.
His mother dies, 27 October 1955.

**Sydney Brenner comes from Johannesburg to join Crick at the MRC Unit, 2 January 1957.**
It is moved to "the Hut" and renamed "MRC Unit for Molecular Biology," 1957.
They are joined by visitors Seymour Benzer, Mahlon Hoagland, and others later that year.
Delivers his lecture on protein synthesis in London, in which he states the Sequence Hypothesis and Central Dogma, September 1957.

**At Harvard University as Visiting Professor, February–June 1959.**
Declines invitation to move to Harvard.
Is made a Fellow of the Royal Society.

**Crick and Brenner debate with François Jacob and Jacques Monod, and the concept of messenger RNA is born, 15 April 1960.**
Becomes Fellow of Churchill College in 1960 but resigns in 1961.
Nirenberg reports that poly(U) codes for Phe at the International Congress of Biochemistry, August 1961.
Is appointed a nonresident Fellow of the Salk Institute, December 1961.
Crick, Leslie Barnett, Brenner, and R.J. Watts-Tobin publish "General Nature of the Genetic Code for Proteins," 30 December 1961.

**The MRC Unit begins to move to its new building on the Addenbrooke's Hospital site, March 1962.**
Telegram from Stockholm arrives, 18 October 1962. Visit to Stockholm, December 1962. Lectures on the genetic code.

His brother Tony dies, 8 March 1966.
   Delivers Royal Society Croonian Lecture, 5 May 1966, and plenary
   lecture at the Cold Spring Harbor Symposium on the genetic code,
   June 1966.

**Account of the successful effort by Maurice Wilkins and Crick to prevent
publication of Watson's book by the Harvard University Press appears in
*The Harvard Crimson*, February 1968.**
   Watson's book, *The Double Helix*, published by Atheneum Press,
   February 1968.

**Gives the Rickman Godlee Lecture at University College London, on the
social impact of biology, October 1968.**
   First paper on embryogenesis published in 1970.
   His model of the chromosome published in *Nature*, 1971.

**Attends conference on extraterrestrial communication at Yerevan, Russia,
September 1971.**
   With Leslie Orgel publishes "Directed Panspermia" in 1973.

**Spends a sabbatical at the Salk Institute and at the University of Aarhus,
Denmark, 1976–1977.**
   Resigns from the MRC, February 1977, and accepts Kieckhefer
   Distinguished Professorship at the Salk Institute, 1977.
   First paper on neuroscience published by *Scientific American*, 1979.

**First meets his future young collaborator Christof Koch in Tübingen, 1981.**
   Their first paper together, "Towards a neurobiological theory of
   consciousness," published, 1990.

**Accepts Order of Merit from Queen Elizabeth II, 7 July 1991.**
   His book on neuroscience, *The Astonishing Hypothesis*, published, 1994.
   Becomes President of the Salk Institute in 1994 but resigns for
   health reasons in 1995.
   Receives sextuple bypass surgery, 9 November 1995.

**Suggests the claustrum as the key structure in producing consciousness.**
   Corrects draft of the paper, 27 July 2004. He dies on the evening of the 28th.
   Family and colleagues hold memorial, 3 August 2004. His ashes were
   scattered on the ocean.
   Public memorial at the Salk Institute held, 27 September 2004.

# Preface

*I'm very happy you got to Cambridge because now you are back in your right environment . . . I hope they let you stay there for always because I am sure there can't be a better place to live and work in than a great University like Cambridge providing of course one has the brains to cope with it and you have more than your share of those so don't let any wandering "lust" get you again because I like to think of you moving in that atmosphere which is most suited to one of your temperament.*

Edie Hammond, 1947

*I must be dreadfully romantic to be marrying an impecunious man like you!*

Odile Speed, 1949

*Oh Francis. I was terribly in love with you. But you really had too many advantages. You were my first lover and a very good one.*

Former Lover, 1956

*No one man discovered or created molecular biology. But one man dominates intellectually the whole field, because he knows the most and understands the most. Francis Crick.*

Jacques Monod, 1970

*He was very urbane, well educated. He reminded me most of Henry Higgins in My Fair Lady. There was quite a resemblance. That kind of style. No shrinking violet, that's for sure.*

Seymour Benzer, 1990

*I feel . . . that a great light, a beacon, has gone from the earth (although its power and prescience will be felt for centuries)—and in a more personal way a sense of being orphaned, for he was a father-figure to us all.*

Oliver Sacks, 2004

*When I first met Odile and Francis, in Tübingen, in 1981, the two seemed like an unlikely couple. Francis was loud, opinionated and hyper-rational, closely resembling the British actor Jeremy Brett, playing the brilliant but arrogant Sherlock Holmes with his mannerisms; Odile on the other hand, remained quietly in the background, representing the warm and fuzzy side of the Crick duo.*

Christof Koch, 2007

"Every great man," wrote Oscar Wilde, "has his disciples and it is always Judas who writes the biography."[1] Fortunately, Francis Crick has been so prominent a figure in 20th century science that we can expect him to have many biographers, not all of them betrayers of the master, nor necessarily disciples or hagiographers. Like many who knew him, I am an admirer, and have been from the time I first met him. Admirers, however, are not necessarily hagiographers.

In 1953 Francis Crick, aged 37, wrote up his "rag-bag" of a thesis, as he called his doctoral dissertation, and prepared to leave Cambridge and Great Britain. He was to spend a year at Brooklyn Polytechnic Institute, and after that? He hoped to return to Cambridge, but the future of his research unit there was not yet decided. Nine years later, and once again in Cambridge, he became a Nobel Laureate. In 1991, after Crick had received (and refused) many subsequent honors, Queen Elizabeth bestowed upon him membership of the Order of Merit.

This remarkable transformation later prompted Oliver Sacks to ponder Crick's life. How much "continuity," Sacks wondered would Crick see between the teenager at Mill Hill School who talked about the "Bohr atom" and the Periodic Table, his "Double Helix" self, and his present self?[2]

In Crick's case we have surviving documents from his early years that suggest a wayward spirit, somewhat flippant and frivolous, with a rather immature sense of humor, and not the world's best listener; in short, not yet the serious and committed person he was to become. Part of that change came in the war years when he worked for the Royal Navy and at war's end became a government scientist at the Admiralty. To flesh out this evolution of his personality we need a more detailed biographical study than is currently available. We also need to appreciate the changing landscape of the biological sciences during the period of Crick's greatest achievements. Great scientists have become great not only because of their qualities of intellect, leadership, experimental skill, anticipation of the future, and dedication to the science, but also because of the state of a science at a given time. Crick emerged from World War II just when new techniques in physical and analytical chemistry, X-ray crystallography, and electron microscopy were finding their way into the biological and medical sciences. This created the potential to better understand the nature of proteins and the nucleic acids. Crick's discovery with James Watson of structure of DNA in 1953 provided an intellectual structure in which to reformulate the relations between these two most important classes of molecules for life. As the fifties gave place to the sixties the immensity of that significance became recognized.

In 1966 when I first met Crick, I was a historian of science working in Oxford. Having published my doctoral dissertation I had been looking for a fresh topic to tackle, and had contacted a friend from my student days, John Prebble, in the Botany Department at Imperial College London. By this time, he was teaching biochemistry and was well aware of the importance of DNA. "Why," he asked, "don't you investigate how Watson and Crick discovered its structure?"

"DNA," I asked myself, "what is that?" For I could not recall any mention of DNA or RNA in our lecture courses. True, the biochemistry lecturer Dr. Russell had explained to us the exciting achievement of Fred Sanger in revealing the amino acid sequence of insulin, surely a discovery as central to molecular biology as the structure of DNA. Indeed, I can still recall the excitement with which Russell worked his way across the blackboard, chalk in hand, drawing squares to represent amino acids, each square linked to its neighbors like the cars of a freight train. Then with plant pathologist Ronald Wood, we had performed experiments on the tobacco mosaic virus, finding, for instance, how the contents of cigarette stubs can cause the disease in tobacco plants. Did we appreciate that the mosaic on the leaves of the host was due to a nucleic acid? Had the significance of this for heredity been appreciated? I cannot now recall. That was in 1953. When, 13 years later, I wrote to Russell for advice on the subject, he responded with enthusiasm, describing how DNA had been first obtained from the pus in discarded hospital bandages in Tübingen and in 1869 had been given the name "nuclein."[3] That was encouraging! But first I needed to establish what work was already under way on this subject.

A chance to find out came on the 17th of May 1966 when Crick was to deliver Oxford University's Cherwell Simon Lecture.[4] This was an opportunity not to miss. The large lecture hall of the University Museum was packed. In came the tall, slim lecturer. He took from his pocket a few sheets of paper with headline jottings on them. His subject, "The Influence of Physics on Molecular Biology," turned into a personal and absorbing account of the discovery of the structure of DNA. James Watson's book, *The Double Helix*, had yet to be published, so the story Crick proceeded to expound was news to all of us listening with rapt attention. His account was brim-full of humor and self-criticism— the famous encounter with Erwin Chargaff, the disaster of their first structure for DNA, and more—in hindsight, so many seemingly improbable events. Many times, Crick had to pause to let the laughter die down.

When I approached Crick after the lecture, he told me that Watson had written an account of the discovery, "a sort of pseudo-novel like, you know, Truman Capote's *In Cold Blood*."[5] Whether or not it would be published was not yet decided. Then there was a history of the origins of molecular biology and the phage group in the form of a soon-to-be-published *Festschrift* for Max Delbrück,[6] and that was it. When I heard from him again in November, he explained that he had "been having arguments with Jim Watson about the desirability of publishing his book," and for this reason had postponed writing to me.[7]

Despite the continuing uncertainty over Watson's manuscript that winter, Crick agreed to help me set to work on *The Path to the Double Helix*. Little did I know then that my proposed project would later be drawn into the dispute between Watson and Crick over the publication of Watson's account of their discovery. Crick's advice to Watson, I later learnt, was that he should scrap the book and "either write a proper history," or put the manuscript aside pending the death of any who object. Meanwhile, he suggested, Watson might make his manuscript "available to selected scholars," and he told him that I was contemplating such an account. Fortunately Watson stood his ground![8]

Choosing, rather, to produce a broad picture of the years leading up to and including the discovery of the structure of DNA in 1953, I labored long to produce *The Path to the Double Helix*, published in 1974.[9]

While I was working on this book, Crick was invited to take part in a conference on "Lives in Science" directed by the Harvard physicist, Gerald Holton, in September 1969. Unable to accept, Crick suggested I go in his stead. It was held at the Rockefeller Foundation's 50-acre estate, Villa Serbelloni—the fabulous Bellagio Study and Conference Center, set on a hill at the divide between Lake Como's two southern arms and surrounded by formal gardens. That meant preparing a biographical essay on Crick for the conference. To my surprise, now, I find that this essay ran to 16,000 words, thus making it the first, albeit brief, biographical essay on my subject![10]

To make this essay into a full biography, I judged, could prove to be a millstone around my neck, and in any case Crick did not want such a book published in his lifetime. Impending retirement from teaching in 1990 brought thoughts of a change, and Crick agreed to my working on his biography, again provided publication would not occur during his life. Owing to teaching commitments until the spring of 2001, it was not until later that year that I began to write this book. Recently I was surprised to learn that it has been assumed in some circles that I had been working on it ever since the 1970s.

I introduce my subject hosting the party held to celebrate the news of his Nobel Prize award and describe what it was like to be at the ceremony. Then Crick's response to the eventual publication of Watson's *The Double Helix* is discussed.[11] These sections address aspects of Crick's life with which many readers are already familiar. My account then follows a chronological path from early life in Northampton, on to high school and university, Crick's experience working as a scientist for the Royal Navy during the war years, the turn to biophysics in the postwar period, the heroic decades of molecular biology in the 1950s and 1960s, the move to the United States in 1976, and his adventures in neuroscience at the Salk Institute in California.

Interspersed within these accounts are sections given to his controversial political and social views, his style as a scientist, the year he spent working in Brooklyn, and 22 years later his emigration to America. Concerning his evolution as a research scientist, I have stressed the importance of his war work, particularly the influence upon him of the colleagues with whom he then worked. Although the years 1947–1949 spent at the Strangeways Laboratory are usually dismissed as of little importance, Crick encountered there the concerns of biologists, the goals they sought, and the means they were using to attain them. It was then that he began to formulate his ideas on protein structure. I seek to show that during his first 3 years at the Cavendish Laboratory (1949–1952), he developed an accurate appreciation of the problems of protein structure determination and, moreover, gained a deep understanding of the theory of X-ray crystallography and the chemistry of proteins.

The 1950s and 1960s were the golden years of the fledgling subject of molecular biology. Those were the days of a beautifully simple picture of the molecular gene, of the power of the genetic code containing the organism's blueprint, and how its message is expressed in protein synthesis. This picture was crafted with rhetorical eloquence by Crick in 1957. Since that time, complexity upon complexity has been heaped upon that simple picture as researchers turned from bacterial viruses to higher organisms. Crick's major effort in this direction was his attempt to understand how chromosome structure may offer clues to the control of gene expression. Although his own ideas here proved to be wrong, it was his effort that led Roger Kornberg and others to develop this field with great success. Meanwhile, others discovered molecular mechanisms for the manipulation of genetic material, leading to applications in medicine, agriculture, and forensic science. Thus, the research of the 1950s and 1960s with which Crick's name is so closely associated came to have more and more salience. Today, with the availability of the genomes of so many species, including *H. sapiens*, "going molecular" is having repercussions in all the biological sciences and outside them in archeology and genealogy. The second half of the twentieth century thus witnessed a revolution felt throughout biology and beyond, the first half having seen the revolution in physics.

This book, then, is an attempt to create a dynamic picture of the remarkable evolution of Crick's career and his role in shaping the new foundations for biology. With his strength of character, forward vision, and dedication, he stepped naturally into this role, his passion for the work fed by hopes of a world committed to scientific humanism instead of religion. He became a crusader, an icon, in whose presence one felt awe at the sparkle of his intellect and the breadth of his knowledge. He was unique and he lived to exert a powerful influence upon the biological sciences at a peculiarly opportune time.

I have sought to understand why Crick was so passionate, and at times, vehement, about the new knowledge of molecular biology. Therefore, I have described the opposition he encountered from certain X-ray crystallographers (in Chapter 11) and virologists and some, but by no means all, biochemists (in Chapter 13). He also saw that this new biology had important implications for society, and assumed the role of preacher and evangelist. His last book closes with "Dr. Crick's Sunday Morning Service," and he described one of his lectures to the Cambridge Humanists as: "A Sermon with book [Monod's *Chance and Necessity*] for a text." Reflecting on how science has already transformed our understanding of our place in nature, Crick repeatedly expressed his conviction of the impact that scientists' understanding of the brain will have on our culture. When that comes, he predicted: "man's whole view of himself will be radically changed" (Chapter 16). This hope kept him working on the brain until his last hours.

Today if you ask members of the general public if they know the name Francis Crick, very likely their answer will be No. Mention the name Jim Watson and DNA, and many would say Yes. With the increasing attention being given to DNA science by the public media today, and the inclusion of Crick in Harper Collins *Eminent Lives* series,[12] Crick's life and achievements may become much better known. I hope *Francis Crick: Hunter of Life's Secrets* will make an additional contribution.

ROBERT OLBY

*19 June 2009*
*University of Pittsburgh*

# Acknowledgments

I owe a great debt to the late Francis and Odile Crick for their encouragement, patience, generosity, and hospitality over the years during which the research for this book was under way. Francis arranged for a copy of his scientific papers to be deposited in the Mandeville Special Collections at the University of California at San Diego for the benefit of American scholars. He also donated his personal papers to the same institution. He introduced me to many of his friends and colleagues and informed me about his early life and his ancestors. He read and commented on versions of the first 14 chapters, and we had many discussions about the subject matter of the later chapters. My understanding of a number of scientific issues, has benefited from his remarkable powers of exposition. Odile shared with me her memories of the early years of her life with Francis. Her brother Philippe Speed and his wife Marguerite informed me about Odile's upbringing in the Speed family and sent me photos of her from those years. I have benefited from meeting and corresponding with Francis' son Michael and his daughters Gabrielle and Jacqueline. I thank the family for allowing me to quote from the scientific and personal papers of their father. I thank Michael for permission to publish an abstract of the full family tree, which he had updated.

On the personal side of this biography, I thank Clive Binfield for sharing with me his work on Nonconformist church history in Northampton and Roderick Braithwaite for providing extracts from his history of Mill Hill School before its publication. I thank, also, four former personal assistants to Crick and Sydney Brenner in Cambridge—Alison Auld, Pauline Finbow, Sue Foakes (née Barnes), and Jennie Maskell—for their delightful and touching recollections. Special thanks go to Kathleen Murray, who served as Crick's administrative assistant for the last eight years of his life. She facilitated my visits to La Jolla and went out of her way to assist me. To the many friends and colleagues of Crick who granted me interviews or communicated by e-mail, telephone, or letter I offer many thanks. Their names are listed separately in the notes section, but I should mention here the special help of Freddie Gutfreund and Mark Bretscher.

The archives of Crick's scientific and personal papers have been absolutely essential to this project. I thank Lynda Claassen, Head of the Mandeville Special Collections at UCSD, for giving me access to these and permitting me to quote from them. Special thanks go to Mandeville archivist, Steve Coy, for his unfailing help, and to Ellen Kennedy, who catalogued the Crick family personal papers. Also, thanks are due to Chris Beckett, Helen Wakely, and Archives Head, Lesley Hall, for their help and permission to cite from the scientific papers at the Wellcome Library; to Colin Harris at the Bodleian Library, Oxford, for the John Kendrew Papers; to Cliff Mead, Head of Special Collections at the Valley Library, Oregon State University, and his colleague Chris Peterson, for the Ava Helen and Linus Pauling Papers; to Allen Packwood, Director of the Churchill Archives Centre, for the papers of Rosalind Franklin, Sir John Randall, Sir Winston Churchill, and O.V. Hill; to Ludmila Pollock, Director of Libraries and Archives at Cold Spring Harbor Laboratory, for the Sydney Brenner and James Watson papers; to Martin Levitt and Charles Greifenstein for the Felix Haurowitz papers; to Judith Goodstein for the Delbrück papers and the oral history of Seymour Benzer; and to Maggie Kimball, Stanford University Archivist, for permission to quote from the Kreisel papers. I thank Jeremy Norman who gave me access to the Jeremy Norman Molecular Biology Archive at the time it was in his possession at his house in California (now deposited at the Venter Institute in Rockville, Maryland). The assistance from staff at the local history sections of the Public Libraries of Cambridge, Northampton, Ely, and Havant in the United Kingdom is also much appreciated. Every effort has been made to contact the copyright holders of figures and photos in this text. Any copyright holders that we have been unable to reach or for whom inaccurate information has been provided are invited to contact me or the Cold Spring Harbor Laboratory Press.

My friend Martin Packer in Birmingham, U.K., has been both a loyal and stimulating correspondent who has been generous with his assistance. He suggested that a call for recollections from those who knew Dr. Crick be made in the Caius College newsletter, the *Caian Alumni*. This brought responses going back to the prewar years. Particularly valuable was the response from Crick's high school friend, John Shilston. I thank Martin for his enthusiasm and demands for accuracy—he pointed out many errors both large and small—and his skill with using the resources of the Web. I have learned much from him. Readers who consult the notes to the chapters will also find that many of Crick's friends and former colleagues granted me interviews. A full list will be made available on my website. I am grateful to them all for their valuable contributions. Thanks go also to those who generously supplied photos of Crick. Their names are given in the Illustration Credits section.

I acknowledge with much pleasure the advice given me in the early stages of the book by my friend, science writer Geoffrey Montgomery, and by the Banbury Center Director, Jan Witkowski, who first expressed the interest of Cold Spring Harbor Laboratory in my manuscript. I also thank my Pittsburgh friends and colleagues, especially Stanley Shostak, Peter Machamer, and James Bogen, for their advice and criticisms. During the work on this project I have greatly appreciated the pleasures and benefits of a faculty position at the University of Pittsburgh. For research assistance I thank Suzanne Hepburn, Stacy McAuliffe, and former graduate student James Tabery. Recent discussions with fellow 1950s undergraduate Martin Cole helped in recalling events half a century ago at Imperial College London.

My visit to the Wellcome Library in 2001 was assisted by a Burroughs Wellcome Travel Grant. Further visits to the United Kingdom were supported by a National Science Foundation award, including my semester as an Archive By-Fellow at Churchill College Cambridge in 2004. I thank the Archives Committee at Cold Spring Harbor for the receipt of the Sydney Brenner Research Scholarship for 2007 to support the final stages in the preparation of this manuscript for publication. I appreciate also the work of the staff of Cold Spring Harbor Laboratory Press: especially Publisher, John Inglis; Project Coordinator, Joan Ebert; Editor, Judy Cuddihy; Permissions Administrator, Carol Brown; and Production staff, Denise Weiss, Mala Mazzullo, Rena Steuer, Kathleen Bubbeo, and Lauren Heller.

Last, but not least, I thank my partner, Ann Ruth, for her love, support, and patience while I struggled to bring this book to completion. As Michael Crick remarked recently, "Glad to hear the glacier is moving into the sea!"

Robert Olby

*19 June 2009*

*University of Pittsburgh*

**Note to the reader**

British readers will note the use of American terms for British ones (e.g., first floor for ground floor, doctoral dissertation for PhD thesis, high school for comprehensive school). All readers will note instances where a term is spelled in both the American and the British ways. My rule has been to use the former except where quoting from a British source—thus hemoglobin (American) and haemoglobin (British).

# The Call to Stockholm

*Crick will need a Nobel Prize which, rumour has it,
he will one day get for his helices. . .*

*We now await the Nobel Prize and Royal Medal.*

*Now for the Nobel Prize![1]*

One October night in 1962, the Cambridge police received a call about a disturbance at one of those tall town houses[2] toward the end of Portugal Place, a little backwater off of Bridge Street and close to St. John's College. Elsewhere, the houses are more modest, and the numerous bikes left on the sidewalk betrayed their role as student accommodation. The name "Portugal" refers to the precious cargoes of port and sherry that in former times arrived on barges and were destined to be consumed by the College Fellows at High Table. Even in 1962, this area of town supported a mixed population of "town and gown." Nearby were a pawn shop and a second-hand clothes shop, the former once patronized by Francis and Odile Crick in time of need. Normally, it was a peaceful part of town, apart, that is, from the annual Fifth of November festivities, when fireworks and bonfires celebrate the capture of Guy Fawkes in 1603 as he prepared to blow up the Houses of Parliament.

It was 18 October, however, and there was no peace on Portugal Place. Mrs. Bessicovitch's dogs at No. 21 had become agitated at the sound of exploding fireworks raining down on the street below. Figures silhouetted against the night sky on the roof of No. 19 were clearly to blame. Clinging to the chimney pots, they were lighting firecrackers and throwing them down to the street. Inside, a lively party was in full swing.[3] A policeman was sent to investigate. As he reached the front door, he might have noticed a metal helix that was painted yellow, attached to the wall above him. If he did, its significance was not obvious to him. Knocking loudly to make his presence felt above the noise within, he was greeted by a tall debonair man with a slight stoop, lively blue eyes, sandy hair, and his dress a touch flamboyant. A friend described him as "a lambent troll of a man."[4]

With the politeness that befits a gentleman, the owner apologized, explaining that he had not been aware of the activity on the roof, this being a big party crammed into many small rooms on five levels. Odile and assistants were hard at work in the kitchen and dining room in the basement or running up and down one or other of the two flights of steep stairs, carrying the refreshments needed to sustain the guests on the many floors. Bottles of champagne were much in evidence, and everyone was having a good time—so much so that when Dr. James Watson put through a trans-Atlantic call to Crick, the latter had later to apologize that he "was incoherent, but there was so much noise," he explained, "I could hardly hear what you said."[5]

The host assured the policeman that the fireworks would stop at once. Then, with a mischievous twinkle in his eye and smile on his face, he asked "Do you know what we are celebrating?" "No," came the answer. "Then you can read about it in the papers tomorrow. But why don't you come in and have a drink?" The officer graciously accepted and, politely taking off his helmet, enjoyed a drink before quietly leaving. The problem had been dealt with. When diplomacy and charm were needed, both the host and the officer could provide it.

Readers of the *Cambridge Evening News* the next day (Friday) found the heading *Nobel Theory was Discussed in "Pub."* The pub was The Eagle, on Bene't Street. The Cambridge scientist thus honored was Dr. Francis Crick and his address was given as "The Golden Helix,"[6] 19–20 Portugal Place. That Thursday, he had learned that he was to share the Nobel Prize in Physiology or Medicine with the young American biologist Dr. James Watson and the New Zealand–born biophysicist Dr. Maurice Wilkins. Their award was for their 1953 discovery of the molecular structure of deoxyribonucleic acid or DNA and, in Crick's case, also for his contributions toward the elucidation of the chemical code in which our genes are written—the genetic code.

By the time the telegram arrived from Stockholm at Crick's office on Thursday, the shops were about to close, for it was early closing day in the town. Fortunately, in the Medical Research Council's (MRC) Laboratory of Molecular Biology (LMB) where Crick worked, the resourceful lab steward Mike Fuller was at hand.[7] He raced round the pubs and pleaded with the Colleges to part with some of their bubbly. Meanwhile, Odile immediately placed an urgent order with Matthews, the wine merchant. Ice was a problem that she solved by visiting the fish-mongers who hammered large blocks of ice into pieces for her.

Among the many guests was the Caltech biochemist Hildegard Lam-from.[8] Always keen to have a lively party, she brought with her a gener-ous supply of fireworks, including rockets. As the party progressed, guests began to congregate on the little roof garden where Hugh Huxley, as legend has it, deftly held each lighted rocket in his hands until the moment was right to release it skyward. Although he failed to score a direct hit on the St. John's College chapel roof, one rocket did land in the College courtyard. Then, four braves scrambled up to the roof, car-rying a supply of fireworks that they proceeded to light and throw to the street. Among the four was Les Smith. He stuffed some fireworks into his trouser pocket and, when they caught fire he burned his hand removing them. Fortunately, his lab chief, the Nobel laureate Fred Sanger, was at hand to take him to Addenbrooke's Hospital.

Memorable as was this party, the reason for the party was a bigger event: the award of the Nobel Prize to Crick, James Watson, and Mau-rice Wilkins. One writer[9] put into words his thoughts as he fantasized about receiving the prize. He could, he wrote,

> hear the *crinkle* of the envelope as I open the telegram from Stock-holm. I can *smell* the leather and velvet scent of the blue box in which the Nobel medal lies . . . I think hard and long of those who are the real thing, whose names will last in the household of the mind . . . I imag-ine myself in their skin of glory—because that is what it is, a skin in which their lives have changed and become luminous.[10]

When a week later John Steinbeck turned on his television at his home in Sag Harbor, Long Island, he was seeking news of the Cuban mis-sile crisis, but discovered that he had won the Nobel Prize for Literature. "What was his reaction?" asked a *The New York Times* reporter. "Disbe-lief," he replied. "Then what happened?" "I had a cup of coffee."[11] No fantasizing here! Did Crick have any idea that he might receive the award? True, Jim Watson had the goal of winning the prize way back in the 50s during his days as a postdoc. He was not surprised. But Crick had been totally unaware of Watson's secret ambition and had not been plan-ning his life with his compass bearing set on a Nobel Prize. Indeed, it was not until 1956, he recalled, that it occurred to him that their discovery of the structure of DNA was Nobel Prize–worthy, due to a casual remark by another scientist.[12] True, the Nobel award would have been a great sur-prise in 1955. But it was not unexpected 7 years later, especially consid-ering he had been consulted by Sir Lawrence Bragg about the matter in 1960 and by Jacques Monod in 1961.[13]

Interviewed on the day of the telegram from the Rector of the Karolinska Institute, Crick was in an expansive mood. He told the reporter from the *Cambridge Evening News* that "the crucial part of this work was done in two months" and that "most of the early work for which he was awarded the Nobel Prize for Medicine was discussed with Dr. James Watson . . . in a Cambridge public house . . . I wouldn't say this was hard work." He confessed, "It was mainly theorizing, there were no long hours of laboratory work—nothing dramatic.[14] Alongside this column, the *Cambridge Evening News* inserted a box entitled "Doctor's Hobby." Taken from *Who's Who*, it reads

> Dr. Crick is 46. He was educated at Mill Hill School, University College, London, and Caius College Cambridge. He has been married twice and has a son by his first wife, and two daughters by his second wife.
>
> His recreation is listed in "Who's Who" as conversation, especially with pretty women.[15]

It was later that he acquired a new hobby—gardening—"to the astonishment," he confessed, "of all my friends."[16]

To reporters from the major dailies Crick was more restrained, but what, one wonders, did readers of the Cambridge paper make of this newly minted Nobel laureate? Clearly, he was not the typical research scientist or Cambridge academic. He was ebullient and frank, and it seems he cared little what others thought of him. He would be provocative, he would be outrageous, make you think, have fun. The anonymous writer, *FRS*, actually the literary critic George Steiner, was later to describe him thus:

> His panache is exactly dated, as is his fine-beaked profile; it belongs to *the belle époche.* His wispy yet far-flung flourishes of hand and torso are vintage 1905. The voice is pitched high but incisive and the mind moves, with an obvious, bewildering celerity, behind a fusilade of rather campy laughter. It is the kind of voice one goes grouse-shooting with.[17]

Our subject, it seems, reminded *FRS* of Oscar Wilde, the author of *The Importance of Being Earnest.* Both men displayed a sharp, even waspish, wit. Both men had a strong sexual drive, but in different directions. Both have rightly been described as brilliant and extroverted, but Crick shunned the publicity that Wilde courted. Despite Crick's raucous laughter, he would have proved no more willing a companion than Wilde to take on grouse-shooting expeditions. But parties, entertaining, socializing at a pub, or going to an art exhibition—such were the times when Crick's exuberance for life, his outgoing nature, and his curiosity could be expressed.

Now, this party was one in a series marking prestigious awards to members of this Laboratory. First came news on 1 October of the Gairdner Foundation Award to Crick; followed on 18 October by the Nobel Prize in Physiology or Medicine to Watson, Crick, and Wilkins; and concluding with the news of the Nobel Prize in Chemistry to Crick's colleague John Kendrew and his colleague and former doctoral supervisor Max Perutz on 1 November. If anything had still been needed to place the LMB on the international map of scientific eminence, this was it. As for Crick's finances, he no longer felt the restrictions of poverty. "He was able to afford some luxuries, buy some toys—a fancy sports car, a yacht, a cottage in the country," recalled Crick's second wife, Odile, with that jovial lilt in her voice.

In December, the prizewinners needed to be in Sweden to accept their prizes and each to give his Nobel lecture. There would be a number of formal occasions requiring specific dress codes for which Odile, herself a fashion designer, needed to devote more than a little attention. This, Crick recalled, kept her very busy, and no wonder. She really splurged, buying an evening dress of oyster satin embroidered with beads and pearls, costing £100, a small fortune in those days. Her daughter, Gabrielle, was amazed at such extravagance!

It would be cold, Stockholm deep in snow, the sun showing its face for only 4 hours before the long twilight of a typical winter day enshrouded the city. This would not be a 1-day commitment, but instead would occupy the best part of a week and would be tiring. There would be much shaking of hands, many introductions, long speeches, short speeches, formal dinners, and receptions. Photographers, television crews, and reporters would be milling around, creating the atmosphere of opening night at a theatre.[18]

All of the Crick family—Francis, Odile, their daughters Gabrielle and Jacqueline, and Francis' son Michael—came to Stockholm to witness the prize-giving. On exiting their plane, they were "blinded by a host of floodlights and heard camera shutters clicking all around." In the airport VIP lounge, they were interviewed by the Swedish Press and then driven to Stockholm's Grand Hotel, where more pressmen awaited them.

After two days of parties and receptions came the day of the award ceremony: 10 December. That Monday morning, the prizewinners were instructed about the formalities of the occasion in Stockholm's Concert Hall. At 4:00 p.m. with the audience seated and the faculties of physics, chemistry, medicine, and literature above them on the stage, the ceremony began. A fanfare of trumpets brought King Gustaf VI Adolf and Queen Louise and their family into the hall to sit at the front of the stage. After the playing of the Swedish National Anthem, the doors at the back of the stage opened to reveal the laureates-to-be, ready to walk to their seats. The orchestra played and the royal party and audience rose. As the laureates reached their seats, they bowed to the King and all were seated. Below them, filling the front row were the Crick and Perutz families on the King's right, and on his left were the Wilkins, Watson, and Steinbeck families as well as the Soviet Ambassador.

The prizes were awarded in the historic order: physics,[19] chemistry, physiology or medicine, and last, literature. For each, the prize's sponsor delivered a "Proclamation" describing the nature and importance of the work that had merited the prize. When it was the turn of physiology or medicine, the sponsor, Professor Arne Engstrom, addressed Crick,[20] Watson, and Wilkins:

> Your discovery of the molecular structure of deoxyribonucleic acid, the substance carrying the heredity, is of utmost importance for our understanding of one of the most vital biological processes. Practically all the scientific disciplines in the life sciences have felt the impact of your discovery. [It] . . . opens the most spectacular possibilities for the unraveling of the details of the control and transfer of genetic information.
>
> It is my humble duty to convey to you the warm congratulations of the Royal Caroline Institute and to ask you to receive this year's Nobel Prize for Physiology or Medicine from the hands of His Majesty the King.[21]

Crick was the first to go forward for the physiology prize. Descending from the stage, he reached King Gustaf, who put into his hands the Prize diploma in its blue leather folder, the "assignment" for the Prize money, and the gold medal carrying a profile of Alfred Nobel. Cameras clicked, a fanfare of trumpets sounded, and from all around came applause from the assembled guests, now risen to their feet. Crick gave a very slight bow to the King, then to the Queen, and to the guests as he returned to the stage. Watson and Wilkins followed. The ceremony ended with the singing of the Swedish National Anthem "Du gamla du Fria."

The laureates were then driven to Stockholm's magnificent City Hall to enjoy, with 600 other guests, the banquet in its Golden Hall. The prizewinners and their families were first introduced to the King and his Court before they entered the Hall together and dined with the Royal family at a table raised somewhat above the rest and running down the center of the hall. From this position, Crick could survey the guests on either side, the men in formal dress, the women in evening gowns. He could take in the sparkle of jewels and the shine of silk and taffeta and enjoy conversation with the pretty Princess Desirée. Odile was seated next to the King and when asked about her family, pointed out her eldest daughter Gabrielle, sitting nearby, and the King then waved to her. To Perutz, Odile remarked, glancing at the Queen, "I had always thought that diamonds were overrated. Now I see how they sparkle when they are really big!" In the speeches that followed the banquet, Watson was delegated to reply for himself, Wilkins, and Crick, the others having declined the task. He recalled the time that they discovered the structure of DNA. He told the assembled guests how they knew then "that a new world had opened up and that an old world which seemed rather mystical was gone."

> Our discovery was done using the methods of physics and chemistry to understand biology. I am a biologist while my friends Maurice and Francis are physicists. I am very much the junior one and my contribution could have only happened with the help of Maurice and Francis. . . .good science as a way of life is sometimes difficult. It often is hard to have confidence that you really know where the future lies. We must thus believe strongly in our ideas, often to [the] point where they may seem tiresome and bothersome and even arrogant to our colleagues. I know many people, at least when I was young, who thought I was quite unbearable. Some also thought Maurice was very strange, and others, including myself, thought that Francis was at times difficult.[22]

It was a frank and personal ending to a speech that had encapsulated a feeling of the enormity of the revolution that was occurring in the life sciences, to which their DNA structure had contributed so much.

We know from correspondence at the time that Crick was very aware of the debt he owed to those who had influenced him and had paved his way. Thus, in the glow of his Nobel award, he wrote to the virologist Wendell Stanley, who had crystallized a virus in 1935, "You were in the field long before any of us and your work and the work of your lab has always been an inspiration to us."[23] Amid all of the fun and champagne of the party held at his house the previous October, Crick thought of the eminent American chemist Linus Pauling. "I realized," he wrote to Linus, "that it was in the same room that Peter and Linda celebrated your own prize eight years ago. I am sure I don't have to tell you, but Jim and I have always be[en] very conscious of the profound influence that your work on the α-helix had on our approach to the DNA structure."[24] And to J. Desmond Bernal, British pioneer of protein crystallography, Crick wrote, "Watson and I have always thought of you as our scientific 'grandfather', fortunately perennially young."[25]

The festivities continued in a hall that Perutz described as "resembling the court of a medieval castle." To open the ball, the students sang and John Kendrew gave a speech of thanks, after which the dancing began. By 2:00 a.m. the last dancers left the floor, but not before the photographers had taken their pictures, which included one of Crick dancing the twist with Gabrielle. So ended a memorable but exhausting day.

The laureates gave their Nobel lectures on the following day. Crick spoke on the genetic code, its principal features discovered only the year before. Watson devoted his lecture to ribonucleic acid or RNA, and Wilkins spoke about DNA. Then, accompanied by their wives, they dined that evening at the Royal Palace. Fortunately, it was not until 13 December that the guests had an early awakening due to the annual celebration known as the "Festival of Lights" or "Santa Lucia, the Queen of Lights." The Cricks were not accustomed to a line of young women dressed in white with red sashes entering their bedroom at 6:00 a.m.! The tall young woman chosen to be Santa Lucia led her singers from bedroom to bedroom, with a wreath with lighted candles on her head, a tray of coffee, gingerbread, and saffron buns in her hands. Following her women carried candles and sang a refined "version of the Italian organ-grinder's 'Santa Lucia' song" in Swedish.[26]

### Becoming Famous with "Honest Jim"

As early as 9 months before the announcement of the Nobel Prize, Crick's progress was being followed by *The New York Times*. At first, Crick had enjoyed the attention of the press, but as time went on, he tried to discourage it. Denying to a *New York Times* reporter that he was "spiraling into fame," he pointed instead to DNA.[27] As he later explained, instead of believing that "Watson and Crick made the DNA structure, I would rather stress that the structure made Watson and Crick. After all, I was almost totally unknown at the time, and Watson was regarded, in most circles, as too bright to be really sound."[28]

Already in early 1962, Crick, not yet a laureate, had become "the constant focus of newsmen" wanting to know what secrets he had wrested from the gene. "But when he learns their questions," wrote a disgruntled reporter, "the door of the Golden Helix slams shut. He hates being asked what he has done and is as near publicity proof as a famous man can be."[29] Crick's attitude to the reporters had changed, and with good reason, for the media so often misrepresent what one says, quote passages out of context, and interpret jokes as serious statements. As for television, he long refused to accept appearance requests because, not being shown live, one has no control over the editing. Then there was the matter of having a private life. "You don't want," he explained, "to be recognized as soon as you walk across a college campus."[30] If you become an icon, you may be spotted in a crowd, no matter where. Become a celebrity and people treat you as if you were some sort of "giraffe."

Did Crick's laureate status change his life and invade the privacy he sought to preserve? Certainly, reporters were more insistent in seeking interviews with him, but his secretaries succeeded in keeping most of them at bay. "His instruction," recalled secretary Sue Foakes (née Barnes) "was to 'get rid of them' "[31]—not just reporters, but even a whole film crew. He was now very well known in the scientific world. Among the general public, though, it was not Nobel status that was to make Crick a household name, but the book his fellow laureate James Watson wrote. This "fresh, arrogant, catty, bratty and funny"[32] account of how Watson and Crick came to discover the structure of DNA sent Crick into a fury. "The whole exercise of preparing this chatty, personal, gossipy, and egotistical account," he fumed to Watson, "grossly invades my privacy" and is "a violation of friendship."[33] As one who was always discreet about personal matters, Crick acted on the principle that one does not say behind a person's back, let alone publish, what one would not say to his or her face. It concerned him that Watson wrote critically about his friends' failings and their personal lives, and he did so while they were still alive.

Writing to Watson in the spring of 1967, Crick admonished him:

> It is not customary to write intimate books about your friends without their permission, at least until they are dead. I would remind you that Bertrand Russell delayed the publication of his autobiography till he was over 90, and that Lord Moran's much criticized account of Churchill's health was not published till after the latter's death. The fact that a man is well known does not by itself excuse his friends from respecting his privacy while he is alive. Only if a person himself either gives permission or discusses his own personal affairs in public should his friends feel free to write about them. The only exception is when private matters are of prime and direct concern, as in the case of Mrs. Simpson and King Edward, and even then the British press wrote nothing for months.[34]

Crick did "not concede that pure scientific research lies in the public realm in the same way that politics or military affairs do." Politicians and military figures should "expect to have their behaviour written about."

> But the point of science is what is discovered, not how it was discovered or by whom. It is the results which need to be brought home to the public. It is quite inexcusable to invade someone's privacy to describe how the structure of DNA was discovered to people who don't even know what it is, nor why it is important. I have no objection to a genuine historical description. It is vulgar popularization which is indefensible.[35]

Crick's concern about privacy was echoed in some of the book reviews of *The Double Helix*. Thus, the French microbiologist Dr. André Lwoff was to remark that Watson's "cold objectivity is applied to persons he likes, as it is to crystals or base-pairing. Very few are spared. May God protect us from such friends!"[36] Another judged the "private world" that Watson revealed as "unbelievably mean in spirit, filled with the distorted and cruel perceptions of childish insecurity. It is a world of envy and intolerance . . . scorn and derision."[37]

So forthright were the objections expressed by Crick and Wilkins to the publication of Watson's manuscript by Harvard University Press that the president of the university, through the Harvard Corporation, overruled the Press, and this was despite the "unanimous support from the faculty body Syndic."[38] During this event, it was the former Cavendish Professor, Sir Lawrence Bragg, and Crick's colleague and fellow Nobel laureate John Kendrew who submitted to Watson a set of revisions needed if Bragg were not to withdraw his foreword to the book. Watson obliged and the more outrageous remarks present in *Honest Jim (A Description of a Very Great Discovery)*, as the manuscript for the book was originally entitled, were removed. Those apart, Watson refused to make further changes. Not even the suggestion from the lawyer acting for his publisher, now the Atheneum Press, moved him. It was that the opening sentence of the book: "I have never seen Francis Crick in a modest mood," be altered to read: "I can't remember ever having seen Francis Crick in a modest mood."[39] Nor would Crick's hope that Bragg would withdraw his foreword to the book be fulfilled. Although Watson had made it clear that he would not be offended, Bragg responded, "It is generous of you to say that you would understand if I wished to withdraw my preface, but, frankly I should be sorry to do so!"[40] The motive behind Bragg's support is clear. It was he who in May 1965 had urged Watson to write up the discovery for publication, fearing that Crick's greater fame would feed the suspicion that it was he, not Watson, who had discovered the structure.[41]

At this point, Watson was encouraged by a letter from Kendrew, telling him that "Francis showed me his last salvo to you about *Honest Jim*—my impression is that he is now giving up the struggle."[42] Crick's salvo had ended with the following statement: "My objection, in short, is to the wide dissemination of a book which grossly invades my privacy, and I have yet to hear an argument which adequately excuses such a violation of friendship. If you publish your book now, in the teeth of my opposition, history will condemn you, for the reasons set out in this letter."[43] But Kendrew wrote Watson that "I had the Braggs staying and helped him redraft the introduction. Hope all now straightforward."[44] Happily, Watson could report back that he had "almost concluded a contract with a serious New York House [Atheneum Press], who know that Francis' moods are not predictable and have accordingly consulted some good lawyers."[45] Wilkins and Crick also consulted lawyers. At first, the firm's specialist on literary works, Robert Montgomery, judged the book to be "libelous and an invasion of your privacy," and he objected to the book as "a vulgarization of the events"[46] but subsequently, he changed his opinion and warned that persisting in their attempt to suppress it would only increase sales. The way now clear, Atheneum quickly published the book under the title *The Double Helix*. It soon became a best seller; it has been translated into some 20 languages, and worldwide sales have exceeded 1 million. Few other popular books about science have been as successful.

Director Ronnie Fouracre approached Crick early in the 1970s, wanting to make a documentary film based on *The Double Helix*. This resulted in two films, one for educational institutions and another for the general public. A version of the latter in collaboration with the BBC, and the commentary by Isaac Asimov, was shown in the United Kingdom in 1974. Ten years later, Crick and Watson were approached by Mick Jackson of the BBC. He wished to make "a 'docudrama'—something between a documentary and a drama."[47] The result was the well-known film *Life Story* shown in the United Kingdom in 1987. The version of this film shown later that year in the United States was called *The Race for the Double Helix* and also *Double Helix*. A version for Germany was also prepared under the title *Wettlauf zum Ruhm* (*Race to Fame*). But Watson's ambition to have a Hollywood-style movie made of the story foundered.

In due time, Crick was to revise his judgment of *The Double Helix* and acknowledge that Watson had told the story in such a way that many readers found they were unable "to put it down."[48] There was suspense, and, for a popular book, Crick now judged, it relays a surprising amount of science. Yet his first reaction had been so different.

"'Who' I asked myself, 'could possibly want to read stuff like this?' Little did I know! My years of concentration on the fascinating problems of molecular biology had, in some respects, led me to live in an ivory tower."[49] And Watson? He is almost more proud of this book than of the discovery it describes, for contrary to the latter, it was his achievement alone—and in the face of fierce efforts to suppress it.

How is Crick portrayed? Surely, he is really the hero of the book. But we are left with no doubt about Crick's alleged failings: immodesty, arrogance, forcefulness, talkativeness, competitiveness, and at times, irritation. Chapter 1 describes these qualities thus:

> I have never seen Francis Crick in a modest mood . . . he was often not appreciated, and most people thought he talked too much . . . he talked louder and faster than anyone else, and when he laughed, his location within the Cavendish was obvious . . . he would not refrain from telling all who would listen how his clever new idea might set science ahead. . . . Already for thirty-five years he had not stopped talking and almost nothing of fundamental value had emerged. . . . The quick manner in which he seized their facts and tried to reduce them to coherent patterns frequently made his friends' stomachs sink with apprehension . . . a stray remark over sherry might bring Francis smack into your life.[50]

Then, too, *The Double Helix* revealed Watson and Crick's dependence on the DNA data obtained in London by the physical chemist and X-ray crystallographer Rosalind Franklin, who had died in 1958. Hitherto, their references to her work had been obscured by vague acknowledgments to the researchers at King's College London. But in writing the manuscript *Honest Jim*, Watson was for the first time specific about the source and nature of these data, admitting that there were those who at the time thought they had "stolen" such information from Franklin. This revelation was commented on by several reviewers of the manuscript, with the senior scientist J.D. Bernal remarking that Watson shows the "mechanism of scientific discovery . . . in a very bad light."[51] At least, one should add, Watson was honest about it—hence his first suggestion for the book's title, *Honest Jim*.

Crick would tell the *Cambridge Evening News* that they accomplished the work in 2 months and that it wasn't "hard work. It was mainly theorizing. There were no long hours of laboratory work—nothing dramatic." Yes, there was discussion in The Eagle pub, at the lab, and walking by the Colleges, but who prepared the excellent DNA (Rudolph Signer), who revealed the ratios of its constituent bases (Chargaff), who discovered its crystalline form (Maurice Wilkins and

Raymond Gosling), who determined the orientation of sugar to base in the molecule (Sven Furberg), and who took the X-ray diffraction pictures of DNA that led to the distinction between two forms—the "wet" and the "dry"? No one at Cambridge—it was Rosalind Franklin with Raymond Gosling in London. In truth, the DNA discovery looks like a confirmation of the microbiologist René Dubos' remark: "Science is not the product of lofty meditations and genteel behavior, it is fertilized by heartbreaking toil and long vigils—even if, only too often, those who harvest the fruit are but the laborers of the eleventh hour."[52] On the other hand, it was Watson and Crick who sought out the data from these disparate sources, mulled over their reliability and likely relations, saw their significance and through the disciplined and persistent exercise of accurate model building discovered the structure. The adage "chance favors the prepared mind" was certainly true of them.

Anatomizing the events surrounding the discovery of DNA, although a crucially important episode in Crick's life, will provide but a lively snapshot from a long, varied, and highly productive life. When we retrace our steps to Crick's early years at school and university, we find that neither was able, in the end, to keep the torch of his curiosity aglow. War work, on the other hand, placed urgent demands on him, and, working with brilliant colleagues, he responded, showing qualities of decisiveness and leadership.

What inborn traits lay hidden and what experiences honed them, finally igniting them? His experience during World War II matured him; his early protein studies established his command of X-ray crystallography and revealed his brilliance. Thus prepared, he could turn to DNA. Yet, solving its structure was but the beginning of his career in what was to become molecular biology. How, then, did he rise to the challenge of the genetic code, grapple with the mysteries of protein synthesis and the structure of chromosomes, and fantasize on the origin of life, before finally plunging into neuroscience? To answer this question, we must explore his uneasy relationship with Cambridge University and its colleges, see how he fought for agnostics and humanists, and assess his influence in the building of institutions such as the Medical Research Council (MRC) Laboratory of Molecular Biology and the Salk Institute in La Jolla, California. Francis Crick's life was long and continuously active. It began in an undistinguished way but ended very differently.

## CHAPTER 2

# A Difficult Act to Follow

 *Apart from my grandfather there was no one of any particular distinction [in the family] although as my father once said, "At least none of them was hanged!"*[1]

### The Family

On 8 June 1916, while war raged in Europe, there was a happy event in the home of Harry and Annie Elizabeth Crick: the arrival of their firstborn, Francis Harry Compton Crick. Annie then persuaded her sister Ethel to carry the baby to the top of the house. This act was intended to ensure that in years to come, her firstborn would "rise to the top." "Amazing," thought Francis in retrospect, but "it did show quite clearly that she was ambitious for me when she knew nothing about me at all, except [that] I was a boy. . . ."[2] Such was our future Nobel Laureate's promising introduction to this world. Yet what a sad world of wholesale carnage it was, for World War I still had 2½ years to run. When Francis was 3 weeks old, the Battle of the Somme alone had claimed the lives of more than 19,000 British soldiers and had injured more than 50,000.

Annie and Harry were married on 8 September 1914, little more than 1 month after the outbreak of the Great War. By the time of Francis' birth, Annie was 37 years of age. She looked forward to great success for her firstborn, and she had good reason. Both the Wilkins and Crick families could claim records of achievement because each had built a thriving business. It was Harry's father Walter Drawbridge Crick and Annie's father Frank Foster Wilkins who had built their respective family's wealth. Young Francis was later told that he "would do well, but your grandfather will be a difficult act to follow."[3]

The genealogy of Francis' family has been traced back to Thomas Crick, born in 1520 in the village of Weston Favell, close to the mid-

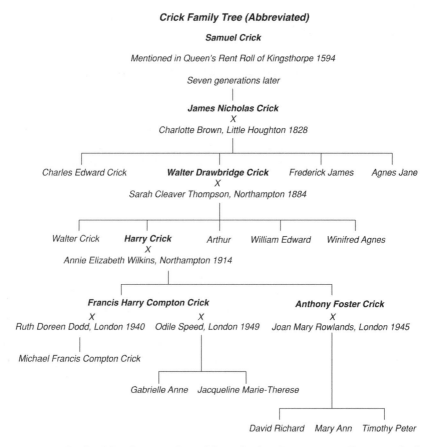

*Figure 2.1* Abridged family tree, adapted from the family tree originally researched by Annie Crick and updated by her grandson Michael Crick.

land town of Northampton (Figs. 2.1 and 2.2).[4] Three generations later, the Cricks had moved to Kingsthorpe and five generations later to Hanslope, where Walter Drawbridge Crick ("Walter Senior" in the future) was born in 1857 at the family farm of Pindon End, some 10 miles south of Northampton. To the west lay the Roman road known as Watling Street—now the busy Route A5—and through the farm ran the London and North Western Railway Company's line to Northampton. Walter Sr.'s first employment was as a clerk in the railroad's freight department. Next, he turned to selling shoes for the Northampton firm of Smeed and Warren. He traveled far, to Scotland, Ireland, and the North of England, and to good effect, for he proved to be an excellent salesman. Meanwhile, his distant cousin William Latimer[5] was trading

***Figure 2.2*** Map showing the villages of Weston Favell, Hanslope, and Crick. Although Crick's family is known to have lived in Northamptonshire for several centuries, no record is known of a member living in the village of that name. (Scale, 1 inch ≈ 10 miles.)

in leather. In 1880, they formed the boot and shoe firm of Latimer, Crick, and Gunn. When his codirectors Latimer and Gunn retired, Walter Sr. directed the business himself under the name of Crick & Co.

In its heyday, the family firm had been a jewel in the crown of Northampton's boot and shoe industry. The factory, on the corner of Hazelwood and St. Giles, was housed in a handsome four-story building, in which some 300 employees worked. Walter Sr. invested in modern machinery that enabled the firm to produce a wide range of products, from sportsmen's boots to glacé shoes "that would make Cinderella envious." The company's specialty was ladies' and children's boots. In the guide *Where to Buy in Northampton* (1890), Messrs. Latimer, Crick, and Company were specifically mentioned among enterprising firms "who have by their mode of manufacture considerably improved the quality of boots and shoes."[6]

The firm flourished under Walter Sr.'s management. The factory occupied a central site in Northampton and the firm had six retail shops located in the London area. He established the family's allegiance to the Congregational Church and sent his sons to the well-known, nonconformist,[7] "public" (i.e., private!) boarding school, Mill Hill School. He possessed qualities that were to be reflected by his grandson Francis—great energy, curiosity, and a love of science. Unfortunately, from 1899 on, his health had been poor due to diabetes. He died in 1903 after suffering a fatal heart attack. He left his wife Sarah; four sons Walter Jr., Harry, Arthur, and William; and a daughter Winifred. At the time, his estate was valued at a healthy £24,909.[8]

The eldest son, Walter Jr., had already left Mill Hill School in 1902 to learn the family business, but his brothers were still in school, running up boarding school fees for three. Had Walter Sr. not died so young, Francis suggested, his sons might well have gone to university and taken up careers in varied professions, for neither Walter Jr. nor Francis's father Harry was interested in the commercial world; both would have preferred a different career. For example, Harry favored the law. "This would have suited his quick penetrating mind and love of detail,"[9] wrote his sister Winifred Dickens, Francis' aunt on his father's side.

Walter Jr. took over management of the company at the tender age of 19. Harry, 2 years younger than Walter, was soon to join him. To their credit, the prosperity of the firm did not decline until the 1920s. Further decline followed the stock market crash of 1929. By 1932, the factory had been sold, and only the shoe shops in the London area remained. The strain on Walter Jr. fueled his obsession about what he judged were the errors of the British government's deflationary policy. The high interest rates, he complained, were reducing liquidity and deflating the economy with ruinous efficiency. Although he became quite fanatical on this subject, the pamphlet *Abolish Private Money, or Drown in Debt,* written with Oxford Nobel Laureate Frederick Soddy, does demonstrate considerable command of the data and the ability for forthright expression.[10] Indeed, Walter Jr.'s outspoken accusations directed at the financiers, and specifically, Jewish financiers, did not escape critical comment. Addressing the Northampton Business Men's Association in 1925, he had boldly declared that "You and I are suffering, and the financiers have so much money they don't know what to do with it. Look at the £7,000,000 loan subscribed fifteen times over in half an hour. The money is in the country, but you have not got it."[11]

These anti-Semitic opinions were not shared by the rest of the family, and "as a child," Walter Jr.'s sister Winifred could remember "Walter talking in this vein and arguing most heatedly, in fact," she recalled, "I was quite frightened of him"[12]—not that she ever entered into the discussions, for she was 12 years younger than Walter and to him, "little girls should be seen and not heard."[13] What lay behind these tirades, it would seem, was the declining success of Messrs. Crick and Company. Indeed, writing to Francis, Winifred commented that according to Walter, it was the Jews who "put Crick and Co. out of business as you probably know."[14] However, it was rumored that Walter Jr.'s management was partly responsible for the company's decline. Many other shoe manufacturers in Northampton survived the recession, including Messrs. Smeed and Warren, for whom Walter Sr. had once worked. When World War II began, Walter Jr. left the company in his brother Harry's hands and moved to the United States where he worked as a shoe salesman. Following the death of this second wife, he moved to an orange farm in California where he died in 1960. His sister, Winifred Dickens, visited the site of his latter years and wrote: "It was quite a climb to this place and I thought it was a sort of museum and he was allowed to live there presumably as a sort of caretaker."[15]

Harry steered what was left of the family business through the difficult years of World War II. He had a "buoyant personality" and was a keen sportsman and loyal supporter of the Abington Avenue Church. Regarding business and social circles in the town, his obituary stated that "he was a consistently popular figure, with a fund of good humour and wit."[16] When Odile Speed, who was to become Francis' second wife, was introduced to him, she found a warm, jolly, and outgoing personality, in contrast to Annie, who was more formal and reserved than he. Harry would play games with his grandson Michael and teach him the art of gardening. Michael adored him. He was also very fond of his grandmother, although her punishment for him when he was naughty was to lock him in a pantry (that had a window). But she had to catch him first; hence this form of punishment became impractical once Michael could run faster than his grandmother.

Harry's passion was tennis, and he pursued it with much enthusiasm, becoming the county's singles champion for several consecutive years. Once, he even played at Wimbledon. Such was his zeal that he would sometimes secretly visit the tennis court on Sundays, for as church secretary, he felt he could not have done this openly. Annie warned Francis "not to mention this to other members of the congregation since some almost certainly would not have approved of such sinful conduct."[17] Evidently, Harry and Annie did play tennis together when Francis was a small boy. Ethel Nutt happened to meet the family on holiday in Filey once and recalled going to watch them on the court and found that "they were two very fine tennis players."[18] Harry enjoyed the competitive side of the sport and sought to instill the same spirit in his sons by offering £10 if either of them could win a set played with him.

In the years following the decline of Crick and Company and the selling of the factory in Northampton, the family lived modestly. Their nanny Ivy Parker had already left in 1924 when she married Frederick Followell, whom she had first met at the Cricks' house.[19] The Cricks parted with their resident housekeeper several years later and this meant Annie had to learn the art of cooking.

As World War II drew to a close, wealth returned to the Crick family. They enjoyed steady income from the shoe shops Harry managed in London—the up-market shop in elegant Prince's Arcade off Jermyn street, where customers had their own shoe forms and were able to buy their shoes on account—as well as the shops elsewhere, for the less affluent. Harry benefited also from the success of the Gordon Woodroffe Leather Manufacturing Company that Annie's cousin had established in Madras, India, as a subsidiary of Gordon Woodroffe of London. During the war, Harry had become its agent in Britain and a company director. Nonetheless, it was not until the last years of his life that the family had a car plus a chauffeur, because Harry was unable to drive due to his heart condition. When he collapsed one night in February 1948 at the George Row Club where he regularly played bridge, it was a fatal heart attack. "At least he went out gloriously," Francis recalled. The attack occurred in the excitement of winning the grand slam![20]

Of Walter's brothers, Arthur steered clear of the family firm. He moved to Kent where he built a very successful pharmacy business, manufacturing the antacid pills that preceded the "Tums" that we know today. He went on to establish what today would be called a health spa. The fine house he built in 1932 in Maidstone was called "The Dell." Thanks to Arthur's financial success, Francis received support as a graduate student in London, and when Francis needed funds to purchase a house in Cambridge, it was Uncle Arthur who stepped in to help.

The third son, William, was considered by Annie to have been the cleverest. Sadly, however, he was one of more than 1700 local men who died on the battlefields of the first World War. He was only 20 years of age and had hoped to become an architect. When he fell in the Second Battle of Arras in 1917, he was a second lieutenant in the King's Own Yorkshire Light Infantry.[21]

Winifred, the youngest, was tall, bright, and forthright. In 1921, she married Arnold Dickens, son of (William) John Dickens, the founder of the family firm Dickens Brothers Ltd., leather manufacturers in Northampton since 1897 and still in business today. From this connection between the two families, one might have expected young Francis to have known about the work of Arnold's youngest brother Frank Dickens (1899–1986), the well-known biochemist who was to become a Fellow of the Royal Society in 1946. However, when Francis' parents sought advice about a career in science for him, they turned to a physicist friend who was visiting Northampton. As Francis explained, "They didn't know themselves and they certainly had very little help in explaining scientific things to me." Yet they were "very supportive," especially Annie. They thought that "science was important and they wanted to encourage me."[22] Annie even had a phrenologist "read" her son's bumps.

Summing up his father's generation, Francis wrote, "I think all the Crick boys were reasonably intelligent. Walter [Jr.] turned into a crank, but my father and my Uncle Arthur were quite bright, although my father's real passion was tennis." However, other than Walter Sr., "there was no one of any particular distinction" in the family, he added, "although as my father once said, 'At least none of them was hanged!' "[23]

### The Wilkins

Similar to the Cricks, Annie's family, the Wilkins, owned a chain of shops under the name of Wilkins and Darking Ltd. The founders were Frank Foster Wilkins and Edward Milner Darking. They called themselves "Hatters and Complete Outfitters." Although mostly located in the Midlands (close to Northampton), one of the ten shops was on the south coast of England at Fareham and another in Surrey, not far from London. Frank managed the business from his home, called Compton House, in Phippsville, Northampton. The obituary in the *Northampton Independent* mentioned "his remarkable gift for organization . . . a natural aptitude for figures and finance . . . Of a rather retiring disposition," the report added, "he took little part in public life, and except among his intimate friends few realised what a really strong personality he possessed."[24] By his death, "Northamptonshire Nonconformity has lost a staunch supporter," declared the obituary, especially his support of the Congregational Church at Victoria Road, where he served as its deacon, trustee, and later as treasurer. He married Elizabeth Perks in the winter of 1877, and 2 days before Christmas 1878 their first child, Annie Elizabeth, was born.

Crick considered both Walter Sr. and Annie's father Frank Wilkins entrepreneurs, who introduced the model of a chain of retail shops in their respective trades. The decline of both enterprises was, not just in Francis' view, a sign of the times but a reflection of the lesser vigor of the descendants. Neither grandfather, he reflected, had married well.[25] When Wilkins offered to sell the shops to their respective managers, all but one responded positively. The manager of the Trowbridge shop, however, had been found with his hand in the till. Frank Wilkins willed this shop to Annie and her sister Ethel. Thus enriched, they were able to aid Francis financially, when he gave up his Admiralty position, by paying for Michael's school fees. Aunt Ethel subsequently left her house, a leasehold from St. John's College, to Francis.

At the time of her marriage, Annie was an elementary school-teacher. She had obsessions regarding matters of hygiene and health and was a teetotaler. Well organized and quite a disciplinarian, she had a strong personality.[26] She was the eldest of a large family that included two sisters, one stepsister, and three brothers. Fortunately, she was blessed with an energetic disposition and was responsible, efficient, and methodical. Francis' school friend Michael Wittet wrote that "she was always such an active and informed woman" and he recalled "the long and interesting letters" she wrote about family activities.[27] Like his mother, Francis was to become a very competent correspondent.

Single until the age of 35, Annie and 27-year-old Harry married in 1914. Annie's younger sister Ethel, however, remained single. Also a schoolteacher by profession, Ethel enjoyed teaching Francis to read, and they both followed his career with the greatest interest. Annie kept a "shrine" consisting of her son's scientific awards, and, after she died in 1955, Ethel kept it at her house. Unlike Annie, Ethel did have some understanding of the nature of her nephew's scientific achievements. Francis was not impressed by his mother's intellectual attainments; rather, he was amused by them. His affection for her, however, was strong, and her death on 27 October 1955 upset him deeply. The next day, he witnessed the signing of her death certificate and then retired to his room, remaining there for 3 days. His grief continued for several months.

### Francis' Early Life

Francis grew up in a world of business, chapel, tennis, and bridge. It was middle class, suburban, and provincial[28]—in short, not an intellectual hotbed for an eager mind! The name "Crick"[29] suggests the distant origin of Harry's ancestors from the village of that name, some 20 miles northwest of Northampton (see Fig. 2.2). Crick was and continues to be quite a common name in Northampton. Indeed, families with that name worldwide seem to have remote roots in or near the Crick village. But Annie found the name Compton of greater interest. Belonging to her ancestors, she traced the Comptons back to Lydia Compton in 1638, hoping to find a connection with the aristocratic family of that name. These aristocrats could be traced back to the 15th century, the Earldom of Northampton to the 17th century, and the title Marquess[30] of Northamptonshire to the 19th century. Everyone in Northampton knew about the Marquess' fine mansion Castle Ashby, dating from the 16th century.[31] Sadly for Annie and her younger brother Eric Wilkins, who drove her around the churches of the County, the hoped-for result was not realized, but this effort did yield the first Francis Crick family tree. Annie's efforts amused Francis, for connection with the aristocracy meant little to him. Michael's comments after viewing portraits of the Compton ancestors at the Castle were that "They look like they are our relatives. It is hard to believe that we all lived within a ten mile area side by side for so long and are not related."[32]

Francis' birthplace was "Holmgarth," the family's house on Holmfield Way on the outskirts of the village of Weston Favell, close to the town of Northampton. The village takes its name from the 11th century overlordship of Weston, comprising land held in the 13th century by Richard de Weston and after him by John Favell.[33] The center of the village comprised the church and, facing that, a group of thatched cottages with stone-mullioned windows. To the south lay the rectory, a red brick house, beyond which the ground slopes gently down to the limits of the parish marked by Northampton's river Nene. To the north, the shoe manufacturer James Manfield built Weston Favell House in 1900 with a small park around it. Here, because the land was 400 feet above sea level, Francis could see the fields running down to the river Nene. Crick recalled the large field at the back of his house.

> I could glimpse the Nene in the distance, with scarcely a house in sight. I recall trying quite unsuccessfully, to dam with mud a small stream running from the smaller lake in Abington Park. . . . One summer I helped get in the hay at Little Billing, on the farm of a friend, using pitch-forks and a wagon drawn by a horse. Several winters I skated on the meadows by the Nene, especially flooded for this purpose.[34]

Weston Favell was a convenient location for those working in the nearby town of Northampton. The Crick family's presence there signaled their financial position in the commercial world of that town. The Cricks were not as wealthy as their friends the Sharmans, who lived nearby in a large house and had a "rather sporty Vauxhall"—a four-seated coupé, nor were they as affluent as the Manfields. But the family had the resources to employ a nanny, Ivy Parker, and a housekeeper who cooked and cleaned; thus the formalities of gracious hospitality were maintained. Crick remembered how

> . . .my mother was "at home" on certain days. If a visitor called at other times the housekeeper often told them my mother was not at home (even when she was in the house) and the visitor would usually leave a calling card which was received on a silver tray. The grocer called on my mother. They discussed her weekly order, the grocer telling her what delicacies were special that week. The goods were then delivered to the house. The doctor, an elderly man with a good bedside manner, made house calls as a matter of course.[35]

The house of Francis' boyhood was intended to be the first half of a duplex, the other half not yet built. Although his parents "bought the other lot," they "only got as far as [building] the garage before the money ran out." That garage was important for Francis. The Cricks had no car, so he was able to use it for "amateur theatricals and various chemical experiments." Late in his life, memories of these experiments were brought back to him when he read Oliver Sacks' delightful autobiography of his childhood, *Uncle Tungsten.* There, Sacks described the kind of experiments that had given Crick such pleasure.[36] But Francis took special pride in the electrical circuits he fitted to glass bottles containing explosive mixtures. Such an arrangement enabled him to explode the bottles remotely. Although they worried, his parents permitted him to continue such experiments, but only if they were conducted under water in a pail.[37] Little did he know then that he would one day be designing electrical circuits to activate mines at sea.

Apart from the bangs and explosions in the garage, how quiet life was then—there was no wireless (i.e., radio) in the Crick household, no television, not even a wind-up gramophone on which to play 78s. But on visits to the Sharmans in their big house overlooking Abington Park, he could listen to their fancy wireless "with glowing valves" and later see the "first experimental broadcast of the [now obsolete] Baird system for television." Street lighting was by gas, and Francis could recall the "gas man on his bicycle coming around every evening to turn the lights on with a long pole." His mother sometimes played their upright piano, and his father "occasionally had long conversations" on their telephone (Northampton 1110) "about the Northampton Tennis team," for he was for 8 years match secretary of the Northampton County Association.[38]

Francis' parents enjoyed the performances of the Repertory Company that played at Northampton's Royal Theatre.[39] They also organized a reading circle with their friends to read plays at their house. But Francis and his younger brother Tony were tucked in bed before the guests arrived. When Francis was old enough to accompany his parents, they took him to the theatre. From such experiences and as a youth play-acting in the garage, Francis began what was to prove a life-long enthusiasm for the theatre. Surely, it was his youthful appreciation of the dramatic arts that would later contribute to his remarkable mastery of the art of public lecturing.

Neither parent ever learned to drive, so when 8-year-old Francis started grammar school, he either walked the 1¼ miles to school or took the bus; later he bicycled. On Sundays, the family often walked to the church on Abington Avenue. There was so little traffic. Indeed, it was not until 1929 that Northampton could boast a single traffic light.

Only once did the family take the train to London before Francis became a boarder at Mill Hill School, and that occasion was not for intellectual culture but to see the tennis at Wimbledon. Other than that, they were known to take a family holiday at the seaside town of Filey, North Yorkshire, or on the French coast.

Looking back on those years in Northampton, Francis recalled the companionship of his brother Tony, 2 years younger than he, and their devotion to tennis.

> I can hardly believe it now, but as a boy I was mad about tennis. I can still remember the day when my mother woke me early and told me (what bliss!) that I could miss school that day as we were going to Wimbledon. My brother and I would sit, sometimes for hours, beside the courts at the local tennis club, waiting for the drizzle to stop and hoping at least one of the courts would become dry enough for us to play on it.[40]

### Science and Religion in the Family

Francis never knew his grandfather Walter Sr., but he admired him for his enthusiasm for science and especially for the fact that he had corresponded with Charles Darwin. Walter Sr., despite his business responsibilities, found the time to pedal on his trusty bike, along lanes, over fields, alongside cuttings, beside cliffs, and into quarries in search of fossils far and wide. He became a prominent member of the Northamptonshire Natural History Society, coauthored some seven papers on geology and paleontology, and was elected a Fellow of the Geological Society.[41] It was science that had attracted Walter Sr. to the charismatic Reverend George Nicholson, President of the Natural History Society and minister of the Congregational Chapel on King Street and subsequently at Abington Avenue.

Walter Sr. must have been a good taxonomist, his chief delights being the mollusks and among them, the *Foraminifera.* Two fossil gasteropods that he discovered were named after him: *Mathilda Cricki, Hudl.* and *Trochus Cricki, Wilson.* That was an honor his grandson did not achieve, for Francis never felt the passion of the collector. On the other hand, he did treasure the fact that his grandfather had corresponded with Darwin;[42] he received a mention in the last brief note Darwin sent for publication at the end of his life.[43]

Walter's faith in the Christian message did not prevent him from appreciating Darwin's great work *On the Origin of Species.* He read it carefully, noting the remark about the dispersal of mollusks to new territory. In the chapter on geographic distribution, Darwin mentioned

Charles Lyell's report of a water beetle that was caught with a "freshwater shell like a limpet firmly adhering to it." Even tiny-shelled animals, it seemed, had their dispersers. Then in February 1882, Darwin received a letter from Walter Crick telling him that he had himself caught a water beetle, its antenna firmly held by the very same species of mollusk that Darwin had mentioned. Walter Sr. sent the specimen to Darwin and there followed a series of letters between the two, the last from Darwin on March 26, 3 weeks before he died. In addition, before that sad event, Darwin sent a note to *Nature* about the subject of dispersal, in which he thanked Walter Crick for his supporting evidence.

Nicholson was the friend who drew Walter Sr. into the dissenting Protestant tradition of the Congregationalists; he became a pillar of the Abington Avenue church in the suburbs.[44] Nicholson's successor at the Abington Avenue Church, the Reverend Charles Larkman, described Walter as "an earnest religious man, but a man with an all-round knowledge of science, a man whose religion was the religion of the scientist—not an emotional religion, but a firm, closely reasoned, earnest faith, which especially appealed to young men."[45]

Although there are parallels between grandfather and grandson, there are also marked differences. Francis never mentioned Walter Sr.'s support of the local church, his role as secretary of the building fund and of the finance committee, his service as a deacon (or lay officer), and his work with the Young Men's Bible Class. In this vein, grandfather and grandson offer a striking contrast—the grandfather perceiving the harmony of science and religion, the grandson their discord.

Harry and Annie were religious "in a quiet way," wrote Francis. "We had nothing like family prayers," but they went to church every Sunday, as did their sons when they were old enough. As registered members and regulars at the services, they had a pew assigned to them. But neither parent impressed Francis as being "especially devout."[46] The Congregational church on Abington Avenue did not preach a fundamentalist gospel. On the contrary, it was, in Clive Binfield's estimation, at the liberal end of the spectrum of Northampton's Congregational churches. Take Charles Larkman, minister from 1894 to 1926. He had "a keen and original mind, and with his remarkable gifts of vivid and pungent expression, he appealed particularly to the thoughtful and the perplexed, and rendered valuable service to an age fearful of the impact of scientific thought on religion."[47]

Did his treatment of the interface between science and religion fall foul of Francis' growing knowledge as he devoured scientific literature? Or was it Larkman's successor George Russell, whose approach to the issue that Francis rejected? Russell succeeded Larkman when Francis was 10 years of age. Considered a fine theologian, he did not stay long in Northampton,[48] but it was during his ministry at the Abingdon Avenue Church that 12-year-old Francis remembered telling his mother that he "no longer wished to go to church." He could not recall just what led him to this "radical change of viewpoint. I imagine that my growing interest in science and the rather lowly intellectual level of the preacher and his congregation motivated me though I doubt if it would have made much difference if I had known of other more sophisticated Christian beliefs. Whatever the reason, from then on I was a sceptic, an agnostic with a strong inclination toward atheism."[49]

Binfield has researched the membership records of the Abington Avenue Church. They included the Cricks' successful neighbors, Mr. Sharman the builder and Mr. Basset-Lowke, designer and manufacturer of those wonderful model trains that schoolboys so admired. Clearly there was enterprise, skill, and wealth in the Church's comparatively small congregation. Its members made handsome gifts to the Church—so much so that it was sometimes referred to as "the income tax church," with such gifts reducing tax liability. But financial success and technical skill do not necessarily signify the ability to respond to the questioning curiosity and demands for evidence of the kind that young Francis made. And living in Northampton, even a 12-year-old would know about Charles Bradlaugh, the freethinker and founder of The Secular Society. Elected to Parliament five times between 1880 and 1886, he was repeatedly thrown out because he refused to take the oath on the Bible. Following his fifth election victory, he finally took his seat. Francis must have walked past Bradlaugh's statue in Abington Square many a time. In this central location, it has been a constant reminder to the citizens of Northampton that the authority of religious institutions in public life can be challenged.[50]

The ethos of the Abington Avenue Church seems to have centered on a concern with respectability; one senses a certain complacency. Financial success brought with it the feeling of having risen in the social hierarchy. The grandeur of the Church and associated buildings on Abington Avenue[51] bore witness to this rise. Binfield surmised that to "a logical agnostic" such as Francis, the "well-heeled provincial liberalisms must have seemed goadingly inadequate and inconclusive." No one measured up to the intellectual standards demanded by Francis, including Larkman, and the fact that his mother admired the minister "because of his upright character" did not alter this judgment. Was it, asked Binfield, "the thoughtless premature rejection by a bumptious boy too sure of his as yet undeveloped abilities? Or was it the (almost fatal) logic of an exceptionally clear and elegant, above all, truthful mind?"[52] His second wife Odile was clear that it must have been the latter.[53]

## Science in an Encyclopedia

Although Francis found something lacking in his family circle at Northampton, the family often found his concentration on scientific subjects hard to handle. There just was no one in the family circle with whom Francis could talk about the scientific topics that fascinated him. Thinking about them, reading about them, discussing them fed his zest for life. From the vastness of space to the minuteness of the atom, he was entranced by it all and chattered away about it—so much and so often that his family, said his aunt Winifred, called him "Craxie," so "lop-sided" did his interests appear.

Uncle Walter did claim some knowledge of science, and he encouraged his nephew in the pursuit of his scientific enthusiasm. He tried to teach him glassblowing and permitted him to use a shed at the bottom of his garden for chemistry experiments. Later, however, Francis was to find his uncle's knowledge "half-baked," especially when as a teenager he had the temerity to correct him on the topic of discussion. Unperturbed, his uncle would stubbornly stick to his own view. Once, his uncle insisted to Francis that light travels in straight lines, and he would not hear of light being diffracted. When Francis showed him a page of a book in which diffraction of light is described, Uncle Walter looked at the book's date of publication and responded "but the book is out of date!" Later, probably when Francis was at the sixth-form level of science (11th or 12th grade), Uncle Walter gave him a problem to determine the quickest route by which to travel, at first beside a desert and then across it, to reach a given point the other side, allowing for the fact that speed in the desert would be much slower than that over land alongside it. When Francis produced the solution that contained root 2 in the formula, his uncle rejected it on the grounds that there was no such thing as root two![54]

Harry and Annie made no pretense at possessing scientific expertise. They realized that Francis' constant stream of scientific questions called for an in-house authority, and they found one in the form of Arthur Mee's wonderful *Children's Encyclopedia*, a multivolume, multiauthor work that Francis always kept. The work had originally appeared in the form of a biweekly serial beginning in 1908 but was published in a print version in 1910. This was the edition that Francis received. Mee claimed that it was "the first attempt that has ever been made to tell the whole sum of human knowledge so that a child may understand" and based upon the "finest ideas of education as set forth by Herbert Spencer." Deploring the fact that "the art of saying things simply has long been dying out," his team of writers had sought to revive this skill and produced a "book for grown-ups and children too—to be read *by* children or *to* children."

The examples set in this encyclopedia by Dr. C.W. Saleeby in the physical sciences, and Edward Step and Ernest Bryant in the natural history section are impressive. The opening sentences of their entries in the *Encyclopedia* awaken the curiosity of the reader. Then, a clear, yet unpatronizing, exposition follows. Titles included "What is water made of?," "How heat works for us," "The making of other worlds," and "Can chemistry build up life?" The latter question was to occupy center stage for Francis some 30 years later when he began to toy with the idea of going into biology. Saleeby's answer to that last question was "No, chemistry certainly cannot build up living matter yet, and perhaps it never will be able to do so. But we ought to know how far chemistry can go in this direction. . . ." Eighty years before this encyclopedia was published, the reader is told, the German chemist Friedrich Wöhler had synthesized urea and, no doubt in the future, "chemists will soon be able to make all the compounds that compose living matter, or protoplasm, and then call the mixture protoplasm. But it will be only *dead* protoplasm, we may be very well sure."[55] Then follows the familiar analogy of the house and its designer. The conclusion: Living protoplasm also needs an architect. But Francis' disaffection with the Church and religion had begun. He was not interested in the opinions expressed, especially those appealing to faith in the unseen creator. He wanted the facts, and scientific information as generally known was there for the taking. But nebulae were still thought to be star clusters, whereas quantum theory and general relativity were yet to come. Nor had Rutherford described his planetary theory of the atom when this edition of the encyclopedia appeared. Nonetheless, the sense of big mysteries to be solved was conveyed to the young as in the following passage: "Chemists are working at the riddle of the ether, for it comes into every scientific question, and is the unanswered problem at the bottom of everything. Yet there is reason to hope that the problem is not insoluble, and its solution will be the discovery of all time."[56]

Francis read the encyclopedia avidly:

> I absorbed great chunks of explanation, reveling in the unexpectedness of it all, judged by the everyday world I saw around me. How marvelous to have discovered such things! It must have been at such an early age that I decided I would be a scientist. But I foresaw one snag. By the time I grew up—and how far away that seemed!—everything would have been discovered. I confided my fears to my mother,

who reassured me. "Don't worry, Ducky," she said. "There will be plenty left for you to find out."[57]

Little did she expect that in the future, his discoveries would add ammunition to his agnostic persuasion. And what irony that Arthur Mee,[58] whose encyclopedia subscribed to the thesis of the harmony between science and religion, fired Francis' search for naturalistic explanations of the mysteries of the universe, but failed to stem his growing skepticism about religion.[59]

## CHAPTER 3

# From the Provinces to the Big City

*He was a highly competent VI former who was expected to do very well, but we had no real expectation of his future brilliance.*[1]

Northampton Grammar School (now Northampton School for Boys) is an ancient foundation; as the School Song has it:

When Thomas Chipsey made this school
In fifteen forty one,
Not for his gain or pride he sought,
But for posterity he wrought
Thus was our work begun.

The school's magazine *The Northamptonian* opened cheerfully in July 1925. Its editor recalled how they had enjoyed "all the delights of a perfect summer." Only a shower of rain on Sports Day reminded them of the fickle nature of British weather. Fortunately, it came during the tea break. Top of the list of summer achievements was, of course, the achievements of the cricket team, who "covered themselves with glory." Also deserving mention was the school's contribution to the town's Pageant, held that year to celebrate Northampton's rich history. The school had put on a dramatic rendering of Northampton in the Middle Ages, with the script written by the headmaster and the production organized by the head of the lower school, P.B. Bascombe.

In the "School Notes" section, the magazine welcomed 14 new boys, among them "F.H.[C.] Crick." He was placed in Miss Holding's class: Form I Lower.[2] Seventy-five years later, Crick still remembered her as "an inspired teacher" who "made everything interesting."[3] Also not forgotten was Mr. Bascombe, known for his brilliant productions of Gilbert and Sullivan operas. Crick's love of musicals and the theatre surely owed a debt to the school's drama program.

As was often the case in those days, the headmaster of the Grammar School was a classicist, William Cooke, a graduate of Queens' College, Cambridge. He was a keen sportsman, fond of a good story, and his "irrepressible geniality," we are told, "was always breaking through."[4] As Crick moved up through the school, he did not again encounter as fine a teacher as Miss Holding and he found the science teaching he received unequal to the knowledge that he had acquired through all of his reading. But the school had a good reputation and a commendable record of scholarships to "Oxbridge" colleges.[5]

Crick enjoyed the freedom of rural life in Weston Favell. He could amuse himself with his chemical experiments, explore the fields near the house, or read voraciously any scientific books at hand. He could visit the family factory in town and watch the machines driven by belts from the ceiling. With his brother, he could go to Weston Way to play tennis on grass courts at a small tennis club or visit the larger club at which his parents played. At Uncle Walter's, he could enjoy glass blowing and making photographic prints, or he could visit his friend who lived on a farm in the village of Little Billing.

At home were formalities to endure and the healthy constraints of Nonconformist styles of living at that time—eschewing extravagance, not displaying the trappings of wealth, not driving a car if you could easily walk or take the bus, not working or requiring others to work on Sundays, and no sports-related activity requiring supervision. There was more than moderation in matters of drink, because Annie was a teetotaler. Consequently no alcohol whatever was found in the house. All was very respectable: a fusion of Victorian waste-not–want-not ideology with ethical concerns of the Nonconformist tradition. Yet, even given these constraints, he had plenty of freedom to occupy himself as he wished. Hence, the suggestion of going away to school did not particularly please him. His parents, however, "were determined," he wrote, "that I should go to a boarding school."[6] They chose Mill Hill School, which his father and his three uncles had attended.

By sending their sons away as "boarders" to prestigious fee-paying schools, the Crick family was following the practice of many of the wealthier middle-class families of England. Most of these "public schools," including Eton, Harrow, Oundle, Rugby, and Westminster, are Anglican foundations. The Nonconformists had followed suit in the 19th century with Eltham College in Mottingham (a South London suburb), The Leys in Cambridge, and Mill Hill School in the North London suburb of that name. These schools broadened the social experience of their students, bringing together the sons of professional and commercial families—families, nonetheless, that had contrasting political loyalties from Fabians to Tories. There, Crick would rub shoulders with boys from different parts of the country—Northerners met Southerners, rural lads met city lads, and they all received a Christian education. No question of the separation between Church and State existed here, for this was England in the 1930s, and these fee-paying schools received no State funds anyway. And girls? Not in a grammar school for boys, please!

Well-known Mill Hillians in the 1930s included Denis Thatcher who was to direct the family firm Atlas Preservatives and marry Margaret Thatcher. More widely known were Patrick Troughton, Shakespearian actor and the second Dr. Who, and Richard Dimbleby, the BBC TV and radio commentator, whose family owned the *Richmond and Twickenham Times.* Less well known was Francis Cammaerts, brave son of the Belgian poet Emile Cammaerts and founder of the "Jockey" intelligence network that operated under very dangerous circumstances in France during World War II.[7]

So Mill Hill it was to be. After special tutoring in mathematics and Latin at Northampton Grammar School, Crick sat for the Mill Hill School's scholarship exam and was awarded a scholarship. Now a 14-year-old who had not attended church services for nearly 2 years, here he was facing regular religious observances: morning service in the school chapel, evening prayers in Ridgeway House daily, and two chapel services on Sundays. How irksome to our young skeptic, but he took it philosophically. "So I would sort of sit through it," he recalled, "I was brought up with a basic knowledge of Christian ideas, rather patchy in some ways, I would say, but surprising some of my friends from time to time by quotations from bits of the Bible . . . I had so much read out to me at one time or another in lessons and at school."[8] But as in Northampton, so at Mill Hill: He could hardly have been fed a diet of fundamentalist Christianity, for the headmaster Maurice Jacks was a Unitarian.

Crick enjoyed singing, and for a while he participated in the Chapel choir. In 1931, he and his friend Harold Fost were cast in the production of *H.M.S. Pinafore* as "Ladies." He had a very pleasant singing voice, a fact that was noted at a party in La Jolla some 60 years later when, as an elderly gentleman, he joined a group that was singing at the piano. (Little did the pianist know that it was Crick.[9]) In short, despite the chapel and prayers, Crick greatly enjoyed life at the school and made a number of lifelong friends whose members were a diverse and lively group. Closest to him at that time were Harold Fost, John Shilston, Michael Wittet, and Raoul Colinvaux. Crick shared a study with Fost, Shilston, and Hamish Hey. Their "House" was Ridgeway, and in the Ridgeway dormitory, Shilston's bed was next to Crick's.

Best known of these friends was Raoul Colinvaux, who became a barrister[10] and the law editor for the London *Times*, and was later to share an apartment with Crick in London. His editions of *Carver's Carriage by Sea* have long been the standard work on the subject.[11]

Crick's love of fun, jokes, and chatterbox habits earned him the nickname "Crackers." Harold Fost recalled what fun he was and how very witty too.[12] Crick was the boy who appeared on both ends of the 1934 school photograph by dint of running behind the group between the sequence of shots taken from one end to the other. Another incident at school concerned the radio that, like many a radio enthusiast in those days, he and Shilston had assembled. Although radios were acceptable belongings at school, listening to them during "prep" in their study was forbidden. This, explained Shilston, "was a period of private study, of one and a half hours every evening, with an optional extra half hour, during which time there was supposed to be absolute silence."[13] But Crick and Shilston spent most of the time "reading and listening to the radio—both strictly forbidden." The teacher, monitor, or prefect[14] on duty patrolled the corridor, listening for unruly behavior and forbidden entertainment. Crick evaded such vigilance by introducing into the circuit that powered their radio a contact that broke when the door opened. Sure enough, when the teacher entered their study, the music stopped. After several tries, the next time the teacher entered their room, he shut the door while remaining in their study. Yet strangely, the music did not come on again. Crick had foreseen this possibility and had placed another contact button in the circuit under the lid of his study desk. This was the moment to press it, thus breaking the circuit. Crick's pleasure at recounting this story 70 years later was very evident! Shilston remembered that they also found it easy to get their radio "into resonance, and by modulating this via a microphone," they succeeded in broadcasting "gramophone music to a nearby study." As for the know-how involved, he reckoned that "it was about seventy-percent Crick—his physics was way above mine."[15]

Shilston described the ambiance of the school as "very liberal for its time . . . no corporal punishment by the staff, for instance. There was a fairly strict dress code; games and the Officers Training Corps were compulsory, but generally the atmosphere was pretty relaxed. Discipline out of school hours and in the House was administered by the monitors and prefects." Of the many sports Crick was "compelled to play," he performed best at tennis, and he played for the school team during his last 2 years.

Unlike the credit system that leads to graduation in American schools, in Britain there has long been a nationwide set of exams called the School Certificate (now called GCSE) that is normally taken by students aged 16. In the 1930s, the majority of students who passed these exams would leave school at this stage. A small number would stay on for an additional 2 years to prepare for the Higher Certificate (now called GCE Advanced Level). Good performance was required in these exams to gain entry to a university. A few students usually stayed on for a part or all of a third year to take the entrance and scholarship exams to the Oxbridge colleges.

Bright students like Crick could be moved ahead of their contemporaries. Entering the school in 1930 as a scholar aged 14, Crick was placed in the Vth Form, where many of the other students were 16 years old (10th grade in the American school system), and would take the School Certificate at the end of that academic year, i.e., June 1931. Crick easily passed these exams and entered the Science VIth Form. Science students at Mill Hill attempted the Higher Certificate exams at the end of their first year in the VIth and usually repeated some of them in their second year to improve their grades. Crick achieved a distinction in physics and in mathematics-for-science, and a pass in chemistry[16] in the first year (1931–1932), which placed him further ahead of many of the other students. Naturally, being moved ahead like this could have a downside for students, Crick concluded. "I was slightly handicapped by being rather young," he recalled, ". . . much younger than the other people."[17]

At this point, Francis' brother Tony entered the school. To be closer to the London shoe shops supervised by Harry and to avoid two boarding fees, the family moved to Mill Hill, and Tony attended as a day student (this was called being a "day-boy"). Francis was allowed to continue as a boarder, for his parents expected that the academic year 1932–1933 would be his last at Mill Hill. When they consulted with the staff, however, they were advised to allow him to stay for another year and to sit for the scholarship exams for the Oxbridge colleges. These required knowledge of either Latin or Greek, so Francis took a Latin class with the senior classics master. The experience did not stir in him "a love of classical literature . . . the language proved too severe a barrier," despite the obvious good nature of the teacher. Crick still remembered one occasion when the classics master announced

> not without subdued irony in his tone, that he had read in the newspapers that the world would come to an end at midday that day. As the hour approached he glanced at the clock and suggested that we should prepare ourselves suitably for the event. He himself stood on his chair and solemnly bowed his head. We were not quite sure where to look, so we looked at him. Finally the clock struck twelve. He

opened his eyes, got down off his chair and, with a wicked twinkle in his eyes said, "There seems to have been some mistake." My heart warmed to him for this theatrical interlude, which he did out of sheer good spirits, but even that was not enough to make me concentrate on my translations.[18]

Needless to say, Crick did not pass the Latin exam. "I cannot recall," he wrote, "that I learnt much in my last year [his third in the VIth Form]. I would have done better to have gone abroad and learnt a foreign language but this possibility did not occur to my parents nor to me." He did benefit from the headmaster's literary appreciation course for the science VIth. It nourished his fondness of English poetry and encouraged his habit of browsing among the literary books in the school library. Clearly, Crick was not one of those who in Britain would be called a "swot," so diligent, so determined to shine in the teacher's eyes. As Shilston recalled, "Neither Crick nor I was particularly diligent in our studies, certainly by modern standards. After I got into the Sixth [form] and moved to a study I didn't do much out of the classroom and the lab. Crick was probably a bit more conscientious, but not a lot. I suspect that by his third year, when he was clearly becoming bored, he did even less."[19]

Crick's failure to win an open scholarship to either an Oxford or Cambridge College is surprising. Had his Latin been his Achilles' heel? This was the second time that he had been coached in the language for an entry exam: first for Mill Hill and now for the Oxbridge colleges. He has described how "hopeless" he was at it, but he just was not interested. One can well imagine him feeling that Latin was no earthly use for the training of a physicist, so why bother with it? Anyway, what kind of an institution was it that demanded expertise in a dead language? It bore the marks of tradition and the establishment, neither of which appealed to the young iconoclast. Yet he knew that great names in physics studied at Cambridge.

His failure in the language exam could not have been the whole story. Other scholars from the sciences had failed the Latin before him and were awarded scholarships, subject to their returning to Cambridge the following spring and taking the "little-go" examination, as did J.D. Bernal in 1919 and others after him. What further reasons could there have been for Crick's failure here? How strongly, one wonders, did the school support his candidacy? Michael Hart, a former headmaster, recalled in 1970 the school's memory of Crick: "We remember vividly all the characteristics (piping shrill voice and laughter) described by his contemporaries at Cambridge today. . . . He was a highly competent VI former who was expected to do very well, but we had no real expectation of his future brilliance."[20] "No real expectation" is revealing, and Shilston regarded his friend of those days in a like manner: "I think it fair to say that in those days he didn't appear much out of the ordinary. He was obviously intelligent, with an enquiring mind, but gave no sign of his future eminence."[21] Adding to this, the school's historian, Roderick Braithwaite, noted that the distinction of being named "Head of School (Maths and Science)" went to Charles Priestley, a contemporary of Crick's who became a physicist,[22] not to Crick. Nor did Crick have the honor of being made a prefect or even a monitor.[23] True, those distinctions required achievement in sports—but Crick's place on the tennis team surely satisfied that requirement. No, according to Shilston, the tennis did not count! Was his prankish humor a deterrent? Was he considered insufficiently mature and not an appropriate role model? Had he sometimes flouted the school's dress code, for instance? Was he perceived to be too flippant and disrespectful of authority? His fellow Mill Hill scholar and close friend Shilston found Crick to be "mildly eccentric by school standards and an extrovert." For instance, he wore suede shoes at school.[24] Boys wore leather shoes, but as Crick must have reasoned, no leather, no shoe polishing! That might have been acceptable in Northampton, the shoe capital of England, but not at Mill Hill. How very trivial, one might say! But it was symptomatic of one who did not feel the need to conform to the norms of those around him.

Spending 3 years in the VIth form meant that much of the same syllabus was gone over twice, some of it three times. The inevitable result for one with as lively and curious a mind as that of Crick's was that, as he recalled, the subject matter that had earlier so enthralled him became stale. Not that the teaching was to blame, because, like a number of other independent schools, Mill Hill had a very good reputation in the teaching of science and mathematics. As Crick acknowledged, "My whole scientific career was grounded in the excellent science teaching I had while at Mill Hill."[25] That did not prevent him from later criticizing the way chemistry was taught in those days—not only at Mill Hill, but nationwide. The method seemed to him to be more like a "set of recipes" than a science with a logical structure.[26] What did attract him was the wonderful way in which the organization of the Periodic Table of the Elements could be understood as the expression of atomic structure, a theme he emphasized in the talk he gave as a VIth former to the school's Scientific Club on the Bohr atom. Doubtless, it was for contributions such as this talk that he was awarded the Walter Knox Chemistry Prize and the Form Prize in physics in 1933.[27] His expository skills, it seems, were already in evidence.

As for mathematics, he enjoyed it. His teacher Herbert Coates was really excellent. Appointed back in 1916, he took a personal and all-round interest in his students. Coates did not underestimate Crick's ability, and it was Coates who followed Crick's subsequent academic career at University College London and lamented Crick's failure to win first-class honors there in 1937.[28] The apogee of his scholastic performance, it seems, had come during the year 1932–1933, his second year in the VIth form, when he took the Higher School Certificate for the second time and was awarded the Form Prizes in physics and in mathematics. Thereafter, he needed the stimulus of fresh intellectual encounters to challenge him; taking the Higher Certificate for a third time would not be the answer! Instead, his competitive spirit found fulfillment in tennis, squash, and billiards. As for reading, he found the works of great novelists and poets that he could read in the school library more to his taste than physics textbooks.

Although Crick was later judged to have an excellent understanding of mathematics, he himself was conscious that he did not have that love of the "discipline of rigorous proof" for its own sake—one that marks the pure mathematician to be. As he put it, "I lacked that single-minded absorption with mathematics as such which a real mathematician needs." He did not feel the excitement that his future friend, the 12-year-old Georg Kreisel, experienced on reading the proof about the angles subtended by a chord of a circle on points on its circumference.[29] Nor for him at this stage in his life was biology of particular interest, and, unlike his brother Tony, Crick had no desire to enter the medical profession. Therefore, he did not study biology at an advanced level, for at that time it was taught only in the Medical VIth form.

### University

Without the Latin and a scholarship, and with his parents on a lower income, Crick had to find an inexpensive alternative to Oxbridge. What could be better than University College London (UCL)? He could live at the family home in Mill Hill and travel daily by bus to Hendon Central station, then take a subway to Goodge Street and a short walk to the College, which in all took "the best part of an hour." This was not the ideal way to enjoy university life, but in contrast to Oxbridge, there was a benefit—none of the religious associations of their colleges was present at UCL. Regarded in its early history by the "establishment" as "the godless institution of Gower Street,"[30] any form of religious teaching or requirement for religious conformity was forbidden in UCL's statutes. Crick could reassure himself of its liberal and utilitarian spirit every time he walked past Jeremy Bentham's clothed skeleton with its wax head in its display cubicle at the South Cloisters. This was the college that in the 19th century Oxford and Cambridge Universities had sought to prevent from receiving the formal status of a university. With some justice, Bentham had called them "the two great public nuisances . . . storehouses and nurseries of political corruption."[31]

By the time that Crick learned of the failure of his attempt to win an Oxbridge scholarship, the time for the entrance exam for UCL had passed. Fortunately, he was accepted provisionally, pending his performance on the entrance exam during his first year in the program. Evidently, he had not expected to find himself in this situation. There were also disadvantages at UCL: The physics course did not live up to his expectations, and the other physics students were not like his school friends from Mill Hill. They would "go home with their little bags every day" once lectures and labs were done. Their goal was to become high school physics teachers.[32] How different it had been at boarding school. There, only the day-boys went home after school, and many had higher and more varied ambitions than these physics students.

Crick was always a good letter writer, and we know from Shilston that they "kept in touch during the rest of the 1930s, mostly by correspondence. I had a number of slightly mad letters from him on a variety of subjects—the usual setting the world to rights at that stage of one's life. I remember one ended 'isn't it nice to think you know everything'."[33] Unfortunately, this correspondence, treasured for its amusing content and greatly enjoyed by Mrs. Shilston, disappeared many years later when they moved to a new house. Shilston later regretted how little of their contents he could remember, but he added that

> They were very "interesting times" from 1933 to 1939, with plenty of opportunity for political comment, which Crick no doubt made. He had a youthful iconoclasm and skepticism, and I think was moving from the Protestant Christianity of his youth—and school—towards his later firm atheism. What I do remember is that he would address his envelopes in a humorous manner, addressed to the postman rather like one from Eliot to Clive Bell, though not in verse. One was to "that miserable hole, Keighley, where I was living at the time."[34]

These were stirring times. Recall the abdication of Edward VIII on 10 December 1936, bringing to an end the voluntary gag by the British press on his relationship with Mrs. Wallis Simpson. The accompanying ribald jokes and cartoons soon appeared. Odile Crick recalled the song "Hark the herald angels sing. Mrs. Simpson has stole our King."

But there was also outrage at the treatment administered to this popular young King by the government—forcing him to either give up his
liaison with Wallis or abdicate. Then followed Archbishop Lang's
address, in which he regretted that the King "should have sought his
happiness in a manner inconsistent with the Christian principles of
marriage, and within a circle whose standards and ways of life are
alien to all the best instincts and traditions of his people."[35] What sanc-
timonious moralizing by the head of the Church of England! Crick's
later skepticism toward the institution of the monarchy and his lack of
confidence in the political judgment of government may well have
been stimulated by these dramatic events. Such a position would have
been natural for one who even then, Shilston remarked, "was more or
less indifferent to public opinion, and did not have a very high regard
for authority."[36]

Crick did not take his work at UCL casually, but he could not recall
that the physics course stimulated him in any special way: The exciting
developments in quantum theory and relativity had not yet permeated
many undergraduate physics courses in England, except at Cambridge
where Arthur Eddington in the department of astronomy, and Ralph
Fowler and Paul Dirac in the department of mathematics were teaching
them in the 1930s. Nor did many universities on the Continent teach the
new physics (Göttingen was the prominent exception). UCL offered
only six lectures on quantum theory at the very end of the course.

Although Professor Edward Neville da Costa Andrade was author
of the fine 1927 book *The Structure of the Atom*,[37] by the time Crick
entered his physics department, contemporary developments in the
field had left it behind. There was only one question in Crick's physics
final examinations on the Bohr atom; the remainder were on more traditional topics, such as "Describe a method suitable for measuring the
viscosity of a gas," "Outline the experiments that would have to be carried out to enable the mass of the earth to be determined."[38] As Crick
later remarked, he was trained "in what would now be regarded as historical physics,"[39] and its bias was experimental rather than theoretical. The character of the department had been shaped by its condition
in 1928 when Andrade's reign began. The laboratories, recalled
Andrade, "were very poorly equipped and funds were scanty, so that
there was no question of experimental research requiring elaborate
apparatus."[40]

There was disappointment at UCL when Crick was awarded only a
good second in the 1937 physics finals in subsidiary mathematics. This
was chiefly due to the fact that the course had not stimulated him. The
kind of physics taught reflected the interests of the professor and the

strong research school that he had established in "the fields of the physics of metals, viscosity, acoustics, brittle fracture, deposition and sputtering of thin films, radiation damage, and sensitive flames." For Crick, gifted with such powers of concentration and rapidity of thought, it is noticeable when his interest in the subject of conversation around him ceases, because he lapses into silence and his penetrating gaze is withdrawn. How often, one wonders, did those competent but unexciting lectures try his patience and dampen his enthusiasm? Did he realize at the time that the teaching was somewhat out of date? No, he answered, because he did not yet have the knowledge to make that judgment.[41] And yet, what exciting years had the first three decades of the 20th century been for physicists. The quantum theory had shattered belief in the fundamental continuity of physical processes, and relativity had reconfigured our notions of space and time. Contrast Crick's apathy for the undergraduate physics courses at University College London (1934–1937) with the American physicist Richard Feynman's delight in what he was taught at the Massachusetts Institute of Technology (1935–1939), both in formal lectures and in the informal special tutorials in quantum mechanics.[42]

Crick's degree results notwithstanding, he was permitted to become a graduate student at UCL, and thanks to Uncle Arthur, he had financial support. He was 21 years old and still living at home. At last, by sharing with his school friend Raoul Colinvaux, he could afford to rent a centrally located apartment in London close to the British Museum on Coptic Street. In those days, his apartment was called a "cold house" (with no running hot water). It was probably Colinvaux who began widening Crick's experience of life by introducing the naïve provincial to London's night life, and, as Odile remarked, he learned fast.[43] Now Crick's conversation with women would very soon turn to the subject of sex.

Crick could also keep up with his intimate circle of Mill Hillians—John Shilston, Harold Fost—and with Shilston's girlfriend Joan Barker, a student from the Trinity College of Music. They would all meet for a "singsong," at Crick's apartment, and Joan played the piano. Joan also met these friends at cricket games and was impressed by them all. "They were a very clever lot," she recalled.[44] Taking a light-hearted attitude to cricket, Crick enjoyed confusing the batsman by bowling with both arms; hence the problem of from which arm the ball would come. Another form of Crick's relaxation during these years was by playing bar billiards. A photograph of him from his UCL days is suggestive of a somewhat affected young man, the pile of books under his arm and cane in hand, suggesting his pride in scholarship.

Crick did not reevaluate his situation following his degree result, but simply took the next step in the academic sequence and became a doctoral student. "That was presumably the next thing to do," he explained." Nor did he debate the topic to research. "I didn't <u>choose</u> a problem, I was <u>given</u> a problem, and it was a very dull problem,"[45] he emphasized. The task was to build an apparatus with which to measure the viscosity of water under pressure at temperatures above 100°C. Since 1930, Professor Andrade had paid special attention to viscosity because of his interest in the liquid state, and he wanted to test a mathematical equation that he had suggested in 1934 to represent the variation of viscosity with temperature. He claimed it "holds over practically the whole temperature range." Applying the formula to water at high temperature, he reported, "has led to interesting results. . . ."[46] Continuation of this work became Crick's research topic.

Thus, for 2 years Crick busied himself building the apparatus. The centerpiece was a sealed copper sphere containing water at high temperatures. When the sphere was oscillated, the damping effect of the water was measured, from which its viscosity was to be calculated. Building the apparatus involved setting up an electromagnetic device to supply the oscillating force (hence the coils, condenser, and wires) and using cooling equipment for temperature control and a camera for recording the oscillations. Some of these parts can be glimpsed in the only extant photo of this apparatus. Regarding his skill in this work, he admitted that he was "no good at precise mechanical construction," but he received assistance from the professor's senior laboratory assistant Leonard Walden, "and an excellent staff in the laboratory workshop. I actually enjoyed making the apparatus, boring though it was scientifically, because it was a relief to be doing something after years of merely learning."[47]

This research problem was typical of the kind that Andrade liked to study. Forming part of his long-standing interest in the viscosity of liquids and the related subject of creep in metals, it called for the experimental tradition that existed so strongly in his department. Crick did not find him to be an appealing character or effective mentor, but he had no choice, because under Andrade's autocratic leadership, other members of the staff were only exceptionally permitted to have doctoral students.[48] Yet the Senior Lecturer and Deputy Director of the laboratories Dudley Orson Wood seems to have been an excellent mentor of students and could surely have directed Crick's graduate research.

The award of the Carey Foster Research Prize in Physics to Crick in his second year and a report that Professor Andrade sent to the Admiralty in 1946, however, tell us that Crick must have performed this work well. Andrade's report said that "He showed great ability as an experimental and theoretical physicist, has an ingenious and lively mind and shaped very well as a research student. He is a very able physicist with plenty of initiative."[49]

Before completion of this research, war was declared and the laboratory in London was closed. Up to this point, Crick had not been adequately trained to follow modern physics, and the years 1940–1947 devoted to military research did not provide an opportunity for closing that gap. To add insult to injury, the apparatus of his doctoral research was to meet an untimely end in 1941 from a German "land mine" that fell on UCL, thus foreclosing the possibility of completing his doctoral research.

World War II was the first in which London and many commercial cities, ports, and industrial complexes were systematically bombed from the air by enemy planes. The impact of warfare was therefore felt keenly at home, the dislocation and the nightly interruptions to sleep from air raids—first the wail of a siren, pitch rising and falling, then another and another. How mournful a chorus, symbolically chanting the message to take cover. Sleep in the subway station, or get into your Anderson shelter.[50] Recall, if you were alive then, the almost total blackouts at night, with only feeble glimmers of light coming from the traffic lights, covered but for "a tiny cross of red, amber or green."[51] Car headlights had to be similarly masked, and all street signs, route directions, and names of railroad stations were removed to confuse the impending invaders. Iron railings were taken away and melted down to make tanks. Food scraps were preserved to feed pigs, allotments opened for gardeners to "dig for victory." This war was to have a profound impact on all who lived through it, especially Crick, for he was to become directly engaged in outwitting the enemy at their own game—sinking ships and destroying submarines by using ever-changing designs of explosive mines. Was this the challenge that at last set alight the fire within him? Would the demands of military conflict do for him what preparation at Mill Hill for scholarship exams and the undergraduate program in physics at University College had not done?

## CHAPTER 4

# War Work for the Royal Navy

*He was ingenious, one step ahead, not blinded by technique and the masses of data. In the hydrodynamic work he showed subtleties in directions he was not familiar with. Among the scientists at Havant, he stood out.[1]*

The third of September 1939 was an extraordinary day. Parliament was in session on a Sunday! Prime Minister Neville Chamberlain was informing the members of the House of Commons about the ultimatum delivered to the German Reich by His Majesty's Government. Having received no response "by the time stipulated," he explained, "consequently, this country is at war with Germany." Addressing the nation on the radio, he expressed his despair with emotion: "Everything I have worked for, everything that I have hoped for, everything that I have believed in during my public life has crashed in ruins."[2]

There had been preparations for war. To combat poison gas attacks, everyone had their own gas mask to carry around in its little box. Sand bags were everywhere, and in the sky above the city those elephantine "barrage" balloons hovered silently and silver-gray to ward off enemy planes. On all major roads, a mass exodus by car and bus was under way. Colleges and schools closed and then relocated far from the southeast of England.[3] Soon, Crick's physics department was evacuated to Bangor, North Wales, and he returned to the family home in Mill Hill, there whiling the time away by playing squash with his brother Tony. Francis, at 23 years of age and a physicist with excellent mathematical skills, was awaiting the call to serve his country.

After these dramatic and fearful days, all was quiet. No air raids, no invasion. What a "phony war" this was! Yet, at sea and at the mouths of our major rivers, trouble was brewing. The German Navy set out to impose a stranglehold on Britain's merchant shipping—an activity so vital to the nation's survival. Not only were there many German

U-boats to fear, but in addition, German mines. Most were the traditional "contact" mines that exploded after contact with a ship's hull. Moored on lines from the seabed, they were suspended beneath the surface at the appropriate depth. Minesweepers knew how to deal with them: Their "sweeps" severed the cables holding the mines and when they rose to the surface, they could be safely exploded. However, 1500 of this inventory were mines of a new design that could not be disposed of in this way. Uncertain of their firing mechanism, the Royal Navy referred to them as "influence" or noncontact mines.

These mines were laid on the seabed by U-boats or dropped from planes into shallow waters where they sank to the bottom; they would later explode under the "influence" of a ship passing over (Fig. 4.1A,B). Soon, merchant ships in the important estuaries of the Humber, Thames, and Stour were sunk—70 of them in November and December alone. Fear of further destruction brought shipping in and out of the eastern coast ports to a standstill, save for one deep channel in the Thames. Mines had traditionally been used as a defensive weapon, but these unmoored mines could be deployed in an offensive role and released over the enemy from the air. Had the German Navy possessed a larger store that had been ready in 1939, Britain might well have been starved into submission.

Winston Churchill as First Sea Lord[4] correctly gave the highest priority to discovering the secret of the "influence" mines. Was the mechanism activated by the sound of a ship's motor, the magnetic field of the ship's hull, or a combination of both? Not until the end of November was a specimen recovered and safely dismantled to reveal its magnetic sensor.[5]

The following month, Crick received a call for an interview at the Admiralty in Whitehall, close to Trafalgar Square, where the administration board of the Royal Navy was headquartered. There, the mathematician Frederick Brundrett appointed Crick to the Admiralty Research Laboratory located in Richmond Park, Teddington, on the western outskirts of London. Mindful of the errors of the first World War, this time the government's policy was to avoid sending scientists to fight at the front and, rather, exploit their skills, especially those of physicists and mathematicians, for analysis, technical innovation, and invention. Aided by the Royal Society, it had set up a "Central Register" that by the fall of 1939 contained the names of more than 6000 scientists that were to be considered for scientific deployment; 10% were physicists.[6] In being assigned to the Navy, Crick was joining a service that employed several academics with University College London (UCL) connections—UCL's professors of chemistry, Charles Goodeve,

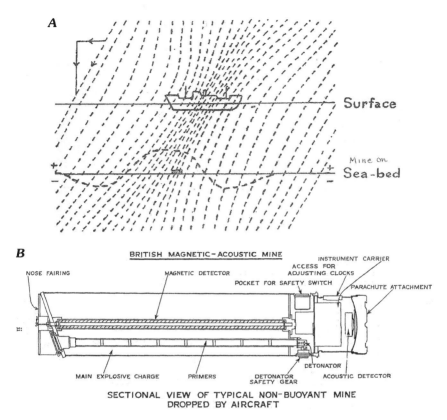

**A**

Surface

Mine on
Sea-bed

**B**

BRITISH MAGNETIC-ACOUSTIC MINE

INSTRUMENT CARRIER
ACCESS FOR
ADJUSTING CLOCKS

NOSE FAIRING     MAGNETIC DETECTOR     POCKET FOR SAFETY SWITCH     PARACHUTE ATTACHMENT

MAIN EXPLOSIVE CHARGE     PRIMERS     DETONATOR     ACOUSTIC DETECTOR
                                      DETONATOR
                                      SAFETY GEAR

SECTIONAL VIEW OF TYPICAL NON-BUOYANT MINE
DROPPED BY AIRCRAFT

*Figure 4.1* (*A*) A ship pictured in the earth's magnetic field, with a magnetic mine resting on the seabed beneath it. (*B*) Magnetic mine in a ship's vertical magnetic field.

and mathematics, Harrie Massey, the latter appointed to UCL on the eve of the war.

Meanwhile, Francis' brother Tony became qualified in medicine at Middlesex Hospital Medical School in 1941. He then joined an ambulance unit in North Africa as a First Army Medical Officer. He went on to see service in ltaly and Greece.[7]

World War II proved to be not only a test of military might but also of invention, innovation, and sagacity. From code breaking, penicillin, and radar to operational research, the achievements of Britain's "boffins"[8] have been celebrated. But little attention has been given to the naval research in which Crick was involved. Although the science was routine, the application of theory could prove challenging. For Crick, these years were of unmistakable importance, especially when, in 1941, he joined a high-powered team dealing with mine design. "In some ways," he judged, this "was the most decisive step in my career." For the first time, he found himself working with "contemporaries who were more experienced and able" than he.[9] Unlike Crick and the experimental physicists he had encountered at UCL, these were exceptional mathematicians and theoretical physicists, most of whom had been trained at Cambridge University. To them, Crick was an experimentalist, but this did not prevent him from playing the game of a theoretician, and in time, they learned to put him "on a more equal footing."

Crick was in awe of their chief, the Australian-born physicist Harrie Massey. With Nevill Mott, he was author of the important 1933 book *The Theory of Atomic Collisions*[10] and was a Fellow of the Royal Society at the youthful age of 32. Here, Crick received his mentoring in the profession of research scientist—he learned to isolate the problem, plan the research, maintain its momentum, explain its nature to a general audience, work in a collaborative manner, and deal with administrators—in this case, naval officers. The urgency of the work and the responsibility, with the lives of so many at sea dependent on it, elicited Crick's nascent powers of efficient organization and authority, and his enthusiasm and confidence swept the work along. The results impressed his bosses, providing him with powerful advocates when he was ready to start a new career after the war.

At the outbreak of war, research on torpedoes and mines was conducted at Teddington and at *H.M.S. Vernon,* the latter not a ship, as the name suggests, but a land-based collection of industrial and research facilities built around Vernon House at Portsmouth on the southern coast of Britian. Within this establishment were many sections, including Mine Design, Mining, and Minesweeping. The bombing of *H.M.S. Vernon* in 1940 and 1941, however, was to cause most of these activities to be evacuated to safer locations. The Mine Design Department (MDD) was relocated to the two estates of Leigh Park House and West Leigh House, both of which were requisitioned by the Navy and located near the small town of Havant, some 10 miles from *H.M.S. Vernon.*

### Crick as Naval Scientist

At the Admiralty Research Laboratory, Crick was a Temporary Experimental Officer, a civilian appointment. Consequently, he and his fellow scientists were not subject to the military rules of dress and conduct. Thus, when Crick was asked to take his hands out of his pockets when addressed by an officer, he replied, "Why should I?" In addition, Crick noticed a difference in attitude between the regular scientific servants, who had been in the Admiralty before the war, and the temporary ones, like himself, who had not. The instinct of the former was to accept what they were told to do and attempt to achieve the stated objective. The instinct of the new boys was to immediately ask "Why should we want this?" Such a question was regarded by the old-timers as improper.[11] In addition, Crick's "youthful sense of humour" annoyed them, but it endeared him to John Gunn and his future wife Betty Russum, the latter addressing him in letters as "Dear Infant." The naval staff, however, "got their own back by taking him out for equipment trials in rough seas."[12]

His first assignment was with the Minesweeping Division, where a group was assessing a plan to use planes as sweeps for magnetic mines. The idea was to equip planes to fly low over the sea and carry a huge coil wound around a steel core. A substantial electric current would be passed through the coil, creating a very powerful magnetic field that would be sufficient to activate the sensor of the magnetic mine below. The scientists needed to determine whether, after detonation of the mine, the plane would then be able to clear the plume of water that resulted from the underwater explosion. Knowing the physical properties of the captured mine's mechanism—its "influence," a scaled-down test was performed in the lab. The scientists placed a sensor beneath an overhead runway built in the lab. A large electromagnet was then sent hurtling down the runway. Crick assisted "while we diligently collected data on the response at various speeds and magnetic field intensities."[13]

Both this analog device and the theoretical calculation indicated that "with luck, the plane would escape damage from the exploding mine." Thus reassured, circular coils 48 feet in diameter, carrying a powerful electric current, were prepared and slung beneath several "retired" Wellington bombers. These "Flying Wedding Rings," as they were called,[14] did detonate the magnetic mines, but the method suffered from the narrow reach of coverage and the uncertainty of its limits. Only on the narrow waters of the Suez Canal did it prove effective. Here, Crick takes up the story:

> As an afterthought I was set the task of suggesting how our own magnetic mines, which had hardly come into service, could be made invulnerable to such a sweep. The obvious solution was to put in a small clockwork mechanism so that the mine had to be activated twice, with an interval of at least 7 seconds or so between activations, since this could hardly be done by a single aircraft. A ship, on the other hand, usually produced a more prolonged magnetic signal and would still be able to blow up the mine.[15]

Successfully solving problems of this kind led to the work he was to do for the rest of the War—designing mine firing mechanisms that would outwit the enemy. Such work was performed by the MDD at the requisitioned stately home West Leigh House, near Havant, close to Portsmouth, on the South Coast. The MDD had been the subject of a 1941 Admiralty report, which judged that the enemy still held the technological initiative in mine design and minesweeping. Progress by the Royal Navy in new mine design was tardy. Waxing testy, the report deplored the "lack of initiative and slow procedure at Havant" and the "insignificant role" given to the Scientific Group. There was too much red tape and control from above. Havant's technical staff complained that they needed "men of really high quality . . . for the technical work of the department."[16] The Advisory Panel noted their plea, and in the resulting reorganization, five scientists—Crick among them—were brought from Teddington in November 1941. Of these, the physicist Harrie Massey, as Deputy Chief, became the boss of what he believed was "a powerful group."[17] At Havant, he recalled, "there was a vital task of raising the morale of the scientists working under very difficult and underprivileged conditions." An injection of talent was required.

Massey may have lacked confidence when dealing with naval authority. For instance, "he would hum and ha before making a difficult phone call," recalled Crick. But he was clear about what had to be done, and under his direction, the scientific section of the MDD was transformed. The problem was how to build better mines. British magnetic mines were coming into service and development of the acoustic mine was under way. But, as Crick explained, all of the mines seemed "to be far too easy for the enemy to sweep. Yet the Admiralty system for devising, developing, testing and bringing into production new types of mines was cumbersome in the extreme . . . what mattered was not just to produce a mine that was reliable but one which also was difficult to sweep yet could be activated by the enemy shipping."[18] The problem was that when a new mine design was introduced, "in no time at all a specimen would fall into enemy hands," its secret discovered, and preventive measures taken. "The only solution seemed to be to introduce new mines, or at least new types of detection, or new combinations of detectors, so quickly [that] the enemy could not keep pace with the rapid changes in design. But how could this be done within the Admiralty production systems?"[19]

This was the substance of Massey's analysis, aided, Crick wrote, "by John Gunn who acted as his principal assistant and [was] prodded by me." In critiquing the existing program, Massey had gone "outside the strict bounds of his terms of reference." But had he not done so, Crick believed, their "efforts would have been ineffectual." Massey's solution was to create a small organization given solely to developing novel kinds of firing circuits and to fit them, with varied combinations of magnetic and/or acoustic sensors, time delays, etc., to the standard British mine.[20] Crick explained:

> The mine circuits, in their special anti-shock box, would be assembled in a small workshop at "HMS Vernon," the land-based establishment at Portsmouth. The workshop would be manned by WRNS (Women's Royal Naval Service) and managed by two naval officers. There would be a small group at West Leigh to design the new mine circuits and to see that they were properly tested for activation safety and resistance to damage. I was made head of this scientific group. The whole operation was referred to as "MX". . .[21]

MX was established in January 1943 in the machine shop at *H.M.S. Vernon,* which was by then no longer a target of enemy bombing. From that time until the end of the war, Crick had, from a work point of view, "a splendid time. I had practically a free hand in choosing what to do. Each new mine circuit had to be approved, but no high level committee sat on top of me on a day to day basis and queried everything I did. We were, in effect, running a private little war of our own against enemy shipping. Naturally this could not have been done without the elaborate organization to produce the main body of each mine and the equally complex efforts of the Air Force in laying them in enemy waters."[22]

The circuits they devised, Crick described as "unbelievably crude" by modern standards. There were strict limitations: The standard components were resistors, capacitors, and small clockwork mechanisms.

> One special job we were given was to devise a circuit to attack submarines coming out of the French Atlantic ports, submarines being high-priority targets. Aerial photos showed that the sub was escorted by a minesweeper which went through the channel ahead of it. It was child's play to make a mine circuit which only came alive if a sweep, with its much stronger magnetic field, first passed over the mine. After a short interval, to allow the magnetic field of the sweep to fall to a negligible value, the main detector was switched on, set at a sensitivity such that it might, with luck, blow up the following submarine.[23]

More difficult to tackle were the large merchant ships converted by the Germans to sweeps that they called *Sperrbrechers* (barrage breakers). Their magnet consisted of old train rails "bound together side-by-side and wound with electric cable." The resulting magnetic field was very high, close to the ship, and sufficiently far enough away to blow up the British mines at a safe distance. In the summer of 1942, aerial reconnaissance photographed a *Sperrbrecher* shortly after it had exploded a mine. Using this series of photos taken during a period of

7½ seconds, the photographic experts were able to estimate the time of the explosion from the spread of the resulting water ripples. The photographs also permitted determination of the *Sperrbrecher*'s position at that moment. The mine, it appeared, "had been detonated between 450 and 525 feet ahead of the *Sperrbrecher*'s bows."[24] But Crick and his colleagues did not consider this sufficiently accurate to determine the *Sperrbrecher*'s magnetic field. It was not until simulated explosions were performed and photographed that they were confident to decide how weak to make a sensor that would still trigger right under the *Sperrbrecher* but not before. To Crick and to Massey, "It seemed completely obvious that such a sweep was a sitting target." Moreover, it was much more important to sink or badly damage a sweeping vessel than to do the same to an ordinary merchant ship, because what mattered most was the ratio of ship activations to sweep activations. The special sweepers were far fewer in number than the other ships. "By putting German sweepers out of action we almost automatically made it possible for our mines to sink more merchant ships."[25] Crick described the alteration that was needed to achieve this result trivial. "So easy was the modification and so dangerous was such a mine to enemy sweepers that I planned it to be one of our first circuits. Imagine my surprise when I learnt from Massey that the Air Force, when they heard of our proposal, at first refused to lay such a mine. 'Will it,' they asked, 'sink an ordinary ship?'"[26]

The answer, of course, was no, because the sensitivity of these special mines was to be reduced so that they would respond only to the powerful magnetic field of the *Sperrbrecher*. Yet, wherever these special sweeps would be sunk, areas of sea harboring Britain's standard mines would be left unswept to blow up enemy ships. Massey had his way, and by the end of hostilities, more than 100 *Sperrbrechers* "had fallen victim to the magnetic mines they had gone out to sweep."[27]

Crick remembered Massey telling him that he had to "bang his fist on the table before he could get the officer to see that, in the campaign as a whole," the *Sperrbrecher* was a more important target than were merchant ships. Looking back on this incident reminded him that "the thinking of service officers in those days was sometimes rather unsophisticated."[28] But when Massey had posed the problem to Crick, the response had been very direct. "At once he had replied with that very clear incisive mind of his, 'reduce the mine's sensitivity.'"[29]

Another problem concerned production delays that were caused by the standard procedure. Here, Crick instituted what he called the "parallel method of development." After thinking of a new circuit for the activation of the mine, he first ordered any new components he might need. The stores for the MX workshop were Crick's responsibility, and he had a staff member look after them. "He was far from being an ordinary store man. His main job was to visit factories in odd corners of the economy and cut red tape by wheedling components out of people. . . . As a hush-hush special operation, we could command a fairly high priority."[30]

Because time is of the essence in war, Crick, in his enthusiasm for quick results, would sometimes persuade his naval colleagues to begin assembling the new circuit before the results of the trials were known, and at least one circuit had to be withdrawn because he had judged its potential too optimistically. On the other hand, Crick could recall one case in which he had an idea for a new circuit at breakfast; then "the whole organization sprang into action and six weeks later some of these new mines were sitting on the bottom of the Baltic, awaiting an enemy ship." Quick results called for flexibility—making do with what was available and not trusting assurances about delivery.

> "'It's on its way, old boy,' I would be told. I soon learnt to reply, 'Yes, I know it's on its way, old chap, but when will it arrive?' At first such direct questioning produced a surprised silence but I soon learnt to exact my authority. This was not always as easy as it might have been because I looked far too junior for my position. On more than one occasion I was mistaken by a visitor for the office boy, purely because I dressed somewhat informally and partly because I still looked young for my years (I was about 26 or 27 at this time)."[31]

Then there was the issue of reliability. The Admiralty had instituted a whole series of tests to ensure that equipment was safe to use and sure to work. But this could delay a new mine by as much as 1 year from prototype construction and testing in a Scottish loch to component ordering and the shock testing of production models in the Severn estuary. Of course, one could introduce a few mines of the new design into the war zone and mark the results, but this was to tempt fate by offering the enemy time to develop countermeasures before the new device could have a major impact. (Indeed, this was the mistake made by the Germans with their first magnetic and acoustic designs.) Crick decided that the best strategy was to wait until several hundred mines of the new design were accumulated and then "lay them all about the same time. With any luck this might paralyze enemy shipping for a period." Thus, in August 1942, 500 of the first British acoustic mines were laid in just two nights and, in addition, in 1943, more than 1000 of the first British hybrid magnetic-acoustic mines were laid. In both cases, the resulting carnage was extensive.[32]

Equally effective was the program of mine-laying in Normandy's coastal waters during the 7 weeks before D-day. Some 7000 were laid, and included among them were two novelties. The first was a low-frequency acoustic mine that could be swept only by using a "specialized and heavy apparatus," and the other was an acoustic mine moored to the seabed instead of resting on it. Together, these mines resulted in the sinking or damaging of approximately 100 German ships.[33] Everyone in the MX organization received a congratulatory letter from Admiral Sir Bertram Ramsay for this achievement. And Captain J.S. Cowie, who was in charge of staff requirements for mining at the Admiralty, underlined the success of MX when he wrote "It may be said without exaggeration that the enemy were to regret the day in January 1943 on which this organization . . . came into being."[34]

When the numbers were totaled at the end of the war, Rear Admiral E.N. Poland stated that "British mines sank almost 1,050 German and Italian vessels . . . nearly twice the number of British ships sunk by German mines."[35] In addition, some 500 of the enemy's vessels were seriously damaged. The 63 "Wrens" (a term used for those who assembled the mine circuits) of MX had prepared 7638 circuits[36] based on 100 different designs supplied by Crick. Such mines, Crick reported, proved to be five times more effective than the standard noncontact mines.[37] Of the 263,000 British mines laid in World War II, one-third of them were of the noncontact type, i.e., magnetic or acoustic.[38]

### Importance of the War Work

Working for the Admiralty brought Crick a salary for the first time in his life. This enabled him to marry Ruth Doreen Dodd, the English literature student whom he had met at UCL in 1936. She was an attractive young woman, tall like Francis, fair, and, like him, born and raised in the Midlands. Her birth in the Warwickshire town of Nuneaton in 1913 put her 3 years older than Francis. Although she attended UCL in 1933, she withdrew from the examination at the end of the academic year for health reasons but returned to the College 3 years later, when the same pattern was repeated. Evidently she suffered from exam nerves.[39] In contrast, crossword puzzles were her joy, and often when socializing with Crick's friends, she could be seen sitting with Crick's school friend Michael Wittet at work on the clues—a welcome change from her employment as a clerk at the Ministry of Labour.

At the time of their courtship, she was living on Windmill Street in Soho, the center of London's night life, and only a short walk from

Crick's apartment on Coptic Street. The marriage was a quiet event at St. Pancras General Register Office on the 18th of February 1940 witnessed by their fathers—Harry Crick, company director, and Samuel Dodd, engineer.[40] There was no honeymoon. On 25 November 1940, while bombs were falling and anti-aircraft guns were firing, their son Michael was born in Middlesex Hospital, on Twickenham Road in Isleworth. The family stayed together through most of the war years, but these were difficult times. War work demanded long hours and, much of the time, Doreen "hardly saw Francis, and when he did come home he would just eat and go to bed." Compared with Portsmouth, Havant did not suffer significant damage. There were the usual accompaniments of aerial bombardment—the sirens wailing and enemy action overhead—and, although they never got hit, Doreen recalled that "they were always finding shrapnel in the garden."[41]

Asked whether they were a well-matched couple, Crick's fellow Old Millhillian, Harold Fost, said that they were.[42] But during this war, when aerial bombardment caused such stress among the civilian population, relationships did fall apart. By January 1946, when Crick's work in Havant ended, the marriage had already broken up and young Michael had gone to live with his grandparents. For a while, Doreen lived in one of the rooms in Crick's London apartment, but in 1947, they were divorced and she emigrated to Canada and married a Canadian soldier she had met while living in Havant. In Weston Favell, Michael had meanwhile become very attached to his grandparents. Crick's work at Havant continued until January 1946. For light relief at lunchtimes, he and his friends formed a male team to play against the women's netball team. A member of the women's team, Mary Clark, described how their "opponents were tall and scored many goals and one was Francis Crick who later discovered the DNA molecule."[43]

From November 1941 to November 1943, Crick had enjoyed daily contact with Harrie Massey in the MDD, but Massey was needed in Canada. Aware of the potential of atomic energy from discussions at Teddington, Crick guessed at the reason—the Manhattan Project. Now he would not have Massey to smooth the ruffled feathers of those officers whom Crick had upset when he set them straight, dissecting their arguments and challenging their decisions. Indeed, Massey had encouraged his team of civilian scientists to be tough with the naval officers. "You had to do it," he recalled, but there were tensions.

Under the necessities of war, Crick had given expression to that inner confidence that was to become a hallmark of his forceful personality. And Massey recalled how "full of enthusiasm" he had found Crick to be, but this was enthusiasm combined with "effectiveness. He was ingenious, one step ahead, not blinded by technique and the masses of data. In the hydrodynamic work he showed subtleties in directions he was not familiar with." In short, among the scientists at Havant, "he stood out." Crick, he explained, "is the sort of person that you get a great deal out of discussing things with. He has an originality of approach on almost anything. I always found a discussion with him interesting and stimulating and often contributing quite a lot. This in itself must have been a very considerable factor in the whole thing."[44]

Massey's successor at Havant, Edward Collingwood, was a very different person—a mathematician of no small accomplishment, although "not as academically distinguished as Massey." An aristocrat, his most famous ancestor was Nelson's second-in-command at Trafalgar, Admiral Lord Collingwood.[45] The family country seat is Lilburn Tower in Alnwick, a few miles from the windswept Northumberland coast. Very much at home in "County" society, he was Northumberland's High Sheriff.[46] His attitude was more "man of the world" than was Massey's, and he "took things more smoothly." He would think nothing of calling the First Sea Lord if there was a problem. Collingwood and Crick became good friends, and Collingwood, 16 years older than Crick, was to prove to be somewhat of a mentor for him. It was he who foresaw a distinguished future for Crick and who also predicted a Nobel Prize for him.

The others in the Mine Design group were David Bates, Massey's doctoral student in Belfast and academic colleague at UCL after the war; John Gunn, educated in Glasgow and Cambridge, and future professor of natural philosophy (i.e., physics) at Glasgow; and Richard Buckingham, Massey's colleague at Belfast before the war and at UCL after the war. All members of this group were to have distinguished careers, but Crick was to become the only Nobel laureate among them. All save Crick, who declined, were knighted: Massey in 1960, Collingwood in 1965, Bates in 1978, and Gunn in 1982. Crick apart, they were all theoreticians and excellent mathematicians, with Crick the only experimentalist. They had been immersed in modern theoretical physics, instead of the "historical" physics with which Crick had been involved. Although the same age as Gunn and Bates, Crick was the "new boy," for Buckingham and Bates had already been working with Massey at Queen's University in Belfast.

Massey influenced Crick the most. In 1939, after 5 years of teaching in Queen's University Belfast, he had been appointed to the Goldsmid Chair of Applied Mathematics at UCL. The Admiralty called him to the position of Temporary Senior Experimental Officer at its Teddington Laboratory, before transferring him in November 1941 to the MDD as Deputy Chief Scientist. Reflecting on his time with Massey, Crick wrote

This was my first opportunity to interact on a daily basis with some-one of considerable scientific distinction. Many years later I listened to Massey give an after-dinner speech at the annual dinner of the Royal Society. He spoke in a more direct and forthright way than is usually customary on such occasions, being incisive rather than witty. What astonished me was how many of the general points he made were ones of which I was particularly fond. With a shock of surprise I realized that he was certainly not repeating anything I had said. Rather I had absorbed these attitudes from him when we were work-ing together, though I had quite forgotten their origins.[47]

Massey's expository skills also impressed Crick when he accompa-nied his boss on a visit to lunch with the "Wrens" who were assem-bling the mine circuits in the restored machine shop at *H.M.S. Vernon.* The occasion called for an impromptu speech from Massey. He used it to explain to the Wrens the design principles of the mine sensors. Using the human brain in an analogy, he explained how the brain must be taken apart and the functions of the separate parts established to determine how the brain performs its tasks. Identifying the sensory mechanism of the mine required a like approach. It was an example of the principle of reverse engineering, which is so important in physio-logical research.

John Gunn was a very able mathematician on whom Massey par-ticularly relied. Following an "outstanding performance in the mathe-matics tripos at Cambridge," he had begun his doctoral research under Ralph Fowler, but the Navy soon claimed him. The same age as Crick, it was Gunn's word that would be taken on theoretical matters rather than Crick's. Thus, the very important theoretical work on a method for demagnetizing ships, known as "degaussing," was undertaken by Massey and Gunn, and its practical implementation masterminded by the geophysicist Edward Bullard.[48] Some 18,000 vessels were protect-ed by degaussing up to D-day in 1944.

Closest to Massey and next to him in scientific accomplishment was David Bates, an Ulsterman who was also Crick's age. Bates' association with Massey went back to 1934 Belfast. Crick referred to him as a "very solid fellow"—"Not a man for risqué parties, one suspects?" "That's right," responded Crick.[49] And what did Bates recall of Crick? He remembered that he was an iconoclast then, and, wrote Bates in 1972, "In spite of your many honours I still think of you as the iconoclast I knew at M.D.D."[50] Bates established the foundations for the modern sci-ence of "aeronomy," i.e., the physics and chemistry of the atmosphere and of the ionosphere. He studied the transient products of photo-chemical reactions such as ozone and drew attention to the relationship between human activity and the global environment long before it became a matter of international concern. Although an experiment of his in the UK Space Program drew press attention, he shunned such publicity, because he was "a thoughtful and retiring man."[51]

Beyond the immediate circle of his colleagues, Crick encountered Georg Kreisel, who was to become an eminent mathematical logician. Kreisel was a member of the "Wave" section of the MDD, housed at West Leigh Cottage. The pressing problem that Kreisel dealt with was to predict how the floating harbors planned for the D-day landing on the Normandy beaches would behave under the impact of waves. This called for the skills that he, as a mathematician, could provide. But he also had skills as a conversationalist and powerful discussant, skills that were every bit as intense as those of Crick. One evening, after working late, Crick had gone to West Leigh's cafeteria and found himself "at dinner with a physical chemist and a young man I had not noticed before. The physical chemist was holding forth and gradually I sensed that both the young man and I did not think much of what he was saying."[52]

The young man was Kreisel, and that was the beginning of their friendship that lasted for more than half a century. Although there were 7 years between them, Kreisel had quite an influence on his older friend. Brought to England at the age of 8 to escape the Nazis, Kreisel graduated at the age of 19 with a mathematics degree from Cambridge's 2-year accelerated wartime program. Called to work in naval research and sent to Havant, Crick found in Kreisel a kindred spirit.

Kreisel has long had a hearty scorn for woolly thinking. Hence, in his presence, it was unwise to speak before you had thought out what you were about to say. Crick was later to remark that when he met Kreisel, "I was a rather sloppy thinker with a taste for wit and paradoxes in the style of Oscar Wilde. Kreisel would tactfully but sternly rebuke me for any careless thinking so that under his influence my ideas became more logical and better organized."[53] Kreisel was precise in his expression of ideas and formulation of statements, for how else, he asked, could conversation result in mutual understanding? As for saying nothing for politeness' sake, he asked "Are people stupid because they are polite, or polite because they are stupid?"[54]

They did not "talk shop" during the hours of firefighting, so Kreisel did not know the details of Crick's work, nor Crick Kreisel's. They did enjoy exchanging jokes—many of them, one suspects, not repeatable in polite company. As iconoclasts, they could outdo each other in tilting at the institutions of Church and State—the antiquated sexual mores and the appeal to tradition. And profanities? No one could match

Kreisel, and Crick did not try. More important, they talked about science. Kreisel raised philosophical issues with Crick and described Ludwig Wittgenstein's project on the foundations of logic. Crick judged that it would not go anywhere, an opinion that Kreisel was later to share. Although Crick was often very disparaging about philosophers, the kinds of fundamental questions that philosophers asked about science intrigued him, especially later when he came to the study of consciousness.

Intellectually, these two were equals, and neither one dominated in the relationship. Kreisel never had the feeling that Crick controlled their conversation or that he just wouldn't stop talking. "I never had the feeling of being inundated by him," remarked Kreisel.[55] In addition, he was impressed with the concentration and speed of Crick's thought. Crick could so rapidly present a problem, rehearse its various components almost unconsciously, recognize the essentials, and astutely select an appropriate way to go about solving it. Collingwood, Kreisel recalled, had a very high opinion of such skills in Crick and had noted how effective they were.

Kreisel was a powerful cross-examiner. "Why are you doing this or that?" he would ask. Such a question was to become a trademark of Crick's presence in a seminar. For Kreisel, the question would often stem from his dislike of wasting time on "trivial" matters. "Trivial" was one of his favorite adjectives, and he was later to chide Crick for the time he spent working on what Kreisel considered dull topics in the theory and methodology of X-ray crystallography.

Then, there was the "Don't worry" rule. There may be objections and evidence counter to your hypothesis. Yet if the hypothesis is founded on reliable grounds and has explanatory potential, hold to it and do not be deflected from developing it. A good example of this principle at work was Crick's attitude toward the unwinding problem with DNA helices; for this, Kreisel encouraged Crick to apply his rule.

By war's end, Massey and Collingwood had come to know Crick well and they were impressed. Crick had been forceful and critical toward existing procedures in the MDD and, as a result, had achieved rapid production of novel mine sensors. He had shown clarity of thought, initiative, drive, excellent organization, and the ability to improvise and make do with whatever resources were available. The drive behind the work was the urge to stay a step ahead of the enemy, to be informed of enemy strategy, and to devise responses to it, always planning ahead. This was the order of the day, but conversation with Kreisel was a welcome relief from the strain and pressure of the work. Looking back over the 50 years of their friendship, Crick believed that he had been "immensely influenced by his powerful intellect. We have [had] many interesting and enjoyable times together. If I had never met him my life would have been very different."[56]

Of a very different nature from his encounter with Kreisel was Crick's wartime meeting with the radio broadcaster Robert Dougall regarding the "Gnat" torpedo. In 1944, the Russians had "fished up from the Gulf of Finland a U-boat complete with the latest snorkel equipment and a torpedo," the latter "fitted with a listening device at the head, so that it homed directly on to the noise made by a ship's propellers."[57] Named "T5 Zaunkönig" (i.e., "wren") by the Germans, the British called it the "Gnat." It had been a problem especially for ship convoys since its introduction in September 1943.[58] Although evasive procedures and a suitable decoy were made available, the Allies were still keen to investigate its mechanism. This called for persistent negotiation with the Russians, and finally Churchill had to "lean heavily on Stalin" before British scientists could do so. Edward Collingwood, who had played an important part in devising a decoy for the Gnat, was asked to go to Leningrad to view the equipment. He took Crick with him. Dougall, acting as translator, met the "high-powered Admiralty scientists flown out from London. One was a tall, sandy-haired young man, who walked with a slight stoop. He obviously had an immense sense of fun, which frequently burst out into a high-pitched laugh more like a bray. I liked him immediately and he seemed amused by me. . . . The young physicist said his name was Francis Crick."[59]

Russian scientists had already made their analysis of the weapon. Now in "the grim Peter Paul Fortress, where the submarine laboratories were situated," Crick studied their conclusions while tracing the intricate electrical circuits involved. Ignorant of engineering terms, Dougall ran into some strange Russian words. "When I came upon the Russian expression *gnezdo holastyakow* I said to Francis 'This can't possibly be right. It says something about connecting up with a *gnezdo holastyakov*, which means "nest of bachelors."' That sounds to me more like the homing of a frustrated female than a torpedo!" Francis gave one of his high-pitched brays and explained that "nest of bachelors" is a term used for a certain type of electrical plug.[60]

After completing the work in Leningrad, they spent 2 more weeks writing up their report for the Admiralty, after which Crick came to the railroad station to see Dougall off to Murmansk. He was then flown back to England. One benefit of this trip was that the local Round Table invited him to give a talk on his Russian experience. He enjoyed the event and realized for the first time that he could perform well in this role.[61]

With the war over, Crick continued at Havant until the end of 1945. In the new year, he was transferred to the Intelligence Section of the Admiralty in London. He considered continuing there and applied to become a permanent scientific servant, but the external committee that was appointed to interview him was an odd lot, he noted. Professors from the provinces, they had little idea of the kind of skills needed. One of them had taught him undergraduate physics at UCL. "At first they were not sure they wanted me."[62] At the end of the interview, they asked whether Crick wanted the job, to which he replied, "After this, I'm not sure that I do!" The Admiralty staff did not accept the committee's decision and a second interview was therefore arranged, chaired by the novelist and former Cambridge physicist C.P. Snow. Crick was subsequently appointed.

One document from this period that has survived is a memorandum Crick wrote, pressing revision to the policy of excluding employment of German spies who during the war had been agents for the Reich in Russia. He judged their knowledge of Russian intelligence would now be invaluable to Britain and believed that they should be actively sought out.[63] In addition, R.V. Jones, a prominent figure in World War II intelligence and author of *Most Secret War*, quoted from a letter that Crick had written to him in 1947. It concerned the need for a central organization for military intelligence, "in contrast to the three disjointed sections that had been thrown together by the Service Ministries."

> I am opening the campaign for a central organization, and have been to see Lee and Admiral Langley. It would really help if they got another angle on all this, especially one founded on some experience. I wondered if you could find time to have a chat with Lee (he lives at the Cabinet Offices) while you were in town. Better still, to see Tizard himself. One point needs bringing out strongly. It's no use reorganizing with the same old gang. We *must* have someone more lively to head the thing.[64]

Jones reflected on the lessons learned in World War II: how "the nature of the weapons, brilliance of our sources and the mistakes of our enemies all weighed the balance in our favour." But, he added, "It may well not remain so in the future. . . . If any one man would have carried my hopes that ability in intelligence would ultimately win the day over the organizational disasters brought upon us by [Patrick] Blackett, he would have been Francis Crick, who was one of the Admiralty contributions to the post-war Scientific Intelligence effort."[65] Crick's no-nonsense approach to organizational matters as well as his influence on those around him to work together and put aside interservice jealousies and rivalries were approaches that he was to display in his future scientific career.

Did Crick want to spend the rest of his life on intelligence relating to military weapons? No, but he considered working in industry and visited the Thorneycroft Company's turbine division. This famous company, founded in 1859, had been a major supplier of equipment to the Royal Navy ever since. But a career in applied science, no matter how good the company, did not appeal to him. He had written to professor Andrade for advice, but received no reply.[66] A friend had suggested a career in scientific journalism. Perhaps he could join the staff of *Nature?* In his own mind, however, he was sure he "wanted to do fundamental research." Instead, the problem was "did I have the necessary ability?"[67] To 25-year-old physicist Freeman Dyson, who met him at this time, Crick seemed depressed about his career prospects in science. "He said he had missed his chance of ever amounting to anything as a scientist. Before World War II, he had started a promising career as a physicist. But then the war hit him at the worst time, putting a stop to his work in physics and keeping him away from science for six years. The six best years of his life, lost and gone forever." After that "it was far too late for Crick to start all over again as a student and relearn all the stuff he had forgotten. No wonder he was depressed." They parted, with Dyson thinking "How sad. Such a bright chap. If it hadn't been for the war, he would probably have been quite a good scientist."[68]

The Admiralty had been understanding. They had given him time off to attend lectures and seminars in theoretical physics, and he had taught himself the basics of quantum mechanics, but he was also reading books on organic chemistry and popular scientific literature on biology. Now aged 31, with a "not-very-good degree," uncompleted doctoral research, and a few short reports for Admiralty Research (but nothing published), why not try a new career? Perhaps this lack of qualifications had its advantages—one could be flexible about the future!

During the 1½ years that he worked in Intelligence, he enjoyed the lifestyle of an Admiralty scientist, his office in the "Citadel" in Central London. Built in the early years of the war, Sir Winston Churchill described it as "A vast monstrosity which weighs upon the Horse Guards Parade." It was intended for use as a military command post in the event of invasion and is protected by a roof of concrete that measures more than 14 feet thick—an ugly building to be sure, looking onto the Parade, where the reigning monarch observes the "Trooping of the Colour."[69]

At this time, it appears that Crick lived life to the fullest, encouraged, one suspects, by Kreisel, who had already been transferred to the Department of Miscellaneous Weapons in Fanum House, near Leicester Square. When many of the senior staff were in Germany at war's end, Crick was to stand in for them. Thus came about his witness of the debriefing of the famous scientist Werner Heisenberg.[70]

The apartment that Crick had rented while working at the Admiralty was on the second floor[71] of 56 St. George's Square, a lengthy square that runs between Victoria Station and the Thames in a part of London called "Pimlico." For a time, Kreisel was also there. Looking for a friend to share the flat when Kreisel left, Crick called Robert Dougall, then living with his parents in Croydon. Dougall recalled:

> One day the telephone rang and my mother said there was someone called Francis Crick asking for me. His cheerful bray sounding over the blower so unexpectedly after our meeting over a year before in Moscow was the pleasantest of surprises. It became positively musical when he said he had a flat in Pimlico and wondered if by any chance I would care to share it with him . . .
>
> Francis had even organized a splendid little Welsh lady, a Mrs. Thomas, to come in every morning to cook our breakfast and clean up for us. She was very pale and wore large steel-rimmed glasses and, for some reason, always burnt the toast. Each morning one woke to the sound of conscientious scrapings in the kitchen, while agreeable smells wafted along the corridor. Mrs. Thomas liked what she called "a good fry." She was admirable in every other way, not least in her impassive inscrutability at finding on occasions unexpected females in the flat. Francis's marriage had broken up and he was waiting for a divorce. I was entirely fancy-free.
>
> We had some tremendous arguments about religion, politics, world affairs, Russia and so on and our viewpoint on almost everything was diametrically opposed. Unlike some scientists I have known, Francis seemed to enjoy the company of people with a totally different outlook from his own. . . . There were two fair-sized bedrooms, and a large sitting-room with kitchen and bathroom at the end of the corridor. In the rather dark passageway Francis' bicycle leaned rustily against the wall.[72]

It was clear to Dougall that Crick was casting off any signs of his conventional home background of Northampton "and seemed determined to shake off any trace of stodginess in his make-up." London may have been drab and war torn, but if you wanted entertainment, it could be readily found among the nightclubs on Old Compton Street.

For theatregoers, it was a dream: So many great actors and actresses walked London's boards—Peggy Ashcroft, Laurence Olivier, Ralph Richardson, John Gielgud, and Sybil Thorndike. Dougall thought that "the war and the dropping of the atom bombs had a profound effect on him. Francis was a great humanist and determined not to work any longer for destructive ends."[73]

## The Gossip Test

There had been relief, both during and after the war, from conversation concerning the applied science of weapons. Crick had entered into discussions about the mysteries of cell division with R.V. Jones. What brings like chromosomes together so that they pair up at the onset of division? Are long-range forces at work? After the war, Crick began to read about penicillin, the new wonder drug, and he came across a short address by a chemist whose name he had not heard before—Linus Pauling—in which the world-famous Caltech chemist stressed the importance of weak attractive forces ("van der Waals forces and hydrogen bonds") that operate between molecules based on their complementary shapes. Some such complementarity—Pauling used the analogy of a lock and key—must surely be the basis for the specific attraction between molecules, such as when an enzyme "recognizes" its substrate (its food) and the antibody "recognizes" the antigen of the invading bacterium.[74] Here was the seed that germinated in Crick's mind, yielding in due time the science we know as molecular biology.

One evening, after telling some of his colleagues at the Admiralty about penicillin, he realized that he really knew very little about such topics.

> It came to me that I was not really telling them about science. I was *gossiping* about it. The insight was a revelation to me. I had discovered the gossip test—what you are really interested in is what you gossip about. Without hesitation, I applied it to my recent conversations. Quickly I narrowed down my interests to two main areas: the borderline between the living and the non-living, and the workings of the brain. Further introspection showed me that what these two subjects had in common was that they touched on problems which, in many circles, seemed beyond the power of science to explain. Obviously a disbelief in religious dogma was a very deep part of my nature. I had always appreciated that the scientific way of life, like the religious one, needed a high degree of dedication and that *one could not be dedicated to anything unless one believed in it passionately.* [75]

After Kreisel had shared Crick's apartment in St. George's Square, he recalled how Crick read a little book with a green cover, entitled *What is Life?* by the famous Austrian physicist Erwin Schrödinger. Based on the public lectures that Schrödinger delivered in Dublin in 1943, this slender little book explained, eloquently and with clarity, the problem that the existence of life presented to the physicist. How, in short, do organisms evade the destructive rule of the second Law of Thermodynamics? This law expresses the universal tendency of order to give place to disorder, for heat to be dissipated, and for organization to give place to randomness, such as a house that is not repeatedly tidied up. "How can we," he asked his audience,

> from the point of view of statistical physics, reconcile the facts that the gene structure seems to involve only a comparatively small number of atoms (of the order of 1000 and possibly much less), and that nevertheless it displays a most regular and lawful activity—with a durability or permanence that borders on the miraculous.
>
> Let me throw the truly amazing situation into relief once again. Several members of the Habsburg dynasty have a peculiar disfigurement of the lower lip ('Hasburger Lippe'). Its inheritance has been studied carefully and published, complete with historical portraits, by the Imperial Academy of Vienna, under the auspices of the family. The feature proves to be a genuinely Mendelian 'allele' to the normal form of the lip. Fixing our attention on the portraits of a member of the family in the sixteenth century and of his descendant living in the nineteenth, we may safely assume that the material gene structure, responsible for the abnormal feature, has been carried on from generation to generation through the centuries, faithfully reproduced. . . . How are we to understand that it has remained unperturbed by the disordering tendency of the heat motion for centuries?[76]

For an answer, he pointed to the robust nature of the forces that hold together the atoms of the genetic material in a molecular structure. That structure is very special—not a boring repetition of the same atomic units, such as in a crystal of salt or calcite—but a unique arrangement, a kind of "aperiodic crystal." "We believe," he declared, "a gene—or perhaps the whole chromosome fibre—to be an aperiodic solid."[77] The gene's "enigmatic biological stability" was to be traced not to what seemed an enigmatic chemical stability, but to a chemical stability that is now understood in terms of "the quantum theory of the chemical bond."[78]

Schrödinger's vision inspired Crick. "It conveyed in an exciting way," he explained, "the idea that in biology, molecular explanations would not only be extremely important but also that they were just around the corner."[79] It also focused attention on the molecular structure of the genetic material. Crick had also been reading the prewar work of John Desmond Bernal. Before war broke out, Bernal and Isidore Fankuchen had obtained beautiful X-ray patterns from crystals of tobacco mosaic virus in Cambridge. Such crystals, when dissolved in water and painted onto a tobacco leaf, infected it, reproducing myriad progeny virus particles. That surely is life—life from a crystal! This, thought Crick, is where to start the search for the secret of life— at the level of the molecules in such crystals. Crick later judged that this is when he first turned his thoughts to biology.

When Freeman Dyson met Crick again in 1946, he found him much more cheerful. "He said he was thinking of giving up physics and making a completely fresh start as a biologist. He said the most exciting science for the next twenty years would be in biology and not in physics." Dyson disagreed. Those years, he believed, would still belong to physics. If Crick were to "switch to biology now," he warned him, he would be "too old to do the exciting stuff when biology finally takes over."[80] Crick was already 30, but Dyson's words did not dissuade him.

CHAPTER 5

# Biology at the Strangeways

 *The only thing that ensures success is a consuming passion either for learning or success or both. Riding and dancing and gossip won't do? Working, even overworking may.[1]*

O n 4 May 1948,[2] the owner of the White Horse Riding Establish-ment could be seen riding his horse along a little road some 2 miles from Cambridge, called "Worts Causeway." Trotting beside him was another horse that he was bringing to a large country house called "The Strangeways Research Laboratory."[3] When the horses reappeared, both now had riders. On the second horse was a tall, slim man. Yes, it was Crick. He used to ride when he lived in Mill Hill. Now in Cam-bridge and working at this biomedical laboratory, he took time on Wednesday afternoons to go for an hour's ride. Such a habit at the Strangeways Laboratory had been unheard of hitherto, recalled Muriel Wigby, one of the former technical staff.[4]

But what brought Crick to this laboratory with its strong medical connections, its expertise in tissue culture, and its promotion of state-of-the-art optical microscopy? It had not been Crick's choice. He had not even known of its existence when it was first suggested to him. No, it was J. Desmond Bernal's research group, working on the structure of viruses, that he wanted to join at Birkbeck College, London. When he visited the College, however, Bernal was away in Sweden, and his sec-retary Anita Rimmel, whom Crick described as an "amiable dragon," discouraged him with the response, "Do you realize that people from all over the world want to come to work with the Professor? Why do you think he would take you on?"[5]

Crick had heard about Bernal during his time in the Navy and from reading C.P. Snow's 1934 novel *The Search*. The character Constantine is too close to Bernal for there to be any doubt about who was Snow's model here. Constantine is bubbling over with enthusiasm for proteins, a subject for which he is "full of facts and speculations . . . happy, exuberantly at home, overflowing with a sort of scientific wit."[6] Constantine was convinced that the mystery of life, as well as its origin, lay hidden in the proteins.

Bernal's research on the structure of viruses attracted Crick as a physicist and atheist. He hoped that by studying the structure of such elemental reproducing molecules, evidence would be forthcoming that would support a naturalistic rather than a supernatural starting point for life. This was the way to go if the mysteries beloved by theists were to be dissolved. But, how was he to enter the field? Had a Master's course in biophysics existed, he could have taken it. But there was none, so he turned to his former Admiralty boss Harrie Massey at University College London (UCL) for advice. Expecting Crick to ask him how to get into atomic energy research, Massey was surprised when Crick explained that he wanted to leave physics altogether. That called for advice from Massey's colleague Archibald Vivian Hill, 1922 Nobel Laureate for his research on the biophysics of muscle contraction and the leading figure in British biophysics.

Writing to Hill to request an interview, Crick explained that he had decided "to return to academic life," and that he felt "a strong, though uninformed inclination to some form of biophysics."[7] When they met, Hill stressed that Crick needed to first learn some biology. Perhaps he might like to work with Hamilton Hartridge in his Medical Research Council (MRC) Unit at Cambridge's Physiological Laboratory. A Fellow of the Royal Society (FRS) and a Fellow of King's College Cambridge, Hartridge had a vacant position for his research on color vision. But as Crick explained later, although he went to see him and was "very tempted," he had already decided that his "main interest was in the biophysics of the individual cell and its constituents, rather than in the special senses. . . ." He assured Hill that he had not forgotten his advice "that one should start by learning some biology." The problem was that to secure the funds to accomplish this, "one has to put oneself under the wing of some well-disciplined Professor, and this is not as easy as it looks!"[8] His experience working under Edward Neville da Costa Andrade had not been great. Now he wanted a committed mentor and was not "completely sure" that Hartridge and he "would get on."[9]

On learning of Crick's decision to study the biophysics of the cell rather than vision, Hill telephoned the Secretary of the MRC, the former Cambridge physiologist Edward Mellanby. Since the war, this Research

Council had been able to substantially increase its funding of biomedical research. Not only had it just funded the new Biophysics Unit at Kings College London, but it was considering taking on the support of the research group working on the structure of hemoglobin and myoglobin at the Cavendish Laboratory in Cambridge. At the time, it was unheard of for the MRC to support a group like this in a physics department.

Hill reported to Crick the substance of his telephone conversation with Sir Edward Mellanby, who had responded very positively and expressed the wish to meet with Crick. Hill added that "Mellanby agrees strongly with my proposition that as a preliminary to future work in biophysics you ought certainly to learn your biology. It occurred to him for example that you might enter as a PhD student at Cambridge and work in the Strangeways Laboratory there while spending a good deal of your time in the study of biological subjects."[10]

Picture Sir Edward, now the powerful bureaucrat, meeting the Admiralty scientist. He explains how difficult it is "to place a man, even of his [Crick's] standing, in the biological world with a decent salary." But Crick, who had been earning over £820 per year at the Admiralty, was not discouraged by this warning and replied he would be "quite prepared to come in with a research studentship, say £350 free of tax, and that his people were willing to give him sufficient money in addition, to live in London." Crick then explained that he was attracted to the application of "physical methods to biological products such, for instance, as Bernal's work on the structure of viruses." Mellanby responded that the Council supported young researchers with training grants, and that if Bernal would apply for a studentship for him "the thing would go through automatically." Mellanby's memorandum of the meeting ends with the comment "I was very much attracted to this man."[11] But there was no mention, apparently, of Cambridge and the Strangeways Laboratory.

At that time, there was no standard application form. Instead, Crick sent Hill his curriculum vitae together with a statement of his plan. He explained:

> The particular field which excites my interest is the division between the living and the non-living, as typified by, say, proteins, viruses, bacteria and the structure of chromosomes. The eventual goal, which is somewhat remote, is the description of these activities in terms of their structure, i.e., the spatial distribution of their constituent atoms, in so far as this may prove possible. This might be called the chemical physics of biology.

Problems as difficult as these will not, Crick warned, "yield to any single form of attack," but instead will require a variety of methods and disciplines. What he would therefore need would be:

1. A general knowledge of these related sciences (biochemistry, bacteriology, genetics, etc.) sufficient to enable one to appreciate the advantages and limitations of their methods, and the significance of current work.

2. A thorough grounding in at least one science. In my case this would of course be physics; but I realise that this would have to be extended in the direction of physical chemistry.[12]

Crick reckoned that this program would require between 6 months and 1 year. Some practical experience in simpler biological techniques was also very desirable. Once this background knowledge had been acquired, he wanted to do research on the structure of viruses using physical methods. Because applications were to be decided that week, Hill prepared Crick's immediately and, pending a decision as to where Crick would be based, a provisional award was granted. There was some rule-bending here, because normal practice required that a host institution be found first.

Before departing for his summer holiday on the Broads,[13] Crick received notice that his application had been successful. On his return at the end of July, he met with Sir Edward again. Reporting to Hill on the meeting, Crick made no mention of Bernal. Instead, he wrote "It is clear that it is essential for me to get some biological background, and desirable to do this at Cambridge, if possible." In typical unstuffy manner, he ended his letter to Hill reporting on what fun the holiday had been. "I got very brown and learnt to sail a small boat. I am now bursting to get down to work!"[14]

There was nothing left to do but to accept the advice given and try his luck at Cambridge. For this, Mellanby had already prepared the way by writing to Honor Fell. Crick, he wrote, "is not only very nice but he is obviously an able man and the Council thought sufficiently well of him to decide to back him."[15]

The account Crick gives of his first visit to Cambridge is not exactly glowing. There was Richard Keynes, who "talked to me as he ate his sandwich in front of his experiment," and Roy Markham, who described his work "in such a cryptic manner . . . that I could not at first grasp what he was telling me."[16] Neither offered him a place. At the Strangeways Laboratory, Honor Fell introduced him to Arthur Hughes, the microscopist. Fell suggested that with Hughes, Crick should study, under a magnetic field, the movement of particles taken into living cells in tissue culture. Hughes hoped that inferences might be drawn from these movements regarding the nature of the semiliquid contents of the cell. Oh—and one more thing—there was a room vacant that Crick could take. It had belonged to the physicist Douglas Lea, but he had died earlier that year.[17] That Lea's death had resulted from an unexplained fall from a fifth floor window in the University Library was naturally not mentioned.[18]

This did not seem to be a very promising program, but Crick reported back to Sir Edward that everyone he spoke with at Cambridge was kind and helpful. He had stayed at Trinity College with his friend Georg Kreisel. Kreisel had gone to Max Perutz, who was working on the structure of hemoglobin at the Cavendish Laboratory, to tell him about Crick, who then visited Perutz. (This, their first encounter, is described in Chapter 6 of this volume.) Crick also met the young biochemist Peter Mitchell. Honor Fell at the Strangeways was "very nice," suggested a project for him, and advised him regarding lecture courses and reading.[19] Fell also sent a report to Mellanby in which she stressed her view:

> No one can have any real insight into living material until he has done some research with it. It has been our experience that physicists in particular are very apt to over-simplify the interpretation of biological phenomena when they first begin work in this field.
>
> I have therefore offered to give Mr. Crick a bench in this laboratory where, if you approve, he can do some part-time work with Dr. Hughes, Dr. Swann and myself on certain aspects of the mechanics of cell division. We have a problem relating to the viscosity of protoplasm for which a technique has already been developed; I think this would suit Mr. Crick quite well and give him plenty of experience in handling and observing living cells. He seemed attracted by the idea.
>
> We were very favourably impressed by him—he seems intelligent and enterprising and has ideas.[20]

Crick accepted gratefully. Many years later, he recalled that he had lacked a deep interest in this research project,

> but I realised that in a superficial way it was ideal for me, since the only scientific subjects I was fairly familiar with were magnetism and hydrodynamics . . . this led to my first published papers. But the main advantage was that the work was not too demanding and left me plenty of time for extensive reading in my new subject. It was then that I began in a very tentative way to form my ideas.[21]

When asked why he had gone to the Strangeways, Crick's reply was very clear: "Because no one else [in Cambridge] would have me!"[22] At least he had an offer from Cambridge, albeit not from the University. Surely, though, London was the center of British biophysics, with Hill at UCL, Bernal at Birkbeck College, and John Randall heading the new MRC Unit at King's College London. Bernal had been ruled out by Hill, although not at first by Mellanby. Hill strongly disapproved of Bernal's efforts to introduce central planning into scientific research, and he disliked the political associations involved. He wanted Crick to come and work on muscle with him, but Crick declined. What of Randall's Unit? Advised by Massey, Crick went to see Maurice Wilkins, Randall's Deputy Director. The two got on famously, and Wilkins found Crick to be "how bright and lively." But when Wilkins suggested to Randall that an offer be made to Crick, "Randall was not keen—he had decided that Crick was rather boisterous and talked too much."[23] Thus, the die was cast—the Strangeways it would have to be. On the bright side, Crick had seen in Honor Fell a good director under whom he could work. She was 40 when Crick first met her. Like Crick, her joy throughout her life was science. She too had a "capacity for friendship" and "a sense of fun" that her colleague the electron microscopist Audrey Glauert saw had "reflected in the brightness of her looks and the warmth of her smile."[24] But she was a cell biologist who used in vitro methods; X-ray crystallography of material extracted from the cell would have to wait.

Hill and Mellanby favored Cambridge for all of the research units clustered there—the Low Temperature Laboratory, the Molteno Institute, and the Strangeways Laboratory—and the University's science departments, including the Cavendish "with its interest in molecular structure." At Cambridge, Crick could learn about the "various techniques necessary for the purpose he has in mind."[25]

Crick resigned from the Admiralty in mid September, said goodbye to his monthly salary check of £68 7s 6d, and prepared to move to Cambridge. But he retained the St. George flat for subletting. His old Naval boss Edward Collingwood wrote him, "Well, well! But congratulations on your success in storming the medical fortress." Collingwood had heard that the Admiralty had been seeking to retain Crick, and he felt sorry for Frederick Brundrett, who had interviewed Crick in 1939 and was still responsible for hiring. "As the centrifuge carries off the lighter particles" quipped Collingwood, it leaves him "with Shaw, Headlow and Captain Wright." Evidently they were not a very bright bunch. Brundrett, on the other hand, would soon become Sir Frederick and in due course succeed Sir John Cockcroft as Scientific Adviser and Chairman of the government's Defense Research Policy Committee. It was surely he who came to Cambridge in an effort to persuade Crick to return to the Admiralty. Crick refused, but as he later remarked, "Looking back, it was absurd because I had [had] a tenured job!"

Collingwood now played the part of mentor and took Crick to task over his financial plans for his future life on a research grant. ". . .if you want to go on how am I (or you) to reconcile an official income of £400 with a just charge of £450 for rent? You will need a very large rent from London [subletting] and a very patient aunt [Aunt Ethel's financial assistance] to make ends meet on those terms; and if they don't meet comfortably you may be sure that the studentship will not work. If, as is probable, you feel like biting my head off, think it over first."[26]

Evidently, Crick did respond by biting Collingwood's head off in a long letter justifying his plans, for Collingwood, fired back

> Your 24 pages don't leave me unmoved. And I suspect that they really leave you unconvinced too. . . . You, as a professional student of riper years, are drawing up an elaborate social budget. As a matter of history and everyday experience it is quite certain that success comes to the student who keeps to his attic and not to the intellectual butterfly who flits from salon to salon to party in search of relaxation or "stimulus." . . . Environmental conditions are never ideal and the only thing that ensures success is a consuming passion either for learning or success or both. Riding and dancing and gossip won't do? Working, even overworking may. Yours ever Edward Collingwood.[27]

### Experimental Cytology

The research on which Crick now embarked followed the tradition of experimental cytology. Most of the knowledge of cell structure had rested on the techniques of "fixing" and staining cells—killing them in order to reveal their contents—whether these were chromosomes in the nucleus or the several organelles in the cytoplasm such as the mitochondria. In contrast to this morphological method, experimental cytologists manipulated the cells: They isolated them one from another, pricked their membranes with microneedles, subjected them to centrifugal forces, and introduced foreign bodies into the cytoplasm, all done without killing the cells.

Fell had found that fine magnetic particles, when introduced into the growing medium for tissue culture, found their way "naturally" from the medium into some of the cells. These cells, it appears, had ingested the particles but continued to function normally. It occurred to Fell and Hughes that one could study the behavior of the cytoplasm—the "living" semiliquid contents of the cell in which the nucleus and cell "organelles" float—by wiggling these entrapped particles using a powerful magnetic field. The magnitude of the magnetic force required to produce these motions in the particles then offers a comparative measure of the resistance by the cytoplasm, which in turn is dependent on its viscosity.

At this time, experimental cytology still harbored elements of the once fashionable tradition of colloid science. Life was attributed to the special state of the semifluid contents of the cell, and colloid science was supposed to offer the appropriate theory for that state. What can we learn about the organization of the cytoplasm from establishing how viscous it is, that is, how sticky or resistant to flow? This is referred to in Lester Sharp's 1943 *Fundamentals of Cytology*, a book that was recommended to Crick. Sharp gave values for the viscosity of protoplasm that ranged from two or three times that of water to values hundreds of times higher. Such changes, he suggested, might underlie the alterations that take place when a cell divides, thus offering clues to the mechanism of cell division (mitosis).

Crick read the literature on the subject and, in a manner that was to become his "trademark," he swiftly dispatched the claims therein. Thus, the German botanist Alfred Heilbronn, who had inserted small fragments of iron into slime molds, then recorded the strength of the magnetic field required to move them, which was in Crick's opinion measuring not the viscosity of the cytoplasm in which the particles were moved, but their adhesion to the bottom of the container. "Whatever else was being measured," wrote Crick, "it was certainly not the viscosity."[28]

After discussing Heilbronn's inadequate methodology, Crick considered the consistency of his results to be "all the more remarkable"—a hint at the fudging of results perhaps? He criticized other researchers for their sloppiness. For example, William Seifriz, an authority on protoplasm, used this term for such a variety of objects, complained Crick, that it "makes nonsense of any discussion of the structure of cytoplasm." Another—actually the tissue culture pioneer formerly at the Strangeways, Robert Chambers—reported his results, wrote Crick, in a manner that "to make the results as they are stated, even approximately quantitative," was impossible.[29] What a wonderful Crickian put-down!

Not only does this 1950 paper supply us with examples of Crick's piercing critique and concern for the quality of experimental evidence, it also offers us a preview of his conception of the relationship between theory and experiment. Without a general theory of the structure of elastic liquids, such as the cytoplasm, he warned, "it is not yet possible to relate features of their behaviour with features of their structure." He suggested in vitro experiments with "crude protein solutions, or possibly a suspension of Claude's microsomes, etc., to see how magnetic particles behave in them." But superficial similarities might prove misleading. "To discover which features of a model are essential to give the properties measured, and which are accidental . . . it is necessary to have a theory, however crude, to account for at least some of the observable features."[30]

The research that occupied Crick from the fall of 1947 to the spring of 1949 involved growing cell cultures in a nutritive medium and setting up the visual and cinematographic recording equipment. He then made a large number of carefully recorded measurements of the responses to a magnetic field of the iron particles engulfed in the cells. Three kinds of data had to be collected: those from "twist," "drag," and "prod" experiments. In the case of the "twist" experiments, he recorded the time taken for the particle to twist through an angle of 90°. To attain more precision, he also photographically recorded these twists using a 16-mm camera. He projected the frames onto a sheet of paper, making a faint outline in pencil of the particle on the frame and comparing this with successive frames to establish the response time. This tedious work in which he studied some 8000 frames proceeded at the rate of 120 frames per hour. For the drag experiments, the magnet needed to be very close to the cells, so Crick devised a minute container that measured 4 mm in diameter to enclose the cells. The prod experiments were not treated quantitatively; their aim was to test the strength of structures such as the nuclear membrane and cell wall.

Nothing of any significance came from this work, although references to it continued for some time. Crick and Hughes concluded that the cytoplasm is "a thixotropic gel, showing feeble elastic properties." This was not a new idea, they admitted, but it was the first time that "the order of magnitude" of its elasticity had been established. As for theories about the structure of the cytoplasm, Crick was pretty dismissive. He rejected the "brush-heap" theory of Seifriz (like a pile of matches) and the "framework" theory of Frey-Wyssling, both of which were popular in the literature. According to the latter, the fibers of the framework had the thickness of a single polypeptide chain. This prompted the retort, "This may be true but there does not appear any evidence for it. The argument really amounts to saying that they cannot be so thick that they can at present be detected."[31]

He was particularly opposed to any suggestions that a fixed structure exists in the cytoplasm, such as the so-called "cytoskeleton" of the Oxford biochemist R.A. Peters, simply because the numerous coordinated chemical reactions taking place there seem to require it. What evidence was there of a structure that he judged to be transient? "If we were to suggest a model," he wrote, "we would propose Mother's Work

Basket—a jumble of beads and buttons of all shapes and sizes, with pins and threads for good measure, all jostling about and held together by 'colloidal forces.' "[32]

Crick shared with Hughes the authorship of this, his first paper. Some passages giving considerable experimental details suggest Hughes' authorship, others remind us of the later Crick, but the critique of the literature described above is vintage Crick. Hughes had set up the phase-contrast microscope for him and taught him how to use it. When a glitch occurred, Crick often had to call on Hughes to sort out the problem. Otherwise, the project was just handed over to him. A second paper on the physical theory underlying the experiments, with Crick's name as sole author, he put together from sources he knew in physics.

How did all of these measurements of little particles help a physicist entering biology? They gave Crick plenty of experience in peering down the microscope at cells and interpreting what he saw. Indeed, this may have been responsible for his distaste at looking down a microscope ever since![33] And there is no question that he executed this work with great care and thoroughness. His dedication to what he considered a project of doubtful value is remarkable. Certainly the literature on the topic confirmed his suspicion that biology has more than its fair share of sloppy experimentalists and muddled thinkers. But the real benefit of these 2 years was the time it gave him to read, meet other scientists, attend meetings of the newly formed Hardy Club (to which he was elected at its foundation in 1949), and go to some lectures. But his preference when entering a new subject was to read exhaustively until he had mastered it, and at the Strangeways he could do just that.

Did the circle of his Strangeways colleagues have an impact on him? If anything, it was the other way around. For example, he managed to effect a change in the social style of morning coffee and afternoon tea. It was their habit to discuss the day's news, the film seen last night, or the sports news. Crick would ask a serious question about science and disrupt the pleasant banter, but in the process he introduced regular discussion of scientific issues—a habit that he continued at the MRC Laboratory and the Salk Institute. His Strangeways colleagues clearly considered him far too speculative and too ambitious about confronting fundamental questions, as can be seen in the cartoon of him that appeared in a brief history of the Strangeways Laboratory (Fig. 5.1).

What was Honor Fell's opinion? Her letter to the MRC recommending a second year for Crick offers a clear answer.

He has made an excellent progress while he has been in Cambridge and shows a real aptitude for biophysical work. He has not only

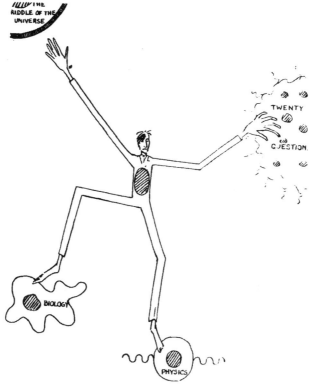

*Figure 5.1* Cartoon of Crick drawn by his colleagues at the Strangeways Laboratory. ("Twenty Questions" was a famous radio panel quiz show.)

acquired considerable biological knowledge from reading, lectures and discussions with other members of the laboratory, but is making good headway with his research. . . . This investigation was begun by Dr. Hughes and myself and we handed it over to Mr. Crick to continue and develop. I think this particular problem has considerable educational value for a physicist, as it involves many hours of close microscopic study of living cells of different types.

I feel that Mr. Crick would greatly profit by another year of study and part-time research. He has to master the elements of an entirely unfamiliar science and no matter how keen and intelligent a man may be, one year is not long enough for this purpose.[34]

Contrasting with Fell's warm support, Hughes' attitude toward Crick as a scientist was not, it seems, so positive. Although he was later to refer a number of times to the magnetic particle experiments done at the Strangeways, no one would be able to gather that it was almost all

Crick's work. Hughes' dislike of Crick's speculative style and the importance Crick and others were beginning to attribute to nucleic acids are evident in the following passages in Hughes' 1952 book *The Mitotic Cycle:*

> It is difficult to resist the impression that the elaboration of theories in cytology is sometimes carried to great lengths. . . . Within recent years, the nucleic acids have proved a favourable subject for exercises of this kind, the scope of which has not been restricted to this planet alone. The question at issue is not whether truths beyond the immediate reach of observation can be apprehended by imaginative inference, but whether such theories have been found to serve as a useful basis for further research, and to stimulate fresh inquiry into particular aspects of cell division. It is doubtful whether a survey of the progress of cytology would uphold such a claim. . . . Anything in the cell could be explained by a general appeal to the nucleic acids, as readily before an experiment as after it.[35]

Who talked during the tea break about nucleic acids and life on other planets or apprehended truths by imaginative inference, if not Crick? However, Hughes did appreciate Crick's mathematical ability, as when he thanked him for correcting the equation for determining the interval between mitoses.[36]

When asked what kind of a man Hughes was, Crick's response was to say, "Well, you see, he was a microscopist." That, it appears, coming from Crick, was a somewhat damning remark, because microscopists have loved the light microscope, an instrument that reveals objects down to 10,000 times the size of an atom, but nothing smaller. In spite of this limitation, the profession was gung ho after the war about the new potential of this venerable magnifying instrument. As one microscopist put it in 1951, "it seems as if we are witnessing the dawn of an inspiring age in microscopy."[37]

The principal cause of the excitement was the phase-contrast microscope, an instrument that was making possible the study of the fine structure of living cells rather than that of cells killed by fixing and staining. Hence, cells could now be viewed as they divided. The instrument uses light rays with a phase difference. This causes interference—a slightly out-of-focus effect—which reveals delicate cell structures that otherwise remain invisible. It does not increase the magnification of the microscope but instead brings those hidden transparent structures in the cell into view. The method had been adapted to microscopy during the war and Honor Fell wanted to use it at the Strangeways. She had urged the release of Flying Officer Arthur Hughes from the Royal Air Force at war's end to make this possible.

On his return to the lab, Hughes adapted the instrument for phase-contrast cinematography to record cell division in action. These films became famous and were shown at many events and scientific meetings in the late 1940s. They showed how in the "dance of the chromosomes" (Fig. 5.2), like members of the two sets of chromosomes in the cell associate in pairs, become arranged on an "equatorial plate," and are then pulled apart and, it seemed, drawn to opposite poles of the cell to become the nuclear content of two daughter cells (Fig. 5.3). This was the dramatic play of mitosis that physicists including Crick had discussed during the war. While Crick worked at the Strangeways, Hughes was often away at meetings, including the International Scientific Film Festival in London in 1948, and in 1949 at BBC Television, showing his films. Their subject—cell division—was the hot number in biomedical science at the time. Hughes also worked with Michael Swann in Cambridge University's zoology department on the arrangement of protein molecules that constitute the spindle between the two poles of the dividing cell.

Hughes began his book *The Mitotic Cycle* focusing on nucleic acids, giving plenty of history, but for any genetic function, he favored "chromosomin," a protein of questionable existence by the time Hughes was writing (1950–1951). Then followed a typical piece of his sarcasm: "In recent years, there has been a tendency to explain every cellular change

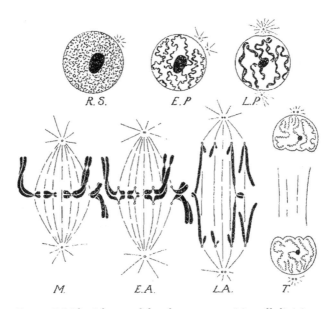

**Figure 5.2** The "dance of the chromosomes" in cell division.

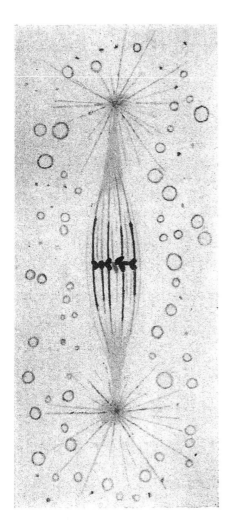

*Figure 5.3* Chromosomes adhere to
the spindle fibers as they move to
opposite poles to form two daughter
nuclei. The thickening of the spindle fibers shown in the drawing is
surely an artifact.

in terms of nucleic acids, much as the Peripateticks once used to interpret all natural phenomena in terms of the four elements."[38]

To whom was that remark directed, if not to Crick? It was Crick
who gave a talk on the nucleic acids while he was at the Strangeways,
and it could well be that Hughes remembered the event or subsequently learned about it. This talk was given in a course for hospital
physicists. Crick's memory of the audience was vivid.

I tried to describe to them what the important problems in molecular
biology were. They waited expectantly, with pens and pencils poised,
but as I continued they put them down. Clearly, they thought this was

not serious stuff, just useless speculation. At only one point did they make any notes, and that was when I told them something factual—that irradiation with X-rays dramatically reduced the viscosity of a solution of DNA.[39]

To such an audience, training for very practical work, the response was surely predictable. But to the Strangeways scientists, the nucleic acids were very relevant because of the focus of much of their work on cell division. Indeed, in 1948 Honor Fell brought a biochemist, Dr. M. Webb, to the Strangeways to study the chemical aspect of cell division. He was to join the Cell Physiology Unit she was forming that would pursue a multidisciplinary approach to the subject. Webb and Jacobson studied the changing distribution of the nucleic acids during the division process, reporting that the chromosomes contain only deoxyribonucleic acid (DNA) at the beginning of the division cycle but later become charged with ribonucleic acid (RNA).[40]

Crick's talk is likely to have touched on the analogies between viruses and chromosomes, both of them nucleoproteins, and to have discussed the chemical basis of specificity. Indeed his impression was that around this time he had begun to formulate the very fundamental thesis that the three-dimensional structure of a protein is determined by the one-dimensional sequence of its amino acids. Moreover, this sequence is somehow encoded in a gene. Whether he went further and claimed that the hereditary encoding is somehow contained in the DNA he could not later recall.

The research environment of the Strangeways was clearly medical. Much of the work was aimed at current pressing problems of the time, for instance, what damage was suffered by the various tissues under radiation treatment? What compounds were most effective in chemotherapy for cancer patients? Honor Fell's priority was with tissue culture techniques that would permit the study of organ development and regeneration. To Crick, she seemed to have something of the gardener about her in her devotion to her tissue cultures. They were, wrote Crick's friend, Maurice Wilkins, her "little dears."[41] The emphasis in the research reflected the major source of the funding and the circle of medical institutions with which the Strangeways had a special relationship. These included the Chester Beatty Institute, Mount Vernon Hospital, the MRC Radiobiological Laboratory in Hammersmith, and the Birmingham Hospital Group. Among these, it was the Chester Beatty Institute that had a strong program of basic research on the chemistry of the nucleus, especially the chemistry of nuclear proteins known as histones.

The applied character of most of the research at the Strangeways was not attractive to Crick, thirsty as he was to confront fundamental questions requiring, he believed, work at the molecular level. Nor was he attracted to the study of cell division. Nevertheless, the Strangeways served him well for 2 years, and by the spring of 1949, he was looking ahead to fulfill his original ambition to enter the field of X-ray crystallography. This time he wanted to work with Max Perutz. Thus, he wrote to A.V. Hill apologetically—seeing as Hill had all along wanted Crick to work with him on the biophysics of muscle.

Dear Professor Hill,                                              7.iii.'49

After much thought I've decided, in spite of your advice to me, to try to join Perutz's group. However much I looked into the possible lines of work I always returned to this one. The problem does really fascinate me, and I feel that after having given up my Civil Service job in order to do work in which I was really interested, it would be better to do this, even if the work is difficult. [42]

Hill replied reassuringly: "I expect you are quite right. If the X-ray diffraction studies of protein are what interest you most, in spite of any deterrent I may have exerted, you can be reasonably sure that your decision is the best one."[43] These two scientists remained on good terms and when Crick became a Fellow of the Royal Society in 1959, Hill wrote to congratulate him. Crick responded: "I was particularly pleased to hear from you, as you were indeed my original sponsor in the new field I was entering. As I remember it, you bothered Mellanby not once, but twice, and it was only because of you that I ever got started at all. I do hope you have forgiven me for not working on muscle!"[44]

### Life Outside the Strangeways

*I must be dreadfully romantic to be marrying an impecunious man like you!*[45]

If the lab was an uneventful place and society rather staid, Crick was never one to let local circumstances cramp his style. There was his son Michael to visit at Weston Favell on alternate weekends. The other weekends were usually spent in London. As one who made friends easily, he could relieve himself of the small world of Cambridge with a quick train ride to the metropolis. There, for instance, was his friend Margaret who was working in the theatre. In October, she complained that Francis had not been in touch with her: "I am astounded beyond measure, that you should attribute your long silence to the intricacies of chicken cells [cells grown in tissue culture]—are they really so absorbing? Indeed I hope not, for it would seem you are doomed to perpetual silence. . . ." Writing again, she appealed to him to "Try and tear yourself away from tissue culture!"[46]

Then there was Edie Hammond, whom he had met during the war while working at West Leigh in the Mine Design Department. She was at the time living in Chichester with her mother, and Crick had been a lodger there after Doreen left him. Edie was a "land-girl," one of thousands who helped the war effort by working on a farm. Crick still remembered one of her asides about pub life in the country: "They don't give you all them drinks for nothing, you know!"[47] She was the kind of spirited lass that Crick enjoyed, and one whose warmth Crick's young son Michael appreciated. She was not as well educated as he, but she thought she would make Crick a good wife. Crick would tease her by testing her general knowledge, for example, with questions such as "Who was Lord Marlborough?" (Answer: "The Duke of Wellington!") But she enjoyed literature, and when she read Oscar Wilde's *Dorian Grey*, she informed Crick that Dorian "kept reminding me of you; not his wickedness of course, but just himself and what he said; I think you must have read a lot of Oscar Wilde in your time."[48]

After Crick's divorce was final in May 1947, he invited Edie to London. Evidently, she accepted, because later she recalled those "halcyon days of July when the skies were blue and the nightingale was singing like a lark in St. George's Square and your heart was somewhat softened."[49] But in the fall of 1948, Crick visited Edie and her mother on a Sunday, bringing with him a female companion. The latter was "looking," wrote Edie, "as completely in possession of you as if she was already your wife. . . . I was devastated by that awful shock I had." The companion was surely Odile Speed. Edie was also appalled by Crick's "sideburns." "You are not a 'Professor' yet," she objected. "'Sideburns' aren't affected by men these days Francis, at least not real men. [They] make you look like a pansy and worse . . . they don't even look like proper 'sideburns,' only like little bits of fluff you have forgotten how to shave off."[50]

In 1949, Crick must have written to tell her about his engagement to Odile, because Edie replied, "Thanks for the bombshell. You seem to derive great satisfaction from dropping them at my feet with remarkable suddenness [*sic*]. . . ." Then she assessed the couple's future.

> I only see the marriage going where the last one went unless your future wife (lucky thing) is prepared to wait on you hand and foot, be patient with your incessant reading and deal carefully with your finances. Still,

I suppose she has all these virtues and many many more or you would never be contemplating marriage with her, since even you must have obtained some degree of level-headedness by now I suppose, although I have never yet met a man quite as frivolous as you were.[51]

When Crick first met Odile, she was a "Wren"[52] officer working in the intelligence section of the Department of Submarines and Mines at the Admiralty. One afternoon in 1945, Crick, who was on a visit from Havant, was chatting with some Admiralty staff when a glamorous young Wren came through the door and her shopping bag burst, liberating the brussel sprouts she had bought onto the floor. At once, up leapt Crick with his characteristic gallantry to assist her.[53] He promptly asked her out to dinner, but she refused so abrupt an invitation. In any case, he looked rather scruffy with his unflattering raincoat hugging his neck. But Crick was not so easily put off and found a way to get in touch with her again. Here were two potential life mates with very different upbringings, education, and interests. Odile was the firstborn of Alfred Valentine Speed and Marie-Therese (née Jaeger). Her birth on the 11th of August 1920 took place at home above the family's well-known shop in King's Lynn, where her father owned a jewelry business at a prime site, number 89 on High Street. He had first met Marie-Therese when she came from Paris to England to learn the language after the first World War had ended. They were married in Paris in 1919.

Regarding class, the Crick and Speed families had much in common—reasonably wealthy middle class and living in the provinces. Just as Francis' grandfather had founded Latimer, Crick, & Gunn, boot and shoe manufacturers, so had Odile's grandfather Alfred Speed founded Speed and Son, jewelers and watchmakers. Making a difference, however, was the fact that Odile's mother was French and Catholic. The Jaegers came originally from Alsace-Lorraine, but Marie-Therese was born in Paris where her mother, Marie Theophile Jaeger, a renowned *couturière*, lived on Rue Lamarck.

Instead of the dissenting tradition of religion of the Crick family, Odile grew up under the strong influence of her Catholic mother. Accordingly, she was sent to the junior section of Our Lady of Walsingham Convent High School, where she was taught by nuns. But at 12 years of age, she was sent to her grandmother and Aunt Madelaine in Paris to continue her schooling there. Two years after having returned to King's Lynn in 1934, she was off to Vienna. The beautiful 16 year old lodged with a family, presumably business friends of her father's (he too was a jeweler). She studied music and art, taking piano and drawing lessons. She also studied with the Viennese sculptor Fritz Wotruba. It was at this time that she also enjoyed sing-songs accompanied by the accordion and decided to learn to play this instrument as well.

When Hitler came to Vienna on 15 March 1938 to celebrate the *Anschluss,* Odile's hosts took her with them to the *Heldenplatz* (Heroes Square) to witness Hitler give a speech. To her amazement, she found herself among several hundred thousand Austrians eagerly awaiting Hitler's appearance on a balcony of the *Hofburg* (palace of the Habsburgs). Teachers had brought their young students and were relentlessly rehearsing the welcome, "Dear, dear Hitler, Come and be nice. Come, show yourself at the balcony." So horrified was Odile at the warmth of the response of the crowd that she telegraphed at once to her parents: "I am not staying here. Take me out of this horrible place."[54]

Marie-Therese came to Vienna and subsequently moved Odile to live with her grandmother in Paris. Soon the political situation deteriorated further and Odile returned to King's Lynn. After living in Paris, it was disappointing to return to provincial King's Lynn. She joined the war effort, but there was boredom and frustration associated with so much volunteer work—for instance, practicing to work as a medical assistant on a train equipped as a hospital for the evacuation of London's casualties should an invasion occur, driving a truck for the Red Cross, but mostly waiting around for something to happen. Not until her mother spotted an advertisement in *The Times* for those fluent in German to apply for employment at the Royal Navy did she find an escape, but into another boring task: listening for German naval radio messages. When Crick met her in 1945, she was translating secret documents from German for the Intelligence section of the Department of Submarines and Mines directed by Commander Ashe Lincoln. In peacetime, she returned to the art world but gave up her fashion design course after the first year. As Crick remarked, "she preferred marriage to further study." [55]

Although, unlike Francis, she was fluent in French and German, she knew no science. When Crick asked her how far gravity extends from the earth, she suggested 10 miles. But this did not prevent their relationship from blossoming. Francis could disparage her ignorance in the sciences and she would cheerfully ignore the put-downs and continue to do so throughout their married life. But she would educate him in matters of fashion, persuade him to take ballroom dancing lessons, and pursue her own skills as an artist. Odile could stand up to Francis' forceful personality, for she was a young woman with a mind of her own who had a clear sense of her own abilities and the beauty to hold Francis' admiring attention. "She had her interest in art," explained Freddie Gutfreund, "and knew that she was admired for her good taste and judgment. She also knew that she did not understand anything about science and did not care that Francis laughed about any comment [on science] she made."[56] Her sister-in-law described Odile as "a great individualist" and that she "loved life, was not especially bohemian [despite the remarks in several obituaries], but truly tolerant of other people and their ways. She was a very natural and talented person."[57]

Odile and Francis both loved the theater and enjoyed partying. Getting together was easy when Francis was transferred from Havant to the Admiralty in January 1946. They were still not far apart when Odile left the Admiralty to study at St. Martin's School of Art on Southampton Row.[58] In 1947, she turned to fashion design and transferred to the Royal College of Art in South Kensington. At this point, Francis moved to Cambridge, but not before making a "brilliant suggestion" to her, for she responded, "What a nuisance you are to go and hide yourself away in Cambridge. As for your brilliant idea that I might follow you there—well I reserve my judgment, but if you were thinking of doing some research in Paris next year well, it would be worth considering!"[59]

Meanwhile, Francis' divorce from Doreen had become final in May 1947[60] and he was now free to marry. As a Catholic, Odile was not. Evidently, her Catholicism had not yet dissolved in the face of Crick's agnosticism, and as her confessor reminded her, she could not marry a divorced man. But she had fallen for Crick's charm, his many winsome ways, and his love for her. There were times when she got angry with him, but he was the first man that she had met in whose company "she never felt bored."[61]

Did she know what was in store for her? Crick's younger brother Tony, who had moved with his family to New Zealand in 1948, hoped she did! He had married Joan Mary Rowlands in 1945 and completed his radiological specialty in 1947, but disliking the shape of the new state-run National Health Service and yearning for the joys of outdoor life, he had emigrated. Addressing his brother by his family nickname, he wrote

Dear Craxie and Odile,

By the time you get this Odile will be an honest woman—God help her—and Joan and I send many congratulations to you both. Craxie is certainly the proverbial lucky man and if Odile does not know what to expect by now she never will. I have always looked on Craxie as the impossible husband but was very shaken when he started to learn French, even if he did get one more mark than I did in School Certificate.[62]

Learning of the engagement, 8-year-old Michael Crick got excited at the thought that he might, like other boys, have a sister in the not too distant future. This set off alarm bells in the mind of his grandmother Annie, who had helped by taking care of Michael for much of the last 4 years. In the forthright manner of a Crick, she wrote to Francis

> Your profession is so badly paid that there are great risks as to the financial position you may find yourself [in], and to saddle yourself with extra responsibilities—needs very careful thought. Babies only last as babies a few years, each year adding considerably to the exchequer, unless you want them to be educated in Council and free schools—you would be wise to budget your finance for the years to come, on a firm basis, not on [a] speculative [one], things in the future change, and often don't work out as one wishes. . . .
>
> In your type of work you won't succeed, if you have heavy domestic worries, and believe me, they will be many on a small income.
>
> Up to now you have had some one to back you, that time has passed and you have only yourself to look to. Forgive my speaking so bluntly. . . .
>
> I wish you every success in your work, and hope it will turn out as you desire.
>
> Yours, A.E.C. [Annie Crick][63]

Annie's concern was understandable. Here was her son with a son to educate, about to remarry, his age 33, his salary £400, and his appointment a 1-year contract with the possibility of renewal. Compare that with Hamilton Hartridge's offer of the MRC staff position that he had turned down!

Now Odile had always been a shrewd judge of character, and a letter she wrote to Crick when they were dating suggests that she had the measure of him in practical terms. He had been back to his family in Northampton for Christmas 1947 and Odile wrote, "Well I must say being in Northampton doesn't improve your literary style," but she wished him "lots of work and entanglements *du Coeur* and may money rain down upon you from the stars and other sources."[64]

Before their engagement, Odile had been concerned to keep the Speed family ignorant of her relationship with Francis, knowing how much her mother disapproved of a divorced suitor. Needless to say, complications arose over when and where to meet. Crick's lodgings on Jesus Lane were too small, and the flat on Hogarth Street in Earls Court that Odile shared with Patsy Lamb was susceptible to visits from her family, especially from her brother. Jokingly, she remarked on the prying "octopus," alluding, one suggests, to family members all seeking gossip on her love life. She wrote to Crick, imagining his reaction. "How my aunts and other relations haunt you! An absolute nightmare—" yes, I was quite fond of <u>her</u>, but she had RELATIONS you know![65]

Finally, in December 1948, Odile broke the news of their engagement to her parents. She reported to Francis that "Mother still rather prejudiced, but it doesn't take much to bring her round and that's really up to you when you meet her."[66] After the event, Odile reported back, "You behaved beautifully . . . *Maman* and dad were charmed with their day in Cambridge."[67]

Crick had taken Odile to visit his parents some time ago, while his father Harry was still alive but unwell. At that time, Harry and Annie were renting Elsmgarth, a beautiful old house in the village of Weston Favell near Crick's birthplace. Harry died in February 1948, so Odile did not come to know him as well as she did Crick's mother.

Odile described Annie as "a very lively person and obviously a very good housewife—very careful and efficient—and [with] tremendous admiration for Francis." Annie had been keen for her favorite son to excel both academically and socially, and to this end, she had taken great care during his upbringing, teaching him politeness, table manners, cleanliness, little courtesies such as writing thank-you letters, considerateness, graciousness, and kindness—in short, she taught him the importance of "behaving as a gentleman," Crick explained. "Annie," said Odile, "was much more concerned with health and intelligence than matters of the heart."[68] She had trained as a teacher and worked in this profession until at age 35 when she married the 27-year-old Harry. Regarding Crick's parents, Odile judged that Francis was more like his mother in temperament and looks, whereas his brother Tony was more like his father, full of fun and warmth. Annie, by this time a widow, invited Odile's parents to Mill Hill to meet her, and Crick invited them to Cambridge.

Now came the problem of finding lodgings in Cambridge. Here, Crick's future boss Max Perutz could help. He and his wife Gisela were moving out of an apartment called "The Green Door." Consisting of two and a half rooms plus a kitchen with bath in "The Old Vicarage," it was centrally located next to St. Clement's Church on the corner of Thompson's Lane and Bridge Street. It was tiny, but the rent was only 30 shillings per week, subsequently raised to 30 shillings and 6 pence.

The apartment occupied the third floor of this ancient stone edifice, its many chimneys towering above the street below. Access to the third floor was via a separate entrance and a flight of stairs in a wooden extension at the rear end of the house. The toilet opened off the stairs, below the level of the apartment. One then entered a passageway with three rooms on the left. On the right side of the passage was clos-et space that Odile would use to store food items—still rationed 4 years after the war's end.

Like so many men, Crick had not thought much about storage space. For him, the chief merits of The Green Door were its proximity to the Cavendish laboratory and the inexpensive rent. Odile, however, was definitely concerned about the "storage space, or rather lack of same." She had only seen the apartment when the Perutz family were still liv-ing there and had not ventured beyond the living room that she reck-oned had measured about 13 feet by 11 feet. She wrote asking Crick to "imagine both our untidy selves wading knee-deep in garments and junk!" Where, she demanded, were the following to be stored? "*a*) all our clothes *b*) household linen and spare bedding *c*) some of Michael's clothes and toys *d*) my drawing paper, chest, paints, canvasses, easel, sewing machine and material *e*) your books *f*) my books *g*) carpet sweep-er and accordion?" This was the first of her concerns, but four more fol-lowed: "Where is the W.C.? I don't object to the bath in the kitchen, you can wallow in the bath while I cook the breakfast. 2. Fuel arrange-ments—are there coal fires, if so where is the coal (probably in the base-ment!) If not, what other form of heating is there? 3. Light—will there be enough light and space for me to paint etc. If not how can I make the most of my prospective leisure? 4. Is there power laid on?"[69]

Odile was clear that they "will need more *lebensraum* [living space] after a few months," so a subletting clause would be advisable, thus enabling them to move elsewhere and rent out The Green Door. It was all very well to enjoy the ambiance of an historic building, but just how primitive would life there be?

On Saturday 13 August, the wedding took place in London at the Marylebone Registry Office. Annie was present with her grandson Michael and members of the Speed and Jaeger families. In addition, there were some 45 other guests, including Georg Kreisel. And what a handsome couple Odile and Francis made, Odile so stunning in her beautiful, tiered dress and hat, both made by her. A reception was held at 96 Cheney Walk, Chelsea.

The honeymoon couple went off to Italy, blissfully ignorant that scientists from around the world were converging on Cambridge for the First International Congress of Biochemistry. On their return, Odile and Francis began their new life at The Green Door. They were poor and their living quarters cramped. To take a bath, one often needed to remove "a miscellaneous collection of saucepans and dishes" before lifting the hinged cover to fill the tub. But the years at The Green Door were very happy ones. As Crick recalled,

Odile and I had our leisurely breakfasts by the attic window in the little living room, looking out over the graveyard to Bridge Street and
beyond that to the chapel of St. John's College. There was much less
traffic in those days, though many bicycles. Sometimes in the evening
we would hear an owl hooting from one of the trees that bordered the
college. . . . Odile luxuriated in her newly found leisure, read French
novels in front of the small gas fire, and attended, informally, a few
lectures on French literature, while I reveled in the romance of doing
real scientific research and in the fascination of my new subject.[70]

The year 1949 was a good one for Francis. He won Odile's hand,
and his wish to study X-ray crystallography was at long last granted.
These two changes in his life set the stage for the major role he was to
have in the science of the 20th century. Odile, likewise, now found herself in the demanding roles that came with marriage to Francis, and she
excelled in all of them.

Was there an inevitability about the move to Cambridge—a kind of
predestination? Surely not. But were it not for the war, it is doubtful
that Crick would have strayed from physics, and one can easily imagine him as a member of the astrophysics community, such as George
Gamow and Tommy Gold. With Odile, he had the good fortune of having a beautiful and steady helpmate, who was shrewd, well organized,
fashion conscious, a wonderful cook, and, as an artist, had her own
interests and circle of friends. Had it not been for employment by the
Royal Navy, he would doubtless never have met her. Nor would he
have met Georg Kreisel, who encouraged him to take the radical step of
moving to biology and then followed his career, helping him with his
mathematical skills when needed. The old adage "It is not what you
know, but who you know" can be important for other reasons than
those seeking preferment through the "old boy network."

## CHAPTER 6

# Helical Molecules at the Cavendish Laboratory

 *I think I have the sort of brain which enjoys puzzles of this sort.[1]*

By early 1949, as Crick's research on the viscosity of cytoplasm neared completion, he was faced with a decision about future employment. There was the possibility of continuing experimental cytology, probably working on cell division, where viscosity changes were still considered possible causes of the chromosomal movements of mitosis. Such research could involve collaboration with his friends Michael Swann and Murdoch Mitchison, who worked in the University's zoology department and who were among his first academic friends at Cambridge. Like Crick, both had had their careers interrupted by war work. They had returned to Cambridge for doctoral research in 1946—Swann to Gonville and Caius College, and Mitchison to Trinity College. Both pursued the mystery of cell division, with Mitchison later becoming a world authority on the cell cycle. Swann, big of frame and friendly of manner, was diplomatic, good humored, and tactful. Crick described him as "very smooth," and he was not surprised by Swann's subsequent stellar career—first with Mitchison building experimental cytology at Edinburgh University and subsequently as its Principal and Vice Chancellor, then as Director General of the British Broadcasting Corporation. But the thought of more biophysics of the cell was far from Crick's mind. He had not forgotten how, on his first visit to Cambridge in 1947, his friend Georg Kreisel had suggested going to see Max Perutz, the protein X-ray crystallographer at the Cavendish Laboratory who headed the

Medical Research Council (MRC) unit for the Study of the Structure of Biological Systems. Kreisel had even prepared the way for Crick by meeting with Perutz himself.

Confronted with the prospect of meeting with Kreisel, Perutz was surprised. "What diffident character needed such a strange Ambassador—an eccentric Austrian mathematician—to pave his way," he thought. In fact, Crick had not asked Kreisel to act thus for him,[2] and when the door opened to reveal the tall Englishman with bushy eyebrows, Perutz recalled, Crick soon "put us all in high spirits by his laughter."[3] Two years younger than Perutz and certainly not shy, the hopeful Crick joked and interacted in a lively manner with the restrained and quiet Perutz. One can imagine this meeting of the voluble and extroverted Crick at 6 foot 2 inches tall with Perutz, a mere 5 foot 6 inches, slight of frame, quietly spoken, dressed in subdued colors, his brown eyes glinting beneath a prominent forehead. How different was the experience from Perutz's expectation!

Now 2 years later, it was time for Crick to repeat his original request to the MRC to fund his research on X-ray crystallography of proteins. Never reluctant to go right to the top, in February 1949 Crick requested an interview with Sir Edward Mellanby.[4] When the two met, Crick did not mince his words. The work he had undertaken on the viscosity of the cytoplasm, he explained, was nearly finished "and he saw no point in continuing it." Mellanby reported that Crick

> discussed many other points which had interested him in biophysics, but his interests were obviously directed more to the purely physical than to the biological side. Out of the spate of words that came forth, I understood that what he really wanted was to go and work with Perutz, who also was very anxious to have him. He seemed to want to use methods of X-ray analysis for determination of the structure of protein and allied molecules.[5]

The interview ended with Mellanby telling Crick that the Council "could not assume the responsibility of looking after him after his training was over, but in the meantime he had better send in an application for another year and see what happened." Another year on a studentship? For 7 days, he considered the situation and then wrote to Hill to explain that he had had

> the opportunity recently to look more closely into the X-ray analysis, and I don't find it nearly so difficult as I expected. Whether it will prove tedious remains to be seen, but with three of us, and maybe more, in the group, things should go at a fine rate. Moreover I think I have the sort of brain which enjoys puzzles of this sort.

It will be interesting to see how far we have got in, say, five years time. Once a few proteins are understood I think my experience of the past eighteen months will be invaluable, but it may, of course, be very many years before we get to that state.

With many thanks for all your kindness to me.

Yours sincerely, Francis Crick[6]

A remarkable letter—grateful, respectful, and modest but firm. "I think I have the sort of brain which enjoys puzzles of this sort," i.e., the shapes and orientations of molecules. He could visualize them, intuitively grasp them, with or without the help of mathematical formulations. Their symmetry elements spoke to him of clues to structure. And Hill's response? He acquiesced regarding Crick's decision but grumbled about how far removed from live tissue was the dead, extracted material studied by the crystallographers.

Now was the time to seek support from Perutz and Sir Lawrence Bragg. Crick had taken care to read the two papers that Perutz had recently published about the progress of his work on the structure of hemoglobin. Thus prepared, he was able to ask what seemed to be intelligent questions, despite the fact that he had no more than a beginner's knowledge of X-ray crystallography. Perutz also had done his homework, reading Crick's Strangeways papers on what Perutz called "this rather meaningless project," and "was impressed with his analogy of the cytoplasm with mother's work-basket." He considered it "a very perceptive observation" and promptly jotted it down in the commonplace book that he kept for sayings that appealed to him. From the start, he took to Crick, whose voluble spirit did not irritate him as it had John Randall in London.

Next came an interview with Sir Lawrence Bragg. Considering the failure of the interview with Randall in 1947, Crick had reason to be apprehensive. Fortunately, perhaps, he was "much in awe" of Bragg because the Cavendish Professor was "already a legendary figure"— holder of the senior physics chair in Britain who had become a Nobel Laureate before Crick was born.[7] Bragg's report to Mellanby of the meeting with Crick was on the cool side, but accepting: "I shall be very glad to have him in the laboratory. I think he would be a useful member of the team, and also that he would be getting the kind of experience he appears to desire. . . . This is merely to say that as far as I am concerned I should welcome the arrangement."[8] Perutz also wrote to the MRC and explained that he had had many conversations with Crick since they first met in 1947. "He has always struck me as an exceptionally intelligent person, with a lively interest, a remarkably clear analytical mind and a capacity for quickly grasping the essence of any problem."[9] Because Crick was an exceptional case, he found it difficult to suggest a figure for Crick's salary. But he included a memorandum from Crick in which the subject was mentioned. Here, Crick explained that he "would very much prefer not to continue on a studentship. This was very reasonable while I was entering biophysics, but it is bad for one's morale to continue as a student for too long, and I should like to feel that I could settle down to some years' steady research, without having recurrent worries about financial matters."[10]

Crick was in suspense for 5 weeks. Then came the news: He had been offered an "unestablished" position in Perutz's unit starting 1 June, salary £700 per annum, no mention of future increments, and no superannuation (i.e., pension).[11] Crick accepted. because the position was "unestablished," the Council could terminate employment when they wished. Usually, these contracts were made for an "indefinite period" but sometimes for a "stated period of years."[12] Hence, Bragg could expect Crick to complete a research project for his PhD and then leave to take an academic position elsewhere.

### The MRC Unit

Crick was joining an unusual research unit among those supported by the MRC. The majority were associated with teaching hospitals and university medical schools but not Perutz's MRC unit; it was based in the laboratory of a physics department. This was not just any physics department, but Cambridge University's world-famous Cavendish Laboratory, where Sir J.J. Thomson had discovered the electron, Sir James Chadwick the neutron, and Sir John Cockcroft and Ernest T.S. Walton "split the atom." Those had been the golden years of the era of Cavendish Professor Lord Rutherford, author of the planetary theory of the atom in which negatively charged electrons circle around a central positively charged nucleus like the motion of the planets around the sun. The Cavendish under Rutherford had been supremely successful at experimental physics, and unlike University College London (UCL) in Crick's student days, it was at the cutting edge.

This era closed with Rutherford's sudden death in 1937. How different was his successor, Sir Lawrence Bragg. Whereas Rutherford had been the king of experimental nuclear physics, Bragg was, with his father Sir William, one of the pioneers of X-ray crystallography and coauthor of the well-known equation that bears their name. Rutherford had tolerated the X-ray crystallographers, but he did not find their work inspiring. He would put his head inside the lab door and teasingly ask "how the stamp collecting was going."[13] Bragg, by contrast, was delighted when in 1938 Perutz showed him his remarkable X-ray diffraction patterns of hemo-globin. Bragg's eyes lit up at the sight of them because he saw their potential at once. For him, proteins like hemoglobin were now a chal-lenge to the very art of the X-ray crystallographer. "I was thrilled by them," he recalled, "and formed the ambition to get out as a final act in my X-ray analyst's life something as complicated as a protein."[14] That life had begun with his enunciation of the Bragg law of reflection in 1912[15] when he was 21 years of age. Three years later he and his father Sir William were jointly awarded the Nobel Prize for Physics.

Perutz was the sole survivor of the group including John Desmond Bernal, who, in the 1930s, had been working at the Cavendish on X-ray crystallography of biologically important organic molecules. Bernal's graduate student Dorothy Hodgkin (née Crowfoot) had left for Oxford in 1934, followed by Bernal to Birkbeck College London in 1937, and Isidore Fankuchen to Brooklyn in 1939 where he taught at the Polytechnic Insti-tute. However, Perutz was not the sole X-ray crystallographer at the Cavendish. Other crystallographers still at Cambridge, such as Helen McGaw and newcomers William Cochran and June Broomhead, belonged to the Subdepartment of Crystallography. They and Perutz's MRC Unit were housed in the Austin Wing, built just before World War II, and locat-ed behind the Cavendish buildings that front onto Free School Lane.

In 1949, Perutz's MRC unit had been in existence for 2 years. Its members included John Kendrew, working with Perutz since 1946 (one year before the unit was formed), and Hugh Huxley, who followed him in 1947. Like Crick, both had served in the war, with Kendrew in oper-ational research and Huxley in radar. Kendrew was 1 year younger than Crick and Huxley 8 years younger. The efficient, suave Kendrew was to find Crick's presence in the group stimulating but at times irk-some. When trouble brewed, however, it would be Kendrew the diplo-mat who was trusted to calm the waters.

When Crick joined the group in 1949, Huxley had turned from hemoglobin to myosin and actin, the motility proteins of the muscle fiber, and Kendrew to myoglobin, the protein that supplies the muscles with oxygen. In this way, they distanced themselves from more menial projects, such as assisting Perutz's hemoglobin research. But all three were keen and expectant about solving protein structures. The mem-bers of the Subdepartment of Crystallography, however, were highly skeptical of such hopes, which was a reflection of the views of the pro-fession at large. This difference of opinion within the Cavendish, in addition to the limited contact between the two groups, was to have repercussions. But as we shall see, Perutz's "cuckoo in the nest" was to benefit from some of the work in the Subdepartment.

Crick started work in the Austin Wing on 1 June 1949. Soon, everyone in the building, including the Cavendish Professor on the second floor, would know when Crick had arrived by the sound of his voice and his laugh as he bounded up the stairs to the second floor. Crick's work habits did not respect traditional hours, so Bragg in his office on the floor above would also know that Crick started some time after everyone else. In contrast to Crick, Perutz, Kendrew, and Huxley were quietly spoken, serious, dedicated to research at the bench, and they kept regular hours. Crick's voice and laughter were soon to reverberate in the Cavendish Professor's ears. As Perutz recalled, Crick "talked volubly, each phrase in his King's English strongly accented and punctuated by eruptions of jovial laughter that reverberated through the laboratory."[16] And on two occasions those working close by knew by the flood in their midst when Crick's apparatus must have been leaking.[17]

The members of the Subdepartment, led by William Taylor, were salaried members of the University physics department and had teaching responsibilities. Taylor belonged to the "old school," formal, highly organized, and his office a picture of neatness. Among the several stars in his group, he was a nonentity. He ran his Subdepartment on the third floor with the utmost efficiency and was, it appears, resentful of Perutz's unit, whose members were not obliged to teach, were answerable to and funded by the MRC and not the University, and, what is more, had the ear of the Professor. For the staff of the unit, however, there was a downside—the uncertainty as to future employment unless given an "established" position. Even then, closure of a unit usually followed the retirement of its director.

### The State of the Art

When Crick entered the laboratory, X-ray crystallography was a young branch of the venerable subject of crystallography. Its own specialist journal *Acta Crystallographica* had only come into existence in 1948 following the formation of the International Union of Crystallography. For the most part under the institutional umbrella of physics, X-ray crystallographers vied with chemists for the solution of chemical structures. Chemists' molecular structures were based on chemical and physical properties, mode of synthesis, reactivity, etc. To represent the attachments of atoms to one another in the molecule, they used the convention of bonds (single, double, hybrid single/double, triple, and ionic). Properties of elements and compounds were explained in terms of the kinds and directions of these bonds, their differing strengths, and the freedom of movement around them. In contrast, X-ray crystallographers revealed images of this molecular world not in terms of arms linking atom with atom, but by gradients of charge—*electron density*—derived from X-ray-diffraction patterns. The peaks in this mountainous world represent the electron clouds around the nuclei of the atoms, the cols between the peaks represent the "bonds" of the chemist, and the valleys the weaker scattering atoms. The distances between peaks could be measured accurately, as could their directions by three-dimensional analysis.

Protein chemists believed that the distinctive properties of each of the numerous kinds of protein were due to their differing composition and constitution, the former referring to the proportions of the particular amino acids they contained and the latter to their spatial arrangement. The concept of a protein as a long chain—the polypeptide chain, so-called because it is composed of a series of amino acids such as glycine, alanine, lysine, and tyrosine, held together by peptide bonds— goes back to Franz Hofmeister and Emil Fischer in 1902. But confidence in the view that the protein molecule in its native state existed in the form of one or more long chains folded in globular proteins such as hemoglobin was debated until the mid 1940s. Skeptics remained into the early 1950s; they questioned whether the chains discovered by chemical investigation were the products of the treatment or whether they also are present in the molecule in its native form. On the other hand, the long-chain configurations for the fibrous proteins of silk, collagen, and wool were accepted by the time Crick entered the field.

In February 1950, when Crick had been working in the unit for just 8 months, John Kendrew gave a talk at the Hardy Club on the structure of the polypeptide chains in fibrous proteins. He summarized the results of a study that he had made with Bragg and Perutz of the molecular conformation in these fibers. Although William Astbury had won wide approval in the 1930s for his model of these chains in wool, calling them $\alpha$ keratin, there was concern that it did not fit modern estimates of the distances and angles between neighboring atoms in organic compounds. Instead, physical chemists had been suggesting that "spiral" models wind around the axis of the molecule. Such conformations are termed "helical." (By definition, a structure is said to be helical when between successive units of the chain [such as steps in a staircase] there is a translation along the axis [movement occuring up or down the staircase] and a rotation around it.) The helices of the crystallographers also retain a constant diameter, thus forming long cylinders, rather than coming to a point as that of a church spire. With a constant diameter, every unit of the chain occupies an identical environment.

Because the polypeptide chain is the very backbone of all proteins, Bragg wanted data on the peptide bonds that link one amino acid in the chain to its neighbors. What freedom of orientation had the atoms in this bond? It was known to be somewhat shorter than a normal single bond, and Linus Pauling at the California Institute of Technology (hereafter, Caltech) in Pasadena, claimed that it was a kind of hybrid single-double bond. The two groups thus joined, i.e., the carbonyl (CO) and the imino (NH), would therefore be "fixed in the same plane" (Fig. 6.1). This "planarity" would place constraints on the flexibility of polypeptide chains (Fig. 6.2A), thus having a decisive influence on the conformation that any such chain could take.

The only evidence on the matter from work in the Cavendish came from H.B. Dyer, a research scholar from South Africa, who worked in the Cavendish Subdepartment of Crystallography. He had studied a dipeptide in which the amino acids cysteine and glycine are joined by this bond. The result was not conclusive. Therefore, Bragg played it safe and with Kendrew and Perutz built models of polypeptide chains, almost all of which had non-planar links. Crick, who read Dyer's PhD dissertation at this time, took the same view of it as had his colleagues.

From Kendrew's talk at the Hardy Club, Crick saw how, by using metal skeletons of the atoms, with their bonds represented by measured lengths of the rods from a set of parts like those of a child's construction kit, the crystallographer could experiment in space. Possible conformations could be explored, with some eliminated and others preserved. Bragg, Kendrew, and Perutz had produced a wealth of possible structures, and the only serious limitation that they accepted was

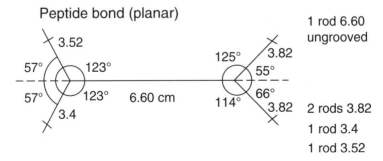

**Figure 6.1** Crick's specifications to the machine shop for construction of the planar C—N bond. Scale, 5 cm = 1 Å.

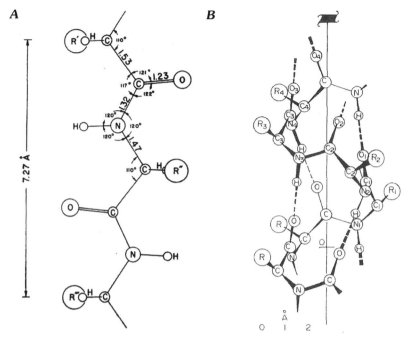

**Figure 6.2** (*A*) Planar peptide bond. (*B*) The Cavendish helical polypeptide chain with planar peptide bond.

that there should be a crystallographic repeat in the helix responsible for the very prominent intensity on the X-ray pattern from the wool fiber (α keratin), supposedly right on the meridian at 5.1 Å. They therefore set the pitch of the helix at this distance. Although they did build one model with an approximately planar peptide bond (Fig. 6.2B), the resulting helix had to be contorted to fit into the repeat distance. "Not surprisingly, *all* their models looked ugly," wrote Crick, "and they were unable to decide which was best."[18]

Crick was not sufficiently up to speed on the structural debate regarding the peptide bond to question the group's work, and apparently nor was anyone else at the meeting. When Crick's three colleagues left Cambridge for meetings later that year, he was given the task of checking the proofs of their article for the *Proceedings of the Royal Society*. Still inadequately versed to act as critic, he could do no more than look for typographical errors.

Opinion in the Cavendish on the subject remained undecided through 1950. Pauling's brief announcement in the *Journal of the American Chemical Society* in November, stating that his group had found new helical conformations for polypeptides by model building, failed to raise concern. He may have been the most renowned chemist in the world, but his note, recalled Crick, "was rather cryptic, and we thought we would wait for the longer papers." After all, "Pauling was not infallible in what he did. He was often right, but he was often wrong. Therefore the fact that he had announced a structure didn't mean that it was right."[19]

When the first of Pauling's 1951 papers appeared in the April issue of the *Proceedings of the National Academy of Sciences*, the details of his proposed α and γ helices had to be taken seriously. The latter structure had not been found in nature, but the parameters of the former (Fig. 6.3) were already supported by data from synthetic polypeptides synthesized at Britain's Courtaulds' laboratories. The pitch of their helices was 5.4 Å, not the 5.1 Å found in wool, and the Courtauld results had convinced Pauling to publish. He took the view that if his model conflicted with the data from wool—the 5.1-Å intensity—then so much the worse for those data. And as it turned out, this prominent but smeared arc of intensity is in reality composed of two arcs that are slightly off the meridian and that have been thrown toward it for reasons Crick was later to discover.

When Pauling's paper reached the Cavendish in May 1951, there was consternation. The α helix soon looked convincing to Crick's colleagues. "Bragg," he recalled, "was quite cast down. He walked slowly up the stairs" to his office. Bragg's long-time rival in the field of crystallography from across the ocean had once again challenged his work. What would be next in the firing line? How would their loyal supporters, the MRC, react? In retrospect, Bragg regarded this episode in their work as "the biggest mistake of my scientific career."[20] Not only had they accepted the 5.1-Å repeat, but they had only built chains with integral repeats, i.e., with the pitch of the chain coinciding with a whole number of links in the chain. Crystallographic principles make no such requirement. This unhappy experience taught Crick the crucial importance of accurate model building and what such an approach can achieve. It was also an example of how confidence in data at a fundamental level (partial double bonds) can outbid the claims of experiment at a higher level.

If the molecular chains in fibrous proteins such as wool and collagen presented problems, discovering their presence and arrangement in a globular protein such as hemoglobin presented a far greater problem. These remarkable giant organic compounds, legion in variety, extraordinarily specific in their action, surely held the keys to the mystery of life itself. Here, the chains, if present, must be folded in char-

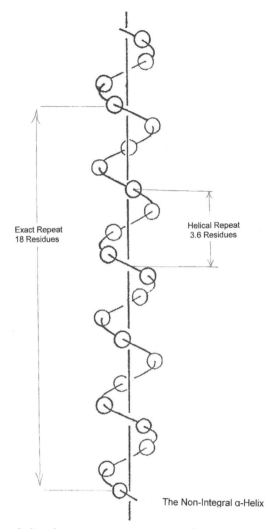

**Figure 6.3** The α helix showing its noninteger pitch and long crystallographic repeat, as depicted by Crick.

acteristic ways. Consider our respiratory pigment, hemoglobin. It combines with oxygen in the lungs and releases oxygen into the tissues, where it gathers carbon dioxide before returning it to the lungs and liberating it. Myoglobin also receives oxygen from hemoglobin and delivers it to the muscles. How can such molecules perform these vital and sophisticated functions? When do they "know" to release and when to take up? Biology teaches us that there is a firm relationship between structure and function, whether at the level of organs, cells, or the molecules therein. What, then, is the structure of these enormous molecules? This was the challenge that occupied Perutz, Kendrew, and Crick while Huxley was concerned with myosin, the muscle protein.

Answering this challenge called for the preparation of crystals and a search for the structure of the constituent molecules within them, because in crystals the molecules are arranged in a regular order and their atoms can be treated as points in a repeating pattern in three dimensions known as a lattice. A lattice in two dimensions can be constructed from repeating elements in any patterned wallpaper or design on a dress. The same applies to the pattern of atoms in a crystal. Optical microscopes, regardless of their design, cannot penetrate to such minute details, but X rays can. Unfortunately, they do not produce images, reflections, or shadows of the molecular lattices. Instead, they produce a diffraction pattern. No lens exists to sort out the X rays in such patterns and bring them into focus. Special procedures are therefore needed to achieve what no lens can. These patterns must be interpreted to yield a plausible structure for the molecules lying hidden in the crystal.

The term "reflection," often used to refer to the spots or "intensities" on the diffraction pattern, had been introduced by Bragg in 1912 when he drew his analogy between optical and X-ray phenomena. He imagined planes drawn through points of the crystal lattice like a succession of mirrors and visualized X rays hitting these planes and being "reflected." The emerging rays then interfere with one another, causing a diffraction pattern, as do light waves passing through a grating. In 1913, in a paper coauthored with his father Sir William, Bragg set out the relationship among the glancing angle of the ray as it strikes the lattice ($\theta$), the wavelength of the X rays ($\lambda$), and the separation between successive planes in the lattice ($d$), namely, $n\lambda = 2d \sin \theta$ (Fig. 6.4). This permits the assignment of individual intensities of the diffraction pattern to particular planes of atoms in the crystal, and hence the determination of the unit cell of the crystal. To proceed further and determine the architecture of the molecules in the unit cell requires more information. The crystallographer can determine the intensities of the spots and thus derive the amplitude of the rays producing them, but not their *phases*. In the absence of sufficient clues from chemistry, however, one needs the phases.

Hence the so-called "phase problem," namely, how far is the crest of one wave of an X ray behind or in front of another, came to be. The phase ($\varphi$) of a wave may be represented by the angle of a vector (Fig. 6.5) where, for example, one wave 360° behind another adds to the intensity, but at 180° behind, the wave decreases or cancels it out.

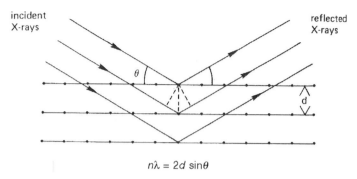

$$n\lambda = 2d \sin\theta$$

**Figure 6.4** The "Bragg reflection" of X rays from a lattice plane.

Knowing the phases, one would not need to guess or to go through lengthy trial-and-error calculations to determine a structure to fit the data. Could there be a direct way to derive the phases from the diffraction pattern? This challenging enterprise, begun in the 1940s, is known as the search for "Direct Methods."

With small molecules and adequate supporting evidence from standard chemistry, the structures that had already been suggested by chemists for a number of small organic compounds were confirmed by the X-ray data. Unfortunately, in solving most of these structures, the crystallographers were not telling the organic chemist something he did not already know. The methods the crystallographers used to overcome the phase problem had been introduced in the 1930s, but early suc-

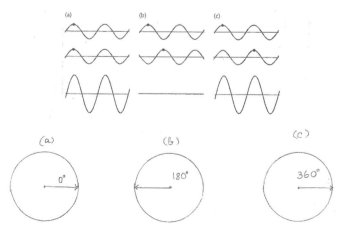

**Figure 6.5** The phase of waves represented by their phase angles (φ).

cesses gained by their use came chiefly in the later 1940s and early 1950s around the time that Crick entered the Cavendish Laboratory.[21]

It was then that Perutz and Kendrew were hopeful that they were overcoming the problems of interpreting hemoglobin's diffraction patterns. They were varying the salt and water content of the crystals and comparing the results to gain clues about the shape and dimensions of the molecule and the phases of the reflections. If they knew both the intensities of the reflections and their phases, they would be able to perform a Fourier synthesis to reveal the structure of the molecules in the crystal. Named after the famous French physicist, Jean Baptiste Fourier, this "synthesis" refers to the way in which it is possible to assemble the several elements present in a physical event, such as the sound of a flute. Using Fourier analysis, one can identify the constituent frequencies and, knowing these frequencies, one can use Fourier synthesis to reassemble these elements and produce the original sound.

There is an analogy between this example and the case of the molecular structure in a crystal. The many spots in the diffraction pattern bear a relation to the structure of the molecules in the crystal. But there is a snag. To recreate the molecular structure from its diffraction pattern requires knowledge not just of the location and intensity of the spots, but of their phases. In small molecules, the phases had often been discovered by the trick of inserting a "heavy" atom that would produce a very prominent effect on the diffraction pattern of the crystal. This atom would serve as a "marker," and the phases of other reflections could then be determined in relation to it. Success with this approach depends on there being no alteration in the structure of the molecule. The substitution of the heavy atom must be "isomorphous." This "heavy atom or isomorphous replacement method" was applied successfully as early as 1937 by J. Monteath Robertson and Ida Woodward in Glasgow, Scotland, "without the need of any chemical assumption," wrote Robertson, and "almost unlimited possibilities for the analysis of complex structures were thus opened up."[22] But this would require the synthesis of new derivatives of the compounds under investigation. X-ray crystallographers would need the skills and knowledge of the chemist, for, as Robertson pointed out, these methods represent "a chemical rather than a physical or mathematical approach to the solution of the phase problem. . . . Phase differences which cannot be measured are thus transformed into amplitude differences which can be measured."[23] Inserting a heavy atom into so huge a molecule as hemoglobin, however, surely would have relatively little impact. So for several years no attempt was made to find one.

## Patterson Synthesis

There remained, thankfully, another approach to the diffraction pattern that does give some evidence of the arrangement of the atoms in the crystal, even without knowledge of the phases. Introduced by the American X-ray crystallographer Lindo Patterson in 1934, the Patterson synthesis involves taking the raw intensities of the "reflections" instead of their amplitudes (the latter being the square root of the intensities) and putting all the phases at zero. This permits the calculation and plotting of what is known as the Patterson map of vectors as a series of contours that resemble an altitude map of hilly country (Fig. 6.6). Peaks on the map then mark prominent vectors. Unfortunately, these maps do not give the actual locations of the atoms in the molecule nor in the crystal. Instead, they give the distances between the principal atoms or groups of atoms that are scattering the X rays and their directions (hence the term "vectors"). They include those distances between atoms in the molecule that are connected by a chemical bond and those not thus connected. Adding to possible confusion is the fact that these Patterson maps are what mathematicians call convolutions of the data, meaning that the vectors are superimposed onto one another. Knowing this, it is hardly surprising that errors of interpretation can occur, especially when the investigator has a good idea what he or she should find! Imagine measuring the distances between all of the trees in a woodland, drawing lines to mark these distances and their directions, and then placing all of these lines to issue from the same point. It then becomes a no-win situation to restore their original arrangement without other information.

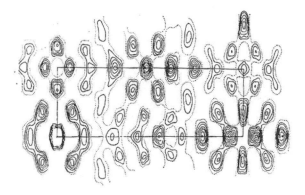

*Figure 6.6* Crick's 1953 Patterson diagram of ox hemoglobin.

At a symposium on hemoglobin held in Cambridge in 1948, Kendrew and Perutz had reported on the progress of their hemoglobin research. To aid their audience, they gave a primer on Patterson synthesis because most of what was known then about the structure of globular proteins relied on this method. They admitted that its physical meaning is "one of the most difficult conceptions in crystallography," and only the "supreme importance which this method has now assumed in the analysis of macromolecular structures," they explained, justified including it.[24]

Among the explanatory diagrams they drew was one depicting two long-chain molecules and, alongside them, the corresponding vectors of Patterson synthesis. They numbered the distances in the molecules and the corresponding vectors in the Patterson synthesis (Fig. 6.7) so that the relationship between the two could be seen. Note that the vectors do not necessarily relate to bond distances; they relate to any distances between two scattering atoms, whether within the same molecule or between neighboring ones. Interpreting these beautiful Patterson maps is thus limited by what seems to the novice to be a sort of scrambling of the data. Hence, it is sometimes referred to as the "poor man's Fourier." The mathematician knows that he or she is dealing with an "autocorrelation function of the electron density."

Nevertheless, Perutz's optimism was infectious, especially when supported by the enthusiastic Bragg, who, hard headed though he was, had retained a boyish sense of adventure that made the pursuit of

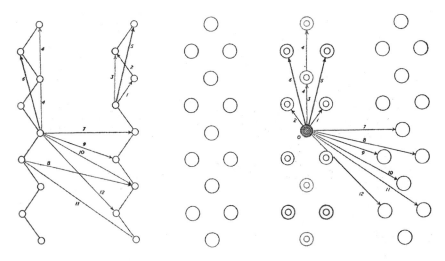

*Figure 6.7* Relationship between molecular structure and Patterson vector structure.

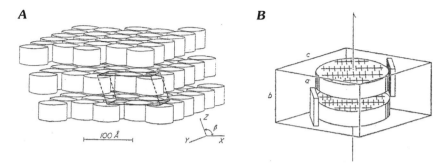

**A**    **B**

*Figure 6.8* (*A*) Packing of hemoglobin molecules in the crystal structure, showing layers of hat-box-shaped molecules separated by liquid. (*B*) Kendrew's model for myoglobin.

hemoglobin so tempting. Perutz could also point to his fortune in discovering that crystals of horse hemoglobin proved to be so well suited for analysis. Their Patterson syntheses revealed vectors that were very suggestive of polypeptide chains running parallel, straight across the molecule, and four layers deep. From his studies of shrinkage and swelling and of the crystals in varied salt concentrations, Perutz concluded that the overall shape of the hemoglobin molecule was rather like a hatbox with the chains sitting like logs in a firebox or like cigars in a cigar box side by side (Fig. 6.8A,B). There was no doubt, explained Perutz, that the structure "happens to be one of those exceptional types . . . where the vector structure bears a strong resemblance to the real one. It was the parallelism of the polypeptide chains in each haemoglobin molecule, and of all the molecules of the crystal, which facilitated the interpretation of the three-dimensional Patterson synthesis."[25] (Fig. 6.9A,B). Had it been otherwise, he admitted, it might well have been impossible to decipher the Patterson diagrams. At this stage, Kendrew, who was working with the much smaller respiratory pigment myoglobin, fitted his data to Perutz's hatbox model, but for myoglobin, a very shallow box consisting of a single layer of parallel polypeptide chains sufficed.

Ten months after Crick's arrival at the Cavendish, Perutz had traveled across the Atlantic to attend the 1950 meeting of the American Crystallographic Association at Penn State College in Pennsylvania. Before leaving, he wrote to the MRC requesting a salary raise for Crick: "Mr. Crick is a man of outstanding ability, who has already made important contributions to every aspect of our work which he has touched. There is no doubt at all that we should like to keep him on as a member of our unit as long as he is willing to stay with us—for I am

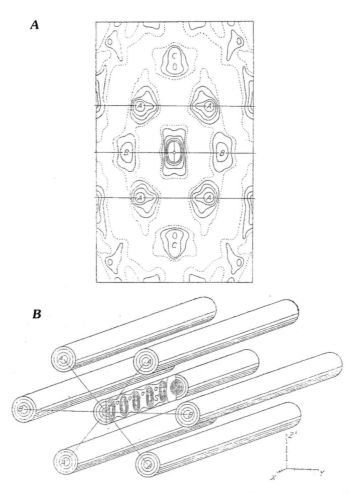

**Figure 6.9** (*A*) Patterson projection of hemoglobin normal to the *x* axis. (*B*) Idealized illustration of rod-like feature in vector structure.

sure he has a brilliant career in science before him."[26] Returning to Cambridge, Perutz found that Crick had still not succeeded in finding a suitable protein of his own. What had he been doing? He had been rereading Perutz's papers on hemoglobin.

To his surprise, Crick had come to the uncomfortable conclusion that "most of the assumptions that had been made in those papers were not substantiated by the facts." So when he was called upon to make a 20-minute presentation at a lab seminar in July 1951, he devoted it to a critique of the unit's work on hemoglobin. At Kendrew's suggestion, the phrase that he used for his title was "What Mad Pursuit" (from

Keats' *Ode on a Grecian Urn*). He hammered home his points without restraint, at times waxing loquacious. "They were all wasting their time," he declared. Going through each method, including the Patterson synthesis and Bragg's very own invention, an optical method called the "Fly's Eye," Crick sought to demonstrate that "all but one was quite hopeless."[27] The exception was the isomorphous replacement method, which had yet to be attempted and involved introducing "foreign" atoms into the hemoglobin molecule by chemical substitution without altering the structure of the crystal. The foreign atoms, chosen to act as strong scatterers of X rays, then serve as "markers" to aid in the search for the phases of the other atoms around them. Contrary to received opinion, Crick had calculated that this approach had some "prospect of success."

As for the hatbox model and the parallel rods inside it, Crick recalled that he had been "very unkind." He described the impact of his talk in the following words:

> Bragg was furious. Here was this newcomer telling experienced X-ray crystallographers, including Bragg himself, who had founded the subject and been in the forefront of it for almost forty years, that what they were doing was most unlikely to lead to any useful result. The fact that I clearly understood the theory of the subject and indeed was apt to be unduly loquacious about it did not help.[28]

Shortly before he died, Perutz recalled the event with pleasure and admiration. Here was the new research student under his supervision telling his supervisor straight that his work is wrong.[29] But it cannot have been a pleasant experience for Perutz at the time. One eyewitness described how Crick had held up a little ball on which Odile had painted the contours found in Perutz's Patterson diagram of hemoglobin, with the ball representing the height in the diagram yielded by the supposed parallel polypeptide chains. But Crick had calculated what that height should be—and it was much larger than this little ball.[30] Crick's notes for the talk have not survived, but around January 1951, before he gave the talk, he had written a draft of a paper on the subject that was entitled "The Determination of the Structure of Proteins by X-Ray Crystallography: Prospects and Methods." Probably because of Bragg's anger at his talk, and Bragg's caution on another occasion when he told Crick not to "rock the boat," this paper remained in manuscript form. When Crick was shown a transcript of it in 2001, he could not remember having written it. Yet its style and the handwriting are unmistakably his and its theme concerns the methods, as did his talk "What Mad Pursuit."

The paper starts with what he called the "Simplicity Postulate," stating that without regularity or simplicity in proteins "it would be a hopeless task to solve the structure" of any of them. What, he asked, is the evidence? Could it be said that "merely on general grounds, the structure is likely to be simple?" This he found to be "a most dangerous argument." Why? "Because we are dealing with part of a biological system. What is 'simple' for the organism, may not necessarily appear simple to anyone looking at protein structure," he explained.[31] Consider the intricacies biochemists have revealed in the pathways in our metabolism. Does the breakdown of carbohydrates (glycolysis) and their oxidation via tricarboxylic acid (Krebs cycle) seem simple to us? Far from it! The latter cycle alone requires a host of different enzymes. Here we see the former physicist arguing like a biologist, a feature uncommon among biophysicists at the time.

Turning to the X-ray crystallographer's methods, his major concern was directed at Perutz's interpretation of the Patterson diagrams of hemoglobin as published in 1947 and 1949. These were constructed from relative intensity values, not absolute values. Absolute values were needed to provide confidence in their interpretation; relative intensity values, on the other hand, could not be used to claim a match between the data and the expectation from calculation. Furthermore, the maps as they stood just did not reveal the degree of regularity and height to the peaks in the contours that Crick calculated should result from rod-like chains running across the molecule in parallel array, only turning, Perutz and Kendrew believed, abruptly at the margins of the molecule. "It is most unlikely," wrote Crick, "that the [Patterson] projection will consist of nothing but rods— viewed end-on of equal weight. . . . The most serious criticism, however, is that no attempt has been made to justify the observed magnitude of the Patterson rods, although this point is central to the whole argument. . . ."[32]

In November of 1951, Crick sent to *Acta Crystallographica* his estimate of what the heights of these peaks should be (i.e., their vector densities) due to such rod-like chains, as Perutz and Kendrew were assuming. His answer was ten times the height of those in Perutz's hemoglobin Patterson diagrams. Among Crick's conclusions was the following:

> It is clear that somehow we must introduce into our simple initial picture sufficient irregularity to throw the vectors away from the central Patterson rod. It is possible that this could be done by making the chain irregular, but the irregularities must be large . . . a globular protein may be more like a three-dimensional framework, and may need a perspective drawing to show its main features.[33]

These were prophetic words, as we shall see. Crick was exposing the myth of the assumption—that proteins forming "single" crystals must be composed of molecules, the atoms of which are in a very regular geometric conformation—not only by the Cambridge group but by others as well. At the same time, his intense study of these Patterson syntheses yielded his intimate understanding of "the poor man's Fourier" and impressed upon him the snares and delusions that lie in wait for the unwary researcher who atempts to interpret them.

Recalling this event, Crick wrote "The main result as far as I was concerned was that Bragg came to regard me as a nuisance who didn't get on with experiments and talked too much and in too critical a manner."[34] From this point forward, Bragg was looking forward to the day when Crick would have his doctorate and leave to work elsewhere. He was now some 13 months into his research at the Cavendish, married for the second time, and the father of an 11-year-old son and 1-year-old daughter, who earned a small salary and had no future contract for employment after completing his PhD More trips to the pawnshop were to be expected. What he needed was a secure university position, and that would involve teaching. But Crick, who did not wish at his age to start at the bottom of the academic ladder, believed that something would turn up. At that time, there did not exist the fierce competition for money and positions that was present later on. All the same, he could not now rely on a warm reference from Bragg.

### X-Ray Crystallography at the Bench

While this drama was being played out, how was Crick's laboratory research progressing? His previous doctoral research had been terminated by the war (see Chapter 3). After the war had ended, he was reluctant to reenter the doctoral program and judged his research topic at the Strangeways to be undeserving of doctoral status. However, both Bragg and Perutz expected Crick to work on a protein that he would be able to call his own, thus achieving original results deserving of a doctorate. Therefore, Perutz had requested permission from the MRC for Crick to register for the PhD "I believe," he wrote, "that Crick is a man of great promise and that it would be well worth while to give him this opportunity to improve his qualifications."[35] At the outset, Perutz had suggested that Crick should try the small protein secretin. And as Crick explained, with a molecular weight of only 5000, secretin "is not only of great interest for its own sake, but also for the light it might throw on the structure of myoglobin, haemoglobin and other proteins already being studied by X rays. This could conveniently be made a subject for a PhD thesis."[36]

Time flew by as Crick began learning the art of preparing protein crystals, and the best part of a year passed before he was persuaded to apply to Cambridge University to enter their doctoral program as Perutz's student. Applying as a member of Gonville and Caius—his friends had "smuggled him into this College"—he explained that he was not as yet a member of the University. His proposed course of research would be the following:

### The Structure of Proteins by X-ray Analysis

The most important requirement at the moment in this field is for extensive data on a few suitable small proteins. It is necessary, however, to find a protein which is not only small but also crystallises in a form suitable for analysis. The initial steps proposed therefore are (1) To make a preliminary X-ray examination of a number of small (or fairly small) proteins not so far studied. (2) to attempt to crystallise certain proteins (which at present give an unfavourable space group) in a better form.[37]

In the British system, graduate students were not required to take a whole series of graduate courses before launching upon their chosen research topic. Unburdened in this way, and relieved of 1 statutory year in recognition of his past research, Crick could hope to complete his research in 2 years, by which time he would be 36 years of age. But he needed to find a suitable protein. Sounds easy, you might say: Are not proteins like hemoglobin legion? They are not when one tries to prepare sufficiently large crystals, because the smaller the crystal, the longer the exposure needed for the X-ray beam to achieve an adequate diffraction pattern. Unfortunately, lengthy exposure destroys the structure of tiny crystals. However, many proteins just refuse to yield crystals that are more than a fraction of a millimeter in size. Then again, some protein crystals have so many molecules in the unit cell (repeating unit of the crystal structure) that all hopes of discovering that structure are dashed by the sheer number of atoms that must be located. Fortunately, the kind of symmetry of the crystal can help. For example, it may reveal that half of the molecule, when rotated into the position of the other half, is identical to the other half. This greatly simplifies the analysis. Hemoglobin obliges here, but many proteins do not.

Crick's notebooks for the fall and winter of his first year (1949–1950) tell the story. This was protein chemistry at the bench for sure: Creating solutions using recipes from the Rockefeller Institute guru Moses Kunitz, grinding up pancreas brought from the slaughter house by Percy the lab assistant, straining the mush through cheese cloth, then rushing the filtrate to the Molteno Institute, shaking it with ice all the way. Whiz the liquid around in the Institute's ultracentrifuge, rush back to the Cavendish with the separated protein, and then have crystals form in the extract. Success came on the last Sunday of October when Crick produced crystals of chymotrypsin.

Crick also tried his hand with hemoglobin, going through a similar procedure and racing back and forth between the Cavendish, the Molteno Institute, and the cold room at the Low-temperature Laboratory (Fig. 6.10). Perutz smelled the first of the two batches that Crick brought back

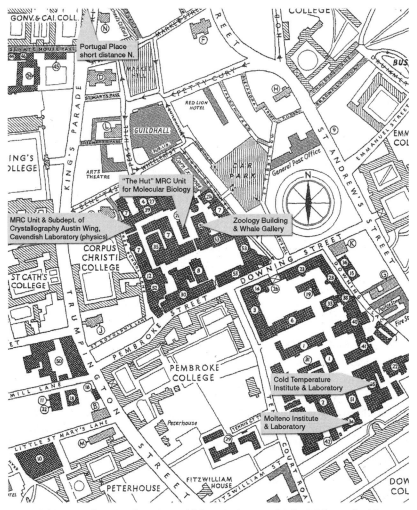

**Figure 6.10** Map showing location of laboratories at which Crick worked between 1949 and 1962.

from the Molteno and asked that he "Throw it away!" Looking in the next batch, Perutz announced that "We have crystals!" Success at last.

Later, Crick worked with other proteins including trypsinogen, trypsin, insulin, and finally trypsin inhibitor, aided by the biochemist Freddie Gutfreund, who, as a member of the Department of Colloid Science, was based at the Low-temperature Laboratory. Together, they visited Michael Green at UCL to learn how to crystallize trypsin inhibitor. Back at Cambridge, Crick succeeded in producing "very tiny crystals." Then, Moses Kunitz sent him material from the Rockefeller Institute from which Crick obtained "quite nice crystals," but when it turned out that 60 molecules are in its large unit cell, he wisely gave this up. Perutz's original suggestion to use the very tiny protein secretin had proved to be too difficult to handle experimentally. Asked if Crick would have had more success if he had persisted, Gutfreund responded

> Of course Francis would have succeeded in getting large enough crystals of the proteins he tried [to work] with if he had not found something more interesting to do! Small proteins that had been crystallized were the obvious choice for trying to get large enough crystals. I think that getting (small) crystals of a protein required a good deal of luck. Getting larger crystals of a protein could be achieved by intelligent planning and Francis could have succeeded. The next piece of luck is when the large crystals have few molecules in the unit cell and a nice space group![38]

While Gutfreund helped Crick over protein crystallization, Crick helped Gutfreund "to set up an optical bench for interference observations of diffusion boundaries. He was always willing to help a friend!" Concerned, moreover, for fellow researchers, he published the results of these studies "mainly to prevent others spending time on unpromising material."[39]

The strengths of Crick's dissertation did not lie in the results of this kind of experimental work, but in his theoretical contributions to the interpretation of diffraction patterns. Unfortunately, this theoretical penchant could lead him into trouble when he insisted on interpreting the experimental results of others or suggesting a method for assigning the sign of a reflection that was, in his view, superior to the one that Bragg had decided to use in his determination, with Perutz, of the hemoglobin molecule's shape. Crick objected to Bragg, an old-fashioned gentleman, that the method he (Crick) had earlier suggested was more reliable and of broader application than Bragg's.[40] When he said he would go back and check them out, Bragg blew his top. The doctoral student was summarily dismissed from his presence and it took the support of Kendrew and Perutz to cool their boss down. As James Watson remarked many years later, "You don't loud mouth your professor when you are a graduate student!"[41]

Crick had also become a convert to the crystallographer's use of model building to help solve the structure of fibrous proteins and synthetic polypeptides. This approach, he repeatedly insisted, is a form of experimental science, and he pointed out that in the case of Pauling's α helix, the results of model building were permitted to take precedence over experimental evidence from X-ray diffraction. Such work was not to be dismissed as mere schoolboy fun, like building Tinkertoy models. This was the great lesson that Pauling, by his example with the α helix, had taught Crick. One does not need to go through the tedium of a Patterson synthesis when one is working with a long-chain molecule with awkward links that can only be coiled into a helix in a very limited number of ways. The Pauling approach was to consider the diffraction pattern produced by the molecule rather than by the crystal lattice, which in the case of these fibers, is not that of a "single" crystal (regular in three dimensions), but of a fiber that is regular in the fiber direction and, to a varying extent, random in the radial direction. "The effect on the X-ray photograph," Crick explained, "is the same as if a single crystallite had been rotated about its fibre axis while the photograph was being taken. Thus all (or almost all) the information is superposed on one photograph, which is rarely an advantage."[42]

How much guidance did Crick receive in his doctoral research? Once or twice a year, he recalled, "I used to go [to Perutz] and say, "Max I demand supervision." In those days, graduate students were often left to sink or swim, especially in a research group that before Crick's arrival had only two graduate students: John Kendrew and Hugh Huxley. Many years later, Perutz told Crick about how he was once "supervised by a graduate student in the art of protein engineering. He set up a gel one day, the next morning I heard that my supervisor got drunk the night before and had fallen off his bicycle, since when I have not heard from him. This should make you appreciate that you had no such trouble with me!"[43] Perutz did suggest the X-ray crystallography textbooks that he should read, but Crick disliked the historical approach these often used, and, although one needed to be able to deal with the algebraic details, he recalled that "I soon found that I could see the answer to many of these mathematical problems by a combination of imagery and logic, without first having to slog through the mathematics."[44] Moreover, his immersion in Perutz's study of shrinkage stages of large hemoglobin molecules meant that he "learned how to deal with diffraction from a single molecule, and only then arranged them in a regular crystal lattice." The conventional approach was to start with the lattice. Crick was to find this alternative approach invaluable later on.[45]

Having failed to find a suitable crystalline protein, Crick lacked a "definite problem" for his dissertation. Taking up one problem after another ended up making a "ragbag" of a thesis. Its title, "Polypeptides and Proteins: X-Ray Studies," appropriately signaled its broad subject coverage. Indeed, he considers it a miracle that he "managed to make a thesis out of it at all." On the other hand, the traditional pocket at the back of the thesis contained his published papers. These, he believed, were "almost enough to make another thesis, including, of course, the structure of DNA!"[46]

### Crystallography with Paper and Pencil

The most important of the theoretical papers that Crick contributed to the professional journal *Acta Crystallographica* in 1952 was the one that he wrote with the Scottish crystallographer William Cochran and the American crystallographer Vladimir Vand on the character of the X-ray diffraction pattern, or transform, produced by a helical chain of atoms. Cochran, a Demonstrator (the American equivalent is Assistant Professor) in the Subdepartment of Crystallography, lectured in the University's crystallography courses. He was a rather shy young man, often depressed, but of exceptional ability. He had resisted Bragg's efforts to draw him into his protein work. Bragg had asked him to work out the transform of a helix, i.e., to determine the kind of diffraction pattern that a helical molecule should produce. But Cochran considered Pauling's $\alpha$ helix a speculative piece of work, and, in addition, he "was convinced that it was impossible to determine protein structure by X-ray methods."[47] It took the arrival of an incomplete derivation of the helical transform to shake Cochran from his disinclination. Vand, working at the time at Glasgow University, had sent his attempt on the problem to Perutz, who passed it on to Cochran. When, as often happened, Crick encountered Cochran in the darkroom, they discussed Vand's paper. They agreed that Vand had correctly derived the answer for a continuous helix, but this did not cover the case of a discontinuous helix as found in a polypeptide chain.

That evening (31 October), Francis and Odile planned to attend a wine-tasting, and because he was not feeling very well, Crick went home from the lab early, became bored sitting by the gas fire, and undertook to work out the helical transform himself. To his delight, he succeeded in solving it. In good spirits, he went to the tasting, only to be bored by the company, and, being unaware that one does not swallow every tasting, he had consumed a great deal of wine! That did not portend the most comfortable of nights. Arriving the next day to the Cavendish, he found that Cochran had also worked on the problem. They had both solved it, using the same mathematical formulae known as Bessel functions (Fig. 6.11), but Crick did so in a laborious manner, Cochran in a shorter and elegant manner. The resulting theory predicted a diffraction pattern with the following four major features (Fig. 6.12):

1. As one moves away from the origin in the fiber direction (*meridian*), the intensities on the X-ray diagram are spaced increasingly far apart.

2. Between these intensities is an absence of spots, the meridional absences.

3. This pattern is repeated at the distances along the meridian that represent the distances between successive units (residues) along the polypeptide chain.

4. Perpendicular to the fiber axis (along the *equator*), the intensities have the form of a hexagonal or pseudohexagonal arrangement.

These features are predicted for the diffraction pattern yielded by helical, long-chain molecules, packed tightly together and lined up in the direction of the fiber, akin to that of a tight bunch of long pencils. Keen to support Pauling's α helix, Crick obtained prepublication diffraction patterns of the synthetic polypeptide poly-γ-methyl-L-glutamate from the Courtauld Laboratory. He and Cochran then used their theory to predict the diffraction pattern that should be produced by this polymer, assuming it to be in the form of Pauling's α helix. Their

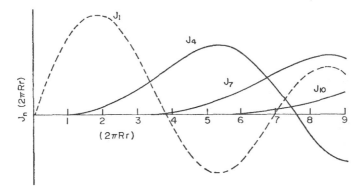

*Figure 6.11* The march of higher order Bessel functions (with $J_1$, $J_4$, etc.).

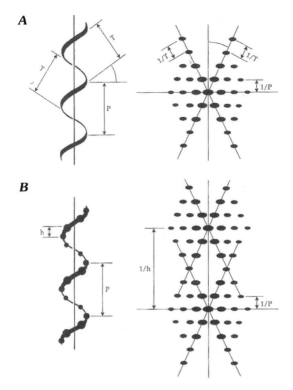

**Figure 6.12** Fourier transform of a helix. (*A*) continuous; (*B*) discontinuous.

predictions were "strikingly borne out." Crick's excitement is evident in the following conclusion:

> We therefore have no doubt that the structure of poly-γ-methyl-L-glu-tamate is based on a helix of eighteen residues in five turns and 27 Å, or a helix which approximates to this very closely. As the structure proposed by Pauling and Corey [α helix] satisfies these conditions and is also stereochemically very satisfactory, it seems to us highly proba-ble that it is correct.[48]

To Cochran working in the Subdepartment of Crystallography, this derivation of features of the diffraction pattern from the architecture of the molecule did not seem to be anything to write home about—not even the fact that they had been able to demonstrate the agreement of their prediction with the data from poly-γ-methyl-L-glutamate. He did not consider his note with Crick to *Nature* "as being of much impor-tance" and he wrote in his diary that "this is the most uninspiring term

I've ever had in Cambridge as far as research is concerned," a laconic response to be sure.[49]

Why was Crick, by contrast, so enthusiastic about this result? First, it meant that for fibers there exists a route to the discovery of molecular structure that bypasses the problem of knowing the phases of the reflections and permits one to omit Patterson analysis. Instead, one could combine the model-building program with testing the model's predictions for the character of the diffraction pattern against the actual fiber pattern reported. Crick also realized that fibers, consisting of long-chain molecules, have to run from one unit cell of the crystal lattice to the next, and the next, for perhaps thousands of cells in the fiber direction, with all of the molecules lined up head to tail. That imposes serious restrictions on the orientations that they can take up, thus reducing the possible conformations, in comparison to the situation with small molecules. In addition, in both proteins and nucleic acids, links between units are sufficiently awkward to add further restrictions on the conformation that the molecules can adopt in the fiber. As a result, the most likely conformation is helical, making the morphology of the molecules cylindrical, and, as has been noted, cylinders tend to pack in an approximately hexagonal array. For all of these reasons, the theory of the Fourier transform of a helix, to give it its technical title, offered a method of moving forward with fibers that are not sufficiently crystalline to permit reliance on the conventional crystallographic approaches used with single crystals. Crick explained:

> Now what happens if you have an array of vertical helices which are equidistantly spaced and parallel but not particularly up and down? The answer is that if you look in projection downwards, it doesn't matter that the helices are "up and down," everything is regular. Now the diffraction that corresponds to that is on the equator. In that case you get spots on the equator, but on the layer lines you won't get spots, because there is no consistent phase relation at all the other angles. Instead you get "bands."[50]

It was 6 weeks from their derivation of the theory before Cochran and Crick could send their short note to *Nature*. Received 14 December, it was published 9 February 1952. *Acta Crystallographer* received their major paper with Vladamir Vand on the transform of a helix 1 week later, and it was also published in 1952.[51] Before either of these papers was sent in, Crick had completed his estimation of the heights of the peaks in the Patterson maps of hemoglobin if the molecule is made up of parallel rods, as Perutz and Kendrew assumed. *Acta* had received his paper on this topic on 15 November. Here, his message was clear: "A discrepancy by a factor of 10 between the vector density is to be expected for a simple model of straight parallel chains and the observed density." Noting that Bragg had entered the fray and, with David Howells and Perutz, had recently arrived at a factor of 9, he could state without causing offence that this "means that the haemoglobin molecule cannot consist almost entirely of straight parallel α-helices." It was not a matter of denying α-helices; rather, he investigated the effects on the peaks of nonparallelism, kinks, tilting, meandering, and corner turning. Most importantly, he detected a "profound asymmetry in the arrangement of the folded polypeptide chains," suggesting that "globular proteins are really three-dimensional in their architecture, and not two-dimensional like the synthetic polypeptides."[52]

Crick was beginning to build a respectable publication record; the derivation of the helical transform and its application to protein fiber structures were original and significant contributions. It was at this point that he traveled to Oxford to tell Dorothy Hodgkin about it. Later, he would recall the importance of helical theory when he would hang a brass helix from his house on Portugal Place. Crick explained that as a single helix, "it was supposed to symbolize not DNA but the basic idea of a helix. I called it golden in the same way as Apuleius called his story 'The Golden Ass,' meaning beautiful." Its yellow color betrays the fact that they never got around to gilding it.[53]

## CHAPTER 7

# The DNA Fiasco

 *It was a complete waste of time.*[1]

Cambridge town center is small, its streets narrow, and if you are pushing a pram,[2] as Odile Crick was one October day in 1951, you might easily bump into a friend. That day, it was Max Perutz. Beside him stood a young man whom Perutz had wished to introduce to Francis Crick—a skinny fellow, tall, with an American accent, and shy. He was the new postdoc for the MRC Lab. His name—James Watson. When Crick came home that day, Odile told him of Perutz's visit with the young American, adding "and—you know what—he had no hair!"[3] Crick realized that Odile had just had her introduction to the hairstyle we know as a crew cut, rarely seen in Cambridge at that time.

The following day, the two did meet, and what a contrast—the garrulous, ebullient, articulate, 35-year-old Englishman meeting the strange, 23-year-old American. As Watson's biographer Victor McElheny described, Watson would look at his shoes as he spoke or, preparing to make a funny remark, "he would suck in his breath through his teeth, and break into an anticipatory smile that was more like a grimace."[4] But Watson was already on a postdoctoral fellowship, whereas Crick had yet to complete his doctorate. Watson knew no X-ray crystallography, whereas Crick was immersed in it. It was the sight of an X-ray diffraction pattern of DNA shown by Maurice Wilkins at a meeting in Naples in 1951 that had fired Watson's enthusiasm to pursue its molecular structure. This meant learning the art of X-ray crystallography, where training in chemistry and/or physics was the form. Undaunted, this young biologist had the determination to persist because he was convinced of the importance of DNA. Hence, it was Watson who brought the problem of the structure of DNA to Perutz's MRC unit in Cambridge. There, he would be introduced to a like-minded iconoclast and agnostic whom he had not before heard of: Francis Crick.

There was strong chemistry between these two scientists. As Watson recalled, "From my first day in the lab I knew I would not leave Cambridge for a long time. Departing would be idiocy, for I had immediately discovered the fun of talking to Francis Crick. Finding someone in Max's lab who knew that DNA was more important than proteins was real luck."[5] Crick's comment on their first meeting was equally enthusiastic: "Jim and I hit it off immediately, partly because our interests were astonishingly similar and partly, I suspect, because a certain youthful arrogance, a ruthlessness, and an impatience with sloppy thinking came naturally to both of us."[6]

It was not that their knowledge of science was mutually complementary, with Crick knowing the physics and Watson the biology. Crick stressed that although he was a member of what has been called the "structural school" of protein X-ray crystallography, he "was always keenly interested in all the genetic work and the biochemical work." Watson, for his part, "had naturally found himself turning to the structural and biochemical side."[7] It is true that when Watson first visited the Cavendish Laboratory, he told Perutz that he knew nothing of X-ray crystallography and was to rely largely upon Crick's tutorial guidance, so much so that Crick toyed with the idea of writing a guide entitled "Fourier Transforms for Birdwatchers"—a reference to Watson's ornithological past. Nevertheless, in Crick's opinion, Watson continued to have difficulty grasping crystallographic arguments. Here, Crick directed and Watson followed. Yet in theorizing and speculating, it was very much a close collaboration, and there was no beating about the bush, no suppression of criticism: Lay out your ideas and assumptions; expect them to be demolished. Two probing minds are better than one. False starts and erroneous assumptions have a better chance of being exposed that way—but not, as we shall see, in their first collaborative venture. Here, collaboration compounded the errors!

At the Strangeways, Crick, like Watson, had been following genetics and bacteriology, including the work of the Rockefeller scientist Oswald T. Avery on the phenomenon of bacterial transformation, work that was to prove crucial to the establishment of DNA's role in heredity. Before Avery, it was already known that when matter from a pathogenic strain of bacteria that is killed by heating is added to the medium in which a nonpathogenic strain is growing, some members of the live strain acquire the pathogenicity of the dead strain. This is due to the production of a protective sugar capsule around the bacterium. It is the transfer of the power to produce such a capsule in the recipient strain that enables it to resist the destructive power of the host's immune system. Moreover, this power is transmitted to its progeny. That this result was not due to a spontaneous reversion on the part of the recipient strain to its former capsular form was clear, because the chemical constitution of its sugar capsule was not that of its own strain but of the donor strain.

Avery, Colin MacLeod, and Maclyn McCarthy had been able to identify the substance transferred as deoxyribonucleic acid, or DNA. How could this be? Could DNA carry genetic specificity, in short, *be* the genetic material? There was skepticism, especially in the Rockefeller Institute, where the biochemist Alfred Mirsky was a champion of proteins. Surely, there must have been traces of protein in DNA that were taken up by the recipient cells. It is the proteins that have the variety of constituents expected of a substance that serves as the chemical basis of heredity—namely, some 20 or so different amino acids. DNA, on the other hand, is composed of sugar–phosphate chains (Fig. 7.1), to which only four different units are attached: the two larger "bases"— the purines guanine and adenine—and the two smaller ones—the pyrimidines thymine and cytosine. A boring molecule, one might say. Yet, besides protein, we find DNA in chromosomes and its sister molecule ribonucleic acid (RNA) in the cytoplasm of all cells, and either one or the other is found in viruses. So when cells divide, viruses mul-

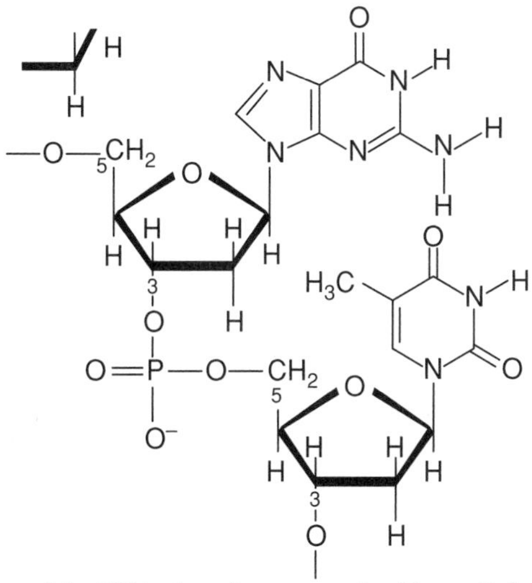

**Figure 7.1** Structure of the DNA phosphate–sugar backbone. Polarity is given by the 3′, 5′ phosphodiester links between successive sugar rings.

tiply, and proteins are synthesized, nucleic acid is present. It was also known that when DNA is damaged, the cell's power to replicate is compromised; hence, the use of agents in cancer chemotherapy that can alter DNA, thereby suppressing cell division. Such agents were in use before the recognition of the genetic role of DNA. However, the consensus had been that DNA was a kind of "midwife" molecule that supervised the replication of the genetic material which must itself be protein. The publication of Avery's group in 1944,[8] however, set in motion further studies of the chemistry of DNA and the phenomenon of bacterial transformation—studies that were to shift the balance of opinion away from the midwife molecule idea in favor of DNA as constituting most if not all of the actual hereditary material.

Crick's view of the matter was cautious. DNA looked like the genetic material, but there might well be some other explanation for bacterial transformation. How confident could he be that what operates for pathogenic bacteria should also operate for higher forms of life that have nuclei and exchange genetic material? He knew that the "chromatinic" structures depicted by Robinow (see Chapter 5) were not established as the nuclear material of bacteria. But the American microbial geneticist Joshua Lederberg had discovered a mating process in the colon bacillus (*Escherichia coli*) and was beginning to construct a linkage map of its genes. Bacteria were not, after all, some primitive form of life lacking a distinctive genetic apparatus. Therefore, results of experiments with bacteria—transformation included—must surely have relevance to higher forms of life as well.

Crick had also read about the biology of bacterial viruses or bacteriophages (phages for short). He knew that some 20 minutes after a bacterium is infected with a phage, the host cell bursts, liberating about 100 phage particles. Clearly, we have here an example of the replication process in which the genetic material of the phage, like the genes on the chromosome, is reproduced and, like genes, it can mutate.[9] Because the phage particle consists solely of protein and nucleic acid, both substances must surely be involved in replication. Yet, although much attention was being bestowed upon the three-dimensional structure of the proteins and their subunits the amino acids and peptides, and on the bases and their nucleotides in DNA, little effort had been directed to the three-dimensional structure of the nucleic acids themselves. Only Florence Bell, a graduate student under William Astbury in Leeds, had produced X-ray diffraction patterns of fibers of DNA drawn from a solution.

That was in 1938.[10] But in 1950, Crick's friend Maurice Wilkins had received the finest quality DNA available, provided by the Swiss biochemist Rudolph Signer at a meeting in London. With help from Raymond Gosling, a doctoral student at King's College, Wilkins obtained a remarkable diffraction pattern of the DNA that was "much sharper and more detailed than any before." There were "dozens of well-defined spots on a clear background"[11] Aided by Wilkins' colleague Alex Stokes, Gosling indexed the spots and identified the unit cell of the DNA as monoclinic—shaped like a shoe box, with one angle between its sides not a right angle. It was this diffraction pattern that Watson saw at the conference in Naples at which Wilkins spoke.

Wilkins and Crick had kept in touch since they first met in 1947. Both formerly physicists, they shared an eager desire to explore the molecular basis to the life of the cell; thus, they had much to discuss. Nucleic acids preoccupied Wilkins, and proteins Crick. Indeed, he once chided or perhaps teased Wilkins by telling him that "what you ought to do is get yourself a good protein."[12] Here were contrasting personalities: Crick directly faced the person with whom he spoke, and Wilkins would turn to one side and in no time one found oneself conversing with his back! A fellow undergraduate described Wilkins as "a rather peevish, slightly old-maidy young man." Later in life, his interaction with Rosalind Franklin was to elicit these features, but with Crick he was enthusiastic, friendly, and outgoing. With Odile, he not only enjoyed conversation about the culinary and visual arts, he was also attracted to her as a woman; had she been single, he would surely have pursued her.[13] Wilkins and Crick would have made excellent collaborators—Wilkins the experimentalist, deft with his hands, and Crick the theoretician, a shade clumsy with his. Both were agreed that DNA was biologically important. Indeed, in 1951, Kendrew suggested to Wilkins that he might like to come and work with Watson and Crick.[14]

Meanwhile, Crick, although absorbed in the study of the three-dimensional structure of proteins, was also very struck by the work of the Cambridge biochemist Frederick Sanger. A highly talented chemist with a gentle spirit, he worked in the basement of the University biochemistry department. Supported first by family money, in 1944 he received a Beit Memorial Fellowship, and, starting in 1951 he was employed by the Medical Research Council. He was revealing the sequences of amino acids in the polypeptide chains of the protein insulin. They followed no obvious pattern. Therefore, Crick believed, such chains had to be laid down on templates, and some form of these templates were then stored and transmitted to the next generation. Protein synthesis was not just a trick mastered by a legion of cunning enzymes. A mold or template that is different for each kind of protein was essential to order the particular sequence of the amino acids.

Here, surely, lay the secret to the hereditary transmission of the remarkable specificity found among the proteins. The gene must contain a specific *sequence* that determines a one-dimensional sequence of amino acids in the polypeptide chain of a specific protein. These chains then fold to form the three-dimensional molecules we know in "globular" proteins such as hemoglobin and the numerous enzymes. Furthermore, when globular proteins are "denatured," the folding is lost and the specificity of action with it. The once elegantly folded chains become long strings, as does the white of an egg when heated in a pan. But the hereditary transmission of the specificity encoded in the genes must be one-dimensional. Recall that in 1944 Schrödinger wrote of "building up a more and more extended aggregate without the dull device of repetition. . . . We believe a gene—or perhaps the whole chromosome fibre—to be an aperiodic solid."[15]

Having thus treated the problem of the storage of hereditary information, there remained the issue of how to express that information. True, it is difficult to conceive of a way in which a three-dimensional molecule such as hemoglobin could be copied in the way a sculptor creates a replica of his work, because the sculptor's mold follows only the external form of the original. Starting with a one-dimensional chain copied from a one-dimensional template that subsequently folds evades the problem. Here, Crick made the assumption that the three-dimensional shape of the molecule is determined by the folding process, and this, in turn, is determined by the nature and arrangement of the amino acids on the nascent chain or chains. Hence, genetic determination could reside in one-dimensional sequences.

DNA was clearly implicated in the hereditary process, but how? If it really is the hereditary material, how does its structure make that possible? Crick's maxim here was that if one is unsure of the function of a substance, one must find its molecular structure. Although he was busy studying proteins with his colleagues Max Perutz and John Kendrew, unlike them he had one ear alert for news of work on nucleic acids and developments on the genetic front. But it would take the arrival of the young American biologist James (Jim) Watson to focus his attention on DNA. As Crick remarked subsequently, "If Jim had been killed by a tennis ball I doubt whether I would have discovered the structure of DNA."[16]

### The Collaboration

Twelve years younger than Crick, Jim Watson had come from a very different background. His father a bill collector and his mother a company secretary and accountant, the family made ends meet as best they could in postcrash Chicago. Growing up in the city's South Side, Watson benefited from that great city's institutions—the Field Museum of Natural History and the library on 73rd Street, with which Crick's Northampton could not compare. Watson came through the educational machine rapidly. Unlike Crick, his career was not interrupted by war, for he was only 12 years of age when America joined Britain against the Axis powers. Nor, like Crick, did he stay in high school until he was 18 years old. Instead of staying on at South Shore High School, Watson moved to the University of Chicago High School, and 2 years later, he was accepted into the 4-year college program for 15 year olds, the brainchild of Chicago University's Chancellor Robert Maynard Hutchins. There, Watson needed to jump over the "intellectual obstacle course" of Hutchins' called "Great Books Program," the course in which he learned to hate philosophy. Yet, he did benefit from some important lessons, as he afterwards admitted. These lessons, he recalled, were to read the original sources, not textbook renditions of them, to value theory above factual learning, and to learn how to think rather than how to remember more and more facts. In retrospect, he realized that he was "acquiring the mental habits" that were to make him acceptable to his graduate teachers and, in due course, to Crick.[17]

Graduating at 19, Watson joined the graduate program at the University of Indiana, where he completed his doctorate under the microbiologist Salvador Luria. His doctoral research proved nearly as inconsequential as Crick's Strangeways project, but it led to his introduction to the informal "Phage Group," composed of those who attended the annual summer course on the biology of phage at Cold Spring Harbor Laboratory on Long Island. Established by two formidable personalities, the Italian-born Salvador Luria and the German-born Max Delbrück, the Phage Group brought Watson into personal contact with those who wanted to unravel the mystery of phage reproduction. Until he met Watson, Crick had no personal knowledge of the members of this powerful circle of individuals.

Before coming to Cambridge, Watson had been working in Copenhagen where he had been sent to learn nucleic acid chemistry. There, with the Dane Ole Maaløe, he tried to establish the identity of the substances that gave genetic continuity to successive generations of phage particles. Not satisfied with the results, Watson turned from the biochemical approach to the possibility of a structural attack on DNA. Changing disciplines in midstream of a fellowship was dangerous, but Watson was undeterred. In early September, he traveled to Cambridge to meet with Max Perutz and Sir Lawrence Bragg, and to request a place for 1 year in their MRC unit. Perutz recalled the event in his charming style: "A strange head with a crew cut and bulging eyes popped through my door and asked me without so much as saying 'hallo': 'Can I come and work here?' I said 'Yes' because I guessed that this must be Jim Watson whom Luria had recommended to Kendrew."[18] Watson returned to Cambridge at the beginning of the semester in early October and it was then that he met Crick.

Before they met, Watson knew only that Crick was one of the protein researchers in the unit. To Crick, Watson was equally unknown—just a surprisingly young American postdoc who was to help John Kendrew with his work on the structure of myoglobin while he learned the trade of X-ray crystallography. Their mutual surprise was that each found that the other was thinking along the same lines about DNA. As Watson recalled, "From the day of our first meeting, Francis Crick and I thought it highly likely that the genetic information of DNA is conveyed by the sequence of its four bases."[19] In fact, Crick was not so confident of his position at that time. In 1968, he said "You ask me what I knew about DNA when I first met Watson and the answer is I really do not remember. . . . I think, in present-day terms, I was not convinced of the overwhelming importance of DNA to suggest that I did experimental work on it rather than on proteins, because I thought proteins were also important and here [in the Cavendish] was the work going on."[20] Thus, it was Watson who brought the problem of the structure of DNA to Cambridge. Had he not come, it would have been left to Wilkins and Rosalind Franklin in London and, very likely, Linus Pauling in Pasadena.

For many years, Crick had been careful not to claim genetic function exclusively for nucleic acids. However, Watson and he agreed that the most likely way to confirm DNA's hereditary role was to discover its structure. Hopefully, that would reveal how it could perform such a role. Then would the chemistry of heredity and the mystery of protein synthesis be illuminated.

These fall days of 1951 were stirring times at the Cambridge unit, its staff still recovering from Pauling's triumph over the $\alpha$ helix. It left a deep impression on Crick. This combination of model building, with attention to precise data on the directions, degrees of freedom, and lengths of the bonds between atoms in the molecule, was very powerful. Executed on the basis of correct empirical data, model building is not only a serious and imaginative process, but a stochastic one, i.e., it is conjectural, serving to create molecular structures and, at the same time, to narrow down the possibilities. Some proposed structures may be in harmony with the X-ray data, but they cannot be built without violating the accepted stereochemistry of the constituent atoms. The structure of the molecule may then be decided by default. Only subsequently does this become proven from unambiguous X-ray data. (In the case of DNA, that was to take almost a quarter of a century.)

Crick realized that with model building, Watson and he might well be able to repeat with DNA the triumph that Pauling had had with the α helix. Had not Pauling conceived of the α helix while ill in bed, using a sheet of paper on which he had drawn a polypeptide chain that he folded to construct a helical model? Crick had seen the diffraction pattern of the crystalline form of DNA that Wilkins had shown at Perutz's Cambridge conference in July 1951 and heard Wilkins describe its likeness to what one would expect of a helical structure. Watson, too, as noted above, had been inspired by the sight of the same diffraction pattern. Their interaction sufficed to bring Crick's thinking about DNA out from "a back recess of his brain."

Wilkins apart, where could X-ray diffraction patterns of DNA be found? There were the prewar patterns from Leeds University Biophysics department. They were republished by Astbury,[21] but were subsequently identified as mixtures of two forms of DNA. Neither Watson nor Crick was aware that a fine pattern of a distinct form of DNA had been obtained by Astbury's colleague Elwen Beighton in Leeds in 1951 but had remained unpublished. Because they had not made a trip to Leeds to investigate further, they were unaware of this development. In truth, they had dismissed Astbury as a pioneer who had already seen his best days. When Watson encountered Astbury at a professional meeting, he was not impressed, for the pioneer was passing the time telling off-color jokes.[22] There was nothing left to do but invite Wilkins to Cambridge in the hope of learning more about the progress being made at King's College.

Wilkins was only too happy to make the trip to Cambridge. For him, it was an opportunity to receive his friends' commiserations for the misfortune of being shut out from the DNA research ongoing at King's. His hosts were expecting to learn about progress in the work, but the expectation was sadly disappointed. The problem that had arisen can be traced back to Wilkins' own suggestion to his boss John Randall that the new postdoctoral fellow Rosalind Franklin, when she came to work at the unit, should be assigned to the X-ray study of DNA, instead of the original topic of proteins in solution,[23] a change that she accepted. Wilkins assumed that she would join the team composed of himself, the theoretician Alec Stokes, and the graduate student Raymond Gosling. Franklin arrived while Wilkins was away, having been given to understand from John Randall that she and Gosling would be the only ones working on this project. "Stokes," wrote Randall, "I have long inferred, really wishes to concern himself almost entirely with theoretical problems in the future," and he continued, "this means that as far as the experimental X-ray effort is concerned there will be at the moment only yourself and Gosling."[24] Wilkins did not learn of the existence of this letter until 25 years later! He then discovered that his name was scarcely even mentioned in it, although it was he who had introduced DNA to the unit in the first place.

Armed with the understanding that she would have DNA to herself, and the undiluted help of Gosling as well, Franklin beat off attempts by Wilkins on his return from his many travels that year to continue his involvement in the work. Now, she had all the "good" DNA that Wilkins had received from Rudolph Signer and brought to the lab in 1950. Although she kept Wilkins at arms' length, she did show him her discovery of a new type of DNA, obtained at high humidity. This underexposed pattern, like the unpublished pattern at Leeds, had fewer data on the X-ray diagram than the crystalline form, but it was an unusual pattern that resembled a Maltese cross, with nothing along the meridian until the prominent reflection at a distance of 3.4 Å. She had also discovered the fact that the crystalline form found by Gosling and Wilkins before her arrival was converted to this new form at high humidity. She called the former the crystalline form (subsequently, A), and the latter the wet form (subsequently, B). All previous published patterns were mixtures of varying proportions of A and B. (Recall that the only other clear B pattern had been obtained in Leeds, but its significance had not been appreciated.) Franklin and Gosling had made a fundamental discovery without which not even a wizard like Pauling could have inferred the structure from the hitherto published patterns. Their study of the distinctive character of these forms and the conditions for their interconversion were necessary prerequisites for the interpretation of either diffraction pattern.

Wilkins, it appears, did not tell Watson and Crick about this important set of results. It seems that he was reluctant to report what Franklin had discovered, but he had missed. The fact that she had reported her success to him with malicious satisfaction gave added force to his reluctance. Using her training in physical chemistry adroitly, she had carefully controlled the relative humidity of DNA fibers to reveal this significant transformation. The only good news that Wilkins had divulged to Watson and Crick was the fact that DNA would be the subject of an internal seminar at Kings in 3 weeks time. It had been planned by Wilkins to facilitate open discussion. Watson was delighted when Wilkins told him that he would be welcome at this seminar.[25]

## Building the First Model

On the 21st of November, Watson took the train to London to attend the seminar on DNA at King's College. Crick had other business and traveled separately, but they were to meet on Saturday to travel together to Oxford, where Crick's destination was Dorothy Hodgkin's laboratory in Oxford University's Chemistry Department. Crick was anxious to talk to her about his work with William Cochran and Vladimir Vand on the derivation of the Fourier transform of a helix before completing their paper for submission to *Nature*. He knew, moreover, that she was one of the few crystallographers who would at once appreciate its importance. Trained as a chemist and a former student of Bernal in Cambridge, at this time she was working on the structure of vitamin B12. She had established the structure of penicillin 6 years before; Crick held her in high regard.

The theme of Franklin's talk, one assumes, was to have been her discovery of the transition of DNA fibers between its two forms, the crystalline and the wet. The X-ray diagram she had created of the latter was rather faint, and although her notes of the talk report the prominent reflection at 3.4 Å on the meridian and the "smears" 40° to the meridian, she did not provide the pattern's other parameters. The drier form engaged most of her attention because of its greater crystallinity, as evidenced from the large number of relatively sharp reflections in the pattern. Prominent in its diffraction pattern was the reflection at 27 Å, as previously observed by Astbury. Unlike Astbury, however, in her notes she interpreted this as due not to a chemical repeat (i.e., of a particular base), but as "one turn of the spiral" (i.e., a structural repeat). Her lecture notes say that both forms consist of cylindrical molecules, each containing several chains (two, three, or four) held together by hydrogen bonds, with the sugar–phosphate backbones being on the *outside* of the cylinders.[26] How many of these conclusions she actually used in her talk remains unknown. Neither Wilkins nor Watson recalled her having said anything about helices, and Wilkins later suggested that such silence would have served to preserve her research distinct from the helical-minded contributions to the colloquium from Stokes and himself. Be that as it may, Watson witnessed the apathy of the ensuing brief discussion from the 15-member audience in the drab, old lecture hall. No stunning news, no fireworks. He left the colloquium somewhat disappointed.

Crick was "in top form" when Watson met up with him at Paddington station to board the train to Oxford. But no sooner had the two seated themselves on the train than Crick began to cross-examine his companion about the King's colloquium. What was in his notes? There were none. He had relied on his memory, a practice that worked for him in a field he knew well, like phage genetics, but now failed him in X-ray crystallography. True, Crick's frequent tutorials during the past 6 weeks had prepared his unmathematical friend to understand some of the talk about helical molecules, but it proved largely irrelevant to the substance of Franklin's talk. The crystallographic jargon proved troubling, and when he recalled the water content as eight molecules per unit cell, the possibility emerged that he might be misleading Crick "by an order of magnitude difference." Franklin's notes for the talk state eight per "structural unit," of which she believed there were about 30 in the unit cell. Crick's annoyance at Watson's trusting of his memory faded away as he set to work with whatever data Watson could remember. Alighting at Oxford station, Crick was in a confident mood. The possible solutions were limited. Just a week of concentrated exploration of possible molecular models would probably suffice to make them quite sure that they "had the right answer."[27]

After returning to Cambridge, Crick set to work to formulate a possible structure. In the shadow of Pauling's success, when he ignored contrary experimental data, Crick reasoned that only the minimum number of data should be permitted to constrain the search for possible configurations of the molecule. It should be helical, and such data that Watson could remember from Franklin's talk should be incorporated. Also accepted was the hunch that the weak attractions of hydrogen bonds, so important in the $\alpha$ helix, were not important in DNA. This conclusion followed from Watson's memory of the water content in Franklin's crystalline DNA—1/30th of the figure that Franklin gave. Moreover, knowing that the bases can exist in two different forms (Fig. 7.2) due to shifting in the location of hydrogen atoms (known as tautomerism), Crick ruled out hydrogen bonding between the bases as a structural feature. In addition, there was the awkward fact that the four bases have different sizes and need to be attached to the sugar–phosphate chains in a highly irregular sequence if they are to serve as genetic material. Therefore, structures with bases facing outward from a central backbone would be easier to build than those with the bases inside an encircling helical backbone.

In no more than 1 week, our two enthusiasts had built several "rough" models based on the fundamental principle of helical conformation—namely, that the repeating elements in the long-chain molecule all occupy equivalent positions. Leaving aside the differing bases in DNA, Crick applied this principle to the sugar–phosphate backbone. The structures thus produced consisted of three such backbones

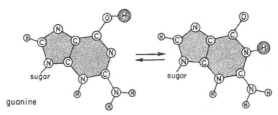

**Figure 7.2** Tautomerism of the bases.

wound around one another forming a tight central cylinder, held together by the electrostatic attraction between the positively charged sodium ions and the negatively charged phosphate groups (subsequently, they substituted magnesium ions for the sodium). This model fitted the X-ray diffraction data that Watson had gathered from Franklin's talk at the King's colloquium, namely, a helical pitch of 27 Å (in the crystalline form), which was a density high enough to require several chains in the molecule. They decided on three chains and the misunderstood value for the water content.

Crick's excitement grew at this result. It was, of course, a "tentative structure." Nonetheless, it looked promising, and if it were true that DNA contained so little water, did contain magnesium ions, and its bases were facing outward and played no part in holding the chains together, then it might have been on the right lines. But this was a big if! Did they check these assumptions with the experts around them, such as Bill Cochran, who had been directing the doctoral research of June Broomhead on the structure of guanine and adenine in the Subdepartment of Crystallography housed in the Austin Wing? It seems not. But Crick, with characteristic enthusiasm, promptly set down a summary of the approach that they used and the features of the molecular structure that resulted. It begins:

### The Structure of Sodium Thymonucleate. A Possible Approach: Summary

*Introduction*

Stimulated by the results presented by the workers at King's College, London, at a colloquium on 21$^{st}$ November 1951, we have attempted to see if we can find any general principles on which the structure of DNA might be based. We have tried, in this approach, to incorporate the *minimum* number of experimental facts, although certain results have suggested ideas to us. Among these we may include the probable helical nature of the structure, the dimensions of the unit cell, the number of residues per lattice point, and the water content. Having arrived at a tentative structure in this way, we have generalized what we regard as the important features and now present these as postulates.

*Postulates*

1. That all residues are equivalent. As Pauling has pointed out, this necessarily leads to helical structures. . . .

2. That the structure will be dominated by the charged atoms. There are no atoms which can *donate* hydrogen bonds except in the basic rings [purines and pyrimidines] and the water [scarce]. Thus hydrogen bonding is unlikely to play the dominant part in the structure that it does in the polypeptide $\alpha$-helix. Electrostatic forces are so big relative to van der Waals forces that we may be confident that the Na$^+$ and the PO$_2^-$ will mainly decide the arrangement.

3. The Na$^+$ and the PO$_2^-$ will all lie very close to the surface of a single cylinder.[28]

Next, they distributed Na$^+$ and PO$_2^-$ in a variety of topological arrangements on sheets of paper that they folded into cylinders. One could then trace links between successive PO$_2^-$ groups climbing up the paper cylinder and estimate the pitch of the resulting helix. This was how Pauling had discovered the $\alpha$ helix in 1948.

At this point, Kendrew urged that Watson and Crick contact the King's group. After all, it was their data that the Cambridge pair was using. Wilkins should be asked to come and see it at once. Crick called them and, according to Gosling, told them that "We've done something rather clever!"[29] The next day brought Wilkins together with Franklin and Gosling to view the structure. Franklin took one look at the model and laughed at them: "Oh look, you've got it inside out!" she exclaimed. She was tickled pink.[30] But Crick was not so easily put down. He proceeded to give his introduction, lecturing his visitors on Fourier theory and helical construction. The more he talked, the less receptive his audience became. At the mention of magnesium ions as the agents holding the three chains together, Franklin positively erupted. Then Watson's error regarding the water content emerged. It was not that he had forgotten, but that he did not understand her terminology. So what in their structure could possibly stabilize it? And that was not all. Had Watson forgotten that at the colloquium she had explained why the backbones must be on the outside of the cylindrical molecules, not the inside, and had he been attending when she had cited support for this point from the much-respected nucleic acid chemist John Gulland? Yet here they were, making this trek from London to see a structure whose backbones would fall apart! Reminiscing later, Wilkins marveled at the sight of "these highly intelligent Cambridge chaps turning up with something which obviously was crazily wrong. The whole thing was inside out."[31]

Crick tried to rescue some merit from the exercise, but even he betrayed a change of mood, that Watson described as "no longer that of a confident master lecturing hapless colonial children who until then had never experienced a first-rate intellect."[32] Meanwhile, Wilkins checked the train times and suggested an early afternoon departure for London. The offer that the Cambridge duo then made of collaboration with King's was turned down. The London visitors had witnessed two mavericks venturing in unfamiliar territory—their territory. Why should they condone such behavior by joining forces with them? As the King's trio departed for the station, Crick and Watson returned to face their colleagues and, in due course, learned the decision of the two lab heads—Lawrence Bragg and John Randall—that DNA was to be off limits to them.

Dutifully, they sent the molds or "jigs" that had been used to make the brass atoms for their DNA model to King's. It was December and time for Crick to send his brief note with Cochran to *Nature* on the Fourier transform of a helix and its support for Pauling's α helix.[33] Shut out from DNA, Watson turned to the RNA virus of the tobacco plant (tobacco mosaic virus, or TMV) in the hope that it might offer clues to the structure of DNA.

Crick, who by this time had been at work at the Cavendish for 2 years, had two publications in press: his note with Cochran to *Nature* and his calculation of the height and character of the peaks in the Patterson diagram for hemoglobin. Using his formula for the Fourier transform of a helix and Pauling's α helix, Crick calculated that such peaks should be 10 times higher than those present in Perutz's Patterson diagrams. It was an important corrective, but how much longer would Bragg tolerate a graduate student who "rocked the boat" and barged in on the research project of another of the MRC's research units, moreover doing so in such an amateurish manner? And to cap it all, he was not progressing with his own laboratory project on a protein he could call his own. To Mellanby's successor as MRC Secretary, Harold Himsworth, Bragg launched a tirade over Crick's behavior and to A.V. Hill, he expressed his concern in a letter beginning:

> There is a young man working here, in Perutz's team, who I believe at one time was a protégé of yours, and advised by you to take up biophysics. This is Crick. I am worried about him. . . . My worry is that it is almost impossible to get him to settle down to any steady job and I doubt whether he has enough material for his Ph.D. which should be taken this year [1952]. Yet he is determined to do nothing but research and is very keen to hang on here. With a wife and family he ought to be looking for a job. I think he overrates his research ability, and that he ought not to count on getting a job with no other commitments. Are you interested in his career enough to wish to discuss it? I should like some help in deciding what line to take with him.
>
> Yours ever W.L. Bragg[34]

Perutz now found himself in a difficult position. To clarify the situation, he decided to write to his contact at the MRC.

> Dear Duncan,
>
> The trouble with Crick is not that he is incapable of independent work—on the contrary, he has much originality. The difficulty is rather that he becomes fascinated by every interesting problem that crops up in the laboratory and throws himself into the fray in an attempt to solve it. He usually makes an original and very useful contribution, but the sum total of his activities, being spread over so many different problems, tends to be different from what is required for the Ph.D.: a single piece of consistent research which is entirely his own.
>
> The research unit as a whole profits from Crick's astute criticism and his many suggestions. On the other hand, for Crick's own career I think it is essential that he gets his Ph.D. May I suggest, therefore, that the [salary] increment be withheld next April, and that Crick shall receive it when he gets his Ph.D. and moreover that you officially inform him of this decision?[35]

The Council's response to this suggestion was forthright. Landsborough Thomson explained, "To withhold the normal increment would be a mark of the Council's severe displeasure and I do not remember that it has ever been done in the case of a member of the scientific staff. In view of your praise of Crick in important respects it strikes us as inappropriate." He expressed the hope that the Council should not have "to use, or even threaten, financial sanctions."[36]

Although Crick did not know that the Council had protected his 1952 salary increment, he rightly feared that his days at the Cavendish were numbered, at least while Bragg remained Cavendish Professor.

## CHAPTER 8

# Two Pitchmen in Search of a Helix

 *I suddenly thought, "Why, my God, if you have comple-
mentary pairing, you are bound to get a one-to-one ratio."[1]*

One day in July 1952, Crick received an invitation with Watson to meet the world expert on nucleic acids—Erwin Chargaff, the biochemist from Columbia University in New York. This eminent scientist was to be lunching at Peterhouse with Crick's colleague John Kendrew. Knowing full well that neither Watson nor Crick had stopped talking and thinking about DNA since the moratorium had been imposed on them, Kendrew invited our two collaborators to meet the great chemist for after-lunch drinks. By way of introduction, Kendrew light-heartedly slipped in the remark to Chargaff that these two "were going to solve the DNA structure by model building!" Kendrew, one suspects, having witnessed their abysmal failure with their DNA model the previous year, judged rightly that they needed exposure to an authority from the world of nucleic acid chemistry.

This proved to be an eventful meeting that the participants were never to forget. It also proved to be a foretaste of the kind of response that some other biochemists gave to the Watson–Crick structure for DNA. Watson was never to forgive Chargaff for the humiliation that his scorn produced. And as we shall see later, Crick avoided rising to any of Chargaff's deliberate taunts in public debates regarding their claims for DNA. Describing this, his first encounter with them, 21 years after the event, Chargaff said, "When I met them first in Cambridge in the Spring of 1952, they did not seem to know much or anything of my work, nor even of the structure and chemistry of the purines and pyrimidines. They were extremely eager to match Pauling's α-helix by something similar in a polynucleotide, and they talked so constantly about 'pitch' that I wrote in a little notebook of mine: 'Two pitchmen in search of a helix.'"[2]

Chargaff possessed the erudition and sophistication of an intellectual from the former Austro-Hungarian Empire. Born in Czernowitz
(now Chernivtsi in the Ukraine), he had earned his "license" as a biochemist, receiving a doctorate in 1928 from the University of Vienna,
the year that Watson was born and Crick a mere 12-year-old schoolboy
in Northampton.[3] In a photo of him taken about that time, Chargaff was
indeed a handsome young man. Working at the Universities of Yale
and Berlin, the Pasteur Institute, and finally, the famous Biochemistry
Department of Columbia University, he identified with the discipline
of biochemistry and was a committed exponent of the techniques of the
organic chemist. In addition, as a product of the broad intellectual tradition of the German-speaking world, he had the erudition that goes
with *Bildung*, i.e., the cultivation of the mind through literature, history, and philosophy. Much of American culture ignited his scorn,
including the world of advertising and the profession of car sales, the
label he was later to throw at Watson and Crick.

It had been in the 1940s that Chargaff read the paper of Oswald
Avery, Colin MacLeod, and Maclyn McCarthy that identified the transforming principle as DNA. It was on the basis of these results that he
decided to devote the major effort of his Columbia lab to nucleic acids
and nucleoproteins. Avery's work had placed a big question mark on
the tetranucleotide hypothesis, which stated that DNA was a simple
molecule composed of a boring repetition of each of its four constituent
"bases" attached to a sugar-phosphate backbone. On analysis, such a
compound would yield equal amounts of the four bases. That being so,
where was the potential for variability in DNA? The chemical data,
mused Chargaff, must surely be suspect.

Adapting the technique of protein chromatography to the nucleic
acids, Chargaff and his postdoctoral student Ernst Vischer showed that
the quantities of the four bases varied from species to species and none
approximated to the 1:1:1:1 ratio of a tetranucleotide, thus dealing the
death blow to that hypothesis. Yet despite the distinctive results of
analyses from different sources, there was an approximate equality
throughout between the (molar) content of the bases adenine (A) and
thymine (T) and between guanine (G) and cytosine (C), expressed as
$A/T = G/C = 1$. No matter how much more G and C there was in a
species, and how little A and T, or vice versa, the ratio between the A
and T and between G and C approximated to 1. Chargaff had described
this work when he was on the lecture circuit in Europe in 1950,[4] but
Watson had only come across Chargaff's papers when browsing in
Cambridge's library 2 years later. Crick had not seen any of them,
although apparently Watson had mentioned Chargaff's work to him.

Now, imagine the scene: Chargaff is introduced to these two apparent nonentities. He is amazed at Watson's appearance—his hair and his accent—this young geek of a postdoc had been doing his best to disguise his American identity by softening his Chicago accent and growing out his crew cut to an extent that made him look ready to take off in the next gust of wind. "Immediately," recalled Watson, Chargaff "derided my hair and accent."[5] Later, Chargaff was to describe Watson at that time as "quite undeveloped at twenty-three, a grin, more sly than sheepish; saying little, nothing of consequence." As for Crick, teasingly provocative and jocular, he had "the looks of a fading racing tout, something out of Hogarth . . . an incessant falsetto, with occasional nuggets glittering in the turbid stream of prattle."[6] Crick and Kendrew as protein boys were saying to Chargaff "Well, what has all this work on the nucleic acids led to? It has not told us anything we want to know!" recalled Crick.

> Chargaff, slightly on the defensive, "Well of course there is the 1:1 ratios." So I said, "What is that," so he said, "Well it is all published!" Of course, I had never read the literature, so I would not know. Then he told me, and the effect was electric. That is why I remember it. I suddenly thought, "Why, my God, if you have complementary pairing, you are bound to get a one to one ratio."[7]

Crick had earlier been discussing possible pairing arrangements between the bases in DNA with the young theoretical chemist John Griffith. The two had gone to have a drink at the Bun Shop, a popular "watering hole" for scientists, close to the Cavendish. They had been listening to a talk by the astronomer Tommy Gold on "the perfect cosmological principle," i.e., the universe is homogeneous in time and space and looks the same at any time and any place. "What is the perfect *biological* principle," asked Crick?[8] Maybe it is the copying of a gene. "Since the bases [in DNA] are flat, perhaps that is so that they can stack on top of one another and attract. Why not work out if adenine attracts adenine, and so on?" With such a means, the DNA of the parent gene could be duplicated in a freshly minted and identical DNA chain constituting the new gene.

Griffith was a quiet fellow, and Crick did not always know what Griffith was thinking. Nor was he aware at this time that Griffith held him in some awe. He knew that Griffith had already been interested in gene duplication, but he had not known that he had devised a scheme of pairing between the bases. All he replied to Crick's request was that he would do as suggested—investigate Crick's like-with-like pairing suggestion. Later, they bumped into each other in the Cavendish tea queue and Crick asked him "Have you done the calculation?" and Griffith replied, "Yes, and I find that adenine attracts thymine and guanine attracts cytosine."[9] Before he had time to say anything else, Crick responded "Well, that is alright, that is perfectly O.K., *A* goes and makes *B* and *B* goes and makes *A*, you just have complementary replication." In other words, the result would be not identical but complementary, and repeating the operation would bring identity again.

It was not until this meeting with Chargaff that he learned about the base ratios and realized "Why, my God, if you have *complementary* pairing, you are bound to get a one-to-one ratio." Now Griffith, it seems, must have known about Chargaff's 1:1 ratios before he met Crick. Otherwise, how did he arrive at these pairs? Moreover, he had been thinking about hydrogen bonding across from one base to the other (Fig. 8.1A), whereas Crick was thinking of electrostatic attraction between the flat surfaces of the bases, one above another (Fig. 8.1B). And, in further discussion, Crick had argued strongly against hydrogen bonding (see Chapter 7). Crick therefore envisioned any structural relationship that would generate the Chargaff ratios to be electrostatic and in the direction of the chains rather than transverse to them.

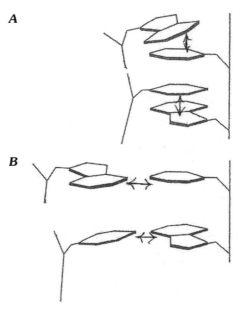

**A**

**B**

*Figure 8.1* (*A*) Electronic attraction between bases in the direction of the polynucleotide chains. (*B*) Attraction by hydrogen bond formation orthogonal to the chains.

Crick now needed to learn whether the Chargaff's pairs were the same as Griffith's pairs. "By this stage," he recalled,

> I had forgotten what Griffith had told me. I did not remember the names of the bases. Then I went to see Griffith. . . . And I asked him which his bases were and wrote them down. Then I had forgotten what Chargaff had told me, so I had to go back and look at the literature. And to my astonishment the pairs Griffith said were the pairs that Chargaff said.[10]

This revelation spurred Crick to devise some experiments to seek evidence in support of base pairing. If he were to find a difference in the absorption of ultraviolet (UV) light by the alleged base pairs when mixed and taken together, and measured separately, that would suggest that base pairing was present, reducing the absorption when the members of each pair were together. He made the same comparison using paired and separate nucleosides. The results were inconclusive, and although Crick considered other possible experiments, he dropped the idea—his expected dissertation completion date was only some 6–12 months away. As it turned out, the plan of the experiment was appropriate, but Crick surmised that the spectrometer cells containing the samples of bases and nucleosides absorbed too much of the UV light.

After the rumpus of 1951, the cloud over Crick seemed to have dispersed. But May 1952 was the time for the Council to review Crick's unestablished post, that had now run for 3 years. Perutz requested a further year for Crick on his unestablished appointment. In support of this request, he outlined Crick's work with the unit since 1949. He explained how Crick had begun by immersing himself in Perutz's horse hemoglobin data to "become familiar with the nature of the protein problem, and to discover whether or not Perutz's conclusions "would stand up to a hard shaking." He confessed that they did not. Crick had shown the need for Perutz to make two revisions and he was just publishing a third criticism—that on the height of the rod vectors in hemoglobin.

Crick's laboratory research was not going too well, but not through any particular fault of his. On the other hand, his derivation of the transform of a helix, and his application of the theory to Pauling's α helix, Perutz believed, "was a feat for which he deserves much credit, as the problem had previously defeated Pauling himself and the very able theoreticians in his laboratory." Perutz suggested prolonging Crick's unestablished position "until we can see more clearly what kind of post would be most suitable for Crick in the future."[11]

The Council's response was to ask whether Crick would be looking for employment outside of the MRC in the future. The Council reminded Perutz that when he had visited the unit, Perutz stated that "Crick would probably leave the lab in about a year from now. Is this arrangement now likely to be altered?"[12] Perutz admitted that "Professor Bragg and I feel it would probably be a good idea if Crick found himself some University job in a year or so, but there was," he declared, "nothing definite about this arrangement."[13] From this correspondence with the MRC, it is clear that Perutz negotiated with the Council judiciously in his effort to keep the way open for his brilliant but trouble-making colleague to gain a secure position in the unit.

### The Coiled Coil

The fall of 1952 was given to work on Crick's dissertation, but both Crick and Watson were keeping their ears tuned for knowledge of Linus Pauling's research activities, for they feared he might decide to turn his skills to DNA. The news that he was busy on a different problem—the coiling of α helices around one another in the wool fiber—was reassuring because it kept open the possibility of success with DNA on this side of the Atlantic, but it unsettled Crick. His thoughts went back to his recent meeting with Pauling who in September 1952 had at long last gotten a visa to travel outside the United States, where, hitherto, McCarthyism had held him within its borders. Thus, it came about that Crick had the pleasure of accompanying Pauling in his taxi ride around Cambridge. They discussed the possibility of Crick coming to Caltech during 1954–1955 and, once arrived at the Cavendish, Crick explained the ideas he had about the structure of α keratin (hair and wool) to his eminent guest.

Remember that the X-ray pattern of this fiber showed that the prominent reflection on the meridian would be in the wrong place if the molecular structure were that of Pauling's α helix. Crick had the idea that if several helices were coiled around one another, then reflections that should have fallen on or near the meridian at the model's predicted distance of 5.4 Å might fall instead at the observed distance of 5.1 Å. A minor inclination of the chains from the axis of the fiber would, he believed, be sufficient to achieve this. The idea first came to him in his work on the orientation of polypeptide chains in hemoglobin. Now, he could apply the idea to the much-studied fiber α keratin.

It was characteristic of him to first ask how much energy would be required. In other words, was the idea feasible? His rough estimate suggested 1/10 of a kilocalorie per residue. "This energy," he added, "increases as the fourth power of the pitch angle, so that the α-helix can easily be bent through small angles, but resists large deformations." He believed that a deformation of about 18° would suffice. In addition, there were packing problems to consider because of the side chains or "knobs," as he called them. If helices wind around one another, would their respective side chains not get in the way? He found it impossible to pack models of the α helix closely side by side, "since a good fit in one place produces a bad fit somewhere else, due to the non-integral nature of the helix. However," he claimed, "it can be shown that by deforming the helices into coiled-coils the knobs can be made to inter-lock systematically."[14] He sent these suggestions in a letter to *Nature* in October, after his conversation with Pauling (see below).

That winter was given to work on the thesis, including further work on coiled coils. Then came the interruptions caused by a second attempt on the structure of DNA, and it was not until 14 March 1953, amid the excitement over DNA, that Crick sent his two technical papers on the coiled coil to *Acta Crystallographica*. The first was an impressive theoretical work deriving the Fourier transform of a coiled coil, aided on the mathematical side by his friend Georg Kreisel.[15] The second was given to the packing together of α helices into two- and three-strand ropes, with a beautiful description of a simple method for visualizing his knobs and holes idea for packing the helices.[16] While at work on these papers, he learned that Pauling was working on just this topic. Had Pauling said anything about this when they were together in Cambridge? Not according to Crick, and because Crick was still a doc-toral student, surely there was all the more reason for the great chemist to tell him about his thoughts and plans. Now, it would have been sur-prising if Pauling had not been at least thinking about how to accom-modate the α-keratin 5.4-Å reflection with his α-helix model. But a search of his surviving papers yielded no evidence that he had set to work on the topic until his return to California. Sure enough, on the plane to Los Angeles on 15 September, shortly after his meeting with Crick, he wrote in his diary:

To be done at once:

Model of 4-residue compound α-helix pitch 60Å radius 1.8Å.

Model of 7-residue B compound α helix for bBb.[17]

Four days later, we find Pauling writing in his diary, "I shall eval-uate the Fourier transforms for some of the compound helixes in alpha keratin proteins (according to my new idea, developed last month)." Within the space of 2 weeks, Pauling's paper was ready. *Nature* received it on 14 October. Crick, meanwhile, as described above, had been working on the mathematics of the stresses involved in inclining the chains so that they could coil around one another. Hence, it was not until 22 October that *Nature* received Crick's letter entitled "Is α-keratin a coiled coil?" which the editor published a month later.[18] Pauling's paper, "Compound helical configurations of polypeptide chains. . ." did not appear until 10 January, 6 weeks later![19] Naturally, when Pauling saw that Crick's paper had appeared first, he objected. The editor had clearly responded to its provenance—the Cavendish Laboratory.

Privately, Crick had been most unhappy about this experience with Pauling. He was sure that he had outlined to Pauling his solution to the problem of the 5.4-Å reflection when with him in Cambridge, but Pauling gave no indication that he had been working on the problem or that he had formed any definite ideas about how to solve it.[20] Moreover, it is remarkable that in November, Pauling gave no indication that he had ever met Crick when he made the following remarks to his friend the former biochemist Jeffreys Wyman: "The trip to Europe this summer was very valuable." He added that his discussions with the

British people working on the problem of α-keratin had emphasized strongly some of the difficulties with the simple structures we had proposed for α-keratin, and after much thought I discovered that the solution of the difficulties consisted in assuming that the α-helixes are not straight, but are coiled around one another. . . .[21]

Could the "much thought" have included conversation with Crick? Is that why the paper Pauling had sent to *Nature* on about 7 October contained an expression of thanks to Mr. F.H.C. Crick, among others, for discussion of the "question of the structure of α-keratin"?

Pauling explained that he was working on cables in α keratin "consisting of six α-helixes twisted about a seventh central one." He added that he had just heard "that a young Englishman, Crick, has submitted the same idea, perhaps not so thoroughly worked out, since the title of his note (which I have not seen) is "Is α-Keratin a Coiled Coil?"[22] His informant was Jerry Donohue, his former postdoc, who was now at Cambridge. Donohue had told Pauling that Perutz had sent Crick's note to *Nature* with a covering letter "requesting high speed publication, so he must think it's pretty hot stuff, although Bragg apparently isn't too keen on the idea."[23] Pauling replied, "I remember that when I was in Cambridge Crick asked me if I had thought about the possibility of α-helixes twisting around one another, and I said that I had. I don't remember that we said any more about the matter."[24] But as we shall see, it was to arise again after Pauling's visit to Cambridge in 1953.

This incident is typical of priority disputes in science—those involved do not remember the incidents in the same way. But the experience taught Crick to be as wary of Pauling's competitive spirit as was Bragg. It also shows that Crick was as sensitive to issues of priority and borrowings as anyone else when he felt he was the victim. True, when Crick compared his approach to the problem with Pauling's approach, it was clear that they had followed separate paths. Accordingly, he stressed this point in a diplomatic letter to Pauling.[25]

Crick continued with his doctoral research throughout 1952. The dissertation, he explained, covered two fields—"theoretical studies on the alpha-helix and on combinations of alpha-helices" on the one hand and "experimental work on two species of hemoglobin" on the other. The former subject dominated the work, but both parts concerned the same kind of problem: the conformation of polypeptide chains and the architecture of protein molecules, both fibrous and globular.

This research laid the foundations of Crick's intimate grasp and critical approach to the analysis and interpretation of X-ray diffraction patterns. He had been teaching himself crystallography "on the job," as Aaron Klug put it, not in the same way as Maurice Wilkins and Rosalind Franklin had, by taking lecture notes when they were Cambridge undergraduates in the physics curriculum. Mathematical skill was, of course, vital to Crick's theorizing in the realms of Fourier theory, and if he ran into a problem, he could turn to his friend Kreisel. Consequently, his mastery of the field was evident as he explained what is important, what to look for, where advances could be made, and how not to go off the rails when interpreting ambiguous data. But like most dissertations, the writing proved to be a tedious chore, and many were the times when he would instead break away to discuss DNA with Watson.

Crick's dissertation makes very clear the crucial importance of his involvement in the hemoglobin studies at the Cavendish for his development as a theoretician in X-ray crystallography. Thus, his ideas about coiled coils and the "knobs" of amino acid side chains fitting into holes in the adjacent chains of a coiled coil of three (Fig. 8.2) came from puzzling over the problem of packing α helices into the hemoglobin crystal. Another venture—the attempt to derive a mathematical expression for the shape of the surface of a molecule in a medium of different electron density—was also due to his interest in Perutz's hemoglobin research. Having trouble with the mathematics, he sent his text to Kreisel, who responded enthusiastically by providing Crick with eight pages of mathematical equations. Kreisel found the work to be "very interesting," and judged that

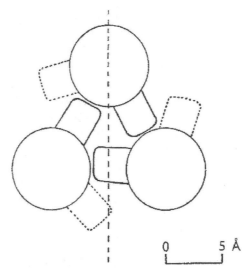

0          5 Å

**Figure 8.2** Schematic illustrating the packing of a three-strand rope—Crick's arrangement of three polypeptide chains in a coiled coil. The figure is a section perpendicular to the fiber axis (*circles*) polypeptide chains, (*knobs*) side chains. Those shown with dotted lines are at a slightly different level from those drawn with solid lines. Outer side chains are not shown.

> there are several professors of mathematics (though not in Cambridge) who made their reputations on less than your letter. I hope I am not mistaken in thinking that the work is new. . . . You might write a research note for the *Proc. Cambr. Philosophical Society*, in which you give the one dimensional case, indicate the three dimensional case. . . . Probably detailed treatment should appear in *Proc Roy Soc*.[26]

Meanwhile. Bragg had come to Perutz's aid by assuming that the hemoglobin molecule had a roughly ellipsoidal shape. Crick recalled how Bragg "made bold simplifying assumptions; looked at as wide a range of data as possible; and was critical but not pernickety, as I had been, about the fit between his model and experimental facts. He arrived at a shape that we know is not a bad approximation to the molecule's real shape . . . it was a revelation to me as to how to do scientific research and, more important, how *not* to do it."[27]

Crick put aside his attempt at a mathematical derivation of the shape based on geometrical figures plus a succession of smoothing functions. Like his unfinished paper on the "Determination of the structure of proteins. . . ," one suspects, both may have been victims of Bragg's displeasure.

## DNA Again

There was a very good reason to discuss DNA in 1952, for there had been a stir among virologists—especially among those studying the bacterial viruses, the phage geneticists. In 1951, they had shown that the phage particle is a tadpole-like object consisting of a protein coat enclosing a nucleic acid interior. If the contents of these particles were expelled, the remaining protein coats looked like "ghosts," or empty tadpoles. Such ghosts could still adhere to bacterial hosts, preventing them from reproducing, and would cause them to burst. But no phage particles could be found in the scattered contents of the host cell. Phage reproduction had not taken place, presumably because the ghosts had been deprived of their nucleic acid contents.

Moreover, other studies showed that the ghosts always remained on the outside of the host bacterial cell. Could it really be that it is the nucleic acid inside the heads of these tadpole-like particles that alone enters the host and causes phage reproduction? Already in 1951, the virologist Roger Herriott had written to his friend Alfred Hershey, suggesting that "the virus may act like a little hypodermic needle" full of DNA. Maybe the phage attaching to the bacterium "cuts a hole through the outer membrane and then the nucleic acid of the virus head flows into the cell. If this is so . . . one should be able to get virus formed by the nucleic acid alone."[28]

Hershey responded that he and Martha Chase had already followed the nucleic acid into the bacterial host using a radioactive "tracer." Like a fish located downstream by the "bleeper" attached to it upstream, the radioactive isotopes of sulfur and phosphorus that were fed to the invading phage revealed the fates of parental phage constituents in their progeny: There was "little or no" radioactive sulfur (the "tracer" for protein) from the infecting phage in the progeny phage, but to Hershey's surprise, 85% of the radioactive phosphorus (the "tracer" for nucleic acid) was transmitted. Ninety percent of the sulfur remained in the ghosts. This is the famous Hershey–Chase experiment, famous because it suggested that hereditary continuity is carried by the nucleic acid, and not the protein. Granted, this experiment was not foolproof, but taken with the evidence from bacterial transformation and from correlation between DNA content and chromosome number, Crick, like Watson, became confident that DNA must be the major, if not the sole, hereditary substance.

Meanwhile, Crick was becoming increasingly impatient with the folks at King's College London. By December 1952, they had had a whole year to come up with some results since the fiasco of 1951. Watson had been marking time by studying X-ray diffraction of RNA and next dove into bacterial genetics. His devotion to this new obsession caused Crick to fret. Watson should be keeping an eye on the literature for new clues about DNA, not burying himself in the biology of the reproduction and genetics of the colon bacillus. And had Wilkins any progress to report? He had informed Crick of his success in the spring of 1952 in obtaining a *Sepia* sperm diffraction pattern that showed helical features. But a lunch that Crick arranged with Wilkins in London that October had not furnished any new information. Should they not, mused Crick, have another go at the structure themselves, despite the ban? Watson declined. True, they would occasionally fiddle with the model atoms they had used the year before, but not with any enthusiasm.

Fortunately, the unit had just received another doctoral student from the United States, none other than Pauling's son Peter. Here was a lively man about town whose agenda included many social activities besides research on myoglobin with John Kendrew. He would often gossip, but one afternoon in December, Peter had serious news to impart. Watson has described how Peter set himself down and, nonchalantly raising his feet and putting them on the desk, announced that the letter he held so prominently in his hand was from his father. In it, Linus reported that he "now had a structure for DNA."[29] How many chains had it?, worried Peter's anxious audience. Were the bases on the inside or the outside of the structure? The letter contained no clues.

"As Watson and Crick passed the letter to and fro their frustrations grew." Crick "paced up and down the room thinking aloud, hoping that in a great intellectual fervor he could reconstruct what Linus might have done."[30] Perhaps Pauling had got it wrong. Watson feared otherwise. Crick was less convinced that all was lost. "Naturally," remarked Crick, "we were apprehensive, but I was more stoical, Jim was more nervy about things." And when they reflected on the chemistry of DNA, the more unlikely it seemed "that even Linus could pick off the structure in total ignorance of the work at King's."[31] There was nothing to do but await further details.

Crick returned to his dissertation and Watson took a holiday. On his return, Peter revealed that his father had written him on the last day of 1952 to say that he and Robert Corey, his coresearcher, had sent in their paper on the structure of DNA for publication and added "would you like to have a copy?" Peter replied, requesting a copy for himself and one for the MRC unit. Revealing was Peter's remark to his father that "for more than a year, Francis and others have been saying to the nucleic acid people at King's, "You had better work hard or Pauling will get interested in nucleic acids."[32] Three weeks later, two prepublication copies were put in the post, one for Peter and the other for Bragg. The covering letter mentioned that the paper was to appear in the February issue of the National Academy of Sciences *Proceedings.*[33]

Watson had gambled his career on being in on this discovery, but now it looked as if he was in the wrong lab; the favorite in the DNA stakes across the ocean had probably won. Before the month of January was out, the manuscript arrived in Cambridge. How did Crick and Watson feel with Peter in the room, the manuscript from his father sticking out of his pocket? Had Linus really discovered the structure? In his forthright manner, Peter told them it was a three-chain helix with the bases on the outside and the phosphate backbones on the inside. So, thought Watson, a year ago they had been on the right track after all, or had they?

His impatience mounting, Watson pulled the paper out of Peter's pocket and began scanning the summary and introduction and then studying the figures that gave the locations of the atoms (Fig. 8.3). Something looked amiss, and as he pored over the illustrations, the truth dawned on him that the phosphate groups were *not* ionized. It was unbelievable! Had this feature been incidental to the structure, it would not have deserved pursuit, but it was crucial. If the phosphates were ionized, the chains would fall apart. Then what would have become of chromosomes? But Watson had read the nucleic acid literature and was confident that the phosphate groups "never contained bound hydrogen atoms. No one had ever questioned that DNA was a moderately strong acid. Thus, under physiological conditions, there would always be positively charged ions like sodium or magnesium lying nearby to neutralize the negatively charged phosphate groups. . . . Yet somehow Linus, assuredly the world's most astute chemist, had come to the opposite conclusion."[34]

Could Watson keep the mistake a secret? No, he had to rush over to the virologists and organic chemists in Cambridge to spread the news and be reassured that DNA is indeed an acid, so of course the phosphates are ionized. It had not been called "deoxyribonucleic *acid*" for nothing! Crick had been equally amazed by Pauling's structure and lost no time in insisting to Kendrew and Perutz that the situation was now very urgent. Once the paper was out, Pauling would learn that his structure was impossible. Indeed, all he had to do was consult his own chemistry textbook! Then he would redouble his efforts and get it right. The shadow of the Cavendish MRC unit's failure to discover the α helix hung over them, and for Crick, there was that uncomfortable memory of his very recent encounter with Pauling as a competitor over the coiled coil.

Watson was already scheduled to go to London in a few days time, so it was decided that he would take the manuscript with him and show

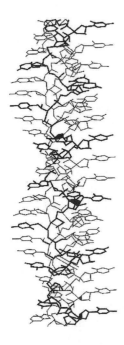

***Figure 8.3*** Linus Pauling's 1953 structure for DNA.

it at King's. It was the last Friday in January. Wilkins was busy when he arrived, so he went to Franklin's lab, and seeing her busy measuring an X-ray photograph, he pushed the door wide open and walked straight in, startling her. Regaining her composure and looking straight at Watson, she "let her eyes tell me that uninvited guests should have the courtesy to knock." After hearing Watson on the subject of Pauling's helical model, she retorted, according to Watson, that "not a shred of evidence permitted Linus, or anyone else, to postulate a helical structure for DNA." She insisted that she had antihelical data to support her doubts about helices, so Watson should know better than to preach to her the gospel of helices. Instead, he should examine her antihelical evidence. She was referring to the apparent asymmetry of some of the reflections. One would not expect this of a helical structure that, like a spiral staircase, possesses radial symmetry. But Watson knew that the previous July, she had mentioned this anomalous datum to Crick when the two encountered one another in Cambridge. Crick, in his detailed critique of Watson's manuscript in 1966, recalled the following.

> In the summer before Linus [Pauling]'s model came out you and I, or at least I, had a longish talk with Rosy in the tea queue at the Zoo[logy] lab at some conference or other in which she firmly maintained the

structure was not helical and I maintained that it was certainly likely to be helical and that she should scrutinize the evidence which appeared to be against it very carefully.[35]

Watson therefore took this opportunity to admonish her to "learn some theory," then "she would understand how her supposed anti-helical features arose from the minor distortions needed to pack regular helices into a crystalline lattice."[36]

Not content with invading her privacy, Watson was now proceeding to insult her by impugning her understanding. And why, she might well have wondered, was he carrying a copy of Pauling's paper, "A proposed structure for the nucleic acids," when she, who had written asking for a copy 3 weeks ago, had not received one? At this point, Watson had the sense to beat a hasty retreat down the corridor with Wilkins, who had come looking for him.

Now Wilkins was a friendly fellow who believed that research should be a shared activity. He had been cautious about telling Crick too much, for he rightly feared Crick's reputation for interpreting other people's data more rapidly than they could themselves. Imparting information to Watson meant it would surely find its way to Crick. Seeing Watson visibly intimidated, however, encouraged him to unburden himself. He had now, he confided, resumed his X-ray diffraction studies of DNA with the help of Herbert Wilson. He went on to report that back in May of 1952, Franklin and Gosling had obtained a remarkable example of their "wet form" of DNA that they were now calling the "B form." Questioned about it, Wilkins went to a drawer in his office and brought out a print. Gosling had only recently passed it on to him in the course of clearing up their lab in preparation for the move to Birkbeck College. For Watson, this was a memorable moment. "My mouth fell open and my pulse began to race," he recalled. The pattern was so much simpler than that of their crystalline form, which they now called the "A form." "Moreover, the black cross of reflections which dominated the picture could arise only from a helical structure. With the A form the argument for a helix was never straightforward, and considerable ambiguity existed as to exactly which type of helical symmetry was present. With the B form, however, mere inspection of its X-ray picture gave several of the vital helical parameters."[37]

Deriving the structure from this pattern might not be such a difficult task after all. More questioning over dinner elicited from Wilkins that the most prominent reflection was on the tenth layer line and represented a distance between repeating elements in the structure of 3.4 Å. This meant that the pitch of the helix was ten times this figure: 34 Å.

The diamond-shaped absences of reflections between the zero and tenth layer lines along the meridian spoke eloquently of a helical conformation, as Cochran, Crick, and Vand had predicted when they had derived the Fourier transform of a helix in 1951 (see Chapter 6). The slope of the helix was equal to the angle made by the prominent "crossways" with the meridian—about 40°. The equatorial reflections suggested a diameter of about 20 Å, an estimate supported by electron microscope pictures of DNA. Watson was almost in delirium, but he completely failed to engender any excitement in Wilkins, who was worrying how he was to fit the bases inside the space within the helix. And Franklin? Watson had infuriated her—why should he, who had neither pulled a fine thread from a solution of DNA nor produced a diffraction pattern from it, be lecturing her? And hadn't she and Raymond Gosling toiled for months to prepare their cylindrical Patterson diagram of the crystalline (A) form of DNA (Fig. 8.4)? Had Watson ever calculated a Patterson synthesis, let alone a cylindrical one? Fortunately he had not tried.

It is now evident that Watson had not absorbed Franklin's description of the two forms of DNA—wet and crystalline—during her talk at

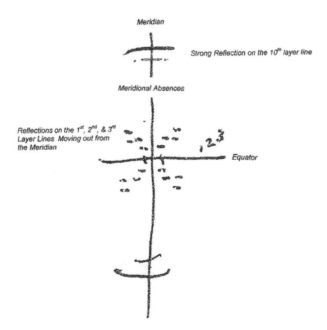

**Figure 8.4** Helical features of the DNA diffraction pattern as Maurice Wilkins described it to Crick in 1952. (Note no dimensions given.)

the seminar in 1951, or he would not have referred to this B pattern as representing a new form. It was no more and no less than a very fine example of the "wet" form that she had already briefly described in 1951. But before this visit on 30 January 1953, only Franklin, Gosling, Wilkins, John Randall, and Herbert Wilson knew of all these data—3.4 Å, 34 Å, and the 40° slope—and that all belonged to the wet (B) form, the diffraction pattern of which was so striking, although possessing few reflections. This revelation of January 1953 shows what a clarification Franklin and Gosling had achieved in establishing the existence and relationship of these two forms of DNA. Hitherto, the helical repeat of the A form at 27/28 Å was associated with the separation between the helical steps of the B form at 3.4 Å.

Before his train reached Cambridge, Watson had sketched this pattern and recorded these dimensions on the blank edge of his newspaper. The following day, Saturday, he would take them to the lab where the one who knew just how to proceed next would be arriving after a pleasant breakfast in the house he had acquired a few months before in Portugal Place—Francis Crick.

## CHAPTER 9

# A Most Important Discovery?

 *It's true that by blundering about we stumbled on gold, but the fact remains that we were looking for gold.[1]*

W hile Jim Watson had been in London probing Maurice Wilkins for information about DNA, the Cricks had been enjoying dinner with their guests, Watson's sister Elizabeth from California and a handsome visitor from Paris, Bertrand Fourcade. The following morning, Saturday, 31 January, Crick read his copy of *Nature*, breakfasted, and set off for the Cavendish. On his arrival, he was greeted by Watson, who declared that even he, a "former birdwatcher could now solve DNA."[2] Crick, suffering a slight hangover, was not amused, nor would he accept Watson's "assertion that the repeated finding of two-ness in biological systems told us to build two-chain models," rather than three-chain models,[3] as they had done with DNA before. For a sound approach, Crick preferred to rely on nucleic acid chemistry rather than biological two-some-ness, and such data were consistent with both two- and three-chain structures. Moreover, biological considerations could just as well lead one to conclude that three chains would be required—"two for replication and one for the cytoplasm."[4] He did, however, accept Watson's argument in favor of two chains based on Franklin's physical data. She had shown that the transformation from the B form to the A form resulted in a shortening of the fiber by about 30%. Taking the density of the A form and its water content led to a figure of about 46 residues in the unit cell. Two molecules occupying the unit cell, each with two chains, would have approximately 40 residues, whereas for two three-chain molecules, the figure would be about 60. Thus, Watson favored the two-chain alternative.

Crick's mild interest changed to concentrated attention when Watson passed on the information that he had gathered from Wilkins and described Franklin's excellent B pattern. Nor did Watson omit mentioning that he had gotten Bragg's permission to make a second attempt on the structure of DNA—the result of describing the B pattern to Bragg and warning him of the threat that Pauling had now posed regarding the hope for a British success with DNA. He had put it to Bragg, he added, that Pauling "was far too dangerous to be allowed a second crack at DNA while the people on this side of the Atlantic sat on their hands."[5] Bragg had been beaten to the goal twice by Pauling, once with the silicate minerals and a second time with the α helix. This could well make a third time, for the King's folk were in disarray, with Wilkins refusing to set to work in earnest while Franklin was still at King's. She, according to Watson and Wilkins, was still fussing over nonhelical data in the A form. Yet just look at her B pattern, obtained—can you believe it—9 months ago, and as far as anyone in Cambridge knew, placed on one side! The distribution of reflections was so striking—that prominent reflection on the tenth layer line at 3.4 Å and the one on the first layer line at 34 Å. Then there were the telltale absences along the meridian creating the remarkable Maltese cross pattern, speaking to Crick and Watson so eloquently of a helical molecule. It could not be so difficult to interpret; even a former birdwatcher (Watson) might be able to do it. Not to attempt to solve it with Pauling on the trail was madness! The entry of the victor of the α helix into the DNA hunt marked "open season" for DNA, and Watson should be permitted to enter the race. Bragg, with the shadow of Pauling falling yet again on the Cavendish, was persuaded.

Was this threat real, or was it bogus—just an excuse for jumping in? Surely, without Franklin's data, neither Pauling nor anyone else could have solved it. But Pauling could call on in-house expertise to produce their own diffraction patterns. True, in April 1953, his postdoc Alexander Rich did just that. Then again, was not Pauling to visit Europe that spring? The King's College Lab could well be on his planned itinerary. Claims that there was no substance to Cavendish fears are retrospective judgments. The fear was real that Caltech would scoop the Brits. Not only was Pauling the wizard at solving structures, but he was also very up to date in the chemistry of DNA's constituents—the 5′–3′ linkage and the β configuration of the glycosidic link—thanks to his visit with his long-time friend, Cambridge's outstanding organic chemist Alexander Todd. None of the Cavendish group seems to have been aware that Pauling had been acquiring important information under their very noses, as we learn from a letter Pauling wrote to one of his financial supporters in 1952:

I talked with Todd about his work a good bit this summer [1952]; we stayed with the Todds in Cambridge. In particular I discussed the significance of his results with respect to the structure of the nucleic acids. Principally as a result of this meeting with him, I have now discovered, I believe, the structure of the nucleic acids themselves. Biologists probably consider that the problem of the structure of nucleic acids is fully as important as the structure of proteins. I think that Dr. Corey and I will probably send in a note on the discovery of the structure of nucleic acids next month. . .[6]

Had Watson and Crick known of the substance of this letter, they would have been even more insistent that the moratorium on DNA be lifted. Fortunately for Watson, Bragg did not need persuading. As he recalled,

I had been feeling very sore at having made such a mess of the [polypeptide] helix, when we had so much right data in our hands, and I remember saying to Perutz so well "We must not miss another boat, ought we not to have a shot at nucleic acid when Cochran's work on the nucleotides, Miss Broomhead's on the purines and pyrimidines, and the formulae for the diffraction by a helix, give us such excellent material?" It was Perutz who insisted that we ought not to venture into this field because Wilkins had been working on it for so long.[7]

On Sunday, 8 February, Wilkins was to come for lunch at Portugal Place, having finally acceded to Crick's repeated invitations. They planned to show him Pauling's paper and he, in return, "would tell you all I can remember and scribble down from Rosie."[8] Meanwhile, Watson had returned to model building and, following his previous preference, had begun to construct backbone-inside structures for DNA, this time two-chain rather than three-chain models. As before, he could leave out the bases and concentrate on the arrangement of the backbones. He started by fixing the phosphate groups and then tried to fit in the sugar rings that are connected to one another by the phosphates. But the sugar rings proved to be so awkward that he just could not satisfactorily accommodate them. As he explained in 1954, "The phosphate groups tended to be either too far apart for the sugars to reach between them, or to be so close together that the sugars would only fit in by grossly violating" accepted interatomic distances.[9] Apparently, he still wanted to have the acidic phosphate groups of the backbone inside, so that the long basic side chains of the protein— which in chromosomes were believed to be wrapped around the DNA—could reach in. A "very feeble reason," retorted Crick. It should be ignored. "Why not," he asked one evening, "put them on the outside?" to which Watson replied, "That would be too easy," and Crick retorted, "Then why don't you do it?"[10] But Watson was afraid of the legion of possibilities that might then be possible for the conformation of the backbones. In despair, he left the lab and went to play tennis with Bertrand Fourcarde. That evening at dinner, Crick returned to the attack on the need to build chains on the outside, but Watson remained reluctant. Thus did the forthright Crick chip away at Watson's assumptions and seek to overcome his reluctance to follow a fresh path.

This was the state of play when Wilkins arrived at Portugal Place for Sunday lunch, looking forward to a merry time to dispel his gloom over the progress at King's. But no sooner had he crossed the threshold than Crick began to quiz him for details of the DNA diffraction pattern. This time, his friend was not forthcoming, because, as he explained, he now had Herbert Wilson working with him, and together they had already replicated Franklin's B pattern, although not with such a striking result. In 6 weeks, Franklin would leave King's for Birkbeck College and then they would begin model building. Wilkins added that she had given a valedictory talk at King's that contained no mention of helices. She devoted the talk to her work on the A form, which she still regarded as nonhelical. In the discussion that followed, Wilkins pressed her on the B form. How, he asked, could her proposed nonhelical structure for the A form "be reconciled with the very good B pattern" (which she did not show to her audience)? "B-DNA," she replied, "is helical and A-DNA is not." To Wilkins, that seemed most unlikely, but being Wilkins, he lacked the courage to pursue his questioning "and simply sat down."[11]

These discouraging reports aside, Wilkins did give them a lead by mentioning that in December the staff of the King's unit had submitted summaries of their work for the laboratory report that Randall gave to all the members of the Committee making a laboratory visitation that month. Among those members was Max Perutz.

Wilkins might not have proffered this information had he known what his friends would ask of him. They showed him the paper that Pauling had sent on his structure for DNA, discussed its errors, and assured Wilkins that once Pauling learned of his mistakes, he would be in hot pursuit of the correct structure. Would Wilkins mind, therefore, "if they started building models again?" The awful specter of competition with them now faced him, and he hated the thought of it. He recalled how he "tried to consider all aspects of the situation . . . but when I assessed the extent of the log-jam in our DNA work at King's, it seemed obvious that I could not ask Francis and Jim to hold off model-building any longer. And it turned out that Bragg was of the same mind: King's had had its chance— more than a year had passed and his moratorium should go."[12]

Wilkins' dream of a carefree weekend was shattered. He was downcast. He "could see no alternative but to accept their position." (Strangely he did not consider having them refer the issue to Bragg and Randall.) He "just wanted to go home, and Francis had the sense not to press me to stay. As I walked out of the house, Jim came into the street and expressed his regrets." Watson had seen what Crick did not see: the dejection written on Wilkins' face.[13]

## The MRC Report

Later that week, Watson and Crick asked Perutz if they could see the report that Wilkins had mentioned. Perutz obliged (an action over which he was subsequently to agonize, faced with the claim that he should first have sought permission). Crick eagerly ran his eye down Franklin's entry. Yes, Watson had reported the essential data correctly— 3.4 Å on the tenth layer line, 34 Å on the first layer line, and a diameter of structure of approximately 20 Å; but Watson had been vague about the slope of the crossways—it was 40° to the meridian. Then there were the details of the change in length and content of water when going from form A to B. These he had remembered correctly.

Watson had reported the information available thus far to Crick following his visit to London. But Crick also spotted what was to prove a highly significant piece of information in Franklin's report that Watson had ignored—that the space group of the crystalline form of DNA is not only monoclinic, but face-centered monoclinic, i.e., C2. This defines the kind of symmetry possessed by the crystal. In other words, the unit cell of the crystal is like a shoe box slightly out of true, due to the fact that one of the angles between the sides or "facets" of the crystal is not a right angle to its neighbors, hence, "monoclinic." C2 refers to the symmetry of the unit cell. It possesses a dyad (the axis of symmetry between paired elements of the structure) *perpendicular* to the long axis of the cell. Furthermore, Crick realized that "examination of the space group implies also that the diads are likely to relate to elements *within* the same molecule, not just the result of some molecules pointing up and others down" and he added that

> The density supported two chains, the shape of the unit cell suggested the diads were at the side. Any reasonable person would have said: "That suggests two chains running in opposite directions." That was the naïve interpretation. In any argument there is always a gimmick by which you can get round it. You just have to decide whether to go along with the argument, and ignore the let-out, or alternatively to ignore the argument because it is not completely fool-proof.[14]

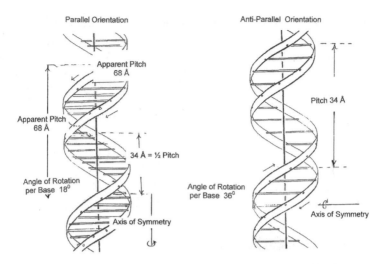

**Figure 9.1** Parallel and antiparallel chains and their importance for the interpretation of the X-ray diffraction pattern.

The possibility of the two chains being either parallel or antiparallel depends on the polarity of the sugar–phosphate backbones (see Chap. 7, Fig. 7.1), not the bases that, by this time, were assumed to follow one another along the chain in any possible (but specific) arrangement. Thus, we can picture the two DNA backbones like a pair of pencils; one, its sharpened end pointing up; the other, down.

Now suppose for a moment that the helices point in the *same* direction. In such a case, rotating them through only 180°, on the fiber axis would bring like-diffracting elements in each chain into positions identical with those vacated by the others. Thus, "from the X-rays' point of view," the upper and lower halves of such a double helix would be identical, and the meridional reflection at 34 Å would therefore represent only half of the true pitch of the helix. The axial separation between bases would then be 180 ÷ 10 = 18° (Fig. 9.1).

Watson had been trying to build chains with the successive residues separated by this small amount, thus making the distances between the steps too small and the structure far too tight. But, as Crick explained, having the chains running in *opposite* directions means that the axis of symmetry lies across the length of the chains. Therefore, 34 Å is the true pitch of the helix, the ten residues on each chain in the helical pitch occupy this distance, and 360° divided by ten is 36° rotation per residue. Nevertheless, Watson persisted model building at 18°. One day, as he left this seemingly impossible task to play tennis, he said to Crick,

"You try it." And while Watson was out, Crick built a chain with the 36° separation and "left a little note for him saying: 'This is it—36° rotation.'"[15] Crick had explained this point to the author in 1967 and 33 years later to Mark Bretscher. At the 2003 celebrations of the discovery, Bretcher had learned from Watson that the C2 symmetry was not used in the construction of the successful model, hence the absence of any mention of it in their four discovery papers. Replying, Crick denied that the C2 symmetry was not used, and he admitted that they should have acknowledged its influence in the detailed paper of 1954, written by Watson, and read and commented on by Crick before publication.[16] Yet the issue of the axis of symmetry and its relationship to the repeat distance of the helix are referred to several times, yet no mention is made of the name for the kind of dyad involved—C2. One wonders why.

Until this point, Watson had done the model building while Crick wrote his thesis. But now, Crick had momentarily taken over and constructed a plausible fragment of the backbone. Watson, he explained, found it very difficult "to grasp the fact that the chains were running in opposite directions." When they had one chain built, "he could not build the second chain the other way up. It was just too difficult! Somehow his mind didn't work that way," recalled Crick. That Crick could make this step was not due simply to his general knowledge of crystallography, because, as he remarked, whereas Watson knew nothing about space groups, he (Crick) "knew very little." After all, there are 230 of them, and he added

> But it so happened that the space group of horse haemoglobin, on which I had been working for some years, is C2. Therefore it happened to be one with which I was familiar. I knew immediately that the space group C2 has diads at the side. Therefore, if these conclusions were correct, and there are ways of weaseling out of them . . . the naïve and as it turns out correct interpretation is that there are diad axes at the side. Jim never liked this argument, I don't think he could quite follow it.[17]

Crick judged that he would not have thought of running the chains in opposite directions "but for the clue of the C2 symmetry." It was "absolutely crucial to have this idea."

So far so good. Watson had at last accepted Franklin's claim made more than 1 year earlier that the backbones must be on the outside of the molecule, and he had concluded from her data that the most likely number of chains in the molecule is two. From her data, Crick had inferred the antiparallel orientation of the two chains and, as a consequence, the repeat at 34 Å, representing the whole, and not half, of the helical pitch. But the bases—How were they to be fitted inside the two helices?

At this point, Watson struck out on his own. He reversed the earlier decision, insisted on by Crick in their failed attempt of 1951, to reject hydrogen bonding as a significant feature of the structure holding the bases together. In a literature search, he found evidence of hydrogen bonding *between* DNA molecules in concentrated solution, and at very high dilution also, where the molecules would be few and far between. This raised the possibility that such bonds exist *within* individual molecules as well.[18] He then consulted the Cambridge dissertation of June Broomhead, where he found illustrations of her structures for the crystallized salts of guanine and adenine (Fig. 9.2A). From this work, performed

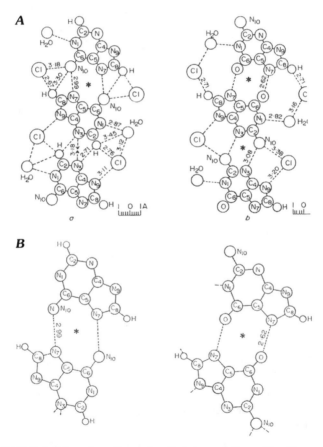

*Figure 9.2* (*A*) Detail from June Broomhead's interpretation of her X-ray data on the purine bases, adenine and guanine hydrochloride. (*B*) Illustration showing how a like-with-like scheme of base pairing, such as that which Watson constructed, could be extracted from Broomhead's figures.

at the Subdepartment of Crystallography, he was able to isolate two locations for hydrogen bonding between pairs of adenine molecules and
between pairs of guanine molecules (Fig. 9.2B).[19] Perhaps such base pairs
could be fitted into the helical cylinder of the molecule, and so could pairs
of cytosine and of thymine. He became very enthusiastic about such a like-
with-like scheme and wrote to his mentor Max Delbrück, expressing his
optimism "since I believe I have a very pretty model, which is so pretty I
am surprised that no one has thought of it before."[20] Unfortunately, as Watson well knew, the adenine and guanine bases are larger (two-ring structures) than cytosine and thymine (single ring). How on earth was he to
pack them into a regular cylinder formed by the two helical backbones
wrapping around them? And how, asked Crick, were Chargaff's ratios to be
accounted for in the model? But Watson did not want to think about them.

Added to these problems, Jerry Donohue told a reluctant Watson
that the formulae he was using for three of the four bases were wrong.
Where Watson portrayed a hydroxyl group (–OH) on thymine, cytosine, and guanine, he should instead place a keto group (=O), and the
imino group (NH) in adenine should be replaced by an amino group
($NH_2$). But changing the bases to the keto forms would make the size
difference of Watson's base pairs even greater, and special pleading
would be needed to bend the backbones far enough to "accommodate
irregular base sequences." On Friday (27 February) he capitulated.
That morning proved to be memorable for Crick. He recalled,

> At this stage, and I remember this very clearly, Jerry [Donohue] and Jim
> were by the blackboard and I was by my desk, and we suddenly thought,
> "Well perhaps we could explain 1:1 ratios [A/T, G/C] by pairing the
> bases." It seemed too good to be true. So at that point all three of us were
> in possession of the idea we should put the bases together and do the
> hydrogen bonding and it was the next day that Jim came in and did it.[21]

Watson explained that he did not necessarily take note of what
would prove to be the important parts of Crick's conversation because he
regularly talked so much, throwing out so many ideas. What Crick
recalled as a precise program—to pair the bases in accordance with the
Chargaff ratios—Watson may have recalled merely as a rejection of his
choice of like-with-like pairings. As Watson felt, Crick was always trying
to organize the others in the office, so they had the built-in tendency to
pursue an alternative plan. As for Chargaff, Watson had been reluctant to
consider his data, yet "It was immediately obvious," recalled Crick, "that
we should look at the base-pairing," but Watson "was not using Chargaff's rules (which he mistrusted)."[22] Thus, Crick's advice for the next
step fell on deaf ears.

Returning from lunch, Watson learned that the metal models of the bases needed for model building would not be ready for several days. With the weekend upon them, that would mean a long wait. His solution was to set to work and cut the shapes of the bases from cardboard. To Crick's amazement, he then left the lab to meet a group at the theatre, where Sheridan's *The Rivals* was playing. Had he appreciated the full impact of Crick's remarks, surely he would have stayed at the Cavendish that evening.

True to form, when Watson arrived the following morning, Saturday, he did not consciously proceed to put bases together in pairs, choosing the members of the pairs to reflect Chargaff's ratios. This meant putting together *unlike* pairs. Instead, Watson persisted with like-with-like pairs, albeit now using the keto forms as directed by Jerry Donohue, for he now respected Jerry's expertise.

### 28 February 1953

When Jerry came in that Saturday morning, Watson recalled

> I looked up, saw that it was not Francis, and began shifting the bases in and out of various other pairing possibilities. Suddenly I became aware that an adenine-thymine pair held together by two hydrogen bonds was identical in shape to a guanine-cytosine pair held together by at least two hydrogen bonds. All the hydrogen bonds seemed to form naturally; no fudging was required to make the two types of base pairs identical in shape.[23]

Calling Donohue over to view his scheme (Fig. 9.3), Watson asked if he had any objections. When the answer was no, "my morale skyrocketed, for I suspected that we now had the answer to the riddle of why the number of purine bases equaled the number of pyrimidine" bases. It seems that Watson had now discovered serendipitously what in principle had been suggested the evening before, namely, mixing the bases in pairs.

When Crick came in, the characteristic skepticism of the collaborator was quickly dispelled. The similar shape of the paired bases was compelling. Caution dictated that there might be other ways to yield the Chargaff ratios, but although Crick tried shifting the bases around, he failed to find any rival pairing positions. Equally important, the symmetry of the bonds connecting the two bases to their respective backbones struck him. They were related by a dyad axis in the plane of the base pairs. In other words, the base pairs could be turned over through 180° to yield the same orientation of these bonds with their backbones, thus confirming the C2 symmetry of the unit cell as recorded by Franklin in

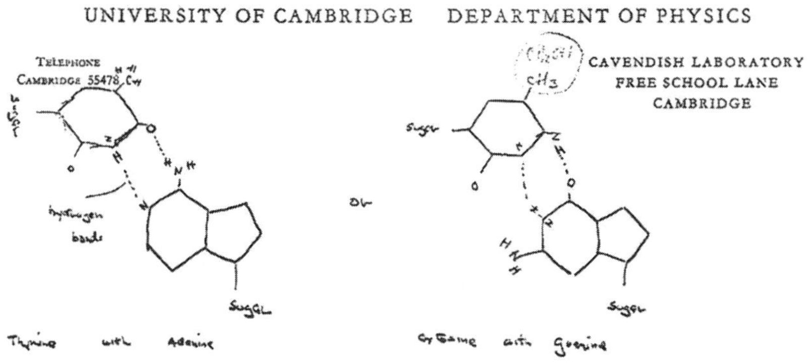

UNIVERSITY OF CAMBRIDGE    DEPARTMENT OF PHYSICS

*Figure 9.3* Watson's scheme of unlike base pairing was discovered on 28 February 1953. These diagrams he drew freehand in a letter that he sent to Max Delbrück on 12 March 1953.

the MRC report. This datum had left its mark on Crick, and again it was proving crucial, for Crick retained a vivid recollection of declaring excitedly to Watson "Look, it's got the right symmetry."[24] Turning the whole molecule upside down, of course, turns the base pairs upside down, but this brings the bonds between bases and backbone on one chain into the position and identical environment formerly taken by the others.

They could now build a model in which the sequence of bases along one chain could be as irregular as they wished, and this would make the sequence on the sister chain equally irregular but determined by the complementary relationship of the pairing scheme. Where adenine was present on one chain, thymine would be present on the other. Likewise, guanine was complemented by cytosine (Fig. 9.4). DNA, despite its plain-seeming composition, would be able to store in the chromosomes the rich data transmitted in heredity, thanks to the legion of possibilities for sequences of the four bases—as Chargaff had suggested 3 years before.[25] And base pairing would replicate the two complementary sequences in accordance with Watson's scheme, thus ensuring genetic continuity from chromosome to daughter chromosomes. Surely, this could be no false trail; there could be no anticlimax after this. It was just so good that it had to be true! So simple, so compact—and going right to the heart of the magic of gene duplication. No dream could improve on this.

Come the lunch break, Crick swept into his favorite Cambridge pub "The Eagle" and could be seen telling "everyone within hearing distance that we had found the secret of life."[26] But Watson was nervous, "slightly queasy." Perhaps because he lacked Crick's intimate grasp of

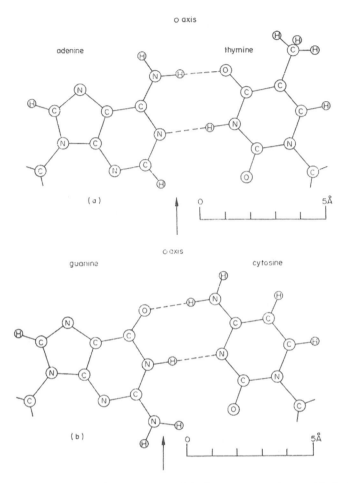

**Figure 9.4** Watson and Crick's depiction of the base pairs, as first published on 30 May 1953.

crystallography, he did not share the confidence Crick possessed from the support afforded to their structure by the C2 symmetry. Having been grievously wrong with their first attempt on DNA, our young postdoc was not yet convinced that they would prove to be right this time either. Many a "discovery" is made serendipitously, but not all such turn out to be great in the fullness of time.

Back in the lab that Saturday afternoon, Crick returned to his thesis, but he could not concentrate. Watson recalled how Crick would "pop up from his chair, worriedly look at the cardboard models, fiddle with the combinations, and then, the period of uncertainty over, look satisfied and tell me how important our work was."[27] Although confident, Crick knew that they must first build a model and show that it conforms with the X-ray data and meets the demands of stereochemistry: The angles and lengths of the bonds between atoms must be acceptable, as must the separations between neighboring atoms.

Monday (2 March) found Crick in the lab before Watson. He had not come to write more of his dissertation. Putting it aside, he was throwing his energies into the model. He was obsessed, and as the work progressed, he became euphoric. Using the cardboard cutouts that Watson had made in lieu of the metal ones ordered from the machine shop that had not yet arrived, Crick checked the placement of the base pairs within the two sugar–phosphate backbones, using a compass and ruler. They fitted perfectly.

Crick gave Perutz and Kendrew a brief exposition when they came to see the model, and Watson went to the machine shop to make known the need for the metal bases. Crick had supplied detailed drawings, and Mike Fuller was delegated to start cutting out the metal shapes. They appeared later that afternoon and serious model building could begin. Here, Crick was driven by his conviction that merely publishing a description of the basic features of their model would be bad policy. They must first submit their idea to the restrictions and demands of stereochemistry. Therefore, all of the distances between atoms in the model as well as the angles of the bonds should be measured. Crick devised a theorem that allowed him to calculate the restraint of one base on the other so they could build a model that consisted of only "a sugar, a phosphate, and a base and just the first atom of the next sugar which you had to put in the right position."[28] That was all that they needed in order to measure the positions of the atoms in the model and find them "reasonably acceptable."

Those early days of March raced by as they attached and detached skeletal metal representations of the atoms using cylindrical collars, such as those familiar to construction kit enthusiasts. It took them several days to measure the positions of all the atoms with a plumb line and ruler. Crick would beat Watson to the lab in the morning, where he could be seen tightening the clamps holding the skeletal model and checking and recording the positions of each atom. Every now and then, a visitor would arrive from the Cavendish to see what all the fuss was about. As physicists upstairs commented, "steam" was rising from the floor below—excited voices, laughter, and Crick's voice as he delivered yet another "buoyant and booming" lecture to the next visitor. This went on all week, ending Saturday morning "by which time," said Crick, "I was so tired, I just went straight home and to bed."[29]

### Wilkins Views the Model

The second week of March brought post from Wilkins, dated Saturday. According to Watson, it was received early Thursday (12 March).[30] Crick opened it at his desk in the Cavendish and read

> I think you will be interested to know that our dark lady [Franklin] leaves us next week and much of the three-dimensional data is already in our hands. I am now reasonably clear of other commitments and have started up a general offensive on Nature's secret strongholds on all fronts: models, theoretical chemistry and interpretation of data, crystalline and comparative. At last the decks are clear and we can put all hands to the pumps!
>
> It won't be long now. . . . Regards to all,
>
> Yours ever
>
> M.
>
> P.S. May be in Cambridge next week.[31]

Crick looked across at the model, sitting there on the table in the middle of the room and thought "was it more a question of laughing or—well, you know, sadness almost. You see. There was the model."[32] Later, Crick joked with Wilkins about the fact that his letter arrived the day that they finished checking their model![33]

Why had they put off the task of informing Wilkins for so long? It was the result of the debacle of 1951, when they had been pressed to inform the King's group prematurely. This time, it was "mum's the word" until the work of checking the model was through. Only when Crick was satisfied with the stereochemistry did they let the secret out. Even then, they preferred not to give the news themselves, although Wilkins was Crick's friend of some years. Instead, the task was passed to Kendrew, whose diplomacy could always be relied upon.

On Thursday, 12 March Watson wrote to Delbrück describing their model. Wilkins' news, which was received, according to Watson, earlier that day,[34] must thus have arrived at the Cavendish as Watson and Crick concluded their checking of the model. At this point, Kendrew called Wilkins to put him in the know. Wilkins recalled how he immediately took the train to Cambridge. Meanwhile, Franklin had begun her new life with Desmond Bernal's group of crystallographers at Birkbeck College, in their bomb-damaged house in Torrington Square, not far from the British Museum.

Arriving at the Cavendish Laboratory, Wilkins was shown the model by Crick, who then proceeded to deliver his explanation of its structure and suggested functions. But he stopped abruptly when he realized that Wilkins was just staring at the model and not attending to his mini-lecture. What impressed Wilkins most, as it had Watson, was the "extraordinary way in which the two kinds of base pairs had exactly the same overall dimensions and shape." Hence, they could be fitted inside the helical cylinder one above another without buckling the backbones. And, as Crick explained, one could imagine separating the members of base pairs and unwinding the backbones, then attaching free bases to each separated chain in accordance with the pairing rule, thus generating two molecules where there had been one. What molecular model for the replication of the gene could be simpler or more eloquent than this? A feeling came over Wilkins that "the model, though only bits of wire on a lab bench, had a special life of its own. It seemed like an incredible new-born babe that spoke for itself, saying 'I don't care what you think—I know I am right.' "[35]

Wilkins' feeling of wonder at the beauty and simplicity of the structure was mixed with deep sadness at the loss of originality that would henceforth attach to the results of his current research program. As he had informed Crick in the letter he sent earlier that week, "I . . . have started up a general offensive on Nature's secret strongholds. . . . At last the decks are clear and we can put all hands to the pumps!"[36] Now, standing in front of the model, he was "rather stunned—the exactness and the replication idea, and the resolving of the paradox that DNA was so regular (even crystalline) and yet contained the complex and irregular genetic message in the sequence of the base pairs. And Francis' sophisticated talk of diad axes of symmetry that had confused me was actually important. . . ."[37]

At this juncture, our two discoverers made the surprising suggestion that Wilkins might like to be coauthor of the paper that they were writing on the structure; he was astounded.[38] The invitation, of course, arose out of their recognition of the extent to which their model depended on information that came from King's College via Wilkins. Well aware of this, and possessive about DNA, to which he had devoted so much of his attention during the last 3 years, Wilkins' response was bitter. The assignment of the DNA research to Franklin had been the first blow to his aspirations, and now, just when he had been assembling his team to return to DNA with Franklin's departure, his friend Crick together with Watson had jumped ahead and discovered the structure. Agreed, he had, albeit reluctantly, given them his permission 1 month ago to start working on DNA again, but had they contacted him since then to tell him how they were faring? No, not, that is, until they invited him to come to view their model. Wilkins' outburst has, perhaps fortunately, not been record-ed, but he has not forgotten how he lost his cool. Crick objected that Wilkins was being unfair to them. They had given the King's group more than 1 year to sort out their problems. They had lent them the jigs with which to form the brass atoms and begin model building. Now, Pauling was hard at work on DNA and had produced a structure, albeit a wrong one. How much longer were they supposed to sit on the sidelines and watch him beat the Brits to the prize? Cooling down and collecting his thoughts, Wilkins responded that he had not been involved in the suc-cessful model building and therefore would be embarrassed if it became the "Watson, Crick, and Wilkins model." He could not accept coauthor-ship. "Crick agreed," recalled Wilkins, but he "explained that Jim had been very concerned that I should not be left out."[39] And Franklin—was there any mention of her work? Apparently not.

Confirmation that Watson and Crick were over the first hurdle came after they had given Wilkins a copy of the paper intended for *Nature* that they had been drafting and redrafting. Wilkins responded in a letter dated 18 March that begins "I think you're a couple of old rogues but you may well have something. I like the idea. . . ." He added that he "was a bit peeved" because he had been convinced the Chargaff 1:1 ratios were sig-nificant "and as I was back again on helical schemes I might, given a little time, have got it."[40] "Peeved" does not convey the state of devastation that, according to Raymond Gosling and Angela Brown, best describes Wilkins at this turn of events.[41] Randall was understandably furious. Some restora-tion of honor was achieved by persuading Watson and Crick to delay sub-mitting their paper for publication until Wilkins could contribute a report on his work to appear alongside the Watson–Crick structure.

Before Wilkins had finished his letter to Crick, Gosling interrupted him with the news that Franklin and he had a paper ready that should be included. Wilkins was not in a generous mood, as one can judge from the following words he added to his letter: "Just heard this moment of a new entrant in the helical rat race. R.F. and G. have served up a rehash of our ideas of twelve months ago. It seems they should publish something too (they have it all written). So at least three short notices in *Nature*. As one rat to another good racing."[42]

Little did Wilkins realize that by this time Franklin had gone beyond the position she had taken at her seminar talk in January when she stated that the B form, but not the A form, is helical. Privately, she now held both forms to be helical, and what is more, she had written a manuscript giving an authoritative interpretation of the B form. She had settled on two chains in the molecule, the bases on the inside, ten per chain in the pitch, and one chain 3/8th out of phase with the other.

Missing from her analysis of the B form was the recognition that the space group C2 means that there is a dyad (axis of symmetry) at the side of the chains, suggesting that they run in opposite directions. Although features of the Patterson diagram of the A form had led her to infer the antiparallel arrangement of the chains in that form,[43] she had not applied this feature to the B form. Crick judged, therefore, that she had not appreciated the anti-parallel feature in "a crystallographic sense."[44] Would she, given time, have embraced the antiparallel arrangement for the B form? Any other alternative seems unlikely. As for the bases, she had not yet worked out how to pack them into her helical scheme. Yes, it was a return to the all-helical consensus of 1951, but it was no mere rehash of her or Wilkins' earlier work. Given a little more time, a month or so, surely she would have solved the structure.

Following upon Wilkins' Cambridge visit, Franklin and Gosling made the journey to the Cavendish Laboratory. Their hosts, aware of the contents of Franklin's report to the MRC, were not anticipating an attack from her on antihelical grounds, at least not for the B form. But how would she react to viewing a model that incorporated data that she must surely recognize as hers? She behaved coolly and gracefully. Betraying no resentment, she commented to Gosling, "It's very pretty, but how are they going to prove it?"[45]

Their structure was appealing, but in her estimation, it was no more than a proposal, and she knew that the best data to support it lay in her work. Neither Gosling nor any others close to Franklin know of her saying that the Cambridge duo must have used her data, yet surely she must have realized that details of her work had reached Cambridge. Crick had no doubts on the matter, but he agreed that in accordance with Franklin's view of scientific research, proposing a structure did not have the same merit and importance as providing indisputable evidence from which it could be derived or established.[46] Watson and Crick had certainly not provided that. At the same time, Franklin must have known that they could not have arrived at their model had she not distinguished and characterized the A and B forms, noted the transition between them, and identified the unit cell of the A form as monoclinic C2.

By this time, Crick had evidently impressed her. At their brief encounter in 1952, he had been rather patronizing, and he later admitted this. After all, at that time, she was talking about DNA in the A form being nonhelical. Now, she could appreciate his recognition of the significance of the C2 symmetry in her data, an inference she herself regretted that she had missed. And Crick had good reason to respect her achievement as recorded in the MRC report. When he subsequently came to London to view the King's data, Franklin was eager to show him her work, and for the first time he saw her 1952 X-ray diffraction pattern of the B form and her data supporting her decision that the backbones are on the outside of the molecule.

At this point, one can note the marked change in personal evaluations that followed the Watson–Crick proposal for the structure of DNA. Before April 1953, Bragg had been expecting Crick to find an academic position and leave the MRC unit. Now, Bragg no longer wanted him to leave. And as we have seen, he was later to express his remorse at having so misjudged him.

Franklin's chief memory of Crick before 1953 went back to 1951 when Watson and he had built that hopeless three-chain model, and Crick had tried in vain to rescue it in the face of her devastating critique. Now the two of them were of like mind about DNA—Franklin had banished her antihelical concerns, she appreciated Crick's grasp of crystallographic analysis, and he appreciated her remarkable skill as an experimentalist. By the summer of 1953, they were in correspondence, and before Crick returned from America in the fall of 1954, Franklin had visited them in Brooklyn. Back in Cambridge, she began to visit the Cricks to enjoy their company and discuss her work with Francis. By April 1956, she had vacationed with them, and in November it was with the Cricks that she convalesced from cancer surgery. Two years later, she was dead. Meanwhile, the important role of her work in leading Crick and Watson to their structure for DNA was obscured, not emerging, until the publication of Watson's little book *The Double Helix.*

### The Aftermath

How did Crick's family take all this excitement? Odile was used to her husband "throwing off ideas all the time and most of them," she added, "seemed to be wrong. One really never knew what was going to come of it." But from the time Watson returned from visiting King's at the beginning of December until they had the structure "there was a sort of suppressed excitement," she recalled, "great sort of goings-on, the whole time they were thinking about it, all the time. We heard about nothing else."[47] His old school friend Harold Fost and his wife Kate visited that spring, and Harold recalled how Francis was frequently quiet, deep in thought, his mind far away.[48]

Crick could not remember announcing to all and sundry in The Eagle pub that they had discovered the secret of life, but he did remember the letter he wrote to his son Michael while successive drafts of their first paper were going back and forth between Cambridge and London:

19 March '53

My Dear Michael, *age 13*

Jim Watson and I have probably made a most important discovery. We have built a model for the structure of des-oxy-ribose-nucleic-acid (read it carefully) called D.N.A. for short. You may remember that the genes of the chromosomes—which carry the hereditary factors—are made up of protein and D.N.A.

Our structure is very beautiful. D.N.A. can be thought of roughly as a very long chain with flat bits sticking out. The flat bits are called "bases." The formula is rather

*like this*

> Sugar ——— base
> |
> phosphorus
> |
> Sugar —— base
> |
> phosphorus
> |
> Sugar —— base
> |
> phosphorus
> |
> Sugar —— base
> ⋮
> *and so on.*

Now we have <u>two</u> of the chains winding round each other—each one is a helix—and the chain, made up of sugar and phosphorus, is on the <u>outside</u>, and the bases are all on the <u>inside</u>. I can't draw it very well, but it looks

*like this*

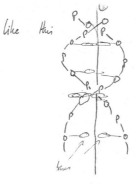

*The model looks much nicer than this.*

The model looks <u>much</u> nicer than this.

Now the exciting thing is that while there are <u>different</u> bases, we find we can only put certain pairs of them together. The bases have names. They are Adenine, Guanine, Thymine and Cytosine. I will call them A, G, T and C. Now we find that the pairs we can make—which have one base from one chain joined to one base from another—are only A with T and G with C.

Now on one chain, as far as we can see, one can have the bases in any order, but if that order is <u>fixed</u>, then the order on the other chain is also fixed. For example, suppose the first chain

| goes | then the second <u>must</u> go |
|------|------------------------------|
| Down | Up |
| A - - - - - - - - - - T | |
| T - - - - - - - - - - A | |
| C - - - - - - - - - - G | |
| A - - - - - - - - - - T | |
| G - - - - - - - - - - C | |
| T - - - - - - - - - - A | |
| T - - - - - - - - - - A | |

It is like a code. If you are given one set of letters you can write down the others. Now we believe that the D.N.A. <u>is</u> a code. That is, the order of the bases (the letters) makes one gene different from another gene (just as one page of print is different from another). You can now see how Nature <u>makes copies of the genes</u>. Because if the two chains unwind into two separate chains, and if each chain then makes another chain come together on it, then because A always goes with T, and G with C, we shall get two copies where we had one before. [See figure on facing page.]

In other words we think we have found the basic copying mechanism by which life, comes from life. The beauty of our model is that the shape of it is such that <u>only</u> these pairs can go together, though they could pair up in other ways if they were floating about freely. You can understand that we are very excited. We have to have a letter off to *Nature* in a day or so. Read this carefully so that you under-

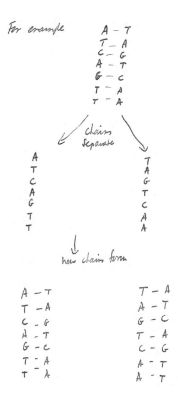

stand it. When you come home we will show you the model. Lots
of love

Daddy[49]

What a historic letter for a 13-year-old to receive from his father!

# Publishing the Model

 *Mad hatters who bubble over about their new structure.*[1]

April 1st 1953 was not just April Fool's Day, but for Bragg a very special day: the date for a site visit from the Rockefeller Foundation's Gerard Pomerat, Assistant Director of the Natural Science Program. At the time, he was one of the most powerful men in charge of foundation funding for basic biomedical science. Bragg wanted to show his important visitor their latest results with hemoglobin, because since 1938, the Foundation had given generous support to keep Bragg's little group afloat. At this juncture, he was hoping for continued support to complete Perutz's project on hemoglobin, now two thirds of the way finished. But Pomerat recorded in his diary that there was "a great air of excitement for Perutz and two of the younger men," who were hoping to show Bragg and Pomerat "what they have been up to in the last week."

> They believe they have really got the structure of nucleic acid from a crystallographic rather than a chemical standpoint. Their clue came out of the beautiful X-ray diagrams produced in Randall's lab and some of the work that had meanwhile been going on in Cambridge. They are just putting the finishing touches on a huge model six feet tall.[2]

Pomerat wrote that these two were "particularly excited" about showing the model to Linus Pauling, who would be visiting that week. Pomerat, however, urged them to first show it to Cambridge chemistry professor Alexander Todd. Also in his diary, written in pencil alongside the phrase about the X-ray data having come from Randall's lab, was Rosalind Franklin's name. Evidently, they had talked about the source of the X-ray-diffraction data. Following these comments was

Pomerat's brief synopses of the careers of the two discoverers. "Both young men," he added, "are somewhat mad hatters who bubble over about their new structure in characteristic Cambridge style and it is hard to comprehend that one of them is an American." How could they not "bubble over"? And who was doing the bubbling? Mostly Crick, one guesses, and that is hardly Cambridge style, where understatement is the rule. Earlier, Bragg on his sick bed had learned that Watson and Crick had "thought up an ingenious DNA structure,"[3] but he had not seen it until Pomerat's visit. He was at once impressed, and his excitement grew as successive biologically important features were pointed out—especially regarding "its potential implications for gene replication." But he, like Pomerat, wanted the chemists to see the model. So Todd came over with some colleagues, approved, and gave it his blessing and the authors his congratulations.[4]

Who should come next but Linus Pauling. The world-famous chemist was on his way to the Solvay conference in Belgium, where he would meet with the other great names in the physical sciences. He walked into the office shared by Watson and Crick, where his gaze turned to the model he had so recently heard about when he was 6000 miles away. Theirs was not a space-filling model such as those at Pasadena; it contained no state-of-the-art equipment, but instead had hand-cut shiny sheets of tin for the bases, soldered to the brass wire arms that represented bonds in a skeletal model. Crick "chattered nervously" about the model while Pauling "scrutinized it" and the copy they now possessed of Franklin's famous B pattern. They anxiously awaited Pauling's verdict—yes, he told them, "they had the answer."[5]

Pauling accepted this defeat gracefully. He was entertained by the Braggs and the Cricks that day and soon thereafter he traveled with Bragg to the Solvay conference. There, he reported on the Watson–Crick model to the eminent gathering, and Pauling responded generously and candidly, admitting that his structure, although only 2 months old, was probably wrong. He added that "Although some refinements might be made," he rated the probability high "that the Watson–Crick structure is essentially correct."[6]

The victor of the α helix was at first amazed. How had this "unlikely team, an adolescent postdoc and an elder graduate student . . . come up with so elegant a solution to so important a structure?"[7] Meanwhile, his erroneous model was out there in the *Proceedings of the National Academy of Sciences* being distributed worldwide for all to see. Obscured by Pauling's gracious and generous response was the scarring that he felt over this turn of events. Truth be told, he had been overconfident, and, in his haste, he had sent in the paper before Robert

Corey calculated the diffraction pattern that it would yield. Only 2 weeks before the paper appeared, Corey and his colleague Verner Schomaker had become convinced that the model could not be rescued by "minor jigglings,"[8] and Corey, 14 years later, was at pains to stress that his own contribution to this model "was essentially nil." Pauling's inclusion of his name was "a generous gesture on his part."[9] It had become a fiasco far more public than the failed first attempt of Watson and Crick on DNA in 1951, which it resembled, with three chains and the bases on the outside. Corey disowned it.

Now Pauling had not forgotten Crick's work on coiled coils in fibrous proteins, and on this occasion, with the DNA structure in front of him, there could be no doubt about Crick's intellectual stature in the field. For his part, Crick had admired Pauling ever since he became familiar with his work. He found Pauling easy to be with, a straight-talking, no-nonsense kind of guy, like himself. Both men had no time for privileged rank and snobbery. Both would admit the errors that they made, but now dinner at the Crick's gave Pauling the opportunity to direct his charm and good humor to Odile, Watson's sister Elizabeth, and a visitor from King's College London, Pauline Cowan, the young X-ray crystallographer who would have collaborated with Rosalind Franklin in 1952 to build models of DNA, had Franklin accepted her help.

Before the end of March, a call was made to Dorothy Hodgkin at Oxford University's Chemical Crystallography Laboratory. Hodgkin and four others—Jack Dunitz and Beryl Oughton[10] from Hodgkin's crystallography group, the chemist Leslie Orgel, and the South African phage researcher Sydney Brenner from Oxford's Chemistry Department—then drove to Cambridge. All were impressed by the model, and for two of the visitors—Orgel and Brenner (about whom we shall hear more later)—meeting Crick for the first time, this visit was of special significance, for both were later to become Crick's colleagues at the Salk Institute. Brenner recalled entering the room in the Austin Wing of the Cavendish Laboratory and seeing the model of DNA.

> Francis was sitting there. This was the first time that I met him and of course he couldn't stop talking. He just went on and on, and it was very inspiring you see. Of course at this stage neither of the two famous *Nature* papers had yet appeared . . . that's when I saw that this was it. And in a flash you just knew that this was very fundamental. The curtain had been lifted and everything was now clear [as to] what to do. And I got tremendously excited by this.[11]

Dunitz also recalled the impact on him of seeing the model:

As Francis explained the various features of the model I could see that everything had fallen into place and that it was obviously right in its essentials. I liked the two strands running in opposite directions, the overall dimensions were approximately correct, and most convincing of all, the base pairing feature of the proposed structure not only gave an immediate explanation of the Chargaff rules that Sydney had described to us but was also revealed as the simple trick for duplicating the double strand. . . . Vaguely, I think I realized that this was a discovery of the first rank, but what I did not see was that it was the beginning of a new science, molecular biology, that was going to change the world within a couple of decades.[12]

## *Publishing the Paper*

After the feverish activity of finalizing the texts of the papers from Cambridge and London and then collating them, Randall and Wilkins arranged with Arthur Gale, the editor of *Nature,* and L.J.F. Brimble, his coeditor, for their rapid publication. We can perhaps picture Jack Brimble in April handling the refereeing of the three papers in his characteristic manner by circulating them among fellow scientists in the Athenaeum Club after lunch![13] One can also imagine with what joy Crick read his copy of *Nature* on Saturday morning, 15 April, over breakfast. He could feel pleased with the brief and nontechnical character of their text and delighted at the admirable support to their structure afforded by the two accompanying technical papers. Had he any inkling of just how vast would be the readership of their little paper, a mere 600 words long? When two decades later he reflected on that question, he admitted to having been surprised at just how much attention the paper had won.

If one were to go to a library today that keeps back issues of *Nature* and turn to volume 171, there would be no difficulty identifying issue 4356 for 25 April—it would be mouse-eared from handling by countless readers, replaced with a photocopy, or missing altogether! Such has been the fate of the paper that was put together by a 37-year-old research student and a 25-year-old postdoc, typed by Watson's sister and the illustration drawn by Crick's wife Odile. Almost a family affair! And what has become of the reprints sent out? Occasionally, one turns up in the auction room. But beware; if you want to bid on one, you may need a deep pocket because you might have to fork over a five-figure sum to win the bid.

There were three Watson–Crick papers on DNA in 1953 and a fourth one published 1 year later. Who of Watson and Crick wrote which paper, or which parts of each? The order of authorship is misleading. Judging

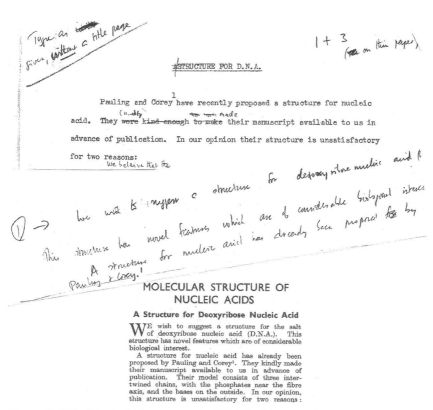

*Figure 10.1* Draft text of "A structure for DNA," with corrections in Watson's hand and an addition in Crick's hand beginning with "We wish to suggest. . . ."

by surviving drafts of the papers, it appears that an early version of the short note announcing the Watson–Crick structure (published 25 April) was written by Watson, but almost all of the corrections, additions, and deletions are in Crick's hand (Fig. 10.1). The first version of the follow-up paper for *Nature* on the genetic implications of their structure was authored by Crick. The earliest surviving manuscript of the third paper, written for the 1953 Cold Spring Harbor Laboratory Symposium on Long Island, is in Watson's hand, but it was very much a joint effort. As Watson recalled, the two authors spent the last 2 weeks of May putting it together, "frequently arguing out the precise language."[14]

Watson returned to Cambridge from the Cold Spring Harbor Symposium in July and wrote "virtually all the fourth paper, but there are minor additions in Crick's hand." The published authorship was decided as follows: Watson had discovered the base pairs and therefore felt that his name should come first on their announcement of the structure on 25 April. The decision for the second paper, written almost entirely by Crick, was decided by the toss of a coin, and Watson won.[15] The third paper was presented by Watson at the Cold Spring Harbor meeting, and as the guest speaker, he felt that his name should be first. That left only the last of the four papers for Crick; hence, his name came first on the paper that Watson, not he, had written for the Royal Society. At the time (July 1953), Crick was at work with scissors and paste, adapting and expanding his publications on protein structure to serve as dissertation chapters, adding explanatory material here and there.

Imagine, now, picking up *Nature* on 25 April, not knowing about X-ray crystallography, ignorant of the importance of the nucleic acids, and alighting upon Watson and Crick's first DNA publication. Entitled "Molecular structure of nucleic acids. A structure for deoxyribonucleic acid," it occupies only the equivalent of one full page of text and begins timidly: "We wish to suggest a structure for the salt of deoxyribonucleic acid (D.N.A.)." This was because the editors would not permit DNA without the periods and they insisted on using the name in full. They rightly feared that many readers would not have seen the abbreviation before and that the authors' names would be unknown to them. If they had encountered the name deoxyribonucleic acid before, however, they might be drawn to read the paper.

We find compacted into these 600 words an account of the main features of the structure, how it differs from previous models, Linus Pauling's in particular, and its novel feature—the manner in which the two chains are held together by the purine and pyrimidine bases, the pairing of which accounts beautifully for the Chargaff ratios. When shown the typescript, the only written comment on the text from Bragg was "Should there not be some figure, even if only diagrammatic? It strikes one as rather hard to follow without one." Crick then gave Odile a rough idea of what they wanted and her stylish, schematic drawing has surely become the most heavily used icon in the biological sciences (Figs. 10.2 and 10.3).

There is no discussion of the biological significance of their structure, apart from two sentences. The first was added by Crick to Watson's manuscript and reads "This structure has novel features which are of considerable biological interest." He had wanted to say more, but Watson was against it. Should the structure turn out to be wrong, objected Watson, they would look silly. Crick also wrote the second sentence that alluded to biology: "It has not escaped our notice that the specific pairing that we have postulated immediately suggests a possible copying mechanism for the genetic material."[16] Georg Kreisel erupted at this much-admired sentence: "God. What a filthy style: 'It has not escaped

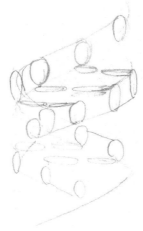

**Figure 10.2** A surviving example of Francis Crick's sketch for his and Watson's DNA structure.

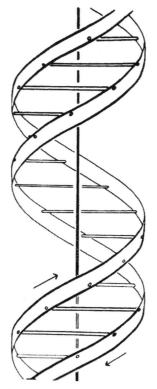

**Figure 10.3** Odile Crick's famous schematic drawing of the Watson–Crick Model.

our notice. . . !!,' " he wrote, and Crick agreed, although they were his words.[17] How portentous, almost as if a member of the royalty was addressing the reader. Others described it as "coy," Crick remarked, "a word that few would normally associate with either of the authors, at least in their scientific work. In short, it was a claim to priority."[18]

By the time this paper had gone in for publication, Crick had digested the contents of the two papers from King's. The data therein were much more supportive of the structure than he had expected. Armed with this additional ammunition, he persuaded Watson to let him write a paper on what at first he called "The molecular structure of the gene." Although Crick's handwritten text covers most of the subject matter of the published version that appeared under the title "Genetical implications. . . ," there are differences in the wording. First, in this first draft he explained how in the structure of DNA "one chain is, as it were, the complement of the other. Such a structure immediately suggests that the D.N.A. is the carrier of the genetic specificity, and that what we are studying is the molecular structure of the gene." Then, he launched into a description of their theory of gene duplication:

> In outline, therefore, we imagine during duplication the two chains unwind and separate. Each chain then acts as a template for the formation onto itself of a new, complementary chain. When each of the original chains has succeeded in forming its complementary chain we shall have <u>two</u> pairs of chains where we had only one before. Moreover, the sequence of the pairs of bases, which we imagine to be the code which carries the genetical information, will have been duplicated exactly.[19]

When it appeared in *Nature* on 30 May, this paper had a greater impact, because, as the French Nobel Laureate Francois Jacob recalled, the first *Nature* paper had not "electrified" him or anyone else in their laboratory at the Pasteur Institute. "I had only skimmed through this article," he recalled, "The crystallographic argument went over my head."[20] Watson and Crick's second paper begins with a sweeping declamatory statement typical of both authors: "The importance of deoxyribonucleic acid (DNA) within living cells is undisputed." Then follow three very carefully worded sentences, probably written by Crick, that neatly express the main reason for the widespread reluctance at the time to accept DNA as the genetic material. "Many lines of evidence indicate that it is the carrier of a part of (if not all) the genetic specificity of the chromosomes and thus of the gene itself. Until now, however, no evidence has been presented to show how it might carry out the essential operation required of a genetic material, that of exact duplication."[21]

This last sentence is the punch line because, as the manuscript goes on to explain, their structure "immediately suggests a mechanism for its self-duplication." The published version then fairly closely follows the draft that was prepared by Crick. The latter begins with an account of the features of the molecule that are of biological interest: It consists of two chains. They are wound "around a common fibre axis" and held together by hydrogen bonds between the bases. But the "conditions for effective hydrogen bonding are restrictive. The only possible pairs are: adenine with thymine and guanine with cytosine (or methyl-cytosine). Given that restriction, any sequence of bases down one chain is possible and the other chain will carry the complementary sequence." Then follows the important inference that ". . .it therefore seems likely that the precise sequence of the bases is the code which carries the genetical information."[22] That word "code," speaking of a secret to be discovered, expressed their hopes for the future.

For the first time, a diagram of the two base pairs is included, and the text then follows Crick's handwritten script with minor modifications. He pictured the two chains of the molecule unwinding, and around them, the four bases (or rather, their nucleotides: base + sugar + phosphate) were freely available. These would then become attached in accordance with the pairing rule to the emerging single chains. If the new series of nucleotides then polymerize, either with or possibly without the aid of enzymes, each single chain would become a double chain.[23] But how the original double helical molecule would unwind without entangling the resulting single chains was a serious difficulty. Still, they judged that this objection to their scheme was not "insuperable."[24]

They speculated on the possible position and role of the chromosomal protein around the DNA and discussed the control of the genetic material and gene mutation—could it be due to the occurrence of "a base in one of its less likely tautomeric forms"?[25] Their paper ends stressing the speculative nature of their scheme for DNA duplication and listing a whole series of questions raised by it, but concludes by rating gene duplication one of the fundamental biological problems for which their model provides a molecular basis; namely, "The hypothesis we are suggesting is that the template is the pattern of bases formed by one chain of the deoxyribonucleic acid and that the gene contains a complementary pair of such templates."[26]

Having sent in this paper, Crick and Watson pretty much shared the writing of their more detailed paper for the forthcoming Symposium in Cold Spring Harbor Laboratory. Crick returned to his thesis, and that fall, Watson sat down to write the longer paper, giving some account of how they arrived at their model. Bragg submitted this to the Royal Society for publication in the Society's *Proceedings*, where it appeared in the summer of 1954. Crick half-jokingly referred to this as "an obscure journal,"[27] and true enough, few scientists have read "The complementary structure of deoxyribonucleic acid" by Crick and Watson.

Crick's preoccupation with his thesis explains how it came about that Franklin's identification of the space group of the A form of DNA, monoclinic C2, was not mentioned in this paper (see Chapter 9). However, Watson did not fail to mention that by trial, they found that "their model can only be built in the right-handed sense. Left-handed helices can be constructed only by violating the permissible van der Waals contacts."[28] He described the base pair, adenine-thymine, as "stereochemically most satisfactory." In contrast, regarding the guanine-cytosine (GC) base pair, they had doubts concerning "the exact structure of guanine," adding their uncertainty as to "whether this pair might form a third hydrogen bond between the amino of guanine and the keto oxygen of cytosine." This gave Pauling the opportunity to present evidence in support of three hydrogen bonds for GC pairs.[29] For the sugar (deoxyribose), Watson and Crick had used the configuration established in Bernal's lab by the Norwegian crystallographer Sven Furberg in 1950.[30] From the work of Wilkins and his colleagues, it later emerged that Furberg's structure (for cytidine) is the configuration found in the A form of DNA, but not the B form. This choice resulted in the sugar rings being further out toward the perimeter of the helix than they should be and the diameter of the helix "a little too large."[31] But as Watson and Crick were aware in 1954, "a better model might result by choosing a different shape" for the sugar ring.[32]

Considering that Watson wrote this paper with very minor suggestions from Crick, it is surely a creditable achievement for a 25-year-old biologist. As he later admitted, this was the first time he had "put together language of the type that Lawrence Bragg, Max Delbrück, and Linus Pauling had mastered so well."[33]

### *The Problem of the Acknowledgments*

As we have learned from Watson's account in *The Double Helix*, he felt some embarrassment regarding the manner in which he and Crick gained access to Franklin's data. Crick, it seems, was not free from some embarrassment either, as his draft of a letter to Randall in 1953 suggests.[34] He considered that they acted appropriately, and, as we shall see, would have been perfectly willing to make any reference to sources that the King's group requested. But from the correspondence between Crick and Wilkins on the matter, it is clear that Wilkins played an important part in dictating the wording of the acknowledgments in their first paper. Successive alterations were made in the references to and acknowledgment of the Kings' work before the papers were sent in. Thus,

> *Watson and Crick*: "As far as we can tell, our structure is roughly compatible with the experimental data. It is known that there is much unpublished experimental material. Until this has been used to test the structure it must be regarded as unproven."

> "We have also been stimulated by the very beautiful experimental work of Dr. M.F.H. [*sic*] Wilkins, and his co-workers. . . ."[35]

> *Final*: "So far as we can tell, it is roughly compatible with the experimental data, but it must be regarded as unproven until it has been checked against more exact results. Some of these are given in the following communications. We were not aware of the details of the results presented there when we devised our structure, which rests mainly though not entirely on published experimental data and stereochemical arguments."[36]

> "We have also been stimulated by a knowledge of the general nature of the unpublished experimental results and ideas of Dr. M.H.F. Wilkins, Dr. R.E. Franklin and their co-workers. . . ."[37]

These changes were due chiefly to Wilkins who had requested the following: "Could you delete the sentence 'It is known that there is much unpublished experimental material.' (This reads a bit ironical.) Simply say 'The structure must of course be regarded as unproven until it has been checked with fuller experimental material'. . . . Delete *very beautiful* and say 'We have been stimulated by the work of the group at King's or something.' "[38]

Wilkins, it seems, was more concerned about not admitting the discoverers' debt to Franklin's work than they were themselves! But recall that he was about to launch his full-scale planned attack on DNA and was now in just the ideal position to provide experimental support for Watson and Crick's model. Admitting a specific debt to Franklin, at this point, would undercut his future claims. Besides, it was hard to accept that the one who had so consistently advertised her antihelical data had mysteriously become the one whose data proved vital to the Cambridge success—a helical success. Of course, had Franklin in 1952 published a short note on the A and B forms and the transition from one to the other, there would have been no question that our codiscoverers would have cited it, and the issue that did concern them—how they knew of her data—would not have existed. (That is not to say that she *should* have published her data then.)

For his part, Crick was happy to modify the acknowledgments, and if Wilkins still did not like the sentences, he offered to "rewrite them until you do."[39] But he drew the line at Wilkins' request to include yet another contribution—this time from a former colleague of Wilkins. This was Bruce Fraser, who more than 1 year earlier had put together a three-stranded model of DNA with its sugar–phosphate chains on the outside and the three sets of bases on the inside, hydrogen-bonded to one another. Wilkins had not thought that Fraser's short note about the model was worth publishing at the time, but now he had changed his mind. Thus, Crick inserted a short paragraph on Fraser's model, ending with the sentence "This structure as described is rather ill-defined, and for this reason we shall not comment on it." To Wilkins' objection, Crick responded that they were "prepared to omit" reference to the structure "entirely, but not to rewrite it." It was, like Pauling's structure, incorrect, and because they had criticized Pauling's structure, so they must criticize Fraser's. He explained:

> Frankly we think you should reconsider the desirability of publishing Fraser's paper at all. As far as we can tell the sole object would be to establish that you had thought of the idea of binding the bases together with hydrogen bonds. Once one has put the phosphates on the outside and the bases inside this idea is so obvious that we can think of no crystallographer so dull as not to have thought of it. The important step is the next one—to put the bases together in a systematic and plausible manner. This Fraser failed to do. We are even doubtful (we have not seen his figure) whether he got the tautomeric forms right. Thus the sole effect of the paper would be to advertise the fact that when he attempted to build models he made a complete hash of it! In writing a fuller paper we should have no choice but to point this out.

> There is a second reason why we think it undesirable that Fraser's paper should be published. It is generally acknowledged that the big step forward in this field was made by Bragg, Kendrew and Perutz in giving precise coordinates for postulated structures. The Cavendish and the California school have been scrupulous in only suggesting structures in a precise form. It is considered acceptable to publish an advance note without actually giving co-ordinates, but only if the structure has actually been obtained in detail first. You will remember that we did not write our letter until we had an <u>exact</u> record of our structure. It is quite clear that Fraser has not done this (has been "deliberately vague") and, more important, <u>cannot do it</u>. We understand the peculiar circumstances which have produced his paper. Nevertheless we think you should seriously consider not publishing it. As far as we are concerned the more incorrect structures suggested the cleverer it makes us appear![40]

Here we witness Crick in the business of directing, making decisions, and demanding standards, much as he must have been when in charge of the firing mechanisms for mines during the war. Between the lines, we can read the intense concern of the writer. "We are not happy," he continued, "about the position of Rosy and Gosling. It is not reasonable for letters to be sent in jointly to *Nature* without having been read by all concerned. We want to see hers, and I've no doubt she wishes to see ours. We think it essential that the whole position should be brought into the open and cleared up as soon as possible." He went on to warn Wilkins

> that it is in your own interests to reach a definite and binding agreement with Rosy as soon as possible. We hear from Peter that she is hoping to see his father this weekend. It is not impossible that she might consider turning over the experimental data to Pauling. This would inevitably mean that Pauling would prove the structure and not you, since he is better equipped in manpower experience and computing equipment than you are. For this reason, therefore, and also because the present position is embarrassing to us, we have written a short note to Randall to suggest a meeting on Wednesday (we could come on Tuesday if the letters are ready by then). We hope that you will have all the final drafts by then, so that we can compare and collate them. As we shall have held up our letter for 10 days we feel that we should see Rosy's letter before ours is submitted, and she should see ours.[41]

With this letter, we learn, was a copy of Watson and Crick's latest draft with Odile's schematic figure of the structure, and apparently what Crick called "the plan of the structure," but no such plan was published at this stage, not even a plan of the base pairs. Evidently, a meeting did take place with Crick, Watson, Wilkins, and Randall (it is not known whether Franklin was also present). Crick came to it very anxious to sort out any problems among the several contributors. It would have been the appropriate occasion for Randall to attend to such questions as the extent of Watson and Crick's dependence on the data obtained in his unit. Watson and Crick, for their part, could have asked whether they would like them to refer to the 1952 MRC report that they had seen before they had succeeded in building their structure.

If the MRC report had been a publication, one would have expected copies to have been readily available at the several MRC units and their staff appraised of their existence. Clearly, this was not the case, and Watson and Crick referred to all of the King's work at this stage as "unpublished," save for the short note by Wilkins and Randall and the earlier work of Astbury, which they stated "are insufficient for a rigorous test of our structure." True, as was later explained, the function of the MRC Committee receiving the report had been "to establish contact between the groups of people working for the Council" in the field of biophysics.[42] But since when did that permit the use without permission, followed by only unspecified acknowledgment of "the *general nature* [author's italics] of the unpublished experimental results and ideas" that it contained? Perutz explained that the report was "not confidential" and added that "it contained no data that Watson had not already heard about from Miss Franklin and Wilkins themselves. It did contain one important piece of crystallographic information useful to Crick; however, Crick might have had this more than a year earlier if Watson had taken notes at a seminar given by Miss Franklin."[43] But Watson, it seems, had not heard it, or if he had, he had not understood or remembered it, and Crick did not have this information (regarding the monoclinic C2 space group) 1 year earlier. This looks like special pleading on Perutz's part.

Watson and Crick did hint that certain specific unpublished data were used in building their model, because they stated that "it rests mainly though not entirely on published experimental data and stereochemical arguments."[44] Later, Crick believed that they should have cited Franklin's contribution to the MRC report in their more detailed paper for the Royal Society's *Proceedings,* written by Watson, because this paper gives some account of the way in which they arrived at their structure.[45] In a footnote to that paper, they did express their heavy debt to the King's College group for information "reported to us prior to its publication," and, they added, "without this [sic] data the formulation of our structure would have been most unlikely, if not impossible."[46] Crick had this report on his mind in November 1966 when at the Genetical Society's London meeting, he advised the author to request a copy. Was this his long memory or the jolt provided by his reading of the early drafts of the manuscript that was to become *The Double Helix,* which contained a reference to it?[47]

When they met at the end of March 1953, both sides, it seems, failed to deal with these aspects of the forthcoming joint publication of the three papers. Randall, it would appear, did not cross-examine Crick and Watson about the source of the King's data that they used. One suspects that his main concern had been to hold up submission of the Cambridge paper until a report or reports of the work at his MRC unit would be ready. Moreover, to have requested more specific recognition for Franklin's contribution than for Wilkins' would have been ill-judged as far as concerns the harmony of the unit. Nonetheless, could Randall not have obtained a statement of the specific data from his unit that proved crucial to their discovery?

As it was, Watson, unlike Crick, became concerned that some would think that they had improperly used the King's College data. This concern lay behind his decision to limit the wider publicity in Britain, desired by Crick, for their structure. Regarding the rights and wrongs of using other people's data, it is clear that Crick was caught up in his excitement with their model. He was focused on one objective— to tell the world of science about their structure and to claim priority for it. It was the same drive that had seen him push through programs in his war work that he now manifested in a different setting. Wilkins was a wounded man. Randall was furious, for here they were in Cambridge treading on his turf again, despite the moratorium. Had Bragg warned him about his decision to terminate it? Had Wilkins advised him of his acceptance of the changed situation? Apparently not. Watson was anxious. Franklin, it seems, kept herself above the affray. But Watson and Crick's statement that the structure they proposed "rests mainly, though not entirely on published experimental data and stereochemical argument," did prove misleading. Its purpose was to disabuse any readers of *Nature* who might assume that they had seen all of the information in the accompanying experimental papers from the King's group that appeared on 25 April, *before* they arrived at their structure. The phrase "not entirely" however, embraced key data that were not available to Linus Pauling or anyone else at the time. This was the data without which Crick and Watson had admitted in 1954 that their solution of the structure would have been "most unlikely, if not impossible." Pauling would have agreed, for consider the following correspondence between Delbrück and Pauling, who had just returned to Pasadena, after seeing the Watson–Crick model in Cambridge: "Dear Linus," wrote Delbrück, "As you can see from this letter I am very much excited about the biological implications of the WATSON– CRICK structure. Every day we seem to discover new ones. We are eagerly awaiting your return to talk these over with you."[48]

In his reply, Pauling wrote

> I was very deeply impressed by the Watson–Crick structure. I do not know whether you know what put Corey and me off on the wrong track. The X-ray photographs that we had, which had been made by Dr. Rich, and which are essentially identical with those obtained some years ago by Astbury and Bell, are really the superposition of two patterns. . . . This had been discovered a year or more ago by the King's College people, but they had not announced it. . . .[49]

But, as Pauling's biographer, Thomas Hager commented, Alex Rich's first DNA diffraction pattern was not taken until 17 April,[50] 3 days before

Pauling wrote the above letter and more than 3 months *after* he had submitted the paper describing his model! Pauling's subsequent suggestion that his lack of detailed data on the structures of the bases also seems strange, seeing that he had been Alexander Todd's guest in Cambridge in 1952; it was Todd who had earlier persuaded Bragg to put the bases on the agenda for research in the Subdepartment of Crystallography in the Cavendish Laboratory. The result was Broomhead's work on adenine and guanine, published in 1948, 1950, and 1951, respectively,[51] and of C.J.B. Clews and William Cochran on pyrimidines.[52]

### *Public Reception of the Model*

After so positive a reception of their model by both the King's College group and by Bragg, Kendrew, Perutz, and Pauling, how many journalists beat a path to Watson and Crick's office? No one came except for a lone photographer, Antony Barrington-Brown. He had been told about the model by a graduate student who was looking for "interesting stories about Cambridge science" for *Time Magazine.*[53] *Time* did not publish an article on the discovery that year or later and Barrington-Brown's photos went unused for 15 years, that is, until Watson put two of the photographer's eight photos in his *The Double Helix,* one showing the two of them posing beside their 6-foot-tall model. In it, Crick is pointing to the model with a slide rule and Watson is staring up at it, his hands in pockets and looking skyward. The other photo shows Crick and Watson relaxing during their morning coffee. Since 1968, the first of these has become a very familiar icon.

It would take a major figure known to the "establishment" to break the ice of journalistic indifference, and Bragg was surely such a one. When he came to London to lecture at Guy's Hospital Medical School on 15 May, the *News Chronicle* published an article under the heading *"Why you are you. Nearer the secret of life."* Written by veteran science journalist Peter Ritchie Calder, it was based on an interview that had taken place the previous day. It mentions the two groups involved in the work, but not Watson and Crick. Bragg told Ritchie Calder that their structure "provides the first rational explanation of how a chemical can reproduce itself." For the study of life, this means "what Rutherford's early descriptions of the nucleus of the atom meant to physics . . . and will open up a vast new field of research into the secret of life."[54] Across the Atlantic in *The New York Times*, two articles appeared following Bragg's lecture. In the second of these, the claim was made that DNA is "as important to biologists as uranium is to nuclear physicists." Nucleic acid had been known, the report added, since 1869. "But what nobody understood before the Cavendish

Laboratory men considered the problem was how the molecules were grooved into each other like the strands of a wire hawser so they were able to pull inherited characters over from one generation to another."[55] Surely a strange and imaginative depiction! Much nearer home, the Cambridge student newspaper *Varsity* did mention Watson and Crick by name and labeled their discovery "the biological equivalent to crashing the sound barrier."[56] But you would not think so if you browsed through volumes of *Nature* or *Science* published during the 1950s[57] or read about the prehistory of what in 1957 became the MRC unit for Molecular Biology and in 1962 the MRC Laboratory of Molecular Biology.[58] In addition, recall that both Crick and Watson were as yet nonentities outside a small circle of those in the know. And after all, their structure was no more than a proposal, but what a proposal it was!

The lack of publicity had not concerned Crick, for he wanted to continue working, both to exploit their discovery and to finish his thesis. Thus, he had remonstrated with Watson when he decided to go to Paris for a week before their paper was written. Where was his sense of priorities with "work of such extreme significance" awaiting him? Meanwhile, Crick established that Franklin's data for her A form—the shorter, more compact structure—could be accommodated by tilting the bases, bringing them into closer contact, and thereby shortening the fiber.[59]

### Doctoral Dissertation

For Crick, it was now back to that tiresome PhD dissertation. Or should he abandon it, now that he had the DNA structure? Well, his colleagues advised him to complete it. Indeed, his old boss, Edward Collingwood advised him that "You will need to be Dr. Crick in the States. As you know it makes academic life easier there."[60] In any case, Crick had no intention of letting Perutz down nor the unit at the Cavendish. So it was back to the tedium of correctly fulfilling the requirements of the University's Board of Research Studies.

Crick was not the only one with a doctoral thesis to finish. Watson also had signed on for a Cambridge doctorate, but he continued to postpone writing it. Then his thesis director Kendrew wrote, asking about his intention, he received the following reply: "I still want the degree and am willing to write the thesis about DNA."[61] He pleaded concern and uncertainty about the draft into military service. But the dissertation was never written. In truth, unlike Crick who was formally still Mr. Crick until 1954, Dr. Watson did not really need another degree. The DNA structure was a far better prize to add to his Indiana University doctorate.

Crick's dissertation consists of 14 chapters, the substance of many of which has already been discussed in Chapter 5 of this volume. At this point, we need only consider the first and the last. The former opens with his view of the importance of proteins and nucleic acids for life. "In general," he wrote,

> we can say that a given chemical reaction occurs in a living cell because of the presence of a highly specific organic catalyst, known as an enzyme, and that all (or almost all) known enzymes are proteins. A cell contains hundreds, probably thousands of different types of enzyme molecules. The central problem of molecular biology is the mechanism of synthesis of these highly specific proteins, and the manner in which the details of this process are passed on from generation to generation. Proteins and the nucleic acids are the key molecules of molecular biology.[62]

Although he placed protein synthesis at the center of his field of enquiry, he also included the nature of the hereditary transmission of the specifications for that synthesis. That, he was certain, involved the nucleic acids. He was deeply impressed by the fact that, despite the variety of proteins in nature, "only about twenty different kinds" of the building blocks—the amino acids—are found in them and that all of these constituents have the same optical property and hence structural orientation—they are *levo,* in that none (save in certain bacteria) has the opposite structural orientation, *dextro*—like having all left hands. He concluded that "This is of the deepest significance, since it shows that structurally proteins are much simpler than one might have conceived, and suggests that the process of protein synthesis is of a simple kind."[63]

Crick's use of the words "deepest significance" reflects his growing conviction, first formed, he thought, before he left the Strangeways Laboratory, that the key to the magic of the proteins lies in the *one-dimensional* sequence of amino acids in the polypeptide chains of which they are made. He recognized that the art of X-ray crystallography was not at the stage where such sequences could be delineated, but in Cambridge, Frederick Sanger was establishing these sequences using the chemical technique he had devised, together with the separation technique of partition chromatography of Archer Martin and Richard Synge. Indeed, it was in the very year Crick moved to Perutz's unit—1949—that Sanger reported his first success in identifying "short sequences of four to five amino acids in insulin.[64] Reports of his progress followed year by year until the completion of the sequence in 1955.[65]

The state of the art of X-ray crystallography could not at that time delineate individual amino acids but it could, on the other hand, yield information about the folding of the chains in the molecule and the manner in which the chains might be packed together. Together, these two approaches formed an effective and powerful approach to the problem of protein structure, for one revealed the *sequence* and the other the *architecture*. Accordingly, it was Crick who would learn from Freddie Gutfreund how the sequence work was progressing and would visit Sanger's lab to see for himself.

There is a frankness in Crick's doctoral dissertation concerning the limitations of his experimental results on hemoglobin. They represent "rather the clearing of the ground for a further attack," and the conclusions are "on rather small points . . . often of a negative kind." The "concrete hypothesis," that follows from them, however, he judged, has "the advantage over earlier models [Perutz's and Kendrew's hatbox, see Chapter 5] in not being in obvious conflict with the experimental evidence"—tactfully put![66] The "concrete hypothesis" was that the chains in the globular proteins are nearly all the Pauling α-helix type, but they run straight for only short distances. Then, he repeated a passage from one of his early papers: "A globular protein may be more like a three dimensional framework, and may need a perspective drawing to show its main features." Four years later, that prophecy was fulfilled when Kendrew unveiled his first three-dimensional model for myoglobin. It was not in the least like a hatbox, but more like the tangle of our small intestines.

On 16 July 1953, Crick sent in his dissertation, signing it "F.H. Compton Crick" as he signed most formal documents at that time. The viva examination took place that summer. Approved in November, it was awarded by proxy on 23 January 1954 while Crick was working in America. After all that has been written about Crick and Watson's failure to cite Franklin's contribution in the 1953 MRC report, it is pleasing to read the acknowledgments in Crick's dissertation. After listing a number of names, he singled out Perutz and Kendrew for special mention and then added that "it is often the manner of attack that matters." On this point, he judged that he had "learnt much from observing the way in which Professor Sir Lawrence Bragg and Professor Linus Pauling have approached crystallographic problems. They don't just go at it using all the heavy resources at hand, but think their way around it and look for the easiest line of attack."[67] Bragg, he later wrote, "made bold, simplifying assumptions . . . and was not disturbed if the data did not exactly fit his model."[68] Pauling, he had noted, was prepared to reject data (from α keratin) that did not fit his model—the α helix.

Thus at 37 years of age, Crick had gained his doctorate, but not based on his work with Watson on DNA. As he explained, that work "was carried out in very intimate collaboration" with Watson, and it would have entailed "recasting the whole plan of the dissertation" to include the two DNA papers that they had published that year.[69] But a reprint of each is tucked in the back of the bound volume, along with five other of his publications. Two of them concerned his work at the Strangeways Laboratory. Much water had passed under the bridge since then and how glad he was to have avoided getting his doctorate for work on the viscosity of cytoplasm.

CHAPTER 11

# Employed by the John Wayne
# of Crystallography

 *I would go home and hold up a dollar for Odile and say:
"That's our last dollar."[1]*

E arly in August 1953, a letter had arrived for Crick on stationery of
the Chemical Congress, which was taking place at that time in
Sweden. It was from none other than Linus Pauling, reminding him of
his invitation to the Protein Conference that Linus was holding at Caltech, 21–25 September. Pauling went on to request that he accept the
status of Honorary Visiting Professor so that he could deliver some lectures following the meeting. "Professor Corey and I," the letter added,
"want you to speak as much as possible during the meeting."[2] At once,
and with delight, Crick wrote to accept.

When he learned that Sir Lawrence Bragg also had been invited to
lecture after the conference, Crick went to see his boss to discuss how
to apportion the subject matter for their lectures. Everyone else knew
that the letter to Crick was a hoax, dreamed up by Jerry Donohue, Jim
Watson, and Peter Pauling. But no one, not even Max Perutz, had the
guts to tell Crick. A week went by before Bragg himself, having been
apprised of the truth by Perutz, revealed the hoax to Crick. At once,
Crick wrote to Pauling enclosing the letter. On reading it, Pauling
thought that he must have written it, that is, until he noted a split
infinitive! (The text had been modeled after phrases found in assorted
letters from Pauling.) Had Watson gone too far this time in his efforts
to repay Crick for his manic enthusiasm—lecturing every visitor in
sight about their structure for DNA? Many, including Odile Crick, felt
that he had. But why was Crick's suspicion not aroused by the phrase

199

"I want you to speak as much as possible"? This was surely because he felt that he had something very important to communicate. Many years later, Watson still found this to be hilariously funny. Watching Crick expound DNA's virtues to all and sundry—eloquent, articulate, and commanding—troubled Watson, and this hoax was a way to get back at him. Yet how hurtful this was at a time when Crick's future was so uncertain.

Going to work with Pauling at Caltech had been a possibility since they met in 1952, but in the winter of 1952–1953, Crick had no future at the Cavendish under Bragg. Consequently, when the X-ray crystallographer David Harker invited him to spend a year with his group at Brooklyn Polytechnic, Crick accepted. He could come once his PhD degree was finished. Subsequently, the eminent protein chemist and long-time friend of Perutz, Felix Haurowitz, had made an informal inquiry about the possibility of Crick coming to Bloomington, Indiana. But no inquiries came from institutions in Britain, Cambridge University included. Looking ahead, Crick guessed that he would probably end up working in the United States for much more than a year and relocating to America was a distinct possibility. He needed immigrant visas and for the whole family.

By the time Crick was due to leave for New York, his success with DNA had altered the situation at Cambridge. Before then, Bragg had been unwilling to tolerate Crick's continued presence in the Unit after his doctoral work was completed. Now, as Crick recalled, the DNA structure had made that concern unnecessary[3] (support for this assessment also comes from Watson). In 1954, Bragg had admitted to Watson how contrite he felt "about his behavior to Francis, saying it was his worst mistake ever in misjudging great talent." Nevertheless, it was not until 1957 that Bragg wrote on this subject to Sir Harold Himsworth, Secretary of the MRC:

> I have always had it on my conscience that when I was at Cambridge I expressed myself rather fiercely to you at one time about Crick. Since then he has been responsible for several really great ideas which have advanced our knowledge of biological structures in a fundamental way. Ructions have not entirely ceased and a good deal of smoothing of ruffled feathers has been necessary from time to time. Crick is a kind of free lance who puts his energy into reading and thinking about other peoples' results which they have sweated away at and then [he] pulls the plum. But he is a good-hearted fellow and he is really clever. I feel ashamed that he has riled me so much—I thought I ought to let you know how much I have come to appreciate what he has done and correct any former impressions I gave you.[4]

In a note that Bragg wrote to Watson, when requesting revisions to the wording of the manuscript of *Honest Jim,* he explained that his main worry about Crick was that:

> Perutz and Kendrew were always saying they could not work because Crick talked so. He also was not producing much. I reported this to Himsworth at the MRC and said I would like to talk about it on his next visit, being doubtful whether Crick ought to go on in the laboratory. I remember the result well. Perutz and Kendrew, as you say, pleaded for Crick, and when Himsworth did come [to the Cavendish laboratory] and said "What about Crick" I was taken by surprise because I had almost forgotten my letter. I at once told him to ignore it, that I quite retracted my suggestion that he might leave.[5]

Unfortunately, early in 1953, before Bragg's change of heart, Francis and Odile had feared that unless Bragg left Cambridge or retired, there would be no position for Francis to return to at Cambridge. But in May the situation changed. Bragg would be out of the Cambridge picture, having just agreed to become the Director of the Davy-Faraday Research Laboratory at the Royal Institution beginning in the new year. In June, Perutz suggested to Crick that he might return to the Unit after the year in Brooklyn, and by month's end, the MRC seems to have agreed to this plan, in principle. Crick then wrote to thank the MRC for agreeing to pay into his pension plan while he was away. But no formal appointment to return was offered, because the MRC needed to review the future of the Unit, now that Bragg was departing. Meanwhile, Bragg had to face the unenviable task of reconstituting the Royal Institution after Andrade's troubled 4-year reign there. Bragg tried very hard to persuade John Kendrew to join him, but he did not invite Crick, who in June was telling a *New York Times* reporter in London that he may leave Britain. In July, he wrote to Felix Haurowitz regarding his suggestion that Crick consider coming to Bloomington. Crick confessed that he had kept Haurowitz's postcard "propped up in front of me on my desk for the past nine months, but until now it has been difficult to write to you definitely about my plans." He explained about his coming year with Harker in Brooklyn and his expectation to return to Cambridge. "My job at the M.R.C. Unit here will be kept open for me," he wrote. He enclosed reprints of the two 1953 papers on DNA and concluded enthusiastically: "It looks as though molecular biology is entering a phase of rapid development, initiated by Pauling's $\alpha$-helix. Such a lot remains to be discovered, which makes it all the more exciting. What a wonderful subject it is!"[6] On the 22nd of August, the Crick family left England sailing on the *Mauretania* to New York, bound for Brooklyn and David Harker's protein structure project.

## The Protein Structure Project

This project was unique in America, because it was dedicated solely to solving the structure of a protein. Perutz had heard about it when he attended the American Crystallography Association meeting at Pennsylvania State University in 1950. He reported to Dorothy Hodgkin the following:

> There is an amount of gossip going around. Langmuire got Harker one million dollars to set up a research laboratory for the X-ray study of proteins, a problem which Harker proposes to solve by application of his inequalities. He went around the U.S. telling people that he has a magical method for doing this, but kept very quiet about it at the Computer Conference. I forced him finally to give us an account of it, whereupon the method revealed itself as pure bootstraps, and all his claims nothing but humbug. But he has the million dollars all right, and Brooklyn proposes to take him after Harvard had turned him down. . . .[7]

No other like institute existed in America. Even at Caltech, Pauling was not studying the molecular structure of a protein, but that of its building blocks, the amino acids and peptides. This left the field of protein structure in America wide open. Armed with support from the Langmuir family trust, Harker secured matching funds from the Rockefeller Foundation and additional support from the Rockefeller Institute, with Brooklyn Polytechnic as host institution, and computing time from the International Business Machines Corporation (IBM). With so ample a fund, Harker could import talent to speed his progress and perhaps yield the first solution for a protein structure—even before Perutz. Most professionals considered hemoglobin an impossible goal, but Harker had chosen a smaller protein, ribonuclease, and would use the "inequalities" in the diffraction pattern to unscramble its data. This was his contribution to "direct methods." If he got there first, what a triumph that would be for American science and for the direct methods he espoused! Beginning with a scientific staff of four, Harker brought Vittorio Luzzati from Paris in February 1953 and Crick in September.

There had been doubts about accepting Harker's offer. According to Watson, Crick feared that the many jokes about Brooklyn must have some substance. They were mostly about pronunciation. For example, you ask for a quart of "earl" (oil) and you chat to the "Oil" (Earl) of Arundel. Were there not more attractive parts of America in which to work? Lacking other firm offers, however, it was a matter of acceptance by default. In the winter of 1952, Crick's future at Cambridge looked bleak, and before the year was out, he accepted the Brooklyn offer. When Harker reported having difficulty finding them an inexpensive three-bedroom apartment, Crick suggested relaxing the price and going, say, to $120 (per week). "If there are still difficulties I think we should not worry too much about the neighbourhood."[8]

## Arrival in America

In August 1953, the Crick family boarded the *Mauretania* for their first encounter with the New World. Seven days later, as they sailed past the Statue of Liberty into New York Harbor, they viewed Manhattan's unique skyline for the first time—so many skyscrapers, so many sizes and shapes, old and faded, young and gleaming, all jumbled together as if they had retreated from the sea to escape destruction. After 7 days at sea, with no buildings anywhere and just the vast expanse of the Atlantic Ocean, the Cricks were once again on terra firma in a city teaming with people. Passing through immigration on 29 August, they were met by Crick's new boss David Harker, together with his wife Katherine and daughter Tatiana. Lacking a car himself, Harker had borrowed a station wagon for the occasion. Tatiana recalled the journey that followed. "We had a flat tire on the way—on the Belt Parkway, I think it is called, and neither my father nor Francis knew how to change the tire. I guess the Parkway police finally changed the wheel."[9] Leaving Manhattan behind, they were at last deposited at their assigned rental apartment.

As they stood on the sidewalk, they were appalled. This was not the old Brooklyn that they had driven through. It was apartment 10 on the sixth floor of 9524 Fort Hamilton Parkway, a street in a distant Brooklyn suburb with no center, no character; it was just nowhere! True, early morning reveille could be heard coming from Fort Hamilton nearby,[10] but after Portugal Place—embedded among spires, leafy yards, and narrow streets—it would be hard to live here on this featureless parkway. The apartment was spacious but gloomy, and the walls were papered throughout with what Odile described as a ghastly design consisting of "tremendous leaves." What they did not know was that Katherine Harker had gone to "enormous trouble" to find an apartment big enough for a family of four, but also affordable. She was understandably "disappointed that the Cricks were less than enthusiastic" about it. Tatiana considered the apartment "rather nice, and everything worked. It just wasn't 'their' apartment."[11] Nor was it their British summer, but a New York heat wave for which there was only minimal air conditioning.

The move to Brooklyn was not to prove a good experience for Odile. She recalled,

We were terribly hard up and I was pregnant and had morning sickness. I had one small child to look after [Gabrielle] who was extremely clinging at two-and-a-half because she had been moved from her home and I couldn't even get anyone to take care of her because she just wouldn't stay with anyone else. And then Michael was with us, very distraught, having been transplanted from his home in Northampton with his beloved "grangran," he was proving difficult. In short, nothing went smoothly.[12]

At a yearly salary of $6000, the offer had seemed generous enough until one figured in rent and a nonresidency tax of 30% of the nominal salary. Actually, clergymen and scientists were exempted from this harsh rule, but you needed to know that fact to claim the privilege. Worrying, too, was the lack of emergency funds, because the foreign exchange regulations of those days permitted British travelers to take so little currency out of the country. Crick recalled how miserable they had been. They arrived with so little money—"And because you're paid at the end of the month, you've laid out everything ahead. So we were often down to our bottom dollar, and I would go home and hold up a dollar for Odile and say. 'That's our last dollar; I'm not paid for two days!' And I always knew when times were hard because she would produce potato pie, which is a delicious thing of potatoes and onions, and that meant that we were getting very hard up."[13]

With the 2½-year-old Gabrielle in tow and all the shopping to do without a car, a kind neighbor took pity on Odile and gave her an old stroller on which to load her purchases. Small wonder that Watson's name was "mud in the Crick household," for he had refused the following pleading from Crick in October: "Do you still feel you can't allow the Third Programme Broadcast? I've yet to find anyone who will say they will object to it, and things have cooled down a bit now. Also, it will bring in $50 to $100, which at the moment I could do with. If you do change your mind I'd be most grateful."[14]

Crick had made the radio broadcast in question on the European Service, and the BBC had wanted to rebroadcast it on Britain's Third Programme that summer, but Watson had refused. Evidently, he did not judge "things have cooled down" enough. Presumably, he was referring to the matter of their use of Franklin's data. Crick lectured him that he "should realize that as a married man with two (+) children I cannot afford to take your detached attitude about money."[15]

Harker's wife Katherine tried to alleviate the Crick family's straitened circumstances by offering Odile sewing work. Odile must have proven her skill, because the blouse that she had made from Tatiana's wedding dress is one that Tatiana still wore four decades later! Tatiana also asked a college friend whose father managed a felt factory to send Odile scraps of felt that could be used to make toys. The Harkers also invited the Cricks for meals. But young Michael enjoyed himself attending the Quaker Friends School in Brooklyn and paying frequent visits on the way home from school to the studio of Sam Magdoff, whose wife Beatrice worked with Crick. Michael enjoyed watching Magdoff create animations for television. This experience was to influence his later career move from neurophysiology to computer games and crosswords, of which he was to become a pioneer.[16]

If domestic arrangements were disagreeable, how then was the project? Was it to Crick's liking? No! Harker expected Crick to be ready to start work at a respectable time, that is, by about 8:30 am. Because the journey to Brooklyn Poly, door to door, like the journey to Manhattan, took 50 minutes by subway, this was intensely disagreeable for Crick, who was accustomed to the 10-minute walk to the Cavendish, where nobody "clocked" you in. Even when Harker made some concession for Crick's commuting time, it was still unpleasant. But Harker was aware of Crick's habits before he came, because after having learned that Crick kept "irregular hours," Harker had written to Cambridge expressing his view on the matter. Crick's response was to break his habit of working late in the lab.

Tall like Crick, Harker was a much more private person. His former colleague at General Electric, Herbert Hauptman, described him as "friendly, courteous and unpretentious, concerned to be helpful" albeit "reserved, almost shy, an old-fashioned gentleman with old-fashioned values." Crystallography was his life and to it "he gave everything—his time, his energy, his total devotion."[17] He would be in the lab early and to have Crick sauntering in at 10 or 11 was unacceptable.

On the strategy for solving protein structures, the two men differed. Harker was one of the founders of the approach to the phase problem by "direct methods," that is, he sought statistical relationships in the data called "inequality relationships" that would permit determination of the phases and then the "reasonable" solution of structures *directly* from X-ray diffraction patterns. Such successes as existed in the 1950s concerned very small molecules with favorable features. Crick was skeptical of Harker's hopes for direct methods when working on such complex and large molecules as proteins. He judged, as did Vittorio Luzzati, that the only possible path to success lay through isomorphous replacement, the route that he had suggested to his Cambridge colleagues and the one to which Harker was becoming increasingly resigned. Harker's optimism about direct methods was based on his expectation that proteins possess a basic pattern in their structure—he called it a "blob"—that direct methods could reveal. But what were these "blobs"? All that came to Luzzati's mind was Harker's "incurable optimism." Harker had been present at Luzzati's talk at Pauling's 1953 conference when he cautioned against the perils of applying a direct method, which was used with success on very simple compounds, to a protein. Did Harker's optimism decrease as a result of this? No, "it could never be knocked," was Luzzati's response.[18] Today, thanks to theoretical developments and powerful supercomputers such as the CRAY T3D at Pittsburgh, the structures of some antibiotics and small proteins have been tackled with success using direct methods—but that is over half a century after the events described in this chapter. In the 1950s, direct methods could not tackle large molecules; as Bragg recalled, they were "outflanked" and their positions usurped by the chemical methods of isomorphous replacement.[19]

### Ribonuclease

The X-ray crystallographer on staff for the protein project was Beatrice Magdoff and Crick was assigned to work with her. The chemist Murray Vernon King had the task of crystallizing Harker's chosen protein, the enzyme ribonuclease. Although he eventually succeeded in producing crystals with 14 different modifications, for several years King was unsuccessful in attaching any "heavy" atoms so that the determination of the phases by isomorphous replacement could be used. No crystals tested by Magdoff and Crick in 1953–1954 revealed a possible case.

Initially, ribonuclease had seemed to be a wise choice, seeing as it was one-fifth the size of hemoglobin. Unfortunately, it proved to be resistant to King's attempts to introduce heavy atoms such as those that Perutz introduced to hemoglobin in 1953 and that Kendrew introduced to myoglobin in 1955. When Crick had been at Brooklyn for 4 months, he wrote a long letter to Kendrew in Cambridge. "Ribonuclease," he reported, "has been going very slowly. First we found that the Patterson had lots of mistakes in it; it has taken some weeks to weed them out and to check it properly. We have just started to plot it. I am only mildly optimistic about it. . . . Low order reflexions," he added, "strongly suggest the approximate position of the molecules in the cell. Dave [Harker] thinks they show the shape, but I have been bitten before over haemoglobin . . . so I am totally skeptical. I think they were surprised by this!" Reflecting on his experience, he judged that when he arrived "they were rather out of touch, not knowing what to believe and what not. Now things are more reasonable. The group as a whole is very pleasant."[20]

Crick was impressed by Beatrice Magdoff's "good sense on all crystallographic problems," and he welcomed the way that "she tends to provide some sanity when hopes get too wild."[21] Likewise, Luzzati respected her competence, and to him she proved to be a "wonderful friend. She, her husband Sam, and their families," he added, "strongly contributed to make our stay in New York pleasant."[22] Odile also recalled their hospitality on a number of occasions.

Sadly, Luzzati was to return to France at the end of January. "He will be a great loss," wrote Crick. "He is very intelligent, and makes a charming colleague. His experience has been in quite different branches of crystallography, and we have learnt a lot from each other." Before Luzzati's departure, Harker held a progress meeting on ribonuclease. Crick described it to Kendrew thus: "Eventually quite sensible, but it took a long time (a whole day). Dave has been pressing me to stay on here permanently. This has been brought to a head by the fact that Odile is returning (to Kings Lynn) in late January, as it is really difficult to cope with a baby here. Odile and I are quite definite that we would prefer to be in Cambridge, and I have just told Dave so."[23] In the absence of the offer of a new contract from the MRC, Crick was tactfully reminding his Cambridge colleague that he wanted to return to the Unit.

The examination of ribonuclease crystals for heavy-atom incorporation led Crick and Magdoff to discover the degree to which protein crystals are sensitive to environmental conditions. Not only did successive crystalline preparations give different diffraction patterns, but smoking a cigarette close to the crystal, as Magdoff did from time to time, caused changes. As Crick explained, "The solvent [for the ribonuclease] was a water/alcohol mixture, very sensitive to temperature changes. These cause evaporation from one part of the tube to another part, producing alterations in the crystal that show up in the diffraction pattern." Drafts also caused sensitivity; they tried blowing cold air over the mounting tube containing the crystal, and this changed the intensity of the reflections by a significant amount. Magdoff and Crick therefore warned that, regarding ribonuclease, "*precise* isomorphous replacement may not be easy to achieve."[24]

Crick was content to have produced with Magdoff and Luzzati a three-dimensional Patterson diagram for ribonuclease (Fig. 11.1) that undermined any assumptions of a simple set of parallel chains as the basic structure. Evidently, assumptions about parallel chains, such as those that Crick had earlier attacked in Perutz's hat-box model of hemoglobin, were still abroad in Brooklyn. Crick did build three trial models to fit the Patterson data, but they were "all very poor, since they either had poor hydrogen bonding or a hole down the middle." Magdoff,

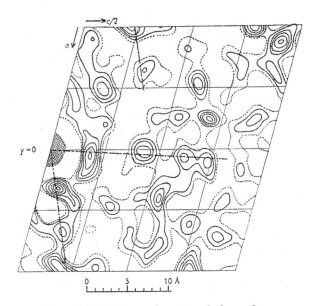

*Figure 11.1* Patterson function of ribonuclease.

Luzzati, and Crick did not rule out the possibility of rods that lie "not parallel, but in a variety of directions," however, they stressed, "It seems to us that these studies only serve to underline the need for the isomorphous replacement method."[25]

Four papers came out of this work, however, as Georg Kreisel was to complain, they were deadly boring, and Crick admitted to Watson, "It is terribly dull, but I feel I owe it to my hosts (do you remember your biochemical work on myoglobin?!)."[26] It was Crick who wrote the first drafts of all of these papers, working regularly on them from 9 am to 7 pm. Evidently, Harker had been impressed by Crick, otherwise he would not have sought to persuade him to stay permanently. Reading between the lines of Crick's letters, there seems little doubt that he had had an impact on Harker's group, with his rigorous and trenchant critique of their approach, and his caution serving to reign in the optimism of their wilder moments. By February, he could report to Kendrew: "The general feeling here is that without a heavy atom we shall hardly get anywhere on ribonuclease."[27] What of Harker's "direct methods"? Crick did not mention them.

Both Harker and his wife came from wealthy families that had lost their wealth. Their manners were Victorian, and Harker "was really very shy and used these Victorian manners to hide behind."[28] Luzzati's impression of Harker helps to explain why Crick felt rather negative about his Brooklyn boss. To Luzzati, Harker seemed like a caricature—a "typical

Far-West American . . . a sort of John Wayne of crystallography . . . very proud of America;" it was for him "the melting pot of all sorts of nations under the obvious rule of honest white guys of British extraction."[29]

Having introduced the isomorphous replacement method into the ribonuclease program, Harker was for many years frustrated by the failure to achieve a successful outcome. This experience contrasted with his earlier experience working on direct methods, to which he had contributed as a pioneer theoretician. He would have been much happier, Crick surmised, had he continued his theoretical line of work. Crick had had battles with Bragg, but he respected him; not so Harker. As director of the project, Harker had tried to go in too many different directions, any caution being drowned by his unquenchable optimism.

When Crick was back in the United States in 1956, he paid a return visit to Brooklyn Polytechnic. After seeing Harker and his colleague Tom Furnas, the instrument designer, he wrote to Kendrew, "I think their progress disappointing, they think they have an isomorphous replacement in the orthorhombic form, but the two different Pattersons are confused (I saw them—and we know this is likely to lead to trouble). Dave seems to think he's neck to neck with you, but in my view you are well ahead (didn't tell him this!). I also didn't discuss Max's theoretical paper, as the subject didn't arise."[30]

This short letter tells of competition and rivalry and is flavored with a sense of the superiority of the Cambridge program. Had he stayed with Harker, Crick might have had to watch the Cambridge successes while frustration would have been growing around him each year that ribonuclease refused to yield its secrets. And Perutz's theoretical paper? It concerned a solution to the phase problem that involved a priority issue with Harker. One can almost see Crick smiling mischievously as he wrote the last two sentences quoted above!

Crick sensed that Katherine Harker had been very ambitious for her husband and pushed him. With $1 million in research money behind him, he took on the challenge to solve a protein structure, hoping to be the first to succeed. Sadly, he was not successful until 1967, 13 years after Crick had returned to Cambridge and 10 years after Kendrew got his first view of polypeptide chains in myoglobin. More important was the fact that even in 1967, Harker was beaten to the structure of ribonuclease by Fred Richards' group at Yale. Their clever chemistry matched by the skill of instrument-minded Hal Wyckoff gave them the edge.[31] When Kendrew visited Harker's group in 1958 and showed them his structure for myoglobin, they were "somewhat surprised, if not amazed, by the unexpected irregular intestine-like folding of the myoglobin molecule and some even wondered if it were correct."[32]

With the ribonuclease program dragging on, Katherine's disappointment grew, and when Crick called to see them years later, he now a Nobel Laureate, "she didn't want to see me; it was too hard for her." In fact, Odile believed that she was suffering from severe depression. As the daughter of the former imperial prosecutor in Russia, Katherine De Savich had already been embittered by the deliberate sabotage of her academic career in economics and "was not an easy person."[33]

A senior professor at Brooklyn Polytechnic, but not a member of Harker's Project, was Isidore Fankuchen. Before the war, he had worked with Desmond Bernal in Cambridge on the structure of the virus that causes a mosaic disease in tobacco plants (tobacco mosaic virus, or TMV). Surely he would appreciate the new advances that were being made in his former field, especially Pauling's contributions on proteins. Far from it! Crick recalled, "Fan claimed there were not enough data to prove the α helix." He even remained to be convinced that the globular proteins in their native state are composed of polypeptide chains, rather than some more complex structure. So when *Science Progress* wanted Crick to write yet another paper on DNA, he wrote instead about polypeptides "to refute Fan."[34]

Here, Crick spelled out the lessons learned from Pauling's success with the α helix: He introduced the noninteger helix to protein chemistry and demonstrated the power of model building for solving structures. Crick judged the building of scale models to be an extremely powerful method, as he explained,

> since it embodies a large amount of data which any successful model must include, but for structures of this type [fibrous proteins] it may well pay to build models without giving much attention to the experimental evidence. It is not going too far to state that, at the stage where model building is usually first attempted, some of the experimental evidence then available will usually turn out at a later date to be wrong, or at least deceptive. There is a case, in fact, for careful model building independent of most of the experimental data.[35]

Fan's resistance to new ideas on helices was nurtured, it seems, by a feeling of jealousy toward the Cambridge Unit that probably traced back to 1950 when Perutz visited Penn State and Fan launched what Perutz called "a bitter public attack on all my work." As Perutz recalled, "After this performance Dorothy Wrinch got up and launched an emotional appeal that I should make all my 'beautiful data' available to the world at large and herself in particular. What with Harker wanting to solve proteins by inequalities and [Martin] Buerger by his implications theory, I felt like being in a madhouse."[36] But Fan, like

Harker, had sympathy for the mathematician-turned-biotheoretician Wrinch, author of the "cyclol" or cage structure for globular proteins. Still grieving for her cyclol hypothesis, she had attacked Perutz's parallel polypeptide chain structure for hemoglobin.[37] Harker, a friend and former colleague of Wrinch, found it "easy to think in terms of helices" in the fibrous proteins, but this, he claimed as late as 1956, did not create "any obligation on the part of the molecules in the globular proteins; these molecules may be built up of units in essentially different arrangements."[38] Clearly, coming to Brooklyn from Cambridge, Crick had entered a scientific environment very different from the one that he had left.

### Social Life

Crick, Magdoff, and Luzzati made a "nice little group" within the larger circle of researchers. Two weeks after Crick had arrived in Brooklyn, he traveled with Harker, Magdoff, and Luzzati to Linus Pauling's Protein Conference in Pasadena. The first talk was given by Pauling's friend Alexander Todd on his brilliant chemical research with Dan Brown into the constituents of nucleic acids and the nature of their internucleotide linkage. This work at Cambridge University's chemistry department, he stressed, relied on the "methods of organic chemistry,"[39] not, as he was surely tempted to add, on X-ray crystallography! Crick spoke next, offering a very confident summary of evidence in support of the Watson–Crick structure for DNA.[40] When Luzzati rose to deliver his talk, he prefaced it by saying "I thought I had some good ideas for this talk before I met Francis."[41] Not afraid to tell a joke mocking himself, he already knew the power of Crick's probing mind, and the "sharp and profound understanding of mathematical problems" he possessed—although Crick's overriding concern was "the structure of biological macromolecules."

Regarding politics, Luzzati found Crick to be unresponsive. The vicious witch hunt for Communist sympathizers conducted by Wisconsin Senator Joseph McCarthy was still ongoing during 1953–1954. Recall the withdrawal of J. Robert Oppenheimer's security clearance in 1954. Although Luzzati discussed these matters with other colleagues, he does not remember "ever doing so with Francis. . . . In these and in later circumstances," Luzzati judged that "Francis seemed to be poorly responsive to political issues."[42]

With Luzzati, and accompanied by Magdoff, Crick discovered America in California. Max Delbrück lent them his "second-best" car so that they could visit Berkeley, San Francisco, Yosemite, and other scenic spots. During the course of these travels, the car broke down in an out-of-the-way place. Crick joked about "second best," and Luzzati remembers them using the disparaging term "jalopy." They were staying at a log cabin well below the main road. The car's battery was dead and there was no way that they could push it up the incline to the road.

Yes, they had a lot of fun on that trip. Once, in a restaurant, a waitress, hearing all the voluble and excited chatter dominated by one particular voice, came to the table. "I can tell where you are all from," she announced. "You are from England," she said pointing to Crick. "You are from New York," she added, pointing to Magdoff, and "Where the hell are you from?" she said, turning to Luzzati. Evidently she had not before encountered English spoken with traces of the influence of three other languages—Italian, Spanish, and French.[43]

But besides the fun, there was serious collaboration. In addition to working together on ribonuclease, Luzzati joined Crick in helping Kendrew to analyze his data on myoglobin (by mail). But most of the time, it was Crick and Magdoff together in the lab, sitting in front of the X-ray tube and disposing one after another of the candidate crystals supposedly altered by isomorphous replacement. Crick was happy working with her, but shortly after Luzzati and he had left Brooklyn, Harker fired her.

The only benefit of Luzzati's departure at the end of January 1954 was that Crick and his son Michael could move into the attractive apartment at 295 Henry Street, vacated by Luzzati. It was in Brooklyn Heights, within walking distance of the lab in an old part of Brooklyn, although not, commented Crick with a laugh, at the prestigious end of Henry Street where Winston Churchill's mother had been born. Back in the lab, Crick found "no one to talk to" about the subjects that really interested him—the structure of RNA that hopefully would give clues to the big mystery and how in protein synthesis DNA determines the specific constitution of each kind of protein.

Luzzati had been very happy to discuss such questions, because he had been a close colleague of Rosalind Franklin's in Paris. He had also visited her in London before he flew to America. Thus, he had seen and "marveled at her amazing X-ray diffraction photos," but for him, DNA was then "just a complex molecule; there was no reason to think that this molecule carried the secret of life"—until he read the 1953 *Nature* papers and met Crick in Brooklyn. These experiences taught him two lessons: First, the structure of DNA "contains crystallographic information that is of supreme importance." Watson and Crick "<u>knew</u> this and it gave them the moral strength behind their conviction." The second lesson was that "the secrets of biology are in the physical special details of molecules," in DNA, the hydrogen-bonded base pairs. That shook the physicist in Luzzati. After this, he felt as if "he had emerged from the pupa like a moth," and he perceived that a physicist's "worse sin is to confuse description with explanation." Many a crystallographer would be content to solve the structure of a molecule. But consideration of its biological function opens up a new agenda. Crick investigated the first agenda in order to reach the second. That made him more than a biophysicist—it made him a molecular biologist. Luzzati's comment here reminds one of the 1953 remark Crick made to Sir Harold Himsworth of the MRC: "My scientific interests are primarily biological, and I only work on molecular structure because I believe it to be the key to the really fundamental biological problems."[44] Yet, despite this "drastically pragmatic attitude," to quote Luzzati, "Francis had a sharp and profound understanding of mathematical problems."[45]

Crick did find intellectual stimulation visiting other centers to lecture and, in the process, making many contacts and sowing the seeds of his future reputation in the United States. Thus, in the fall of 1953, he gave a talk to Wendell Stanley's virus group in Berkeley, California. Pauling's Protein Conference in Pasadena brought him together with 50 other X-ray crystallographers, among them many pioneers and British colleagues. Other visits included the National Academy of Sciences in Washington, D.C., Brookhaven National Laboratory on Long Island, and the Institute of Cancer Research in Philadelphia—where he met Lindo Patterson (author of the method that bears his name) and gave a lecture on DNA; "a very jolly trip," he reported. Then there was the American Association for the Advancement of Science (AAAS) meeting in Boston and a visit to speak at Massachusetts Institute of Technology (MIT). Before returning to the United Kingdom, his itinerary included Cold Spring Harbor Laboratory on Long Island, the Marine Biological Laboratory in Woods Hole, Massachusetts, and, finally, the Gordon Conference, the latter intended for the discussion of cutting-edge science. There, the emphasis is on discussion, so no proceedings are published, and an invitation to one of these meetings is a mark of recognition by the scientific community.

Before Luzzati left for Paris, he and Crick made a much-enjoyed visit to Erwin Chargaff. Recalling how Kendrew had introduced Watson and Crick to the famous biochemist in 1952 (see Chapter 10), Crick wrote Kendrew, "This bit of news will amuse you. Vittorio and I visited Chargaff! He was really very friendly—we were asked to lunch, also to come again. He says our (Jim's and mine) second *Nature* article stinks, but I pointed out that if we hadn't written it somebody else would have done. Otherwise, both being reasonably critical people, we were essentially in agreement. . . ."[46]

Chargaff took this opportunity to give Crick his *early* references to the DNA 1:1 ratios, which Crick and Watson had not yet cited—a tactful hint. And as Crick wrote, "he did it very nicely," not like John Randall who had written him a "nasty letter" about another matter. Crick's affable visit with Chargaff underlines the way in which he was able to put behind him the experience of their first meeting, something that the younger Watson could not do.

### *Another Matter*

The nasty letter mentioned above concerned the structure of collagen, the protein of our cartilage and tendons. Its author was Randall—and, yes, it was again a matter of disputed territory. Randall, together with Pauline Cowan, Tony North, and Stewart McGavin at King's College London, had been working on the X-ray crystallography of collagen since 1951 and began publishing in 1953. It was the second venture into X-ray crystallography in Randall's MRC Unit. After being upset that so much of the credit for DNA was going to Watson and Crick in Cambridge, Randall felt very sensitive regarding the expression of recognition that he felt his group should receive for their work on collagen. This time, it was North's much improved pattern of collagen. Last time, it was Franklin's wonderful B pattern of DNA. But unlike the King's work on DNA, the collagen study had very much been a team effort, using a variety of investigative approaches.

This work was reported at the Discussion Meeting held by the Faraday Society on 26/27 March 1953, just a week after Crick had written to his son Michael that "we think we have found the basic copying mechanism by which life comes from life." The meeting was masterminded by Randall, and 27 March was his day. The King's team showed their much improved diffraction pattern and described the direction in which their model building was going. They were working on helical configurations. This time, hopefully, they would be left to find the structure themselves. The discussion period that followed was curiously silent until a participant identified as Mr. Crick—not yet Dr. Crick—rose and said,

> I did not intend to say anything, but having seen that extremely beautiful photograph which the workers here have produced I do not think that the occasion should pass without one who is in crystallography commenting upon it. To the crystallographic eye it is really something

quite overwhelming. The other photographs of collagen shown earlier could mean anything. I do not have to say it here, because it was in King's College that Dr. Stokes first worked out the theory of helical structure, but the new picture is just exactly what you expect: it has the two characteristics of a helix; absence of a lot of the spots, which means the structure is defined by a few parameters, and an open space in the center without reflections. I do not think perhaps the rest of the audience were aware that this means, with any luck, that the structure of collagen is likely to be worked out in a short period. I make this comment because the picture is rather an important one.[47]

This sounds like a warning shot from the wings. Sure enough, as the shades of winter spread over Brooklyn, Crick began to ruminate over collagen. The result was that by the new year Randall found in his mail a letter from Crick enclosing his paper on the structure of collagen that he had devised. Excitedly, he had also sent copies to Pauling and to Perutz. Randall's notes on Crick's paper are dated 4 January 1954.[48] No copy of Randall's angry letter has survived. But clearly, it had been so strongly worded as to upset even Crick, a man well used to taking straight talk. Randall was furious; was it not enough that Crick had benefited so richly from their DNA work, yet here he was again entering *their* territory? Having done so, he was honor-bound to give adequate recognition for their contribution.

To Watson, Crick wrote "I am very unpopular at King's, because of the collagen note, and I have had a very rude letter from Randall, objecting that I have not given them sufficient credit in my acknowledgments. The whole story is very sad, as I tried hard to do them justice, but their ingenuity at reading implications (or talk of them) into my words has defeated me."[49] He also sought sympathy from Kendrew, but to little avail, because Kendrew had only just visited Wilkins, as we learn from his letter to Watson:

> I suppose you know that Francis has done it again. His collagen structure has reduced King's to a state of fury—Randall has written him a letter beginning "You will lose the respect of your scientific colleagues unless. . . ." Maurice threw off steam to me and to Murdoch [Mitchison] (separately) for a total period of 4½ hours on a single day when we each happened, independently, to visit him; and Pauline [Cowan] threatens to leave the protein field. You may consider yourself well out of this. I do.[50]

Truth be told, it was not as if Crick had blithely ignored the rights and wrongs of entering this field, for in his letter to Kendrew he explained,

> The origin of the structure is quite amusing. One morning over break-
> fast, I had a discussion with Hugh [Huxley], who was staying with us,
> on the question "would it be ethical to try to solve collagen." We
> decided it would be. I said I had an idea for the structure. After Hugh
> had gone I tried it out. It was quite hopeless. But the dilemma was so
> acute that within two hours I had arrived at the one I've written up![51]

Atomic models had to be made because none existed at the lab.
When they arrived, they were constructed of steel that could be bent
with a pair of pliers, "very necessary for my collagen structure!" he
joked to Kendrew. Undismayed, Crick built the model, measured
dimensions and angles, wrote it up, and sent it to *The Journal of Chemical Physics,* where it appeared February 1954. His structure consisted
of two helical antiparallel strands, hydrogen bonded at right angles to
the fiber axis, reminding us surely of these features in DNA (Fig. 11.2).
But unlike the DNA structure, this, his first collagen model, disappeared from the literature almost without a trace. In December 1953 he
was already describing it as "a crazy structure" and subsequently "a
disaster." Truly, his exuberance and confidence had gotten the better of
him regarding collagen as it had regarding the first of the DNA structures. At the time, Tony North, working at the MRC Biophysics Unit at
King's College London, had felt that Crick's collagen model was "highly dubious from the outset," but "Crick thought that it would be simple to construct a model with the right geometry and therefore thought
that we were being singularly incompetent in not producing one rapidly—my impression is that he thought that he would show us up by
devising one."[52] Pauling gave a good description of Crick's model to his
son Peter, noting "serious steric hindrance" between certain of the
hydrogen atoms. He seems not to have written to Crick about the
model. Clearly, he was not impressed.[53]

Now, it was not as if the King's College diffraction pattern was
unpublished; it had appeared in the Conference proceedings. In any
case, collagen was being studied at a number of centers scattered
around the world. Pauling, for instance, felt about collagen as he did
about DNA—it was fair game for him. On the data front, both Richard
Bear at MIT and, in 1954, G.N. Ramachandran in Madras, India were
conducting diffraction analyses. A year later, when Alex Rich came to
Cambridge to collaborate with Crick for 1 month that stretched to 6,
these two found key evidence leading them to the correct model. That
evidence came not from collagen itself, but rather from their study of
the synthetic polymer polyglycine—a close relative because some 30%
of the amino acids in collagen are glycine.[54] Previous to this,
Ramachandran and G. Kartha in Madras had come very close to the

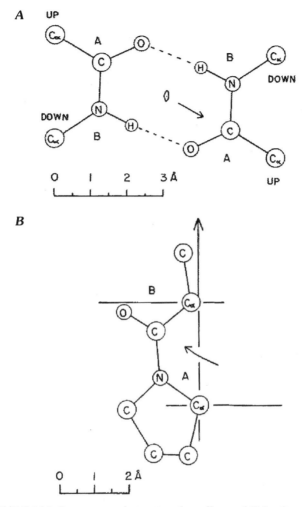

*Figure 11.2* Crick's first proposed structure for collagen. (*A*) Section of the structure of collagen. (*B*) Projection of part of one chain of the structure in the direction shown by the arrow in *A*.

correct structure.[55] Rich and Crick, then aided by their knowledge of polyglycine, modified a small part of Ramachandran and Kartha's three-chain structure to produce the model accepted today.[56] Nevertheless, it is clear that the major contribution came from Ramachandran, a judgment with which Crick agreed.

Crick could operate like this because, as Bragg remarked, he was "a kind of free lance." He could study the work of others intensively "and then he pulls the plum." Thus, the excellent postdocs Pauline Cowan (Harrison née Cowan) and Stewart McGavin at King's College London had established the threefold, left-handed character of the helical structure of collagen based on their study of poly-L-proline. They had told Crick about it before Rich and Crick arrived at their structure for polyglycine and applied the knowledge thus acquired to the data on collagen.[57] But in Cowan's opinion, polyproline had been more instructive than polyglycine.

Reflecting on this episode, Crick judged it to have "all the elements that surrounded the discovery of the double helix. The characters were just as colorful and diverse. The facts were just as confusing and the false solutions just as misleading. Competition and friendliness also played a part in the story. Yet nobody has written even one book about the race for the triple helix. This is surely because, in a very real sense, collagen is not as important a molecule as DNA."[58] Tony North believed that "despite Randall's ill feelings about Crick, other members of the King's lab were less reticent about talking to him, though I don't think we knew how far they [Rich and Crick] had got with collagen. . . . Of course, where we were daft was not realizing that the original Indian model only needed a relatively slight 'tweak' to rectify its shortcomings."[59]

This foray into collagen's structure serves to remind us of the continuing strength of Crick's interest in protein structure. Ribonuclease had been Crick's assigned task in Brooklyn, but his letters to Kendrew reveal the extent of his involvement together with Luzzati in discussing the problems of interpreting the growing data on myoglobin and hemoglobin in Cambridge. One of Crick's long letters is full of advice as well as warnings about false leads and the best way to proceed.

During the year in Brooklyn, Crick had been able to visit many American centers and give several talks and make new friends. Freddie Gutfreund observed how during the winter of 1953–1954 what proved to be a "dramatic change in his fame" had been "quietly developing." Returning to America in the summer of 1954, Gutfreund was met by Francis and Beatrice Magdoff at the docks in New York. "It was a very different Francis from the one I left in September 1953—[he was once more] his usual boisterous and optimistic self. He came and stayed with me in New Haven during the summer and was received with respect by all my friends there."[60]

It was September before Crick returned to the Cavendish. On the 18th, he gave a progress report on his work in Brooklyn. His notes for the occasion show that he was not optimistic for an early solution to the structure of ribonuclease, and as for the anticipated presence in the

molecule of straight chains or "rods," his notes are clear: "No <u>rods</u> anywhere." Instead, "sequence of lumps in the c direction." Then he asked
puckishly, "Mad pursuit? Not entirely pointless, but likely to be pro-
longed."[61] This flashback to his first talk at the Cavendish that had
aroused Bragg's fury 3 years earlier (see Chapter 6) was surely not lost
on his audience! If Kendrew had been there, we could picture the two
of them exchanging at least a wink. And Perutz? He would have been
quietly happy that the Brooklyn team, in Crick's estimation, was
unlikely to beat the Cambridge Unit to the first three-dimensional
structure of a protein.

**1** *Annie and Harry Crick*

**2** *The factory of Crick & Company on the corner of Hazelwood and St. Giles, Northampton, superimposed on company notepaper.*

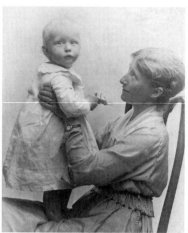

**3** *Annie holding Francis, her firstborn.*

**4** *Tony (left) and Francis (right) as a toddler. "She [Annie] thought I looked like an archbishop."*

**5** *Tony (left) and Francis (right) on a seaside holiday.*

**6** *(Top) Students housed in Mill Hill School's Ridgeway House. (Right) Enlargement of a portion of the top photo, showing Crick (center).*

**7** *Francis Crick (back row, second from right) and Harold Fost (back row, far left) as "ladies" in a 1931 performance of Gilbert and Sullivan's* H.M.S. Pinafore.

**8** *Crick (third from left) in Mill Hill School's Cadet force, c. 1932.*

**9** *Crick (back row, far right) in Mill Hill's tennis team, c. 1934.*

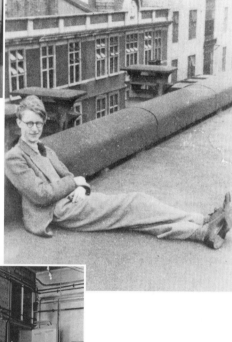

10 *Crick as a physics student at University College London, c. 1937.*

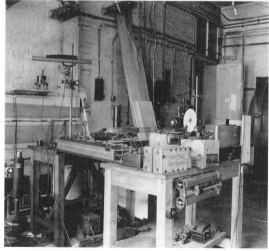

11 *Crick's viscosity apparatus at University College London, 1939.*

**12** (Left) *Betty Gunn (née Rossum) and John Gunn; (top left) Francis Crick; (bottom left) David Bates; (near right) Richard Buckingham; (far right) Harrie Massey.*

**13** *Doreen Dodd, Crick's first wife, c. 1939.*

**14** *Crick with Michael on the lawn of Elmsgarth, c. 1942.*

**15** *Crick's certificate of employment, dated 24 August 1943.*

**16** *The Strangeways Research Laboratory, Cambridge.*

**17** *Dame Honor Fell*

**18** *Georg Kreisel*

*19* Edie Hammond, Crick's girlfriend, in Chichester, c. 1945.

*20* The MX Wrens

*21* Odile as a Wren officer, c. 1945.

**22** *Crick's National Registration Card for 1945–1949 showing "The Green Door" as the current address.*

**23** *The bridegroom and bride, August 1949.*

**24** (Top) The 17th century Vicarage showing the wood-clad staircase
to the third-floor apartment called "The Green Door." (Bottom) The
Vicarage viewed from above, surrounded on its south aspect by
trees. To the right is St. Nicholas Church, on Bridge Street, and in
the foreground are the St. John's College buildings.

**25** *Crick with daughters Gabrielle on his right and Jacqueline on his left.*

**26** *The Crick family punting in Cambridge, c. 1957. Odile is punting while Francis and Jacqueline enjoy the ride.*

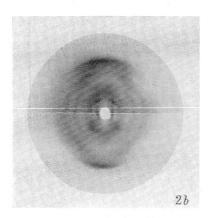

**27** *One of Florence Bell's diffraction patterns from 1939.*

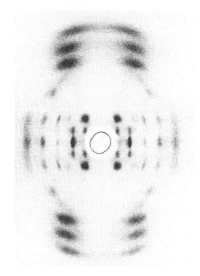

**28** *Crystalline form of DNA obtained by Maurice Wilkins and Raymond Gosling in 1950.*

**29** *Frederick Sanger*

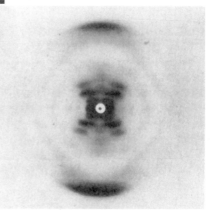

**30** *Elwin Beighton's excellent 1951 diffraction pattern of DNA showing the distinctive features of the "wet" B form of DNA (not published until 1974).*

**31** *Rosalind Franklin*

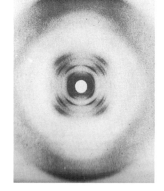

**32** *Erwin Chargaff, discoverer of the base ratios of DNA.*

**33** *Wilkins' X-ray diffraction pattern from oriented squid sperm.*

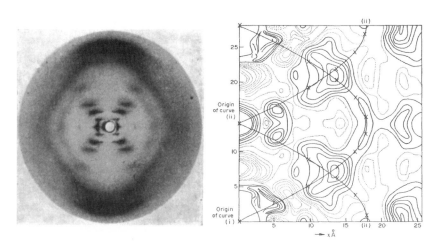

**34** *(Left) Franklin and Gosling's diffraction pattern for the "wet" form of DNA, May 1952 (photo B51); (right) their Patterson function for the "crystalline" form of DNA.*

**35** *Lord Alexander R. Todd (left) and Linus Pauling (right), Cambridge, England, 1948.*

**36** *The nearest to the original DNA model with bases represented by plates of tin.*

**37** *Jacques Monod*

**38** *Max Perutz holding a model of a polypeptide chain as he lectures.*

**39** *Sir Lawrence Bragg*

**40** *Francis Crick and James Watson in Cambridge, 1951, during a walk along the "Backs." Behind them is King's College Chapel.*

**41** *The Golden Helix, 19–20 Portugal Place. Note the brass helix hanging above the door. It represents the concept of a helix and marks Crick's achievment in 1951, the year that he derived the features of the diffraction pattern produced by a molecular helix from theoretical considerations.*

**42** *Telegram from Sten Friberg, rector of the Karolinska Institutet (Stockholm, 18 October 1962).*

**43** *Crick lecturing*

**44** *Maurice Wilkins, Max Perutz, Francis Crick, John Steinbeck, James Watson, and John Kendrew receive the Nobel Prize from King Gustaf VI Adolf, 10 December 1962.*

**45** *Crick beside Princess Desirée at the Nobel Banquet.*

**46** *Crick dancing the twist with his eldest daughter Gabrielle following the banquet.*

*47 Seymour Benzer (left) and
Sydney Brenner (right)*

*48 Crick at the BBC,
broadcasting on
"Living Matter,"
14 October 1960.*

**49** *Gonville and Caius College*

**50** *The Crick Window in Gonville and Caius.*

**51** *Francis and Odile Crick with Leslie and Alice Orgel, about to enter the church for Freddie Gutfreund's wedding.*

**52** *David Hubel (left) and Torsten Wiesel (right) when they won the Nobel Prize for medicine or physiology in 1981.*

**53** *Crick and Christof Koch*

**54** *The Pool at 1792 Colgate Circle, La Jolla.*

**55** *The Cricks' house in the Anza Borrego Desert.*

*56 Francis and Odile celebrating completion of the house.*

*57 The Cricks at 1792 Colgate Circle in 2003.*

*58 Crick with his former personal
assistant, Pauline Finbow,
when she visited the Cricks in
La Jolla in 2002.*

*59 Robert Olby and Odile Crick*

***60*** *Composite photograph of Crick lecturing.*

# The Genetic Code

*It is better to use one's head for a few minutes than a computing machine for a few days!*[1]

Recalling his year in Brooklyn, Crick remembered one winter's day in December 1953 when a visitor appeared at Brooklyn Polytechnic bound for the office Crick shared with Vittorio Luzzati. Tall, approaching 50, and with an accent betraying his Russian origin, the visitor was George Gamow, a well-known physicist and supporter of the "Big Bang" theory of the origin of the universe. Gamow's sense of humor, love of fun, wizardry at card tricks, and sheer ebullience and intelligence made him an admirable addition to Crick's circle. The point of the visit concerned Gamow's brave stab at the nature of the genetic code written in the sequence of bases in DNA; the previous July, he had written to Crick and Watson on the subject. Now, Crick and Gamow met for the first time.

Luzzati was in the room. "It was amazing," he recalled. "These two spirited men debated, argued, and fought their way through the subject of the code disposing of issues, one after another, in their exuberance their voices rising to shouts." This intellectual fireworks display continued until "all the fundamentals were there. In that one meeting, lasting little more than two hours, they had the principles all laid out."[2] Looking back on the occasion, Crick did not find this surprising, because he had been thinking about the code on and off ever since he had received Gamow's first letter 6 months earlier. That had been shortly before he left Cambridge for his year in Brooklyn.

More sustained concentration on the subject had not been possible while he was working on the structure of ribonuclease, because he had accepted the Brooklyn offer before Watson and he had arrived at their structure for DNA. Moreover, a multitude of other concerns had been on Crick's mind about their DNA structure, such as how to unwind the

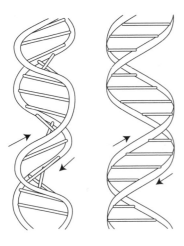

**Figure 12.1** Paranemic (*left*) and plectonemic (*right*) coiling.

double helix to permit duplication. Without unraveling and opening it up, there was no way that free nucleotide bases could be hydrogen-bonded onto each strand to generate two double helices where there had only been one. But unwinding would cause all those super coils that we know so well when wrestling with that willful garden hose. The DNA of even a virus might well need 20,000 untwistings, whereas in higher organisms, the number could be as high as 20 million. Consider, too, the rapid rate at which the duplication process occurs. Unwinding seemed impossible, and Watson's mentor Max Delbrück had made a bet with Crick at Linus Pauling's conference in September 1953 that the two chains in DNA could not be intertwined (i.e., be plectonemic) in the manner of their structure. They must instead exist in the double helical molecule, much like two coil springs that can be slid into each other and withdrawn again (paranemic) (Fig. 12.1).

Of greater interest to Crick was how their structure for DNA could throw light on the function of genes, i.e., how DNA can direct the synthesis of a protein that is the direct product of the gene—an enzyme—that will catalyze the synthesis of a particular pigment, such as blue or brown in the pupil of the eye, that being the *trait* determined by that gene. The original stimulus to explore this topic had occurred in the summer of 1953 as a result of Gamow's "zany" letter, handwritten in printed style, the text peppered with strangely spelled words; this "unknown hand" had intrigued Crick.[3] It began, "I am a physicist, not a biologist, and my interest in biology can be justified, if anything, only by my recently published book 'Mr. Tompkins Learns the Facts of Life'. . . . But I am very much excited by your article in May 30th *Nature*, and think that this brings Biology over into the group of 'exact' sciences."[4]

This book belongs in the series of popular science titles written by Gamow, beginning with *Mr. Tompkins Explores the Atom*. Through the medium of a series of dreams in *Facts of Life*, Mr. Tompkins sees inside himself and learns about the immune system, the chromosomes, and the brain. Mr. Tompkins describes how the genes "sit tight in their chromosomes" in the nucleus, but each "has numerous daughters, all of them hard-working girls, who go out into the cytoplasm and direct chemical processes. . . ."[5] This book was completed in 1952. But it was not until the following summer that Gamow's attention was drawn to Watson and Crick's second paper on the "genetical implications" of their structure, and he became excited about the future of biology as an exact science. Anxious to meet the authors, he first wrote them to ask some questions.

"If your point of view is correct," he began, "and I am sure it is at least in its essentials, each organism will be characterized by a long number written in quadrucal (?) system with figures 1,2,3,4 standing for the four bases. . . ." Disinclined to accept the notion of precise locations of the genes on the chromosomes, he thought of organisms characterized by the entire sequence of numbers (representing the bases) such as a Fourier series. "This," he declared, "would open a very exciting possibility of theoretical research based on combinatorix [*sic*] and the theory of numbers!"[6] i.e., using combinatorial mathematics. Given the four bases in the nucleic acid language, how do they translate into the 20-or-so amino acid language of the proteins? What combination of the former could be used to spell the latter? There would not be a unique solution but there were physical constraints on the possibilities.

Gamow began by taking the important step of deciding which amino acids are used in protein synthesis under the control of genes and which are their derivatives that result from subsequent metabolic activity. For protein synthesis, only the former would constitute the "language" of proteins, whereas the bases in DNA would be the language of genes. The problem, then, became to discover the *code* for translating DNA language into protein language.

Taking Gamow's letter of 8 July 1953 to The Eagle pub, it dawned on Crick and Watson that they had "never actually counted the exact number of different amino acids found in proteins."[7] Gamow had arrived at his list of 20 by a very rough procedure. He began by assuming that proteins are synthesized directly on the DNA of the chromosomes in the nucleus. He thus imagined amino acids finding their way into "cavities" among the sugar–phosphate backbones of the double helical DNA molecule, each fitting into its appropriate cavity, rather as a child finds the objects that fit into the variety of "mouths" in a toy mailbox. He arrived at the figure 20 for the number of different cavities. Then, he

**A**

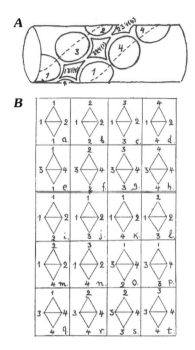

**B**

*Figure 12.2* Gamow's model and coding table for a genetic code.

listed the amino acids found in proteins in order of their prevalence and drew a line under the 20th. All those falling above the line became members of Gamow's "standard" set, to use Crick and Watson's term.

How Gamow arrived at this figure he explained in his letter to Crick that he sent on 22 October 1953, before he had met with Crick in Brooklyn, but he did not publish his formula until the following February. To begin with, he used a triplet code—the cipher for each amino acid would be a sequence of three bases. With four different bases, the number of possible triplets would be $4^3$ (64), enough for each amino acid to be coded by more than one nucleotide triplet. Hence, the code would be "degenerate." From here on, his theorizing was dominated by his picture of protein synthesis taking place on the surface of the DNA double helix, with different amino acids fitting into different cavities on its surface. Next, he estimated the number of differing cavities and arrived at the number 20. Although his reasoning here is rather obscure, it can best be understood by refering to Figure 12.2A, top. The two dotted lines represent the two backbones of the double helix. Between them is a groove giving access to the bases within. The paired numbers 1,2 and 3,4 represent the paired bases—guanine and cytosine and adenine and thymine. The cavities in the helix into which the amino acids are supposed to fit can be visualized in the groove between the two chains. Each is bordered by four bases: 1 3 1 (4) and 3 2 4 (1) in the diagram. Owing to the pairing rules, the base (4) on the left side of the figure "located across the axis of the cylinder" is determined by base 3 across from it, and similarly, the fourth base of the next cavity (1) is determined by base 2 across from it. These members are therefore determined by their partner and cannot be varied. The other three can be any combination of the four bases. Gamow therefore concluded that "each hole is defined by only three of the four nucleotides forming it,"[8] i.e., the bracketed numbers (1 and 4) in these examples play no part in the coding of these cavities. In addition note also that they belong to the sister strand carrying bases 1,3,1 and 3,2,4. The horizontal lines in the diamonds in Figure 12.2B represent this pairing limitation on the fourth base of each cavity, numbers 4 and 1 in Figure 12.2A.

Furthermore, he assumed that neighboring amino acids would be sitting in successive cavities sufficiently close to one another around the DNA molecule to permit their being linked together to create a polypeptide chain. That meant that one of the members of a triplet of bases would always be shared with its neighboring triplet. Thus, base 3 is shared between the first cavity on the left and its neighbor on the right, and base 2 with its neighbor (not shown, off the right side of the figure). Hence, Gamow expected "a partial correlation" between neighboring amino acids in proteins, thus restricting the variety of possible neighbors that an amino acid might have. When he imposed these restrictions on his scheme, he was left with just 20 different cavities, as illustrated in his coding table (Fig. 12.2B).

Evident in this letter is Gamow's thoughtful and enthusiastic response to Watson and Crick, but his illustrations of the idea of a code seemed very simplistic. For example, he suggested that "an animal will be a cat if adenine is always followed by cytosine in the DNA chain, and the characteristics of a her[r]ing is that guanines always appear in pairs along the chain. . . ." Moreover, just how the genes operate in determining the products of protein synthesis he had earlier visualized picturesquely but inaccurately. The analogy of the gene's "numerous daughters . . . who go out into the cytoplasm, and direct chemical processes" omitted a stage. The "hard-working girls" are not enzymes, not even proteins, but instead, as they were known by the 1950s, nucleic acids—specifically, ribonucleic acid (RNA). And the enzymes to which he refers are created in the cytoplasm under the direction of the RNA. Missing the middle term—RNA—it was then natural for Gamow to imagine the protein product of the gene being synthesized directly on the DNA molecule.

Crick was not put off by Gamow's apparent naïveté, for he respected Gamow as a physicist, and as he reminded Watson on another occasion, he had a "partiality for physicists."[9] Moreover, he had a clear appreciation of what one should expect of someone coming fresh to a subject. Characteristically, he was cautious. Thus, before his meeting with Gamow, Crick confided to Watson, "He is trying to see how DNA can synthesise protein, but I shall know better after he's been. I don't think he has a good grip of the problem—yet."[10] After the visit, Crick became more positive: "Gamow was quite fun, but nothing really new yet, although he is learning fast. He is trying to find cavities in our DNA structure to fit the amino acids. I explained all the doubts and difficulties, but encouraged him to go ahead. After all, one never knows."[11]

Crick did note that Gamow's 20 amino acids left out two important members: the amides (–NH₂) of glutamic and aspartic acids (called glutamine and asparagine). Contrariwise, his list included what Crick called "local exceptions," that is, amino acids that are found only in one or a very few proteins, such as hydroxyproline, which is found only in collagen. After making revisions in line with these objections, they were surprised to be left with just 20 in their standard list. No need to follow Gamow and draw a line at 20! They had reasoned their way there (Table 12.1). Such a "standard" list was not to be found in any biochemical textbooks of that time. Instead, there were lists of commonly occurring amino acids numbering in the low 20s and a further listing of rare ones. Crick's deep knowledge of the protein litera-

*Table 12.1.* The magic 20 amino acids found in proteins.

| | |
|---|---|
| Glycine | Asparagine |
| Alanine | Glutamine |
| Valine | Aspartic acid |
| Leucine | Glutamic acid |
| Isoleucine | Arginine |
| Proline[a] | Lysine |
| Phenylalanine | Histidine |
| Tyrosine | Tryptophane |
| Serine | Cysteine[b] |
| Threonine | Methionine |

[a]Strictly speaking, proline is an imino acid.
[b]Cystine is excluded from this list because of its formation from two cysteine molecules.
(Table from Crick [1958]; also see Crick [1955].)[12]

ture lay behind this most important insight, without which the idea of a code would have made no sense. Watson and he were looking at the problem of protein synthesis from the point of view of the source of the specificity or "information" bestowed upon the product, rather than the energy or enzyme machinery required.

Gamow's intervention into this subject had also stimulated Crick in another way. He could join Gamow in thinking about the genetic code in an abstract manner. One could initially put aside tiresome biochemical experiments or ingenious genetic analyses and concentrate on possible coding relationships between the two languages. This had been the spirit of Gamow's first letter, and their open-minded response reflects the fact that they were entering unknown territory. In addition, it was an example of Crick's willingness to test ideas, the precise form of which he rejected. After all, some scheme like it might operate through RNA instead of DNA. Assuming that RNA can create diamond cavities by folding up in some as yet unknown way, Gamow's approach could simply be transferred from DNA to RNA. Yet here again, some amino acids are very similar to some others. Crick doubted that Gamow's cavities would be capable of distinguishing among them. More serious was Gamow's assumption that one out of the four bases from each diamond cavity was shared with a neighboring cavity. This means that any code based on his model would be an overlapping one and, hence, not of the "easy-neighbor" kind.

An eye for current developments in chemistry was needed here. Fortunately, by this time, Crick's knowledge of protein chemistry was, as he recalled, both "detailed and extensive." He had read widely on the subject and talked frequently with the biochemist Freddie Gutfreund. Lunches with Gutfreund at The Eagle pub had been keeping Crick abreast of Sanger's stepwise progress in establishing the sequence of amino acids on the insulin molecule's polypeptide chains. As he learned of each addition to the results, Crick received confirmation of the claim that insulin and indeed all proteins have distinct and fixed sequences. So the question was: Do we find the restrictions of an overlapping code reflected in the amino acid sequences being revealed? And if, by mutation, we alter one amino acid in a chain, do we find cases where at least one neighbor is also changed? If not, Gamow's code is surely a nonstarter.

Thus, while in Brooklyn, Crick continued to follow Sanger's insulin results, but also to work on plant viruses. In this work, mutagens (agents causing mutation) were applied to alter the viral RNA and then to look for chemical changes in the protein of the viral progeny. Such data showed that neither Gamow's diamond code nor his subsequent "triangle" code could be right. That winter at Brooklyn Polytechnic, Crick, assuming that the code is "universal," i.e., identical in all organisms, succeeded in disproving "all possible versions of Gamow's code."[13]

### RNA Structure: Key to the Code

While these excursions into combinatorial schemes were ongoing, Pauling's former postdoc, the X-ray crystallographer Alexander Rich, aided by Watson, was trying to discover the structure of RNA. Their hope was that just as the secret of gene replication was discovered in the structure of DNA, so would the secret of the genetic code be revealed once they found the structure of RNA. "It is hard to believe," Rich said, "that naively for six months we were trying to fold up a model of RNA to create amino-acid sized cavities in it—hoping that a specific fit between such cavities and specific amino acids would lead to the solution of the genetic code."[14]

Watson and Rich's efforts in the previous year had so amused Gamow that he wrote to Crick in Brooklyn describing them. "They have a model of RNA, big and nice looking, but they do not believe in it very much themselves (except Alex Rich who conceived it). It has trapezoid holes formed by two bases and two different 'sugar edges.' And there are twenty different holes."[15] Watson went on to claim that there are two forms of RNA in the cytoplasm; one was formed of single helices such as those extracted from tobacco mosaic virus (TMV), the base composition of which does not follow the Chargaff ratios. The other was composed of double helices that should give the ratios. Perhaps, Watson speculated, one chain of a DNA double helix passes into the cytoplasm where it is converted into single-stranded RNA. There, it unites with another single strand to form double helical RNA. Protein synthesis then occurs, in accordance with the Gamow idea of amino acids fitting into the holes in a double helix, this time composed of RNA.

Watson had been searching the literature on base ratios in an effort to establish whether RNA in the cytoplasm is present in the form of double helices. After 3 weary days, he found the evidence he was seeking in a paper by the Glasgow biochemists R.M.S. Smellie and Norman Davidson,[16] after which he wrote to Crick. His letter breathes excitement:

> We have a structure [of the one stranded form] which is neither very ugly nor very cute but which seems to vaguely agree with the structure of the two stranded beast. This may be hasty since we have no x-rays [diffraction patterns] but is worth the effort since this may be the beast which grows proteins. In any case we now visualize the mysteries of life as follows [Fig. 12.3].

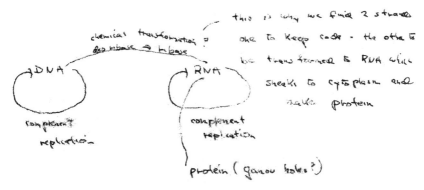

**Figure 12.3** Watson's scheme for protein synthesis, in which he notes "this is why we find 2 strands one to keep code—the other to be transformed to RNA which sneaks to crytoplasm and makes protein."

> All of this is slightly mad but as it is cute. I think it is correct. Your comments seriously desired. Have convinced [Richard] Feynman and slightly [Max] Delbrück. The others are not convinced but are not intuitive creatures, and so do not count. I do not believe the base ratios could be so good if they did not imply self replication. . . . It would be strange for DNA and RNA to be so similar and yet have different replication schemes. I think this idea is worth letter to *Nature* which I am now writing. Do you agree? . . . Would feel best if you did not pass on contents of this letter until note to *Nature* is in. Jim[17]

True to form, Crick responded enthusiastically, even addressing Watson more intimately than usual:

> My Dear Jim, Most exciting! Your letter arrived the day before I was due to go to M.I.T. and give a talk on "The Central Problem of Molecular Biology"! so I was hard put to say nothing about your ideas—but I didn't. . . . Let me say straight away that I like this idea very much. It still leaves a lot of questions open, but it seems to make more sense than all previous schemes. I could be easily convinced if the RNA analytical data were really very good.[18]

Then came Crick's misgivings. "On the basis of this hypothesis, why do not RNA base ratios correspond with DNA base ratios? And if one DNA chain is lost to making an RNA chain there should be a 50% loss of DNA per generation. Perhaps the DNA chain simply acts as a template and is not converted into RNA?" Crick then addressed the issue of publication.

> Re. letter to *Nature*. Why not? Would strongly suggest that the letter takes the form of primarily reporting your experimental results i.e.

(1) That the X-ray reflections are sharp (how sharp are they ?) which shows there must be a repeating structure.

(2) The same for two (or more?) sources of RNA therefore probably a structure, not a host of structures.

(3) Preliminary fiber diagram perhaps stating it is helical?

I should soft pedal any model which is not <u>obviously</u> correct. The great strength of our position on DNA was that the model fitted everything so well. You have no competitors, so can afford to wait till you are certain. You could merely say you have a promising model, and further work is in progress.

Re. bases. Suggest you put it as a question. Are the base ratios really 1:1 for some types of material and not others. Mention experimental error of the determinations (Davidson) and ask for further accurate work. Then put "RNA controlling RNA" as a possible deduction from 1:1 ratio if established. This allows you to get priority for the idea, without sticking your neck out too much. . . . The present position reminds me very much of the period for DNA when we realized that the base ratios meant complementarity, but hadn't got a model.

One could certainly expect the one stranded beast to control protein synthesis and naturally one hopes that its structure will give strong indications of this. Don't overlook this old point. Even if your structure is not very beautiful, and your X-ray data not very good a[n] RNA structure which explains why there are only 20 common amino acids in proteins <u>will be accepted</u> at any rate by me. I have horrid suspicion from time to time that our thinking here is too simple but we shall see. . . . Write again soon. Am extremely interested in it all.

Yours ever Francis[19]

Here we see Crick, like an elder brother, mentoring Watson in a kindly but searching manner. Crick's well-known attention to the evaluation and interpretation of experimental data are to the fore. His concern that Watson should protect his priority, yet do so "without sticking your neck out too much," was wisdom born of earlier events.

This letter also suggests that at this stage (February 1954) he was not ruling out some form of matching between a nucleic acid double helix and specific amino acids. It was Gamow's holes in a double helix structure again, this time for RNA not DNA. But RNA was proving puzzling. Although it appeared to be a DNA-like structure, Watson told Crick that he was "without any solid idea how to build one (we are now even imagining branches!). The whole thing is puzzling and paradoxical (for could DNA be wrong)." He admitted that he was being "slowly driven to despair and to loath nucleic acids. . . . Am quite anxious to talk with you. Alex [Rich] is not interested in really understanding [the] picture and Leslie [Orgel] is not a crystallographer. It is obviously a much tougher puzzle than DNA."[20]

Nevertheless, the news spread that Watson and Rich were about to make history. Back in February, Watson had looked ahead in the expectation of another triumph and asked Crick, "What do we do after protein specificity is solved? I really do not want to do genetics or watch birds, or think about phospho lipids."[21]

### The Armchair Approach

Cryptography is such an enjoyable occupation. No messing with crystals and X-rays, or running chromatograms and centrifuges late into the night. And if one likes playing with computers, it's possible to have so much fun feeding in the data from protein sequencing and calculating the odds for this or that kind of code.[22] But, as Crick was to note, there are simple ways to test a code that do not involve the use of computers. "It is better to use one's head for a few minutes than a computing machine for a few days!" he quipped.[23] The physicists and mathematicians in Gamow's circle, however, continued to pursue the different kinds of codes, with Gamow processing data at Los Alamos using the MANIAC computer. Crick used pencil and paper together with the chemical "bible" of amino acid sequences.

There was another dimension to this work; the sociable and fun-loving Gamow had the idea to form a club that would keep all the cryptographers in touch. He called it "The RNA Tie Club," because the attention of the decoders was now focused on RNA. Membership, limited to 20, was to be by election, and each member would have one of the 20 amino acids as his club name. Among the club's officers, Gamow (alanine) was the synthesizer, Watson (proline) the optimist, and Crick (tyrosine) the pessimist. Also elected to membership were some of Crick's friends, including chemist Leslie Orgel (threonine) and phage geneticist Sydney Brenner (valine). Gamow brought in physicists Richard Feynman (glycine) and Edward Teller (leucine). By circulating papers not intended for publication, the club could promote speculation and discussion among the scattered members (Crick and Orgel were in England, Brenner in South Africa, and the others on the western and eastern seaboards of the United States).

The choice of amino acid club names was made by each member on a first-come first-served basis. Crick, whose name had not appeared on Gamow's original list, was assigned the official role of the club's pessimist, which he filled with distinction, but he also proved to be the most

imaginative and creative thinker of the group. His pessimism was founded on his underlying hunch that many kinds of schemes can be formulated based on combinatorial mathematics, but the growing data on the
amino acid sequence in proteins such as insulin and the hormone β cor-
ticotropin would probably eliminate most of them. By the winter of
1954–1955, returned to Cambridge, Crick had been preparing a manuscript for the club entitled "On degenerate templates and the adaptor
hypothesis," which he sent out in January. From the insulin and β corti-
cotropin amino acid sequence data, he showed that the limits set by
Gamow's overlapping code were well and truly breached by 10 amino
acids that had at least eight different nearest neighbors. He also pointed
out that the sequence on one of the sheep and bovine insulin chains differed by a single amino acid midway along its length. Gamow's code
would require at least one neighboring amino acid also to be different.

Crick's most fundamental objection to Gamow's scheme was the
absence of any distinction between the directions of the sequence of
bases. How does the mechanism for transcribing the sequence know to
proceed in, for example, the direction of threonine, proline, lysine, and
alanine rather than that of alanine, lysine, proline, and threonine?
"There is little doubt," he judged, "that nature makes this distinction,
though it might be claimed that she produced both sequences at random,
and that the 'wrong' ones—not being able to fold up—are destroyed. This
seems to me unlikely."[24] He then turned to the chemistry and stated,
"Now what I find profoundly disturbing is that I cannot conceive of *any*
structure (for either nucleic acid) acting as a direct template for amino
acids, or at least as a specific template. In other words, if one considers
the physical-chemical nature of the amino acid side chains we do not
find complementary features on the nucleic acid. . . . I don't think that
anybody looking at DNA or RNA would think of them as templates for
amino acids were it not for other, indirect evidence."[25]

What, then, do DNA and probably RNA show? According to Crick,

A specific pattern of hydrogen bonds, and very little else. It seems to
me therefore, that we should widen our thinking to embrace this obvious fact. Two schemes suggest themselves. In the first small molecules
. . . could condense on the nucleic acid and pad it suitably, and the
resulting combination would form the template. . . . In the second,
each amino acid would combine chemically, at a special enzyme, with
a small molecule, which, having a specific hydrogen-bonding surface,
would combine specifically with the nucleic acid template. . . . In its
simplest form there would be 20 different kinds of adaptor molecule,
one for each amino acid, and 20 different enzymes to join the amino
acid[s] to their adaptors.[26]

Discussing this idea with Brenner, Crick agreed with his friend's suggestion to call it the "adaptor hypothesis." It implies, he added, "that the actual set of twenty amino acids found in proteins is due either to an historical accident or to biological selection at an extremely primitive stage, otherwise the genetic code would not now be universal."

Crick had spent time trying out a variety of codes, including "easy-neighbor" ones, i.e., codes with complete neighbor freedom for the amino acids specified. He was surprised to find that he could devise such a code, only to have his hopes dashed when he found that it did not allow him to do what nature does and alter just one amino acid in the middle of one of the insulin chains without causing base changes elsewhere. He also tried "combination codes," in which the order of the three bases in each triplet is irrelevant. Thus, ABA, AAB, and BAA would code for the same amino acid. Again, he ran into trouble with the limitations on nearest-neighbor frequencies.

Putting these setbacks to one side and keen to appreciate Gamow's contributions, Crick made the following points: (1) He introduced *degeneracy* into the subject, i.e., "several <u>different</u> base sequences can code for the same amino acid," (2) he devised overlapping codes, a type not favored by Crick and Watson but one that they could test, and (3) his scheme was "essentially abstract." Crick described the evolution of Gamow's approach as originally paying "lip-service to structural considerations, but the position was soon reached when 'coding' was looked upon as a problem in itself, independent, as far as possible, of how things might fit together. . . . Such an approach, though at first sight unnecessarily abstract, is important."[27]

The structural considerations were Gamow's imagined cavities in the DNA helix. Crick's introduction of adaptor molecules now distanced amino acids from any such cavities, whether in DNA or RNA, thereby opening up the range of possibilities for coding schemes—they became legion. "Such a point of view," he explained, "discourages a purely structural approach to the problem, at least for the moment, and throws us back on 'coding' which, it is important to note, still remains a problem even with this new approach."[28] Crick ended this remarkable unpublished paper in a mood of discouragement: "In the comparative isolation of Cambridge I must confess that there are times when I have no stomach for decoding."[29] Watson reported the impact of Crick's paper in a long letter that included the following: "Am not so pessimistic. Dislike adapters. We must find RNA structure before we give up and return to viscosity and bird watching."[30] Crick, in short, was proving his worth as the official pessimist for the club.

However, less than 1 year passed before Orgel and Crick together with John Griffith devised a new code that overcame the problem of how to read the base sequence—the *punctuation problem*—and yet would permit any amino acid to have any other amino acid as neighbors. It was an *easy-neighbor* code. In May 1956, Crick circulated their note on "comma-less codes" among members of the RNA Tie Club. Across the title page, Crick wrote the following quotation: "It cannot be that axioms established by argumentation can suffice for the discovery of new works, since the subtlety of Nature is greater many times than the subtlety of argument." Francis Bacon.[31] This action was prompted by his concern that the only evidence they had for their code was their arrival at the "magic number" of 20; there was no independent evidence.

The punctuation problem concerns how to "read" an apparently continuous message as a series of triplets. Crick explained the problem that arises when dealing with a nonoverlapping code.

> How did we know where the triplets began? Put another way, if we were to imagine that the correct triplets were marked by commas (for example ATC,CGA,TTC, . . .), how did the cell know exactly where to put the commas? The obvious idea that one started, at the beginning (whatever that was) and went along three at a time, seemed too simple, and I thought (quite wrongly) that there must be another solution. It occurred to me to try to construct a code with the following properties. If read in the right phase, all the triplets would be "sense" (that is, stand for one amino acid or another), whereas all the out-of-phase triplets (those that bridged the imaginary commas), would be "nonsense"—that is, there would be no adapter for them and thus they would not stand for any amino acid.[32]

The paper on this subject, published in 1957, was an expansion of the 1956 manuscript. Its title, "Codes without commas," referred to the absence of any signal in the base sequence to indicate the base that represents the first letter of a triplet.

They began with a classification of the types of code, with the letters A, B, C, and D standing for the four bases in DNA (Fig. 12.4).[33] Next, they assumed that some triplets, but not others, could be associated with a given amino acid. "Using the metaphors of coding, we say that some of the 64 triplets make sense and some make nonsense."[34] Representing a sequence of bases by the numbers 1–11, they distinguished among those readings of the bases that made sense and those that did not as follows:

Using the letters A, B, C, and D for the four bases, they then showed that they could devise a code in which the only correct readings of the

**A**

Thus, while overlapping codes seem highly unlikely, partial overlapping is not impossible. At the moment, however, nonoverlapping codes seem the most probable, and these are the only ones we shall consider here.

|  | B C A C D D A B A B D C |
|---|---|
| Overlapping code | B C A<br>  C A C<br>    A C D<br>      C D D |
| Partial overlapping code | B C A<br>    A C D<br>        D D A<br>           A B A |
| Nonoverlapping code | B C A<br>      C D D<br>          A B A<br>             B D C |

The letters *A*, *B*, *C*, and *D* stand for the four bases of the four common nucleotides. The top row of letters represents an imaginary sequence of them. In the codes illustrated here each set of three letters represents an amino acid. The diagram shows how the first four amino acids of a sequence are coded in the three classes of codes.

**B**

We further assume that all possible sequences of the *amino acids* may occur (that is, can be coded) and that at every point in the string of letters one can only read "sense" in the correct way. This is illustrated in Figure 3. In other words, any two triplets which make sense can be put side by side, and yet the overlapping triplets so formed must always be nonsense.

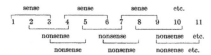

The numbers represent the positions occupied by the four letters *A*, *B*, *C*, and *D*. It is shown which triplets make sense and which nonsense.

It is obvious that with these restrictions one will be unable to code 64 different amino acids. The mathematical problem is to find the maximum number that can be coded. We shall show (1) that the maximum number cannot be greater than 20 and (2) that a solution for 20 can be given.

**Figure 12.4** (*A*) The "comma-less code." (*B*) "Sense" and "nonsense" reading of a sequence.

64 possible triplets amounted to the magic number 20, as in the code below.

|  |  |  |  |  |  | A |
|---|---|---|---|---|---|---|
|  |  | A | A | A | A | B |
| A | B |  |  | C | B | B | D |
|  |  | B | B | C | C | C |
|  |  |  |  |  |  | D |

*Number of triplets,*    2        6        12

Thus, *ABA, ABB,*    *ACA, ACB, BCA,* . . .*give*   *2 + 6 + 12 = 20.*[35]

Georg Kreisel was delighted by this paper; it was just so neat and uncluttered. It also delighted members of the RNA Tie Club. The authors had admitted the possibility of an alternative explanation, namely, starting at a fixed point and proceeding in threes—as later proved to be the case—but their comma-less offspring was so neat. Pressed by correspondents wanting to quote the paper, Crick agreed to its publication, but the authors expressed their caution in the following words: "The arguments and assumptions which we have had to employ to deduce this code are too precarious for us to feel much confidence in it on purely theoretical grounds."[36] It served to underline Crick's and Kreisel's conviction that an economical and logically attractive theory is not necessarily the right one.

### Winter in Cambridge

In November 1955, both Alex Rich and Jim Watson were in Cambridge with Crick. Watson was hoping to return to the structure of RNA with Crick's help at hand. True, Crick's adaptor molecules—if they existed— would distance RNA from the amino acids, making the structural approach to the code through RNA less promising, but Watson had yet to accept that such molecules existed. Rich was in Cambridge to work with Crick on the structure of collagen and was startled after receiving a letter from Max Delbrück that read,

> Here of course everybody is buzzing about Rundle's RNA structure (*JACS* last issue). It seems the most fantastic thing that a complete outsider from the Middle West should have hit the jackpot, and that he should have kept it to himself during all the time the paper has been in press. At first everybody was skeptical and critical of various interatomic distances but yesterday I discussed it with Corey [Pauling's coworker] who had already gone over it in great detail and seemed completely satisfied. Anyhow, since the biochemical meaning of the structure is almost more obvious than the Watson–Crick structure it would be hard to convince anybody that the structure is wrong.[37]

To all three, this was a thunderbolt. Could they check out the report in the *Journal of the American Chemical Society*? No—the latest issue, by definition, would not have reached the United Kingdom yet. But the news was not to be ignored, for Robert E. Rundle was a respected X-ray crystallographer from Iowa. "Our frustration," recalled Rich, "was immense. We worked frantically for several hours attempting to form cavities in a skeletal model of RNA that would accept amino acids, until Donald Casper, sitting on a stool and reading the letter, suggested it might be a joke." Well, perhaps they had been fooled! This was the time to break established habits and dare to make a transatlantic call. Rich called his friend Jack Dunitz at Caltech. In his "clipped Scottish accent," Dunitz mentioned the great expense of transatlantic calls, suggesting that "I should speak very rapidly into the telephone. He would tape my call and play it back later at a slower pace so that he would get the whole message in a most economical way."[38] This suggestion was so absurd that it became clear they had been the butt of a practical joke. The letter had been composed with much zest by those two masters among practical jokers, George Gamow and Max Delbrück. They had "carefully orchestrated it," admitted Rich, "and we fell right into it!" If this did no more than create hilarity, it clearly put the final nail in the coffin of RNA structure as the key to the code. But they had become wary of cryptography, so where else could they turn? Would they have to resort to the "wet" approach of biochemistry at the bench? Crick's experience with combinatorial codes and his hypothesis of adaptor molecules had made that plan increasingly likely to become the only way forward.

Even so, the failure of cellular RNA to throw light on the code did not rule out the potential of another kind of RNA—namely, that found in RNA viruses. Watson had worked on the structure of the TMV particle in 1952–1953. Aided by Hugh Huxley on the technical side and by Crick on the interpretation, Watson had suggested that each rod-shaped particle of TMV was composed not of a single lengthy protein molecule, but of a helical array of look-alike protein subunits, with a repeat occurring every 31st unit.[39] Unfortunately, he could not locate the very small content of RNA in the specimen. Now back in Cambridge, he could revisit this problem. Meanwhile, Rosalind Franklin had confirmed and modified Watson's suggestions for the structure of TMV and went on to locate the RNA embedded in the protein subunits (see Chapter 13). Others had crystallized spherical viruses and found their unit cells to be cubic. Here, Crick's ease with making inferences from symmetry data came into play, because considerations of symmetry indicated limitations to the possible ways of fitting together units that are asymmetric (all left-handed) to comprise a structure with cubic symmetry. The symmetry characteristics of these spherical viruses, he argued, must be composed of 12, 24, or 60 subunits. Watson and Crick concluded the following:

> We assume that the basic structural requirement for a small virus is the provision of a shell of protein to protect its highly specific packet of nucleic acid. This shell is necessarily rather large, and the virus, when in the cell, finds it easier to control the production of a large number of

identical small molecules rather than that of one or two very large molecules to act as its shell. These small protein molecules then aggregate around the ribonucleic acid in a regular manner, which they can only do in a limited number of ways if they are to use the same packing arrangement repeatedly. Hence small viruses are either rods or spheres.[40]

This conclusion is typical of Crick's penchant for deploying the properties of symmetry and considerations of packing to arrive at simple principles that have wide application. Rosalind Franklin, Aaron Klug, Kenneth Holmes, and Don Caspar were to develop this agenda, leading to their illustrations of the remarkable concept of the self-assembly of macromolecular structures. Structures as complex as plant viruses or bacteriophages can assemble from their constituents spontaneously (see Chapter 17). Crick also hoped that the spherical viruses might serve as a model for discovering the structure of microsomal particles in the cytoplasm judged to be the site of protein synthesis. In addition, he had hopes that some properties of the genetic code might be gathered from the proportions of the protein and nucleic acid in the plant viruses (see Chapter 13). It was a matter of exploring any avenue that might lead to the unraveling of the secrets of the genetic code and throwing fresh light on the mechanism of protein synthesis.

The "comma-less code" marked the end of Crick's involvement in Gamow's "combinatrix" program. It had lasted from September 1954 to May 1956. Meanwhile, in 1955, Crick had begun to collaborate with the chemist Vernon Ingram in an attempt to demonstrate sequence differences in the protein lysozyme from different varieties of the hen (see Chapter 14). In September 1956, Ingram's celebrated success in demonstrating a sequence change in sickle cell hemoglobin was announced. From this time forward, the Cambridge MRC unit studied the code at the bench. Behind that experimental agenda, however, lay the lessons of the "combinatrix" program and Crick's critique of it: The code must be degenerative, nonoverlapping, and of the easy-neighbor kind.

# Preaching the Central Dogma

 *I visited Hoagland on my way back to England and am now bursting with ideas about protein synthesis.[1]*

On the 3rd of April 1956, listeners to BBC Radio's *Third Programme* heard the broadcast of a talk by Crick entitled "More about the Chemistry of Heredity." He was excited because in a recent issue of the *Journal of the American Chemical Society* "there appeared a very important scientific paper."[2] He explained:

> It is not really even a paper, but a letter to the Editor, less than a thousand words long, written by Dr. Fraenkel-Conrat, who works in the Virus Laboratory in Berkeley, California. I doubt if many people, casually turning over the pages of the *Journal*—more than six thousand pages of it come out every year—have realised the significance of this short little note, but those of us who are interested in the chemistry of heredity have read it with enormous interest.[3]

It concerned experiments in which particles of the tobacco mosaic virus (TMV) were separated into two parts—a nucleic acid core and a protein coat. Fraenkel-Conrat then reassembled particles consisting of the core[4] from one strain of the virus with the coat from another and infected the host tobacco plant with these reconstituted "hybrid" particles. What, he asked, would the progeny virus particles be like? "You might say that since the plant always copies the virus, the new virus it produced would be exactly like the one you put in. This is a reasonable idea, but it is not the one I would have put my money on. For . . . there has grown up the view that what really determines the making of a protein molecule [the coat] is not a previous specimen of that protein molecule . . . but a long piece of RNA [the core]. That is, the synthesis of one sort of molecule—namely protein—is directed by a chemically quite different molecule, the RNA."[5]

In other words, "one complicated molecule can carry the information necessary to synthesise a quite different complicated molecule." This "has been a dogma to us for some years now," he explained. "Us" meant those pursuing the implications of the Watson–Crick structure for DNA. Those who were like-minded with Crick rejected the assumption that proteins just make more of themselves, for in the nucleus of the cell, they claimed, DNA contains the template for their synthesis. In the DNA-containing viruses it should also prove to be DNA, and in the RNA-containing ones, RNA.

Establishing that this is in fact the case called for clear and decisive experiments. In 1955, Crick had identified as his "favourite experiment" the kind of "TMV reconstitution experiment" published 2 years later by Fraenkel-Conrat. Unfortunately, he told Brenner, "I couldn't persuade the people here [in Cambridge] to try it."[6] That year he had also written to Watson:

> . . . have thought of a beautiful experiment, but don't know who to get to try it. It seems to me possible that if the RNA of TMV is got out very carefully, and if virus-associated protein (i.e., the stuff found in the leaf) is added to this, and the pH lowered slowly the protein might form up round the RNA and make intact protein again. If (this is a much bigger if) this then proves infective (and Schram's recent *Nature* letter encourages this hope), then a beautiful experiment becomes possible i.e. to use the RNA of one strain and the protein of a different strain, and to see what enzymes alter infection. Do you know anyone who might try this? I can't spark up much enthusiasm here among the virus boys.[7]

Although the results reported by Fraenkel-Conrat were as Crick predicted, the task remained of debating and publicizing their significance—hence the importance of attending and speaking at conferences. Now that Crick's skill on the lecture platform was well known, he received many conference invitations. In truth he was becoming a traveling speaker; he played the role of molecular evangelist for the new science, at home, on continental Europe, and especially in the United States. At home, the BBC exploited his style of man-to-man public speaking in which he conveyed his enthusiasm and his optimism about the discoveries that lie just round the corner.

In 1956 Crick gave talks at two conferences in which the "dogma" that one kind of molecule carries the information for the synthesis of a different kind of molecule was vigorously discussed: one in March to the virologists at the Ciba Foundation in London and one in June to the biochemists and geneticists in Baltimore. The Ciba event comprised a selected international group of 34 researchers, among them the seasoned virologists Norman Pirie and Frederick Bawden from the Rothamsted Experimental Station, England's historic agricultural research institute. They faced the interlopers from physics and chemistry: X-ray crystallographers, Crick, Rosalind Franklin, Donald Caspar, Aaron Klug, and the electron microscopist Robley Williams. Bawden and Pirie were skeptical of the claims made by the biophysicists. Thus Pirie was sure that the products of viral reproduction are determined to an extent by their hosts, and therefore one cannot simply consider the virus particle itself as the genetic material, and certainly not just its nucleic acid. But Crick cornered Pirie with the question "Do I understand that you expect that it will eventually be possible to show that the protein part alone is infective?" Pirie responded warily that when writing the paper for the conference, "That was a guess." He was thinking of host–virus interactions that might cause the viral protein to *become* infective. Crick responded preferring the "simplified view that the genetic part, which you must get into the plant, is RNA." It was time to "concentrate on one thing: is pure RNA infective? If we can answer this question most of the other questions are irrelevant."[8]

In the opening paper of the Ciba conference, Crick had discussed "some general ideas about the structure of viruses." This, he admitted, "is a hazardous undertaking," especially in a subject lacking unquestionably true principles. But he would draw on ideas from the study of protein crystals and "make some use of that powerful but dangerous weapon, the *principle of simplicity.*"[9] One such idea was the subunit construction of possibly all viruses. Take the lengthy TMV particle with a molecular weight of more than 36 million. It has a very long thread of nucleic acid (RNA) and around it are packed more than 2000 look-alike protein subunits. In Franklin's studies undertaken with Klug and Kenneth Holmes at Birkbeck College, University of London and with Caspar at Yale, the details of this arrangement were being established (see Fig. 13.1).

Now, if the nucleic acid determines the nature of the protein, one might expect that the nucleic acid/protein ratio would follow the coding ratio (e.g., how many RNA bases code for one amino acid—probably three) (see Chapter 12). Therefore, one might expect to find three times as much nucleic acid as protein. Unfortunately all viruses so far studied had much more protein than nucleic acid, in TMV more than 46 times as much! But if, as Watson had suggested in 1952, each virus particle is constituted of very many identical subunits, this expectation would not hold, because the same RNA "gene" sequence could be used over and over again. It would direct the synthesis of the same protein

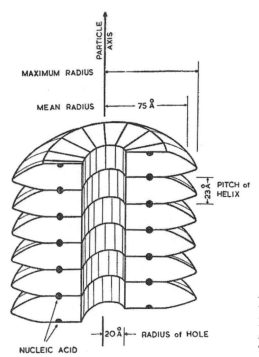

MAXIMUM RADIUS

MEAN RADIUS — 75 Å

PITCH of HELIX
23 Å

PARTICLE AXIS

20 Å ← RADIUS of HOLE

NUCLEIC ACID

*Figure 13.1* Schematic representation prepared by Rosalind Franklin, Aaron Klug, and Donald Caspar of a short length of the TMV particle.

in all the subunits. The coding ratio derived from total RNA to one protein subunit, however, yields a ratio of around 50:1—too much RNA by far! Clearly, then, the coding ratio could not be used to support the claim that the RNA is there to code for the protein.

On the other hand, there was evidence for the dogma from the study of mutant virus strains that Crick described as at least "suggestive." For the strongest case, Crick turned to Fraenkel-Conrat's "hybrid" virus reconstitution experiment that he had described in his radio talk. He told the Ciba audience, it is "one of the most important [experiments], not only for virus work, but for the whole field of protein synthesis, and we look forward to it being repeated and extended by other workers so that its apparent conclusions can be established beyond doubt."[10]

The second conference, entitled "The Chemistry of Heredity," was held at The Johns Hopkins University, Baltimore, in June 1956. Crick now found himself in the presence of Erwin Chargaff again. Two years had passed since Crick's happy visit with Chargaff in New York, but the doyen of the chemistry of the nucleic acids was irritated by the goings-on of the code breakers. Now he used his lecture to chastise Crick and his like in the RNA Tie Club. "Is nature," Chargaff demanded, "nothing but an interior decorator's dream?"[11] Recalling the mischief done by the now discredited tetranucleotide structure for DNA with its regularity of sequence, he suggested that lesson should discomfort "our easygoing cryptographers." Only the base ratios, he reckoned, offer a clue as to how "a concept of biological information can be transcribed in chemical terms. . . . The precise features of this transfer," he judged, "can hardly be discerned at this time; the unbelievable amount of male handicraft spent on getting inept models out of difficulties appears to me to be wasted." For his part, Crick must have felt sympathy for this judgment. Had he not opened his RNA Tie Club paper of 1955 with a quote from the 11th century Persian author, Kai Kā'ūs ibn Iskandar, "Is there anyone so utterly lost as he that seeks a way where there is no way."?[12] The paper closed with his confession that "there are times when I have no stomach for decoding." Yes, Chargaff was on target.

The Baltimore meeting was truly a proving ground for the Watson–Crick model and the associated speculations about its functions. Crick played his hand deftly, not responding to scornful remarks, remaining quiet in discussion until an issue of major importance arose. One such was the existence in DNA of unusual bases. If the key to the structure were the pairings between adenine and thymine and between guanine and cytosine, where do the unusual bases fit in? Take 5-methylcytosine found in some of the phages. Crick explained that it appears simply to replace cytosine, pairing normally with guanine. But why does 5-hydroxymethylcytosine, which also occurs in some DNA, seem always to occur next to guanine? Crick suggested that when DNA replicates, the precursors are not single nucleotides but dinucleotides or longer chains. Thus precursors containing the bases guanine and cytosine would then end up side-by-side in the DNA. Chargaff objected that this would require the presence of "a Maxwellian demon in each cell" to insure guanine and 5-hydroxymethylcytosine are neighbors—"a sort of a higher cellular [*sic*] intelligence unlikely to find outside of Cambridge."[13]

Another difficulty was that a question mark continued to hang over the frequency of breaks in the DNA chains. Naturally, if the breaks were frequent enough, the problem of how to unwind the double helix for its duplication would be overcome. This "unwinding problem" had been of great concern to Max Delbrück since 1953. Now he discussed the subject at length in a paper with the Berkeley molecular biologist, Gunther Stent,[14] who had built a side-by-side, like-with-like model of DNA in which the chains did not need unwinding because the structure was paranemic. Crick rated it "not impossible," but his skepticism was evident.[15] Such were the doubts and questionings he encountered, but mercifully also unexpected support came from a biochemist at Washington University, St. Louis, Arthur Kornberg. Already famed for his skills at the bench, discoverer of many enzymes, and mentor for such future Nobel Laureates as Paul Berg, Kornberg had begun work on the synthesis of DNA in the winter of 1954. The Watson–Crick model for DNA had not been his inspiration,[16] but when he reported significant progress 6 months later to the scientists gathered in Baltimore his attitude had changed. He had found that a DNA "primer" was needed together with all four of the building blocks—the nucleotides of adenine, guanine, thymine, and cytosine—to achieve growth of the new DNA chains.[17] This augured well for the Watson–Crick model, for that is just what one would expect if the synthesis involves copying a template, not just straight polymerization. The primers, it seemed, were really acting as *templates* and a juice was acting as a polymerizing enzyme. The meeting brought Crick and Kornberg together. The St. Louis biochemist was surprised to find the Cambridge molecular biologist, unlike many of that profession, appreciative and respectful of biochemists.[18]

Also supportive of Crick was the talk "Nucleic Acids and the Synthesis of Proteins," given by phage geneticist Sol Spiegelman from the University of Illinois in Urbana. Rejecting the biochemists' *multienzyme* model with no template for protein synthesis, he elaborated the *template* model by arguing that "the key to the resolution of the basic issues of biologically specific syntheses is to be found in the interrelations existing within the *triad of macromolecules RNA, DNA, and protein.*"[19] He concluded on an optimistic note: "The problem of protein synthesis has been brought to the point where further questions must be posed in terms of the chemical structures and reactive interrelationships amongst RNA, DNA, and protein. . . .the era of the direct attack has arrived. The crucial experiments have not yet been executed. However, the systems for their performance are with us, or close at hand. The outlook is depressingly bright for the quick resolution of many interesting problems."[20]

The records of these meetings show the science was in ferment. These events were thermometers recording the stirrings among virologists, geneticists, and biochemists. Although support for the new paradigm was evidently growing, more research threw up fresh difficulties. In Baltimore, R.E.F. Matthews from New Zealand asked, assuming DNA makes RNA and RNA makes protein, why do cells that manufacture very different proteins contain RNA of very similar composition?[21] Moreover, how could it be, asked two researchers from Oak Ridge National Laboratory, Elliot Volkin and Lazarus Astrachan, that in bacteria infected with a DNA virus, they found RNA with a composition close to that of the virus and unlike that of the host?[22] These were clues that were not picked up at the time. Brenner and Crick were to grasp their significance 4 years later (see Chapter 15).

Before returning to England, Crick had to give a talk in July at the University of Michigan in Ann Arbor. His mind, he told Brenner, "is at the moment fixed on protein synthesis (partly the result of having to give a lecture on it here)."[23] As we shall see, this subject was to engage his attention again in the fall in Cambridge and intensively the following year.

### Return to Cambridge

Crick had been away from Cambridge for almost 5 months of 1956. First he and Odile had traveled to Madrid where he attended the International Congress of Crystallography in April, afterward holidaying with Odile in southern Spain accompanied by Rosalind Franklin. Then came the Baltimore conference, and his 10-week spell as Visiting Scientist at the Public Health Service of the National Institutes of Health in Bethesda, where he worked with Alex Rich. He lectured on small viruses at Madison, Wisconsin, to an audience of about 200, although he had been told to expect about 15! His last engagement before returning to Cambridge was the series of lectures he delivered at the University of Michigan at Ann Arbor. He spent a lot of time, he wrote, "preparing the lecture on protein synthesis. The last part was all about microsomal particles: it included the postulate 'the microsomal particles are the only (cytoplasmic) site of specific protein synthesis.'" His conception of protein synthesis at that time was "1. The microsomal particles are inert templates. 2. RNA and protein are normally synthesised there together from common intermediates. 3. The new RNA is often broken down rather quickly and reused. This fits a lot of results."[24]

Back in Cambridge there must have been a mountain of work to catch up on, but in October, with protein synthesis still much on his mind and knowing he was to deliver a review talk on the subject at a symposium on "The Biological Replication of Macromolecules" a year hence, Crick fell to sketching his ideas on how proteins are synthesized. As he explained, it gave him "the opportunity to sort out and write down my ideas, most of which had been formulated earlier."[25] It was in October that he wrote the brief manuscript entitled "Ideas on Protein Synthesis." In two pages of double-spaced typing this remarkable document[26] states the two axioms of what we know as the foundations of classical molecular biology (Fig. 13.2). One prominent phrase therein: "The Triad DNA, RNA, and Protein" reminds us of

### Ideas on Protein Synthesis
### (Oct. 1956)

*(from the manuscript)*

The Doctrine of the Triad.

The Central Dogma: "Once information has got into a protein it can't get out again". Information here means the sequence of the amino acid residues, or other sequences related to it. That is, we <u>may</u> be able to have

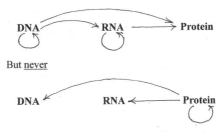

But <u>never</u>

where the arrows show <u>the transfer of information.</u>

*Figure 13.2* Crick's manuscript "Ideas on Protein Synthesis" (October 1956).

Spiegelman's lecture in Baltimore in June when he had referred twice to this "triad."[27] Crick's document also reminds us of the closing summary of that meeting by its organizer, the veteran geneticist, Bentley Glass. There Glass had used a religious simile: "It has become virtually a biological dogma within the past decade to attribute the specificity of biological syntheses to the trinity, RNA, DNA, and protein. The nature of the interrelations between these three, despite all the theorizing, remains veiled in a mystic aura."[28]

Crick's 1956 document and the lecture that followed can be seen as lifting that veil. The document is also important in establishing what he meant by these axioms and what his intentions were. Only then can we rightly appreciate his later response to apparently contrary evidence.

The 1956 manuscript begins with "The Doctrine of the Triad" (i.e., DNA, RNA, Protein). Then follows:

Requirements for Protein Synthesis.

(a) a passive template i.e. one which does not turn over in the process. This can be RNA <u>or</u> DNA.

(b) mixed intermediates of <u>ribose nucleotides and amino acids</u> . . . [i.e., his adaptor molecules].

DNA makes DNA by a special process not involving RNA and only involving proteins in a non-template manner (e.g., as enzymes or as structural supports). Presumably the Kornberg system.[29]

Crick went on to emphasize that "none of the detailed 'information' is in the histone [proteins] of the chromosome, or in the protein of the microsomal particle in which protein synthesis occurs." Moreover different microsomes making different proteins, he assumed, contain different RNA constituents (the information), although they have the same protein. His requirement for the mixed intermediates referred to his adaptor molecules, which, he suggested, bring their amino acid charges to the right place in the sequence of the nascent polypeptide chain. He concluded this short note happily with the words: "This scheme explains the majority of the present experimental results!" (Crick's exclamation mark). That included giving a reason for why the information (i.e., sequence) in RNA that in principle ought to be transferable back to DNA was not so represented. It was that any RNA not associated with the microsomes soon becomes degraded; hence the dotted line from RNA to DNA. Given appropriate conditions, he implied, this transfer was conceivable, but had not yet been observed.

In January of the following year, Crick was again in the United States where he met with the Boston biochemist Mahlon Hoagland. Hoagland was writing up his account of the experiments performed with Mary Louise Stephenson and Paul Zamecnik demonstrating the existence in the soluble RNA of the cell of a form of RNA that attaches to amino acids. Their work indicated that this RNA (now known as *transfer* RNA) brings the amino acids to the site of protein synthesis. Recall that Crick, who had speculated about the existence of such "adaptor" molecules earlier (see Chapter 12), was very enthusiastic. He wrote to a friend: "I visited Hoagland on my way back to England and am now bursting with ideas about protein synthesis."[30] These were honed and sharpened during the year and expressed in the now famous lecture, "On Protein Synthesis" before the audience attending the 12th symposium of the Society for Experimental Biology. There he repeated the substance of his October 1956 document, delivering a masterly but demanding lecture. He did run overtime, and felt that he "didn't get it over very well." The ideas were new to many in the audience, although not to those close to Crick. "It was propaganda, you see," he told science writer Horace Judson,[31] and to be sure, Crick was no stranger to the value of rhetoric.

It was the third week of September 1957, and the location was the Haldane Lecture Room of University College London. The 41-year-old

Crick laid out his vision of the "broad general problem of his subject, emphasizing well established facts which require explanation," and seeking "to relate the problem to the other central problems of molecular biology—those of gene action and nucleic acid synthesis." He had in mind, he explained, "the biologist rather than the biochemist, the general reader rather than the specialist."[32] These preliminaries signaled to the reader not to expect a traditional biochemistry lecture. Rather his aim was to raise fundamental questions and build disciplinary bridges looking for solutions.

It was 10 years since the Society for Experimental Biology held its first symposium, devoting it to the nucleic acids. At that time there was much talk of spirals in chromosomes, bacteria, and plant cells, and W.T. Astbury and Florence Bell reported on their studies of the nucleic acids.[33] Crick deployed his keen analytical and rhetorical skills to persuade his audience that it was time to take a fresh look at the relation between the proteins and nucleic acids. "In biology," he declared, "proteins are uniquely important . . . the most significant thing about proteins is that they can do almost anything." But their main function "*is to act as enzymes,*" whereas the function of nucleic acid in the form of genetic material "is to control (not necessarily directly) the synthesis of proteins." Accepting this "central and unique role of proteins," he declared, "there seems little point in genes doing anything else."[34] No other class of biological molecules will be found comparable to proteins for in them "Nature has devised a unique instrument in which an underlying simplicity is used to express great subtlety and versatility; it is impossible to see molecular biology in proper perspective until this peculiar combination of virtues has been clearly grasped."[35]

The sequence of the amino acids in a protein, he explained, is all-important. For a particular protein it is quite specific. Yet there are differences in parts of the sequence in the *same* protein taken from *different* species, as shown in Sanger's pioneer research on insulin. But the star evidence just then available came from the important work of Crick's young colleague Vernon Ingram, who had shown that sickle-cell hemoglobin, known to be the result of a genetic mutation, possesses just one and only one amino acid differing from that of normal hemoglobin—a valine instead of a glutamic acid. "It may surprise the reader," Crick remarked, "that the alteration of one amino acid out of a total of three hundred can produce a molecule which (when homozygous) is usually lethal before adult life, for my part, Ingram's result is just what I expected."[36]

The last remark here was not a case of arrogance. Rather, it reflected his memory of having to press Ingram to repeat the work because initially it had not given this clear result![37] Thus corrected, the Cambridge lab had eloquent support for the claim that it is the nature of the sequence along the polypeptide chain that determines the identity and function of the molecule. And if hemoglobin can be considered an "honorary enzyme," this claim may hold for enzymes too.

The essence of the problem of protein synthesis, Crick explained, can be "set out under three headings, each dealing with a flux: the flow of energy, the flow of matter, and the flow of information."[38] Biochemists had been concentrating on the first two. He would concentrate on the last. This was not the information of the information theory buffs, who, following Claude Shannon, treated information as the degree of order expressed statistically—the obverse of the second law of thermodynamics. Applied to such an example as language, that version is purely a syntactic concept—for instance, what is the probability of a *c* being followed by an *h* in English or an *s* by a *z* in Hungarian? The higher the probability, the lesser the randomness in the distribution of letters, but that does not give us meaning. Crick's use of the term, however, referred to cryptography and the Morse code. Here the semantic aspect of language was in his mind. How was he using the term "information"? "I mean," he explained, "the specification of the amino acid sequence of the protein." (In this context, treat letters as the analogs of amino acids, then words—very long ones—are the analogs of the polypeptide chains of proteins.)

He began by disposing of the problem of how the chains are folded to yield a globular molecule in proteins like hemoglobin and insulin. Although traditionally this had been treated as a separate process following on the synthesis of the polypeptide chain, he preferred "the more likely hypothesis," that "the *folding is simply a function of the order of the amino acids. . . .*" Where, then, was the need for a master gene to direct the folding, as the geneticist George Beadle had suggested in the 1940s?

Naturally Crick was aware that the folding could be changed by circumstances after synthesis. Thus a marked change of temperature or pH can cause a folded protein to unfold completely. So he added to the above claim the phrase "provided it takes place as the newly formed chain comes off the template." Even then, he admitted there might well be exceptions, especially among the γ-globulins of the immune system and the so-called *adaptive* enzymes.[39]

Now for any code there has to be a limit to the number of symbols in play, and in 1954 Watson and Crick had decided that only a single "standard set" of 20 amino acids can be incorporated in protein synthesis. Moreover, "*the amino acids must be joined up in the right order*. It is this problem, the problem of 'sequentialization,' which is the crux of the matter."[40] It involves disposing of the long-held belief that proteins are synthesized like carbohydrates and fats simply by the aid of specific enzymes, i.e., by other proteins (plus a primer to start the chain building). This, as the biochemist Alexander Dounce had realized,[41] leads to the dilemma of an infinite regress, because these other proteins have to be synthesized by yet other proteins and so on ad infinitum. Now, suppose the nucleic acids were to carry, coded in their chemical structure, the sequentializations of the proteins. Then there is no dilemma.

### A Hypothesis and a Dogma

Accepting the "rather meager" evidence from both DNA and RNA viruses that the nucleic acids are in some way responsible for the control of protein synthesis, how is the process achieved in a higher organism where the DNA remains in the nucleus, but its instructions must find their way to the protein factories in the cytoplasm? Five years before, Watson had expressed this journey in informational terms when he taped a sheet of paper to the wall above his desk in the Cavendish Laboratory on which he had written "DNA → RNA → Protein." The transporters of the information from nucleus to cytoplasm were RNA molecules. Their destination was the particles in the cytoplasm known at the time as "microsomes," described by Crick in this lecture as "rather uniform, spherical, virus-like particles." Hence his scheme required at least two kinds of RNA in the cytoplasm: (1) "template" RNA, synthesized in the nucleus under the direction of the DNA, and transported to the site of protein synthesis inside the microsomes of the cytoplasm; and (2) "soluble" RNA, synthesized from template RNA, liberated, and broken down to form the adaptor molecules ("transfer" RNA) needed for protein synthesis.

Turning theoretical, Crick then offered "an outline sketch" of his "ideas." "My own thinking," he began,

> (and that of many of my colleagues) is based on two general principles which I shall call the Sequence Hypothesis and the Central Dogma. The direct evidence for both of them is negligible, but I have found them of great help in getting to grips with these very complex problems. . . .Their speculative nature is emphasized by their names. It is an instructive exercise to attempt to build a useful theory without using them. One generally ends in the wilderness."[42]

The first of these, to which he granted the status of an hypothesis, "in its simplest form assumes that the specificity of a piece of nucleic acid is expressed solely by the sequence of its bases, and that this sequence is a (simple) code for the amino acid sequence of a particular protein."[43] It "unites," he explained, "several remarkable pairs of generalizations." On the one hand we have the "central biochemical importance" of the proteins, on the other the "dominating biological role of genes, and in particular of their nucleic acids." It also relates two examples of linearity—that of the polypeptide chain that constitutes a protein and that of the succession of sites lying within the "functional gene" on the chromosome. His friends Seymour Benzer at Purdue (Indiana) and Guido Pontecorvo in Glasgow (U.K.) were "running the genetic map into the ground" so that the order of genetic sites on the chromosome could be matched with the sequence of amino acids on the protein it determined.[44] Benzer's minutest genetic distances on the bacterial virus chromosome had by this time come to about the equivalent of 12 DNA bases. The hope was, then, to place on one side the genetic map and alongside it the "chemical map" of the amino acid sequence of the protein determined by that gene—that is, to establish "colinearity" between them. Crick advised his audience that "work is actively proceeding in several laboratories, including our own, in an attempt to provide more direct evidence for this hypothesis." Indeed, a race was on and it was international.

Later Crick realized that he had not expressed his Sequence Hypothesis "very precisely." As written it "rather implies that *all* nucleic acid sequences must code for protein," which, he explained, was not what he meant. "I should have said that the only way for a gene to code for an amino acid sequence of a protein is by means of its base sequence. This leaves open the possibility that parts of the base sequence can be used for other purposes, such as control mechanisms . . . or for producing RNA for purposes other than coding. However, I don't believe anyone noticed my slip, so little harm was done."[45]

Clearly, in concentrating on this aspect of informational transfer he was setting aside two questions about the control of gene expression—when in the life of the cell the gene is expressed and where in the organism. *But these are also questions of an informational nature, although not falling within Crick's definition.*

Crick introduced his second general principle under the unlikely title "The Central Dogma. This states that once 'information' has passed into protein *it cannot get out again.* In more detail, the transfer of information from nucleic acid to nucleic acid or from nucleic acid to protein may be possible, but transfer from protein to protein or from protein to nucleic acid is impossible."[46]

"Information," he cautioned, "means here the *precise* determination of sequence, either of bases in the nucleic acid or of amino acid residues in the protein." In this way he was excluding from his Central Dogma alterations in the nucleic acid brought about by proteins, such as adding something (e.g., a methyl group) to a base in the DNA. Modified forms of the bases had posed a possible problem for the Watson–Crick model, so Crick was well aware of it. He realized, too, that proteins—principally histones—surely play a part in controlling the activity of the genes, but such functions *do not involve the translation of an amino acid sequence of a protein back into what would be the corresponding sequence of the bases in DNA.*

Now why call his second principle the "Central Dogma"? Clearly, as mentioned already, the term "dogma" had been in circulation since the previous year. But it was to cause Crick trouble, and when he ruminated on this question years later he explained: "I had already used the obvious word hypothesis in the sequence hypothesis, and in addition I wanted to suggest that this new assumption was more central and more powerful." "The direct evidence for both of them," he admitted, "is negligible," but by naming them hypothesis and dogma he thought he had "emphasized" their "speculative character." In his mind, "dogma" referred to "an idea for which there was *no reasonable evidence.*"[47] When his friend Jacques Monod later chided him for using a word that refers to something "a true believer *cannot doubt,*" Crick was greatly amused. In his mind the meaning of dogma had been shaped by his youthful agnosticism growing up in a Christian, church-going family. Had he looked in the *Oxford English Dictionary* he would have found some support for his understanding: "an opinion, belief, principle, tenet," *one* meaning being "tenet or doctrine authoritatively laid down by a particular church." Or had he looked in *Webster's College Dictionary*: "a settled opinion . . . a point of view or tenet put forth as authoritative without adequate grounds." But Monod, seeking confirmation for his view in a *Larousse* (1952) would have found: "Fundamental point of religious doctrine: the Catholic dogmas. Opinion given as certain."[48]

The Central Dogma was important for a number of reasons. First it assumed the truth of the sequence hypothesis concerning the *determination of sequence* in the products of protein synthesis. Wherever he used the word "control" in the paper, he meant the control of sequence. At the same time the Central Dogma proposed a framework of permitted and disallowed transfers of information between the nucleic acids and the proteins. "It was realized that forward translation involved very complex machinery. Moreover, it seemed unlikely on general grounds that this machinery could easily work backwards. The only reasonable alternative was that the cell had evolved an entirely separate set of complicated machinery for back translation, and of this there was no trace, and no reason to believe that it might be needed."[49]

Therefore he stressed the negative character of this claim—the directions in which information cannot be transferred (i.e., from proteins to other proteins and from proteins to nucleic acids). He made no mention of the support this negative statement gave to the doctrine of the noninheritance of acquired characters. Indeed, it was not until the cytologist Cyril Darlington pointed this out to Crick in 1955 that Crick so much as made the connection.[50] But in 1957 Crick had mentioned "the dominating biological role of genes," alongside the thinking behind the kinds of experiments that were ongoing in the 1950s—seeking to demonstrate that changes in DNA cause changes in an amino acid sequence. Then, too, where was the cell's available machinery for a two-way system? Last, there was a consensus that most of the amino acids are each coded by several different groups of bases. There is *redundancy* in the code. Translating an amino acid sequence back into its nucleic acid sequence would not therefore assure recovery of the original nucleic acid sequence.

This Dogma was "saying that transfers [of sequence] from protein do not exist." One might ask if this is almost as universal a statement as the Second Law of Thermodynamics—namely, energy cannot flow from a lower temperature to a higher one without an energy source to achieve it (like the refrigerator in the kitchen). Not so, for here Crick knew he was dealing with a system that is the product of the historical accidents of evolution on Earth under natural selection. On another planet in which some form of life may yet be discovered based on different chemistry, the Central Dogma might not hold.

To be sure, there was little in the way of detail known about the cell's machinery for *forward* transfer of the information in DNA, let alone for backward transfer. Recall that the forward flow of information is represented by DNA → RNA → protein. The content of that information in the gene is a *cipher* written in the sequence of the bases in DNA. The code is then the set of relations between a sequence of bases in DNA and/or RNA and the corresponding sequence of amino acids in the protein synthesized. *Transcription* (i.e., the sequence transfer from DNA to RNA) was assumed to be governed by the base-pairing relation. *Translation* of the RNA sequence of bases into the corresponding amino acid sequence of the protein should be governed by the genetic code or chemical dictionary.

RNA must be synthesized in the nucleus when a portion of the double helix is opened up to reveal the base sequence of a gene. The bases for RNA would hydrogen-bond with the exposed DNA chain, following the pairing rules established for DNA, except that in RNA the base, uracil, takes the place of thymine in DNA. This would generate a complementary copy of the gene's sequence that Crick referred to as "template RNA." This would pass into the cytoplasm, where it would become the major constituent of the microsomal particle.

Now suppose this template RNA is a molecular structure with a series of cavities having the kind of geometry envisioned by George Gamow for DNA. Then the translation from an RNA sequence into an amino acid sequence might prove after all to be simply a problem in three-dimensional geometry. But as Crick reminded his audience, all attempts to configure RNA to achieve such cavities had failed. Instead, Crick had introduced the idea of a set of "adaptor" molecules each acting as an interface between the nucleic acid sequence of the genetic message—template RNA—and the amino acid specified by that sequence. In his imagination he pictured these adaptors each having the service of a specific enzyme, and it would be the enzyme that attaches the appropriate amino acid to the appropriate adaptor. Diffusion would bring the adaptor, thus loaded, to the microsomal particle in which the formation of hydrogen bonds would attach the adaptor to the complementary base sequence (Fig. 13.3).

How, then, had Crick in principle resolved the problem of the lack of complementarity in shape between the *template* RNA (the presumed transporter of the information in DNA) and the 20 amino acids that have to be selected and brought each in its turn to the site for synthesis of the protein? He did it by inventing 20 enzymes, each of which attaches specifically to its appropriate RNA adaptor and to the relevant amino acid. In other words, in Crick's view there cannot be recognition between the individual amino acids and a specific nucleic acid sequence of bases, but these enzymes, now known as "synthetases," *being proteins with specific three-dimensional architectures, do have this power.*

This lecture had presented a bold and enticing framework for approaching the secrets of protein synthesis. He had distilled the major claims from unformulated assumptions of the inner circle of molecular biologists, whereas the details of the mechanism of protein synthesis relied heavily on his inventive mind. As he remarked three decades later, "it was a mixture of good and bad ideas, of insights and nonsense. Those insights that have proved correct are the ones based mainly on general arguments, using data established for some time. The incorrect ideas sprang mainly from the more recent experimental results, which in most cases have turned out to be either incomplete or misleading, if not completely wrong."[51]

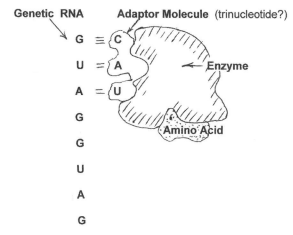

*Figure 13.3* Schematic drawing of an adaptor molecule (1957).

No sign of vanity here. But what of those who were in the audience? Klug, who was at the time a postdoc working on virus structure with Franklin at Birkbeck College, was there. He knew Crick by sight, having been present at the Ciba Foundation meeting in which the subject of Crick's talk was of course not new to him. Listening to Crick probing the mysteries of protein synthesis, in contrast, opened his eyes to a new world. The experience was memorable and he came away enthralled.[52] Even today, reading through the printed version, you can feel the excitement of entering unknown territory, identifying the most significant problems, and throwing aside the clutter of subsidiary concerns to focus on the transfer of information between the nucleic acids and the proteins—the subject of his two-page document of 1956 and the framework for his subsequent lecture.

Also much impressed was the Pasteur Institute geneticist François Jacob. Meeting Crick at the Pasteur for the first time in 1955, Jacob realized that Crick was no mere appendage of Watson, as was generally assumed in Paris. Although handicapped by the effort of giving his talk in French, Crick still impressed. In London 2 years later, he was for Jacob the "star turn" of the Society of Experimental Biology Symposium:

> He talked incessantly. With evident pleasure and volubly, as if he was afraid he would not have enough time to get everything out. Going over his demonstration again to be sure it was understood. Breaking up his sentences with loud laughter. Setting off again with renewed vigor at a speed I often had trouble keeping up with. A formidable intellectual machine. . . . On this difficult subject [protein synthesis] Crick was dazzling. He had the gift of going straight to the crux of the

matter and ignoring the rest. Of extracting, from the hodge-podge of the literature, the solid and the relevant, while rejecting the soft and the vague. His talk was based on hypotheses that, although so far not well supported by direct argument, helped one to grasp this highly complex problem. . . . To go straight to the point, not to worry about details, at least to begin with: such struck me as the lesson to be drawn from Francis's talk.[53]

In this lecture Crick had presented a bold and enticing framework for approaching the secrets of protein synthesis. It was truly a defining moment of the fledgling subject that was acquiring the name "molecular biology."

### The University and the Colleges

The picture of protein synthesis described by Crick in 1957 reminds us forcefully of the excitement and the controversy of that period. How did Cambridge University and the Colleges react to the lively presence of Crick and the MRC Unit in their midst? Of course, it was not the first research unit to be located at or near the University. Indeed, there has been a long tradition in the United Kingdom for research establishments in the sciences to be located close to centers of learning, and Cambridge, having long enjoyed close connections to government, has accumulated many. This has brought an influx of research scientists to the city, some among them wishing to be included in the social life of the University. Inevitably there has been resistance from the University over the use of land and buildings by research establishments not under their control, especially in central locations. From the Colleges, too, there was reluctance about providing dining rights and fellowships to those not involved in any teaching or administrative duties for them. Crick's experience with the Colleges and the University needs to be seen in this context.

Crick was given limited dining rights at his College, Caius, in 1954. Twenty-two years later he was made an Honorary Fellow. When his friends nominated him for a fellowship at King's College in 1957, the response came back that sufficient funds were not available. Although Harvard's widely respected physical chemist John Edsall had written a very supportive and just assessment of Crick's achievements, the appointing committee judged DNA research as a "flash in the pan."[54] Crick's friend, the Cambridge neurophysiologist Horace Barlow, reported back to Crick: "The scheme to get you elected a Fellow was put up to the Fellowship electors during the vacation, and has, I'm afraid, come to nothing. Your case was not considered on its merits at all; they simply said there wasn't enough money. Some of us don't really believe it would be necessary to sell the chapel to provide money for another fellowship, so the matter may come up again . . . keep quiet about it. . . . But I'm afraid the chances of success are down to about 0.01."[55]

In 1960 Crick accepted an "Extraordinary Fellowship"[56] from Churchill College, but a year later he resigned over the issue of building a chapel (see Chapter 16). Subsequently he relented and accepted an Honorary Fellowship. The other Honorary Fellows were Sir Winston Churchill and Barnes Wallis, the aeronautical engineer.

Max Perutz, who held a lectureship in biophysics at Cambridge University from 1953 until 1957, finally received an honorary fellowship at his college, Peterhouse, in 1962. Contrast this with John Kendrew who became a Fellow of Perutz's College in 1947 within a year of arriving in Cambridge. And then Sydney Brenner, brought to Cambridge by Crick in 1957, was made a Fellow of King's College in 1959. But Kendrew had won a major scholarship to Trinity College in 1936, making him a "member of the club." Brenner had no such qualification, but like Kendrew, he was prepared to be involved in some teaching. Moreover, neither of them was, like Crick, regarded as a militant atheist. But as Brenner recalled, "we all had an ambivalent attitude to the University."[57]

In 1957, the Arthur Balfour Chair of Genetics in Cambridge became vacant because its incumbent, Ronald Fisher, had reached 67, the recently extended age of retirement. His choice to succeed him was his friend in Caius College, Francis Crick. Crick did apply. Although his 7-year appointment still had 4 years to run, the opportunity to rejuvenate the University's genetics department and build the science on molecular foundations was a challenge he judged well worth considering.

The external member on the appointing committee for the genetics chair was the Sherardian Professor of Botany in Oxford, Cyril Darlington.[58] In 1955 he had had a confrontation with Crick at a meeting of the British Association. As a pioneer of cytogenetics and a strong personality, he commanded authority. He had performed research on the nucleic acid of the chromosomes and believed the relationship between the specificity of the proteins and the nucleic acids was a reciprocal one. As for helical DNA, he had described what he had called the "molecular spiral" of the chromosomes in 1937,[59] and he felt this premonition of what was to come deserved recognition. Apparently he gave the appointing committee the grounds for concluding that Crick could not claim to be a geneticist. Crick was not appointed.

The committee turned instead to Glasgow's professor of genetics, Guido Pontecorvo, or "Ponte" as he was known. His experience with

Cambridge University is relevant to the issue of the relations between the University and the MRC Unit, because by this time Sydney Brenner was established at the Unit with Crick and together they were working to make genetics an important feature of the Unit's program (see Chapter 14). Ponte had recently turned down an offer from Harvard, but he was attracted to the offer from Cambridge. On experiencing the response to his list of needs, however, he withdrew. Next he unburdened himself in a long letter to Crick to explain his actions. The chief reason for his change of heart, he wrote, "was . . . the feeling that no one in the University wanted me; at least no one offered a word of support. Nobody mentioned College affiliations and Adrian [the Vice Chancellor], when asked, did not even try to explain what the difficulties were." Pontecorvo told Crick that the facilities for genetics were awful, "that intrigue is rampant and that there is a large amount of ill will against genetics accumulated for various reasons by the two previous Professors [R.C. Punnett (1910–1938) and Ronald Fisher (1943–1957)]." Finally, he added, "I phoned Gray[60] and told him he should be ashamed that, with minor exceptions, anything notable in biology at Cambridge is done by physicists and chemists outside the University. I also said that if they have any sense left they should appoint you (if you still want it!)."[61] The committee did not, and Crick in retrospect considered that a blessing, for he had escaped from a future that would have included academic administration and the demands of teaching.[62]

Had Crick been appointed, the plans being developed for the expansion of the MRC Unit might well have been slimmed down, for Brenner would probably have left the Unit to join Crick. But Crick stayed and with Brenner played a vital part in the formulation of an expanded research program that would include the chemistry group of Fred Sanger in addition to genetics. It was Crick's approach to Sanger that led Sanger to suggest to the MRC that his group might be included in the new laboratory.[63]

Meanwhile William Bragg's successor, Nevill Mott had taken up his appointment in 1954 and, in Kendrew's words, was "trying to throw us out of the Cavendish."[64] Fortunately he was persuaded to relent, and after a further 3 years the group found a temporary home in the former Metallurgy Hut in the yard behind the entrance to the Austin building. This move served as the occasion to change the name of the Unit to the "MRC Unit for Molecular Biology." Soraya de Chadarevian has drawn attention to the significance of this change of name for the Unit. She has also described the protracted negotiations between the MRC and the University that took place between 1957 and

1959 over a site on which to construct a new building for the MRC.[65] Finally a location was chosen 2.5 miles from the center of Cambridge alongside the new Postgraduate Medical School by Addenbrooke's Hospital. At this point the Unit became the MRC Laboratory of Molecular Biology—a title justified by the inclusion not only of molecular genetics led by Brenner and Crick, but of biochemistry led by Sanger.

Crick had at first resolutely opposed the suggestion of a site for their Unit out of town, but when all efforts to find a central location failed he acceded. His relation with the University had in any case been cool. Given the choice, it appears he preferred to give talks to the Cambridge Humanist Society than to lecture on science in the departments and colleges. But when asked to give a series of lectures in the University, he responded with a note of his fee, suitably reduced from the payment he had received at Harvard for such lectures. The result was no reply and no lectures. Used as he was to the efficiency of the MRC, he wished to continue working under them, especially because he preferred their management style to the bureaucracy of the University.[66] Hence it was a relief as much to him as it was to Perutz to see that the MRC rebuffed attempts by the University to exercise control over important aspects of the management of the new MRC Laboratory.

## CHAPTER 14

# Crick as Experimentalist
# Attacking the Genetic Code

*Eventually we were able to show by purely genetic meth-
ods (using phase-shift mutants and a neat cross devised
by Sydney), that the code was almost certainly a triplet
code and that most, but not all, of the triplets probably
stood for one amino acid or another.[1]*

I f the DNA structure of 1953 had as great an impact as has been
widely claimed, why did Crick not promptly revise his agenda, give
up the pursuit of the structure of the proteins, and turn to questions
arising directly from the structure of DNA? First, as we have seen he
was already (in January 1953) committed to spending the academic
year (fall 1953–1954) working with David Harker on the structure of
the protein, ribonuclease (see Chapter 11). Next came his new contract
with the MRC in Cambridge (fall 1954–1960) intended for his continued assistance with the program under Max Perutz's direction on the
structure of hemoglobin. Granted, he often strayed from his official
role, making his position "anomalous," as he admitted to his future colleague Sydney Brenner in 1955. But as he explained, he did not want
to "draw attention to it" before he needed to.[2]

These considerations apart, his loyalty to his friend Maurice
Wilkins meant that work on the structure of DNA was to be left to
Wilkins at the King's Unit. On the other hand, Crick could not resist
planning to explore the genetical implications of the structure. Had he
not written the first draft of the 1953 paper on that subject and critiqued
the efforts of the code breakers? Then came his formulation of the
Sequence Hypothesis and the Central Dogma in October 1956 (see
Chapter 13). But he had not forgotten how in 1954 the Polish geneticist
Boris Ephrussi had suggested to him provokingly that genes in the *cyto-*

*plasm* may determine the amino acid sequence of the protein and "all the *nuclear* genes do is to fold up the protein correctly?"[3] Ephrussi's question, recalled Crick, "made me realize that we first needed to show that a *single* mutant in a nuclear gene altered the amino acid sequence of the protein for which it coded, probably changing just a single amino acid. On returning to Cambridge I decided that this was the next important step to take."[4] It would require work at the bench. Hence our theoretician would require collaborators and laboratory space.

Initially Crick needed only enough information from the genetics to show that there had been a mutation, not how in chemical terms the DNA had changed. The goal would be to demonstrate a change in the composition of the protein determined by that gene. This should be observable in an alteration to its amino acid sequence encoded by the mutated gene. As he wrote, "We might, with luck, pick up a change as small as an alteration in just one amino acid."[5] He settled on the small, easily crystallized protein found in tears and eggs, lysozyme. For assistance on the chemical side he turned to the skillful chemist, Vernon Ingram, who since 1952 had been pursuing isomorphous substitution in hemoglobin for Perutz. Also assisting Crick was Muriel Wigby. She had come from the Strangeways Laboratory where she had worked with K.B. Jacobson, the one colleague there besides Honor Fell who appreciated Crick and expected him to do well (maybe even win a Nobel Prize).

The most difficult part of the work would be to characterize the protein—hence their choice of lysozyme. It is small, easily crystallized, and readily obtainable from tears and from egg white. They searched for mutants in the eggs of chick, guinea fowl, turkey, duck, goose, and lesser black-backed gull. Their simple tests revealed differences between the lysozyme from chick, guinea fowl, and human tears, but not one mutant was found among the 12 strains of birds that a chicken geneticist provided. This source should have proved ideal because of the detailed knowledge of the genetics of the breeds.

For a source of human tears, Wigby would hold a slice of raw onion below one of Crick's eyes and she would catch the tears with a little pasteur pipette. They also tested the tears of a half a dozen people at the lab, but all yielded the same lysozyme. Crick thought it a good idea also to use the tears of his 18-month-old daughter, Jacqueline, "but Odile would have none of it. 'What!' she exclaimed, 'Use her precious baby for an experiment!' I was sternly forbidden to attempt it."[6]

The only surviving document on this work is Crick's account of his attempt to crystallize the lysozyme from species other than the domestic fowl by the direct method in egg white. It did not work. "The chick," he noted, "produced copious crystals, the other species none."[7]

To Brenner he wrote:

> Lysozyme. Attempt by Vernon Ingram and myself to find two hens with different lysozymes so far completely negative. . . . It is all rather discouraging. Even if we find a difference we shall still have to show it's due to amino acid composition, and also do the genetics (which may mean doing the tears of cocks!) . . .[8]

At this point the work was overtaken by the sight of more promising material—the hemoglobin that Perutz had received, taken from a patient suffering from sickle-cell anemia. Known to be a genetic disease, and rightly attributed to the unusual behavior of the hemoglobin molecule by Linus Pauling,[9] Crick's expectation was that sickle-cell hemoglobin must differ from normal hemoglobin in its constitution. Could Ingram demonstrate this? Ingram combined two techniques—chromatography and electrophoresis—so that he could achieve a very clear separation between the hemoglobin fragments yielded from enzyme digestion. He then established that the difference between the normal and sickle-cell hemoglobin amounted to the substitution of a single amino acid (a valine for a glutamic acid). At one point, recalled Crick, Ingram reckoned two different amino acids were changed, but Crick and Watson demurred, for that was the outcome predicted of an overlapping genetic code, and they had rejected such codes. "Jim and I were brasher then," wrote Crick, "and refused to believe this. 'Try it again, Vernon,' we said, 'you'll find there's just a single change' and so it turned out to be."[10]

The impact of this demonstration was considerable. Hitherto even Pauling had continued to maintain that the gene by its shape determines—like a mold—the manner in which the polypeptide chain of the newly formed protein is folded. Now it was clear that simply by suffering an alteration to the *sequence* of amino acids in one of hemoglobin's chains, the molecule must have folded up differently, impairing its function. That alteration had nothing to do with the gene acting as a three-dimensional mold as Boris Ephrussi and before him George Beadle had suggested.[11] The three-dimensional shape of the hemoglobin molecule according to Crick is determined by the *one-dimensional* sequence of its amino acids. For Crick and Ingram to have reached this point with their study of lysozyme would very likely have taken several years. They had judged well to drop it, for hemoglobin gave them what they were seeking.

### The Question of Strategy

As we have seen in Chapter 12, several approaches to solving the code were explored post-1953. Watson had chosen to seek clues from the structure of RNA, but unlike DNA, RNA had proved unyielding of its structure. George Gamow and other members of the RNA Tie Club continued to explore combinatorial solutions but without success. Both approaches can be described as *indirect*. Adopting a *direct* approach, however, would mean studying protein synthesis. More attractive to Crick was the possibility to use genetics and, where possible, some chemistry. Applying Frederick Sanger's sequence analysis for proteins, one could hope to establish the change in the amino acids of a protein caused by a given mutation. If one also knew the kind of change produced by the agent causing the mutation (the mutagen), the results could at least be suggestive of the base sequences involved. That approach called for a collaborator already versed in the art of mutagenesis (creating mutants) and well versed in phage genetics—the bacterial viruses that were at the cutting edge of the subject.

The ideal person to fill that role was the South African phage geneticist, Sydney Brenner, who had met Crick briefly in 1953. In the summer of 1954 they became better acquainted, but it was not until 1957 that resources were available to bring Brenner to Cambridge, and he was free to come. In the interval Crick had been keeping Brenner abreast of developments in Cambridge with his detailed and enthusiastic letters.

However, with Brenner still far away in South Africa, Crick's thoughts turned to a direct approach through biochemistry, but using a shortcut that would not require the actual synthesis of a protein. It concerned the adaptor molecules he had invented in 1955, and the Harvard biochemist Mahlon Hoagland had independently discovered nearly 2 years later.[12]

Hoagland had the special skills of the biochemists at Massachusetts General Hospital Laboratory, the "Huntington," where the "cell-free system" was developed. This is a system composed of extracts from living cells in which attempts were being made both to simulate what the cell does so well—synthesizing proteins—and to unravel the machinery involved. It was the "wet" nuts-and-bolts approach that in the end provided the route to breaking the genetic code. It was time-consuming and often exasperating, and biochemical skill was de rigueur. Hoagland had discovered in the soluble RNA fraction of the cell a kind of RNA that binds to amino acids and brings them to the microsomal particles (of the kind later called "ribosomes") in which proteins are synthesized. Little did he know then that Crick had previously predicted the existence of "adaptor" molecules that should do just that (see Chapter 12). These adaptors were initially called sRNA, referring to their source in the soluble fraction of the RNA in the cytoplasm of the cell. Subsequently they were renamed "transfer RNA" or tRNA because they transfer the amino acids to their appropriate place in the nascent polypeptide chain.

Hoagland was enlightened by a December 1956 visit from Watson in Boston. He was "bowled over" and also

> a bit deflated and miffed at having the theoretical framework for our discovery foisted on us by an outsider—indeed, by a molecular biologist—after we had revealed transfer RNA and correctly interpreted its significance! A vision rose before me: we explorers sweating and slashing our way through a dense jungle, finally rewarded by the discovery of a beautiful long-lost temple—and looking up to see Francis, circling above our goal on gossamer wings of theory, gleefully pointing it out to us.[13]

Crick was for a while reluctant to accept that Hoagland's RNA molecules could be the adaptor molecules, because they were so large. He had been thinking of molecules as small as trinucleotides (three bases, three phosphorylated sugars), just sufficient to hydrogen-bond to the complementary triplet of bases that he envisaged should represent the genetic "message" for a specific amino acid (see Chapter 13). Moreover, small molecules could move through the cell rapidly.

In January 1957, a very excited Crick had visited Hoagland, and in their correspondence thereafter Crick invoked a mechanism that would break up the RNA molecules Hoagland had discovered, yielding the trinucleotide fragments he had expected to be the true adaptors. Hoagland countered by suggesting that one of his RNA molecules is large because it may well code for several amino acids,[14] an alternative that Crick had already rejected, because he reckoned such lumbering molecules would, like overweight postmen, transport their charges too slowly.

Suppose, however, that each adaptor, although large, is specific to just one amino acid, as Crick had originally suggested for his trinucleotides. It would follow that such adaptors could offer Hoagland and Crick a "chemical handle" with which to crack the genetic code. Here was a possible *direct* approach, for each kind of adaptor RNA molecule, whether large or small, has to "post" its specific mail (an amino acid) to the correct "mailbox" (the end) of the growing polypeptide chain. No wonder Crick had invited Hoagland to spend a year in Cambridge. Suppose they could separate out the 20 hypothesized tRNA adaptors, discover which amino acid is attached to which adaptor molecule, and then locate the bases on these adaptors that attach to the template RNA (later called the anticodons). This would be no easy task given the limited analytical tools then available. In particular, there was as yet no technique available for establishing the base sequence of a nucleic acid. The prize of success, however, would be the confirmation that the sequences of the bases in DNA do determine the sequences of the amino acids in the proteins; Ephrussi's challenge would have been met and the *rule book* for the translation of the former into the latter would be revealed. They would discover the genetic code! Unfortunately, biochemical skill would be needed, and Crick had had enough trouble already crystallizing proteins in his doctoral research to risk attempting this kind of work without a skilled collaborator. Who better to invite than the one who had discovered the adaptor molecules (tRNA)?

This project required laboratory space and facilities of the kind that a physics laboratory like the Cavendish could not provide. Crick set about searching among the nooks and crannies of the University's jumble of science buildings packed so tightly into the old town center. Being a well-known face at the Molteno Institute from his doctoral years, he persuaded the director, David Keilin, to let him use a small room upstairs that was vacant at the time. A lab was set up there for Hoagland's arrival in September. Then the difficult work began of establishing the cell-free system in which the 20 hypothesized adaptor molecules were to be sought. Fortunately a technician with 5 years' experience working on the extraction of RNA at the Strangeways Laboratory, Muriel Wigby, had recently been appointed to the Unit. With her help and the supply of rat liver extracts from her former boss at the Strangeways Laboratory, K.B. Jacobson, they established that there are different adaptor molecules. However, they were unsuccessful in separating them effectively. Crick later commented that they probably had them, but his colleague John Smith, the well-respected nucleic acid chemist in the group, must have run the adaptors off the chromatogram. Ruefully, Crick remarked, they should have repeated the experiment until they found them.

Recalling his collaboration with Crick on adaptor molecules, Hoagland described how

> Francis still wanted to get his hands dirty, to handle some tRNA and to taste experimentation, beginning modestly with isolating tRNA from rat liver and making some preliminary efforts to separate it into different amino acid-binding species. We set up a lab in the nearby Molteno Institute with the help of my friend and colleague John Littlefield, who was then visiting in Cambridge, too, and we started to work. We would do an experiment and get some variation in results that Francis felt obliged to analyze and ponder at length. I would assure him that the variations were very likely an error—we would not find them if we repeated the experiment. Or, we would try some

new technique for separating different tRNAs, and things would go wrong. I was used to all that, but it soon became apparent that the great theorist was getting bogged down in trivia. I felt he was feeling a wistful nostalgia for juggling big ideas, not test tubes. Watching Francis in this context was fascinating. He had an uncanny ability to analyze and criticize, in *detail*, the experiments of others, but at the bench he became mired in the day-to-day messiness and inconclusiveness. The raw face of biochemical experimentation was not right for his mountaineering spirit. The morning I arrived to find him on his knees on the floor trying to find a white rat that had turned gray with dust while eluding efforts to catch it was the moment we both acknowledged that his experimental days were over."[15]

(Fig. 14.1). In fact it was not to prove the end of Crick's days as an experimentalist, as we shall see. Hoagland returned to the Huntington Laboratory nearly empty-handed in the summer of 1958, although "with mixed feelings." He had warm memories of "hobnobbing with the nobility in the court of molecular biology at the peak of its greatest glory." He recalled "spending many pleasant evenings with Francis and Odile and their friends at their delightful home at Portugal Place, with its golden helix over the front door, eating well, talking, and singing songs" supported by Odile on her accordion. That was the year when Sanger received his first Nobel Prize for his sequencing of the protein insulin, and a party was held at Portugal Place. "At the height of the festivities, someone set a rocket off from the roof of the house. Its ill-planned trajectory caused it to lodge, still smoldering, in the tower of the nearby church." The fire brigade had to be

*Figure 14.1* Crick at the Molteno Institue chasing a rat and Mahlon Hoagland looking on.

called.[16] We can sense the lively state of the scientific activity, too, from Crick's report to Watson in November, 1957:

> Everybody here is very busy. Seymour [Benzer] is studying a model of DNA. Sydney [Brenner] and George [Streisinger] are hard at work on phage tails. Leslie Barnett and Alice Orgel are learning phage techniques. Mahlon is working full steam with Muriel in the Molteno. Tomorrow John [Kendrew] talks about his 3-dimensional Fourier of myoglobin. Vernon [Ingram] has just left for the States. The bacterial flagella look very promising."[17]

### Collaboration with Sydney Brenner

The success in identifying the amino acid alteration with sickle-cell hemoglobin in 1956 was a landmark toward establishing the relation between genetics and biochemistry, and it ruled out Ephrussi's suggestion that the sequence of a protein is not determined by the chromosomal genes, but by genetic determinants in the cytoplasm. More importantly, it confirmed Crick's hunch that a mutation sufficient to cause a serious pathology may be due to an alteration in only one amino acid on a polypeptide chain. This hunch was also supported by the data from the plant virologists' research into the action of mutagens. All the mutants they produced by the action of nitrous acid and by hydroxylamine involved a single amino acid alteration, leaving their neighbors unchanged. But the dictionary for translating genetic sequences of the bases into amino acid sequences of the proteins had yet to be established. Turning to the mutagenesis of phage, however, Crick and Brenner reckoned might show the way.

Brenner is the son of poor immigrants from Eastern Europe, who had settled in Johannesburg where his father set up as a shoemaker. Advancing rapidly through school, Brenner matriculated just before his 15th birthday. Then supported by a bursary from his hometown, he entered the University of the Witwatersrand as a medical student. At 21, with all of his medical courses completed, he was too young to qualify for medical practice. While marking time, he studied in the Department of Anatomy and began research in histology and cytology leading to his M.Sc. in cytogenetics.

Deciding he wanted to go abroad, Brenner applied to Cambridge University's Professor of Biochemistry, A.C. Chibnall, and also to Chibnall's colleague, Joseph Needham. "My main interests," Brenner explained, "lie in the borderline fields between genetics, biochemistry and cytology." As his research interest he wrote "problems related to the specificity of protein synthesis."[18] When neither replied, he turned to Oxford where the physical chemist Cyril Hinshelwood accepted him. There he began his research in phage genetics, and with others visited Crick and Watson in Cambridge for a day in the spring of 1953 to see their DNA model (see Chapter 10). That meeting had been brief, and it was not until the two men got together at Cold Spring Harbor and more especially at the Marine Biological Laboratory at Woods Hole, Massachusetts, that they became personal friends. Before his return to South Africa, Brenner stayed with the Crick family at the end of 1954. There was one more visit, he explained, "because we were talking the same language." The return to South Africa was a condition of his scholarship from the Royal Commission for the Exhibition of 1851. Although sought by several American groups, he had to refuse their job offers. But he and Crick had found out how much they liked to be together. On the eve of his departure over Christmas 1954, he wrote to the Crick family: "I hope to see you all soon, soon; but now am furiously concentrating on leaving this sceptred isle. It is gray here in Southampton; I feel rather gray myself. It is all slipping away and I can do nothing about it. Au revoir Goodbye. Sydney."[19] The "au revoir" was for Crick's daughter Gabrielle, who had become bilingual.

Thus was the seed sown for the remarkable collaboration between these two in Cambridge that was to last from 1957 until Crick left for a sabbatical at the Salk Institute in 1976 (see Chapter 18). Throughout that period, Brenner and Crick were to discuss ideas and plans for experiments day by day. They rarely collaborated in the sense of doing experiments together at the same time; mostly they would be working on different although related topics, but they discussed their work. They would "say anything without having to justify it up to the hilt." Problems were clarified in this way, because Crick insisted on their formulation being clear. They would interrupt each other to continue in dialogue or in duologue, the ideas tumbling helter-skelter to be met by relentless questioning. The idea was no good after all—throw it out! And Crick was "as ruthless with himself as he was with others."[20] There they would be at their whiteboard in the office they shared writing furiously, rubbing out and starting again. They were the opposite of those people, Brenner remarked, "who will not say anything until they've got it all worked out. I think such people are missing the most important thrill about research—the social interaction, the companionship that comes from two people's minds playing on each other. And I think that's the most important thing. To say it, even if it's completely stupid!"[21] Mark Bretscher supported this view, explaining that the reason why Crick and Brenner

made such a wonderful pair, was that Sydney would pour out all sorts of rubbish, all sorts of crazy ideas. It was almost as though he was on LSD or something. But he wasn't. They were all bits of crazy, making crazy connections, and Francis would laugh and say: "Come on Sydney, give me a break," or things like that. Francis would listen to them and sometimes he would see that there was actually an interesting thought in there somewhere. So Francis acted as the sift to filter out from the noise, thoughts that would jumble out of Sydney's head.[22]

"Sydney and I," wrote Crick, "had discussions almost every working day—using several large white boards—but he also spent long hours in the lab and considerable time reading the literature. He was much better than I at thinking up novel experiments. My role was more that of a critic and clarifier."[23] Crick described his memory as "rather fallible."[24] But Brenner has an amazing memory—he is a traveling encyclopedia on a cornucopia of subjects from medieval history to paleontology and computer science. Two areas of expertise of relevance to his future career were histochemistry, in which he learned about the acridine dyes during his Johannesburg years, and, more importantly, the genetics of phage, the subject of his doctoral research in Oxford. This work he continued on his return to South Africa. He also liked nothing better than to work furiously at the bench, whereas for Crick such work did not normally prove enjoyable. As his interest in the phage program grew, however, so did his involvement, and when Brenner was away in the summer of 1961, it became full-time.

Brenner loved to initiate new projects, but when it came to writing up the work his enthusiasm would depart. "I really resent writing," he remarked, "and in fact had Francis not locked me up in a room during our career together I wouldn't have written as many papers as I did."[25] This explains why the famous *Nature* paper on the $r_{II}$ mutants that appeared in December 1961 bears the unmistakable marks of Crick's drafting. Thus these two scientists functioned admirably because they complemented each other in so many ways. Moreover, they shared wit, a great sense of humor, and a healthy iconoclasm about church and state. Crick recalled that: "Collaborating with Sydney not only made all the difference to my ideas and my few experiments, but it was all such fun. It says much for his tolerance and good temper that there was never an angry word between us. Happy days!"[26]

By the summer of 1955, Crick, feeling the time for Brenner's move to Cambridge was getting very ripe, began pressing the issue. At first, the funds were not available, but the financial situation changed when Crick's colleague Hugh Huxley left the MRC Unit, having had an affair with Kendrew's wife Elizabeth. Also at this time, the Rockefeller Foundation made a handsome grant to the lab. At year's end, a firm offer to start in 1957 was on its way to Johannesburg. Brenner accepted. All through 1956, Crick kept him informed of events in Cambridge. By August he was writing excitedly to tell Brenner about the progress of Ingram's work with sickle-cell hemoglobin. Assuming that work goes well, he added, "we should then, I think, press on to phage. This means obtaining the tail protein pure and I think we should start on this, as a long term programme, fairly soon."[27] The idea was to establish the colinearity of the amino acid sequence of the tail protein with the corresponding genetic map.

The strategy to be followed was like that for lysozyme, but the idea had occurred to Brenner in 1954 when he first met the phage geneticist Seymour Benzer. Benzer, originally a physicist, had turned to genetics under the influence of Max Delbrück and the Phage Group. He had pioneered the fine-structure analysis of the genetic map of phages, demonstrating the one-dimensional nature of the map and its divisibility down to distances close to that of a single base in DNA. In the region of the genetic map of phage T4 known as $r_{II}$ he was able to locate two distinct regions that he called the A cistron and B cistron.[28] In each of them he found a series of mutations that have the effect of altering the rate of destruction of the host bacteria (Fig. 14.2). This change affected the character of the plaques, or clear spots, formed by the phage on the bacterial lawn. Wild-type plaques ($r^+$) are small with a clear center but are surrounded by a large halo. Easily distinguished from these are the *rapid lysis* ($r$) mutant colonies with their large clear centers and their sharp edges. A specific inherited biochemical alteration in these minute phage particles was yielding an effect visible to the naked eye. "I was carrying around a whole lot of amino acid sequences," recalled Brenner, "and he [Benzer] was carrying these four or five mutants mapped in a single line. I immediately saw from what he had that the classic theory of the gene could now be broken," meaning that the gene could be dissected into many parts each corresponding to the amino acids of the protein that it determined. As Benzer recalled: "The future seemed to us quite straightforward—isolate the $r_{II}$ protein from various mutants, then establish the colinearity of alterations in amino acid sequence with the locations of the mutations in the genetic map. If we could somehow identify the DNA bases, we could solve the genetic code."[29]

The plan was for Benzer, at the time senior research fellow of the National Science Foundation based at Purdue University, to come to Cambridge for a year to work on the problem with Brenner.[30] Arriving in the fall of 1957, he brought his postdoctoral student Sol Champe with him. A month later another American phage geneticist, George Streisinger from Cold Spring Harbor Laboratory, came over to join

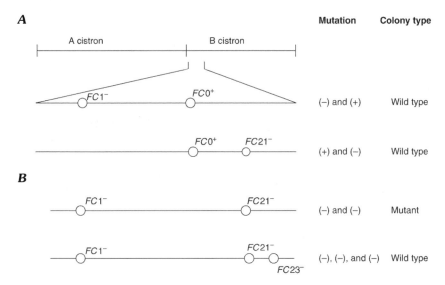

*Figure 14.2* Schematic representation in descending order of the increasing levels of genetic analysis. (*A*) Genetic region of the $r_{II}$ region of phage T4, as determined by *Eschericha coli* strain B as host. (*B*) Single, double, and triple frameshift mutants used by Leslie Barnett and Crick to recreate wild-type offspring in 1961.

them. He wanted to work on the tail proteins of phage. Benzer's arrival in Cambridge brought the number of scientists sharing Crick and Brenner's office to seven. Benzer recalled what it was like:

> It was great, because Crick and Brenner were arguing all the time. We would read Crick's mail that would be sitting on his desk. It was a tremendously stimulating environment. We could always tell when he was in the building; you could hear his "Ha,ha,ha,ha" down the hall. He was tremendously brilliant and stimulating—his mind was working all the time. My wife was very impressed with him. She said, "Closest thing to pure intellect." But that didn't preclude other things. He was into all kinds of adventures on the personal level. He was very urbane, well educated. He reminded me most of Henry Higgins in *My Fair Lady*. There was quite a resemblance. That kind of style. No shrinking violet, that's for sure.[31]

Then there was the typical day at the lab that drove Benzer "a bit wild":

> You'd come in, in the morning, about 10:30 or so, and there would be coffee time. Then there was a little time to work; then it was lunchtime. Then you'd come back after lunch and before you knew it,

it was tea time. And then you were home. Almost nobody would come back in the evening, except the crazy Americans, who were driven types. So, to get back into the Cavendish Lab in the evening, I would have to ring the bell for the concierge to allow me through the gate.[32]

What Benzer did not mention was that he liked to sleep late and work late, a habit not suited to the traditional working day. In view of the cramped conditions, Crick had persuaded the zoology professor Carl Pantin to let him have the use of the former Zoology Museum in the Anatomy Building, a long room without plumbing. There the phage experiments were performed where once a whale had been displayed. The dishwashing facility was accommodated in the office shared by Brenner and Crick, together with seven desks, five of them for the American visitors.

At this time the genetics of bacterial phages was very much at the cutting edge of the subject because of the simplicity of the experimental setup, the minimal requirements for space, and the rapidity of the reproduction of the phage particle. It was especially the combination of miniaturization and large population size ($10^8$ phage particles in a single Petri dish) that enabled molecular distances to be detected from events visible to the naked eye on a little dish! Phage genetics, wrote Crick,

has the advantage that experiments are rather fast, once everything is set up. It does not take long to carry out a hundred crosses, since the manipulations are easy and an actual cross takes only about twenty minutes, this being the time for the phage to infect the bacterium, to multiply inside it (exchanging genetic material in the process), and to burst open, thus killing the cell. The results of the cross must then be plated out on petri dishes, to which a thin film of bacteria has been added. Then the dishes have to be incubated, to produce a lawn of bacteria. Where a single phage has landed and infected a cell, a colony of phage will grow, killing the local bacteria as it does so, forming a clear little hole (called a plaque) in the lawn of growing bacteria on the surface of the plate. This process takes a few hours, so one has a brief respite while it is going on. The petri dishes have to be taken from the 37° incubator and examined to see whether they have plaques or not and, if so, of what type. Interesting plaques are then "picked"—that is, a few phage are picked up with a little piece of paper or a toothpick; grown further and the process repeated a second time to make sure the phage stock is a pure one.[33]

Whereas the genetics of higher organisms was researched by cross-breeding, the genetics of phage depended on the use of cross infections. For both, mutagenic compounds were being used to enrich the supply of hereditary variations. The degree of genetic recombination in mixed infections of phage then provided the data for a map of the genes on the phage's chromosome, reaching even down to molecular dimensions.[34]

## Two Classes of Mutation

At Purdue University, Benzer and his postdoc Ernst Freese had experimented with two types of chemical mutagens—base analogs and acridine dyes. The former created mutants by altering bases in the DNA and caused reversions to wild type by reversing the base change. Both events—the *substitutions* and the *reversions*—happened at the same locus on the genetic map of the chromosome. The acridine dye proflavin likewise both caused mutations and reverted them. Yet neither group of mutagens could restore function to mutants produced by the other group of mutagens. Freese attributed this outcome to different chemical changes in the bases of the DNA involved. One kind he called *transitions*—purine to purine, pyrimidine to pyrimidine, which are chemically easy; the other he called *transversions*—purine to pyrimidine and the reverse, which are chemically difficult. A transition-type mutagen (5-bromouracil or *BA*) could not correct the work of a transversion-type mutagen (proflavin or *PF*) and vice versa. Yet Freese judged both kinds of mutations involve substitutions that occur at the same locus.

When Benzer and Champe arrived in Cambridge, Brenner and Leslie Barnett joined with them in exploring the distribution along the genetic map of these two kinds of mutation. Crick followed their progress with much interest. Of these two mutagens they were surprised to find that one [*BA*] seemed "to act on a set of sites quite distinct from the other [*PF*]." This "absence of any coincidences" they considered "highly significant."[35] It made them skeptical of Freese's assumption that both types involved base substitutions. Moreover, there was another difference between the two groups of mutants revealed by growth on different hosts. Whereas the wild type grew happily on a strain of bacteria known as *K* and on one known as *B*, the *BU* mutants, although growing successfully on *K*, had "leaky" plaques on *B*, meaning they showed only some residual function. The *PF* mutants, on the other hand, did not grow on the *B* host at all. Why was this?

In the summer of 1958, the American visitors said farewell to Crick and Brenner and returned home. During the stay in Cambridge, Vernon Ingram had taught Benzer the fingerprinting technique that permitted identification of amino acids in small quantities of proteins, and Crick had taught him model building in the drafty tower room in the Cavendish. Streisinger was happy to have succeeded in identifying the lysozyme of phage as the most promising protein for his project to establish colinearity of gene and protein. Unfortunately, Benzer and Champe had no luck in locating the genes determining the head or the tail proteins. After their departure Brenner and Barnett continued the work on the head protein and reported some progress in 1959.[36]

Their Cambridge hosts, meanwhile, were as skeptical as ever about Freese's substitutions. Nor, apparently, were they all attracted to the suggestion that some mutations are caused by additions or subtractions of bases from the DNA chain, an idea put forward by Bruce Alberts and Jacques Fresco. On receiving their paper in June 1960, Crick replied that he did not find their "discussion of mutations very illuminating. What you say is plausible enough, but it doesn't solve any of our problems, such as 'How does proflavine act as a specific mutagen?' or 'What causes hot-spots?' "[37]

That year Crick and Leslie Orgel had been speculating on an idea they called "loopy codes." The RNA containing the code, they thought, might "fold back on itself to form a loose double helical structure":

> The idea was that some bases could pair whereas others, which did not match according to the pairing rules, would loop out. The "code" would then depend either on the looped-out bases or the paired ones, or some elaborate combination of these two possibilities. The idea was really rather vague, but it made one important prediction. A mutant at one end of the message might, in theory, be capable of being compensated in its effect by another, toward the other end, that paired with it [Fig. 14.3]. Thus some mutants should have distant "suppressors", as they are called within the same gene.[38]

"I got so excited about this," said Crick, "that I thought: 'I'll do the experimental work myself,' and [in February 1961] I started to work on the $r_{II}$ gene."[39] Here was the ideal experimental system: no tiresome failures to crystallize a compound, no long wait in the night for the product to sediment, and collaborative work. Leslie Barnett would instruct him. For a while he would be sharing bench space with Alice Orgel (the wife of his friend Leslie Orgel), who was working on phage for her doctorate. But his attempt to support his looping-out idea failed. According to that model, many of the suppressors should be at a fair distance from the original mutation and distributed in a symmetrical manner, but they were not. His looping-out idea was knocked on the head.

Meanwhile Brenner, long familiar with acridine dyes, was pondering their mutagenic action. Debating the issue with Crick in The Eagle pub one day, he suddenly realized that you could get deletions and additions "if you stuck proflavine into the DNA—if it went between the

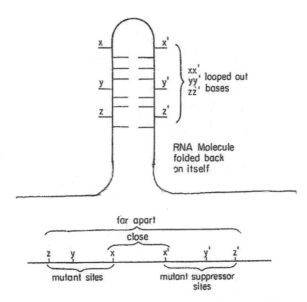

**Figure 14.3** Loopy code scheme for the suppressors of mutants in bacteriophage T4.

base pairs. . . . It was called intercalation." Leonard Lerman, Senior Research Fellow from the University of Colorado Medical Center in Denver, produced evidence for such events while working at the MRC Unit in 1959–1960.[40] So Brenner speculated that during replication the cell could insert an extra base on the other strand that would match the inserted acridine dye on the "mother" strand. Or it could make a compensating deletion, thus either lengthening or shortening the chains of the double helix and causing a mutation or suppressing an existing mutation.

In December 1960, Brenner and colleagues felt confident enough to send a preliminary note to the *Jounal of Molecular Biology,* expressing their "doubts" about the detailed theory of mutagenesis put forward by Freese (in 1959) and to suggest an alternative. "Under this new theory," they explained, "an alteration of a base-pair at one place *could* be reversed by an alteration at a different base-pair, and indeed from what we know (or guess) of the structure of proteins and the dependence of structure on amino acid sequence, we should be surprised if this did not occur."[41]

Working with Benzer in Cambridge in 1957–1958, the group had reckoned there must be at least 150 proflavin mutations scattered within the B section of the $r_{II}$ region of the phage chromosome, of which they had located 58. Further work raised this estimate to more than

200. Crick became very interested at this point. Could he find mutations located closer to the original mutation than he had been searching? These might restore the correct reading of the base sequence—assuming, of course, that the cell has the means to identify where reading of the relevant gene sequence should start? Earlier Crick had judged that "obvious idea . . . too simple, and I thought (quite wrongly) that there must be another solution."[42] Now, it seemed, that might be just what the cell does.

Crick started with mutants of the strain T4B isolated by Brenner, Barnett, and Benzer. He discovered that his initial mutant

> had not one but several, distinct suppressors, all of which mapped fairly close to the original mutant. I decided that I would have to call them all by a distinctive name. I often worked through the weekend, taking Monday off so that our laboratory kitchen (which did all the washing up and also prepared petri dishes for our use) could catch up. It happened that it was a weekend when I needed a new name, and nobody else was around. Mutants were usually called by a letter, followed by a number. Thus P31 meant the thirty-first mutant in the P series, probably produced by proflavin. Unfortunately I could not remember for certain which letters had already been used, so I decided to rename my mutant *FC*0, since I was quite sure no one had used my initials to name mutants. The new suppressives [*sic*] were then named *FC*1, *FC*2, and so on. This use of my initials suggested to some people that I must be conceited, but the real explanation was that I have a rather fallible memory.

> The new suppressors all seemed like good, nonleaky mutants. So why not, I argued, see if they too had suppressors? And indeed they had. I even went a step further and found suppressors of suppressors of suppressors.[43]

These results could be explained in terms of a string of bases that are read triplet by triplet from a starting point on the map, and assuming a nonoverlapping code. To establish this, Crick had first to classify his mutants based on their functional relation to *FC*0. So he assumed that the mutant *FC*0 has added a base. Therefore he classified it as plus. Any mutant that suppresses *FC*0 would therefore be a minus (Fig. 14.2B). This notation was arbitrary and future research could reveal *FC*0 to be minus and its suppressors plus. No matter, at this point, all that was needed was consistency of classification. Crick explained:

> . . . this addition of a base at the *FC*0 site will mean that the reading of all the triplets to the right of *FC*0 will be shifted along one base, and will therefore be incorrect. Thus the amino-acid of the protein which the *B* cistron is presumed to produce will be completely altered from

that point onwards. This explains why the function of the gene is lacking. To simplify the explanation, we now postulate that a suppressor of *FC*0 (for example *FC*1) is formed by deleting a base. Thus when the *FC*1 mutation is present by itself, triplets to the right of *FC*1 will be read incorrectly and thus the function will be absent. However, when both mutations are present in the same piece of DNA, as in the pseudo-wild double mutant *FC* (0 + 1), then although the reading of triplets between *FC*0 and *FC*1 will be altered, the original reading will be restored to the rest of the gene. This could explain why such double mutants do not always have a true wild phenotype but are often pseudo-wild, since on our theory a small length of their amino-acid sequence is different from that of the wild-type.[44]

We can represent these events by a series of letters for the bases, but confining ourselves to three out of the four different bases for simplicity and using for them the letters A, B, and C. Then we have the situation shown in Figure 14.2C. Such alterations to the genetic sequence became known as "phase-shift" or "frameshift" mutations, thus distinguishing them from base "substitutions." Note that, as the figure shows, the researchers did not know whether the deletion or the addition came first in the sequence. All they knew was that one of the mutants was of the opposite sign to the other. In one direction the initial frameshift would be one place to the right, in the other direction, one place to the left.

Crick had begun testing this theory in May 1961 by classifying the suppressors he had discovered. On the 8th of May he wrote: "Decided to see if one r [$r_{II}$ mutant] had many *different* suppressors."[45] He called those restoring function to *FC*0 minus and those suppressing the minus suppressors were called plus. But there was an interruption in the last week of June when he was due at the annual reunion of the Société de Chimie Physique devoted to DNA. Who would have turned down so inviting an invitation as this, located at an isolated hotel at the Col de Voza, elevation 6000 feet, "near the limit of the Alpine forest, and dominated by the ice of the Dôme du Gouter and of the Aiguille de Bionnassay"? Crick accepted and before leaving with Odile for France he dictated a hurried note to Delbrück telling him:

> I have been hard at work doing phage genetics; in particular this question of suppressors. I believe <u>Dick Feynman</u> has done something similar, but I have only heard rumours. NB! I would be most interested to know what he found, and whether he is going to publish. We have an ingenious theory for our results which, if true, would be very important for decoding, but it needs much more work to establish it. If only we had a protein![46]

The conference proved memorable. Recalling it later, the President of the society, Raymond Latarjet, waxed eloquent. He noted the "atmosphere of critical and constructive enthusiasm which pervaded those days." On the third day there was an excursion. "That morning, after two days of bad weather, the clouds vanished. At the Aiguille du Midi and at the Col du Géant, the high mountains covered with fresh snow offered us their most beautiful visage. The wind had fallen; the sun gently warmed the rocks. For a moment, DNA itself was forgotten."[47]

Crick was not giving a talk, but when Streisinger's presentation on the lysozyme of phage T4 ended, Crick rose excitedly to add a stunning footnote. Supporting Streisinger's remarks about the distinctive nature of acridine mutants, he went on to outline the chief results of the Cambridge work, and then added:

All these results are compatible with our suggestion that the action of acridines is to add or delete bases. An attractive additional hypothesis is that the code is read in short groups, starting from one end of the gene. The exact starting point is supposed to determine which group is read. The deletion of a base would then alter the active reading from this point onward. The double mutants produced by the reversion of acridine mutants would then, on this hypothesis be altered not just in two, separated, amino acids, but in a short stretch of amino acids in sequence. If this were so it would be very important for decoding.[48]

### The Fifth International Congress of Biochemistry

After the meeting, the Crick family were booked for a 5-week holiday in Tangier, Morocco. There in a large villa with two Arab servants to wait on them, they relaxed. While Gabrielle and Jacqueline made for the beach, Odile and their German au pair girl, Eleanora, headed for the Arab market. Crick, meanwhile "spent the day on the terrace, in the dappled shade of the palm trees."[49] Thus restored, Crick's next commitment was the International Congress of Biochemistry in Moscow in the second week of August.

Not having been in the Russian capital since February 1945, Crick was keen to see it again. This time it was summer and the drabness of wartime was gone. For Crick the highlight of the meeting came when Matthew Meselson drew his attention to a talk that had been given in a remote seminar room by a biochemist from the National Institutes of Health, Marshall Nirenberg by name. Next Crick recalled "running into Nirenberg and having heard that he had *done* it [taken the first step in breaking the genetic code]." When they talked Crick recalled:

I was so impressed that I asked Marshall to take part in a much larger meeting, of which I was chairman. What he had discovered was that he could add an artificial message to a test-tube system that synthesized proteins and get it to direct some synthesis. In detail he had added polyU—the RNA message consisting entirely of a sequence of uracils—to the system and it had synthesized polyphenylalanine [phe]. This suggested that UUU (assuming a triplet code) was a codon for phenylalanine (one of the "magic twenty" amino acids).[50]

Nirenberg and his German colleague Heinrich Matthaei had thus brought about the first synthesis of a polypeptide using the biochemists' cell-free system and had broken the first letter of the genetic code.

Returning to Tangier, Crick brought his family back to Cambridge, where he soon plunged into more phage experiments. The task now was to test the predictions from the theory, namely all (+ +) and (− −) combinations should be mutants, and all (+ −) and (− +) ones should be wild type or near wild type. Annoyingly, not all of the latter two obeyed expectation, and their number was too large to be ignored. In the face of their beautiful theory, the moment had surely come to apply the "don't worry hypothesis." Brenner explained that under such conditions "the theory should remain. And it was wise of us to take all these exceptions, which showed no relationship amongst each other, and put them on one side. We didn't conceal them, we put them in an Appendix." This appeared in their 74-page paper of 1967 in the Royal Society's *Philosophical Transactions.* There they introduced their concept of "boundaries" between mutants that accounted for many of the anomalies.[51] But in 1961 they merely advanced the simplest suggestions: "the shift of the reading frame produces some triplets the reading of which is 'unacceptable'; for example they may be 'nonsense,' or stand for 'end the chain,' or be unacceptable in some other way because of the complications of protein structure."[52] They also pointed out that the message from the gene sequence would differ depending on whether the frameshift was to the left or to the right. Both might spell "nonsense," but their locations on the genetic map would be different.

The next and most crucial step was to manufacture triple mutants of the form (+ + +) or (− − −). This was a laborious task, but Barnett was at hand to assist. Meanwhile Brenner left for a spell in Paris, but not before leaving a list of suggested mutants to use. First they constructed one (+ + +) triple mutant. Crick recalled how, after placing the Petri dishes in the incubator, they left the lab. Returning after dinner to inspect them,

one glance at the crucial plate was sufficient. There were plaques in it! The triple mutant was showing the wild type behavior (phenotype).

Carefully we double-checked the numbers on the petri dishes to make sure we had looked at the correct plate. Everything was in order. I looked across at Leslie. "Do you realize," I said, "that you and I are the only people in the world who *know* it's a triplet code?"[53]

Crick went on to stress how remarkable was this result:

Here we had three distinct mutants, any one of which knocked out the function of the gene. From them we could construct the three possible double mutants. Each one of these made the gene nonfunctional. Yet if we put all three together in the same gene (and we did separate experiments to show that they had to be in the same virus, not some in one and the rest in another separate virus), then the gene started to function again. This was easy to understand if the mutants were indeed additions or deletions and if the code was a triplet one. In short, we had provided the first convincing evidence that the code was a triplet code.[54]

This event must have occurred around the end of September, for on the 9th of October Crick reported: "We now have convincing genetic evidence that the coding ratio is 3 or a multiple of 3. It is not clear that we can prove that it is 3 rather than 6, but we are trying."[55]

Barnett and Crick then constructed more (+ + +) triples and one (– – –) triple and found them all to be wild or pseudo-wild. Although they were confident about their conclusion, Crick was careful to express it in a rigorous form. The identifications + and – he warned "must not be taken to mean literally the addition or subtraction of a single base." Instead he expressed it as:

+ represents + $m$, modulo $n$

– represents $-m$, modulo $n$

where $n$ (a positive integer) is the coding ratio . . . $m$ is an integral number of bases, positive or negative.[56]

Thus if each mutation were due to two additions or deletions, $n$ would be 6. Clearly the success with the (+ + +) triples did not rule out the possibility of six-member codons.

Suggestive evidence that the coding ratio is 3 came from certain acridine and hydrazine mutants falling "at or close to" the *FC*9 and the *FC*30 *sites.* Unless these mutagens were deleting or adding an *even* number of bases, these data supported the ratio of 3. The two mutants at *FC*30 were *X*146 and *X*225, both suppressors of *FC*30. Next to the data in the laboratory notes identifying the signs of these two Crick wrote "L1 and L2 had equal number of wild and small sharp plaques. These are almost certainly the double, thus *X*225 and *X*146 are negative *therefore coding ratio is 3.*"[57] (See Fig. 14.4.)

True to form Crick wasted no time in writing up this remarkable research. The style of the paper they sent to *Nature* in November is clearly his. Although the text starts with "*we this*" and "*our that*," at the end it is "*I was startled* by the announcement of Nirenberg. . . .*"

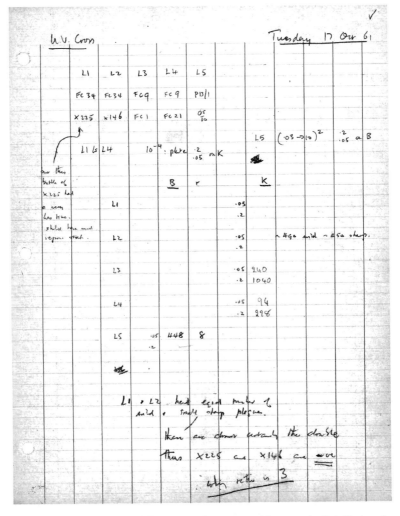

***Figure 14.4*** Laboratory record for 17 October 1961 with notes in Crick's hand concluding X225 and X146 are minus [mutations] and therefore the coding ratio is 3.

Furthermore it is organized in the way a theoretician would prefer. First come their conclusions on the nature of the genetic code:

> . . .it is of the following type:
>
> (a) A group of three bases (or, less likely, a multiple of three bases) codes for one amino acid.
>
> (b) The code is not of the overlapping type.
>
> (c) The sequence of the bases is read from a fixed starting point. This determines how the long sequences of bases are to be correctly read off as triplets. There are no special "commas" to show how to select the right triplets. If the starting point is displaced by one base, then the reading into triplets is displaced by one base, and thus becomes incorrect.
>
> (d) The code is probably "degenerate"; that is, in general, one particular amino acid can be coded by one of several triplets of bases.[58]

An account of the results of experiments undertaken follows, after which he wrote: "Our explanation of all these facts is based on the theory set out at the beginning of this article." Then he set out their predictions based on the theory—for 14 double like-with-like mutants and 28 unlike doubles—and found them correct. "We regard this as a striking confirmation of our theory," he declared.[59] Then he added a typical Crick remark: "It may be of interest that the theory was constructed before these particular experimental results were obtained." Another Crickism is the passage beginning "It would not surprise us if it were eventually shown that deletion 1589 produces a protein which. . . ."[60] The last sentence, too, is vintage Crick, full of optimism: "If the coding ratio is indeed 3, as our results suggest, and if the code is the same throughout Nature, then the genetic code may well be solved within a year."[61] Such enthusiastic optimism could only have come from the pen of Crick, who also had added a note about Nirenberg's and Matthaei's triumph. "This," he wrote, "implies that a sequence of uracils codes for phenylalanine, and our work suggests that it is probably a triplet of uracils."[62]

Crick added what he called additional "weak experimental evidence" for the triplet nature of the code in his *Scientific American* essay of October 1962. This was the group's study of phages containing a deletion of a stretch of the chromosome lying between genes *A* and *B,* called "1589," already discussed in the *Nature* paper. This deletion, it appeared, had the effect of combining the two genes so that they were read as one. By combining deletion 1589 with variable-length deletions in the *A* gene, they could alter the reading frame in the *B* gene. Crick explained: "Now, if deletions [in the *A* gene] were random in length, we should expect about a third of them to allow the *B* function to be expressed if the message is indeed read three bases at a time, since these deletions that had lost an exact multiple of three bases should allow the *B* gene to function. By the same reasoning only a sixth of them should work (when combined with 1589) if the reading proceeds six at a time. Actually we find that the *B* gene is active in a little more than a third."[63]

This is as near as they came in 1960 to establishing that the codon consists of only three bases. Asked if these experiments had been logically planned, Crick replied: "I don't think I could honestly call it 'logically planned.' I think I'd call it 'logically improvised.'"[64] The most remarkable feature of this work is that Crick and Brenner were able to make such significant claims about the nature of the genetic code in molecular terms without knowing anything about the assumed protein product of the genetic region they studied, let alone what was the nature of its amino acid sequence. Nor did they know which of their plus and minus mutations represented the addition or subtraction of bases. They even lacked the means to establish that each mutation involved the subtraction or addition of only one base, although they had come close to doing so.

By exploiting Brenner's frameshift idea, they had been able to use genetics to establish the "general nature" of the code. It was undoubtedly a very ingenious piece of research. But as Brenner remarked,

> . . .it was a real house of cards theory. You had to buy everything. You couldn't take one fact and let it stand by itself and say the rest could go. Everything was so interlocked. You had to buy the plus[es] and the minuses, you had to buy the barriers, you had to buy the triplet phase, and all these went together. It was the whole that explained it, and if you attacked any one part of it the whole thing fell apart. So it was an all or nothing theory. And it was very hard to communicate to people.[65]

Looking back, Brenner considered it "one of the most beautiful, aesthetically elegant experiences of my life, in which just by doing these little operations, you landed up with this detailed description of the molecular structure of living matter."

If this marked the end of Crick's experimental work at the bench, it was a glorious end, and he had so enjoyed carrying out the experiments himself. He found he could get through running two successive sets of experiments in one day providing he started in good time and came back after dinner for an evening session. He did not find this "unpleasant," he recalled, and Odile told him "she had never seen me so cheerful as during the period when I did experiments all the time." This was not the trying world of biochemistry and "for weeks on end, all the experiments seemed to work perfectly," and he did not have to kill a single rat!

Certainly he was keen to register the Cambridge lab's priority for the claims they would see published in *Nature* at year's end. That November he wrote to the President of the Col de Voza meeting and to Nirenberg:

> *To Latarjet (President)*: You may recollect that I reported our basic idea in the discussion at the Col de Voza conference, though not the bit about the coding ratio being 3. At the end of the conference I handed in a short written account. Could you tell me when and where it is likely to be published? It is the only simple means I have of establishing that we had the idea <u>before</u> Nirenberg's astonishing discovery that poly U codes for polyphenylalanine.[66]

> *To Nirenberg*: I enclose an account of our genetical work which we have submitted to *Nature*. We had the basic idea in the summer, before your epoch-making discovery, and reported it at the Col de Voza DNA meeting in June, but we only got the triples after I returned from Moscow. Your PNAS paper arrived here the day before we sent off our MSS, so I was able to add the reference. . . .[67]

On the last day of the year 1961, a Sunday, *The Sunday Times* and *The Observer* printed substantial accounts of the phage work of the MRC Unit. *Sunday Times* Science Correspondent, Tom Margerison wrote an article that appeared under the title "Scientists Have Cracked the Code of Life."[68] The *Observer's* science correspondent, John Davy, reported: "A major advance in unraveling the secrets of life appears imminent. A paper published in *Nature* yesterday, by four Cambridge scientists, suggests that the 'genetic code' is about to be cracked. . . . [They] believe they have established what type of code is involved. This, as all cryptographers know, is the most important step towards cracking any code. . . . This work is developing at such a pace that the layman cannot be expected to keep up."[69]

The reporter also drew attention to the discovery of Nirenberg and Matthaei. But Crick became concerned at the fuss made by the British press over their *Nature* paper. He wrote to Nirenberg reassuring him "that it is your discovery which was the real breakthrough."[70] Yet the publicity continued. Across the ocean *The New York Times* published a full-page article on protein synthesis in February with a double column on Crick, the "Hunter of Life's Secrets."[71] The issue of priority was still concerning him when *Scientific American* was to publish review articles by him[72] and by Nirenberg in that order. Crick stressed to the editor that "it should not appear to anyone that we wish to claim more than is our due. However ingenious and elegant our experiments are it must be realized that it is the biochemical work on the cell-free system which will be crucial."[73] Crick also explained that the basic discovery of Nirenberg and Matthaei came *before* Crick had the triple mutants in Cambridge that clinched their conclusions.

At this juncture Crick had become the focus of much attention. Housed in the hut outside the Cavendish Laboratory and awaiting the move to the new building by Addenbrooke's Hospital, reporters were surprised at his makeshift accommodation. The reporter from *The New York Times* was confronted by "An Owlish Sanctuary." Crick's corner looked "somewhat like an owl's aviary. Model helixes hang from the roof on strings," he wrote, "and scholarship is apparent only from books and untidiness. To intimates this unraveler of the coils of life is elegant in explanation and with his fondness for sideburns and Italian suits, somewhat Edwardian in manner."[74] How incongruous this description of Crick is in the setting of accommodation that looked more like a bicycle shed than the home of a prestigious research unit!

Behind the flamboyance, however, one found a dedicated scientist. Long ago Crick had judged that to make a significant contribution one had to be passionate about the work, and passionate he was. He demanded high standards. Authors of slipshod or erroneous work were called on to publicly renounce their claims or if awaiting publication to withdraw them or delay publication until they had more evidence. True, the Cambridge group's 1961 results did have some errors in mapping that they discovered on extending the research afterward. But they published their corrections.[75] Nor did they fail to pursue all the anomalous results that had been put on one side in 1961. They are described in 1967 in their 74-page paper in the Royal Society's *Proceedings*, together with exhaustive details—including the 27 pages of appendices giving the signs (+ and −) of 234 mutants. Anyone wishing to pursue the subject further would thus be able to do so. Brenner went on to conduct a number of ingenious experiments and Barnett took on the laborious task of making quadruple, quintuple, and even sextuple mutants, showing that only the multiple of 3 yielded the wild type. Crick and Barnett then "worked hard at tidying up all the loose ends." This led to evidence strongly suggesting the existence of a third chain terminator to polypeptides at the ribosome—namely, UGA (uracil, guanine, adenine).[76] Proudly they listed the predictions made in 1961 that were fully confirmed over the following 5 years: (1) the genetic message is read from a fixed point in nonoverlapping triplets; hence colinearity is achieved between gene and protein.[77] (2) The code is highly degenerate. (3) For their double mutants (+ with –) "a string of amino acids would be altered, corresponding to the region of the polypeptide chain between the two mutants."[78] As they checked through the final manuscript, Crick pondered its size and its destination in the august pages of the Royal Society's *Transactions*. They were "burying" it like he and Watson had buried their detailed paper on the structure of DNA in the Society's *Proceedings* in 1954. Crick commented:

> I told Sydney that I supposed that he and I would be the only people in the world who would ever read through it carefully. For fun we decided to add a fake reference, so at one point we put "Leonardo da Vinci (personal communication)" and submitted it to the Royal Society. One (unknown) referee passed it without comment, but we had a phone call from Bill Hayes, the other referee, who said, "Who's this young Italian working in your lab?" so reluctantly we had to take it out.[79]

Authorship of the 1961 and 1967 papers numbers five. Among the names, one deserves special note—Leslie Barnett. When she came to the MRC Unit in 1956 as a technician she began assisting the crystallographers with their computing, but Brenner's need for additional hands brought her to the phage work in 1957. John Finch described her as "one of the laboratory's repositories of phage folk lore which she taught to numerous visitors."[80] The American, Eric Miller, who spent 1987 in Brenner's laboratory, praised her "tough-nose" and her "spirited nature." He judged that

> many of the seminal discoveries in molecular biology would not have occurred without the careful hand, watchful eye and steadfast focus that Leslie brought to the bench and their (Crick and Brenner's) experiments. Although she was officially "retired," I saw first hand in 1987 what Leslie was capable of doing in the execution and development of Sydney's ideas. One could clearly envision her meticulous and spirited endeavors of thirty years earlier as the genetic code and the nature of mutation and mRNA were revealed.[81]

## Conclusion

Over the eight years that had passed since the discovery of the structure of DNA, the armchair code breakers and their critics—including Crick—had had their day. Crick's foray with Hoagland into a biochemical attack on the code had come to naught. Undismayed, he with Brenner, Barnett, and colleagues established the principal features of the type of code to which it belongs. Their ambition to "break" the code by matching a genetic sequence of bases with the amino acid sequence of the gene product, however, eluded them. Biochemists Nirenberg and Matthaei, meanwhile, used the direct approach through polypeptide synthesis in the cell-free system to be the first to break a letter of the code. This was Crick's last foray into experimental investigation at the bench, and he had enjoyed it.

## CHAPTER 15

# The Excitement of the Sixties

*We can now confidently look forward to placing increasing areas of biology on a molecular basis.*[1]

S uch was Crick's confident message to the Royal Society in 1966. Behind his confidence lay the remarkable progress with the genetic code since the findings of Marshall Nirenberg and Heinrich Matthaei in 1961. Another crucially important achievement was that of the French school at the Pasteur Institute in Paris. There the American visitor Arthur Pardee and the future French Nobel Laureates François Jacob and Jacques Monod performed their "PaJaMo" experiment. It concerned the manner in which the colon bacillus switches on and off the synthesis of the enzyme β-galactosidase for the digestion of lactose (milk sugar). To their surprise they found how very rapidly protein synthesis of this enzyme began once the appropriate gene had been introduced into the bacillus.[2] This was followed by the discovery by Pardee and Monica Riley of the reverse situation—destroying the gene led to a rapid decline of the synthesis.[3] Yet were not the ribosomes supposed to carry the RNA templates for a specific gene product—here the β-galactosidase enzyme? Naturally it would take some time before sufficient ribosomes *of that kind* could be produced and protein synthesis of that gene product detected. Equally, the decline of synthesis would follow the gradual decay of the existing supply of ribosomes; it would not end abruptly. In 1959 Jacob suggested that an unstable species of RNA—a "tape," or "X"—must be involved, but the full significance of this remark and these experiments did not come until a year later.

The occasion was a gathering in Sydney Brenner's rooms in Gibbs' House, King's College on Good Friday (15 April) 1960. Crick was quizzing Jacob about their PaJaMo experiment. Jacob described that work and the as yet unpublished work of Pardee and Riley, from which it became clear that either the gene had to be continuously active for protein synthesis to continue or there must be an unstable, and therefore difficult to detect, intermediate—an undetected species of RNA—present and supplying the required template. Crick recalled how

> . . .at some point Brenner and I both suddenly realized the following thing: [Elliot] Volkin and [Lazarus] Astrachan [working at the Oak Ridge National Laboratory] had previously shown that there was an unstable RNA in phage infection, which was DNA-like. . . . We suddenly realized that the Volkin–Astrachan RNA was *the message.* All the experiments were done later by Brenner and his co-workers with the heavy isotope. They were planned that evening at a party in my house. Half the people were enjoying the party and half the people were in little corners trying to devise all these experiments which were later to show that messenger RNA really exists.[4]

Jacob also recorded this eventful moment: ". . .once the gene was destroyed, all synthesis stopped. No gene, no enzyme. . . . At this precise point, Francis and Sydney leapt to their feet. Began to gesticulate. To argue at top speed in great agitation. A red-faced Francis. A Sydney with bristling eyebrows. The two talked at once, all but shouting. Each trying to anticipate the other. To explain to the other what had suddenly come to mind. All this at a clip that left my English far behind."[5] Jacob's RNA "X" and Volkin and Astrachan's phage-like RNA must be functioning in the same way.

Speedily and overflowing with enthusiasm, Crick composed a note for publication under his and Brenner's names:

> We now boldly combine the conclusions of the two experiments [PaJaMo and Astrachan] and generalize them to produce the following hypothesis.
>
> (1) genetic RNA has the same over-all base-ratios as genetic DNA.
>
> (2) it passes into the ribosomes, but it is only a minor component (10–20%?) The major part of ribosomal RNA is *not* genetic RNA.
>
> (3) genetic RNA is (at least in some circumstances) "unstable," that is, it may have only a limited life. . . .
>
> We thus arrive at a picture of the action of ribosomes that differs in important respects from the one we used to hold. On the old picture a ribosome made a single kind of protein and continued to do so while it remained intact (though possibly at varying rates). On the new picture a ribosome may be making one protein at one moment, and a quite different protein a few minutes later.

Although the *combination* of ideas presented here is new . . . the ideas arose during discussions at Cambridge with Dr. François Jacob. . . . It is because it has so radically altered our own thinking that we have thought it worth while to put this particular formulation onto paper.[6]

As the archives show, Crick wrote a second longer draft of this memorandum adding the epitaph: "Is he in heaven? Is he in hell? That damned elusive Pimpernel. . . ." (referring to this hypothetical form of RNA that we know as *messenger* RNA or mRNA). Although it was intended for the RNA Tie Club, he never sent the paper out.[7] Justly he left the subject to his French colleagues, who, after all, had provided the stimulating experimental results and the suggestion of an RNA "X."

In Crick's estimation, the belief that the ribosome (formerly called a microsome particle) was the message, proved to be "the one big howler" in molecular biology, because it had lasted nearly 7 years. He added: "The only thing one can be thankful for is that it wasn't all done by someone, as it were, outside the magic circle, because we would all have looked so silly. As it was, nobody realized just how silly we were."[8] Yet what relief this brought. Because the ribosome was a kind of reading head and no more, its base composition did not need to reflect that of the DNA. Moreover, because most of the RNA in the cell is in the form of ribosomes, the troubling data of A.N. Belozersky and A.S. Spirin could be attributed to the ribosomal RNA having a standard base composition that does not reflect the distinctive DNA base composition of the species' genetic material. This clarification of the mechanism of gene expression encouraged the idea that introducing synthetic RNA polymers into the cell-free system might bring about protein synthesis, thus opening the way to a direct approach to the code, as described in the last chapter.

The presence of this newcomer among the different kinds of RNA and its subsequent experimental confirmation by Brenner, Jacob, and Matt Meselson[9] and by François Gros in Watson's Harvard group[10] in 1961 cleared the way, as we have seen, for the biochemists' success in breaking the code. But there were many skeptics, the young Henry Harris among them. He later became well known in cancer research for his work on cell fusion and the discovery of tumor suppressor genes, but throughout the 1960s he remained a skeptic concerning mRNA. His own work led him to distrust the results of the pulse-chase technique by which the existence of the messenger had been confirmed. In the early 1960s he was invited to present a talk at a Biochemical Society international colloquium in Cambridge. Following talks by Crick on the code and Brenner on messenger RNA it was Harris's turn to speak.

Crick on his way to the lecture room to hear Harris, took Brenner's arm telling him: "I'm going to sit next to you to see you behave and don't get out of control at question time." And Harris recalled:

> I marshaled all the evidence I had for the presence of a short-lived RNA that was made and broken down within the nucleus and for the absence of any such RNA in the cell cytoplasm; and I ended saying that I doubted whether short-lived messengers existed in higher [eukaryotic] cells. The large auditorium was crammed with people, and when I had finished the incredulity was palpable. Francis Crick took up the cudgels on Sydney Brenner's behalf. Francis can be devastating in debate, and I think I gave as good as I got. For Francis it was all good, clean fun; for me it was a matter of life and death.[11]

Come question time, it had been Crick, not Brenner, who needed controlling! On the issue of mRNA in higher organisms not being short-lived, Harris' skepticism was in due course vindicated. The short-lived RNA he had found in the nucleus also, as he suggested, proved not to be the product of structural genes, but nonetheless important. Meanwhile his skepticism over the messenger put him in a "rather hostile world. . . . In some quarters I was relegated to the lunatic fringe which, in my experience, is a distinctly uncomfortable place for a young man to be. . . ."[12] However, Brenner and Crick took his critique sufficiently seriously to make a special visit to Harris, at that time the Head of the Department of Cell Biology at the John Innes Institute at Bayfordbury. Harris went through his objections to mRNA with them in his office—a less public and less "gladiatorial" event than at the Colloquium. Subsequently Crick conceded that Harris was right in questioning the alleged short-lived character of mRNA in higher organisms.

### Crick's Reputation

Notwithstanding persistent problems, the field of research becoming known as "Molecular Biology" was increasingly attracting attention, as was Crick's salience therein. This is clear from the record of his awards and prestigious lecture invitations from the 1960s. During that decade he received many awards, some shared, others singly—the Albert Lasker Basic Medical Research Award (1960), The Research Corporation (1961), the Charles-Léopold Mayer Prize of the Académie des Sciences de l'Institut de France (1961), the Nobel Prize in Medicine (1962), and the Gairdner Foundation International Award (1962). He also gave many prestigious, named lectures, like the Herter Foundation Lectures in Baltimore (1960), the Vanuxem Lectures at Princeton (1964), and the John

Danz Lectures in Seattle (1966), plus some dozen other named lectures in Europe and the United States. From these and the many unnamed public lectures he gave in the 1960s, Crick became very well known among American scientists, and his overseas earnings grew to match his salary from the MRC. Then, mindful of the invitations for Crick to join the staff at the National Institutes of Health, Oregon State University, and Harvard University in the 1950s, there was a fear among British scientists that he might one day leave the MRC and the United Kingdom.

The invitation he received to give the plenary address at the Sixth International Congress of Biochemistry in New York (1964) was a mark of his acceptance by the biochemical community. Harvard students had first come under his spell in 1959, when he accepted a Visiting Lectureship in the Department of Chemistry, and again in 1962, when he was Visiting Professor in the Department of Biophysics. The invitation to join the faculty there proved a difficult one to refuse. His Visiting Lectureship at The Rockefeller University in 1959 proved valuable in spreading the gospel of molecular biology within the scientific community of New York City. *Scientific American* had already brought his name to the attention of its worldwide readership in the 1950s, and in the 1960s he contributed three articles on the genetic code to this widely read monthly journal. In Europe he was becoming well known through his BBC broadcasts on the European Service and on the Third Programme. In these carefully prepared scripts, Crick still managed to convey his enthusiasm about the science—a sense of being at the cutting edge and of seeking a way through unknown territory. It was "man-to-man" talk. Take the following examples in which he introduced the relation between the gene and the chromosome; later in the same program he described Seymour Benzer's attempt, using genetics, to dissect the genetic map of phage down to the individual bases of the DNA.

> Chromosomes are long, thread-like bodies which lie inside the inner part—the nucleus—of the cell, and the units of heredity—the genes—lie in a linear order along the chromosomes. There is no doubt about this. I could show you them in a phase-contrast microscope. I could also show you how in many cases a particular section of a chromosome—that is, a gene—has been identified as responsible for a particular characteristic of the organism. But how does it do it?[13]

> ...the experimentalists may not have reached the limits of resolution. Next year, perhaps, we may be down to half-a-dozen base pairs, or even (dare I suggest it?) one pair of bases. In other words the biological unit is getting down towards the chemical unit, which only contains about forty atoms. We are on the threshold, in fact, of molecular genetics in its widest sense.[14]

Or take Crick's broadcast on "Cracking the Genetic Code," for the BBC's European Service. He began explaining how minute the sperm is, its head but one-hundredth of a millimeter across. The genetic material within "weighs only a few times a millionth of a millionth of a gram. However the information is recorded, it must be written on a very small scale if so much is to be got into such a very tiny space." And he ended the broadcast with characteristic cautionary remarks followed by optimism:

> We still don't know whether the code is universal. The same twenty amino acids are used in proteins throughout Nature, from virus to man, but it is not yet certain that the same triplets code them in all organisms, although preliminary evidence suggests this is probable. If so, we shall have the key to the molecular organization of all living things on Earth. But on Mars, I wonder? Will there be life, or the remains of life, on Mars? And will it be DNA and RNA and protein all over again? The same languages perhaps, with the same code connecting them? Who knows?[15]

Crick's ability to put himself into the situation of the audience and address them at the appropriate level resulted in numerous requests for his service as a lecturer both to large and mixed audiences and to experts in other scientific disciplines who lack the specialist knowledge of the molecular biologist. His clarity of exposition, uncluttered by jargon and inessential detail, made him the ideal speaker to spread abroad the gospel of the "trinity"—DNA, RNA, and protein. And he was fascinating to listen to. "The words flow on in paragraphs," recalled John Maddox,

> not mere sentences, each of them well constructed, yet one can tell from the sound of the voice when a new idea, or a joke has come into his mind. So you wait a second or two, until the paragraph is over, and out it comes, the idea or the joke. To listen to Crick is, odd though he might think it, like listening to a practiced but Nonconformist preacher.
>
> Crick's enthusiasm is from the same vein. He is anxious to be sure that you understand, so that his conversation is peppered with phrases such as "It's like this" or "D'you see?" It is not, however, an old man's trait but a boyish one. He seems to want to share his own excitement at what may be going on inside his head.[16]

Maybe Crick's Nonconformist upbringing exposed him to the rhetorical and personal style of the evangelist, because he surely earned that title in the world of biology. He also had the performer's touch, as Maddox witnessed at the British Association meeting in 1955. Toward the end of an hour-long talk on DNA, Crick suddenly produced a roll of paper about the size of a toilet roll and flung it across the room like a football hooligan would. But the roll of paper was painted with the structure of DNA. The point was to show that DNA molecules are long. "It was flamboyant," wrote Maddox, "but he made his point."[17]

That was in 1980, but 10 years later Crick could still cast his spell over an audience. Witness his "outreach" Dorcas Cummings Lecture at Cold Spring Harbor Laboratory. The neighbors and friends of the Laboratory, keen to hear Crick speak, had packed the auditorium and began to fill a neighboring one as well. John Inglis, Director of the Laboratory's Press, recalled the occasion clearly. "In came the speaker, a tall and slightly stooping figure, his white hair swept back, his jacket a mulberry-color. His topic was: 'How we see.' There was no patronizing tone, no apologies to the scientists present for the elementary level of presentation. Instead he motioned to the back of the hall where they were seated and explained to the rest of the audience: 'They know what I am going to say. They are only here to find out whether the old actor can still remember his lines.' What followed was a fascinating, lucid and elegant exploration of the subject. It was unforgettable."[18]

Afterward, Watson wrote Crick: "I'll remain perpetually indebted to you for your marvelous Dorcas Cummings Lecture."[19]

Aware of his quick mind and readiness of wit, there were also efforts to persuade him to participate in televised debate and to join the famous BBC program *The Brains Trust*. His friends Julian Huxley and Noel Annan often appeared on this very popular question and answer radio program.[20] His response to the first invitation was "I never debate," and to the second, a refusal. Requests to review books brought the brief response "I only in exceptional cases review books." Such exchanges fill many boxes of his "refusal correspondence" now in the Crick archives.

It was the pressure of all this correspondence—from the sane to the crazy—that caused him to invent his refusal card (Fig. 15.1). When he realized how their impersonal character might offend, he often wrote on the blank side of the card a personal note, but by the time he left the United Kingdom for America he had stopped using them.[21] In truth, becoming a Nobel Laureate attracted media attention and Crick was aware that he had to protect himself from the numerous demands that befall Nobel Laureates. Becoming a celebrity, mixing with the high and mighty, opening public events, sitting on committees, gracing fashionable banquets, and the like were not for him. He wanted to continue exploring vistas into the unknown that were opened up by the 1953 structure of DNA and to spread the word about this remarkable

From:
M.R.C., *Laboratory of Molecular Biology, Hills Road, Cambridge.*

---

Dr. F. H. C. Crick thanks you for your letter but regrets that he is unable to accept your kind invitation to:

| | |
|---|---|
| send an autograph | read your manuscript |
| provide a photograph | deliver a lecture |
| cure your disease | attend a conference |
| be interviewed | act as chairman |
| talk on the radio | become an editor |
| appear on TV | contribute an article |
| speak after dinner | write a book |
| give a testimonial | accept an honorary degree |
| help you in your project | |

*Figure 15.1* Crick's refusal card.

progress, not only to the scientific community but beyond to the world at large. There was no time for celebrity stuff, but young children who wrote asking about how to become a scientist received sympathetic and helpful replies.

## Crick's Nobel Lecture

In December 1962, Swedish scientists in Stockholm heard Crick's Nobel lecture on the genetic code, and the following year readers of *Science* found an adapted version of it in the February issue. Crick had used the occasion to conduct an assessment of the state of the subject. He reported the demise of all overlapping and all combination codes, frequent degeneracy in the code, and few "nonsense" codons (defined at the time as those that do not code for a protein). He applauded the biochemists' initiative with protein synthesis using synthetic polynucleotides because it had "opened the way to a rapid, although somewhat confused, attack on the genetic code." His mention of the epithet "somewhat confused" referred to the initial finding that all DNA syntheses of polypeptides seemed to need a polynucleotide containing uracil. This result seemed unlikely to be true. Nor did the introduction of "random" sequence polynucleotides of mixed constitution solve the problem, for in such experiments the sequences of the template polynucleotides being copied were not known. These results therefore "suggestive, rather than conclusive." Between the biochemists' results and those gleaned by the virologists studying alterations in amino acid sequences following from mutagenesis he found a "fair measure of agreement."[22] After summarizing the work of his group on frameshift mutants and the triplet code, Crick issued a challenge to the biochemists: "It has yet to be shown by direct biochemical methods, as opposed to the indirect genetic evidence mentioned earlier [see Chapter 14], that the code is indeed a triplet code."[23]

Notwithstanding the excitement of the time, Crick showed caution on certain points. Although he thought it likely that "most if not all of the genetic information . . . is carried by nucleic acid. . . ," at the same time, he warned that "whether genetic information has any other major function we do not know."[24] Neither was it yet clear whether "some triplets may code more than one amino acid—that is, they may be ambiguous." Nor could he yet say whether the genetic code was universal. These doubts apart, he concluded on a typically positive note:

> Finally, one should add that in spite of the great complexity of protein synthesis and in spite of the considerable technical difficulties in synthesizing polynucleotides with defined sequences, it is not unreasonable to hope that all these points will be clarified in the near future, and that the genetic code will be completely established on a sound experimental basis within the next few years.[25]

Between 1962 and 1966, Crick delivered some 50 lectures on the genetic code at 30 different locations as far away as Vancouver and Hyderabad. His ceaseless travels and frequent correspondence during those years kept him in touch with the results coming from the new experimental procedures of the American biochemists. Marshall Nirenberg and Philip Leder at the National Institutes of Health had introduced their triplet binding test, and H. (Har) Gobind Khorana at the University of Wisconsin, Madison, had devised a method for synthesizing RNA polynucleotides with defined repeating sequences—a significant improvement on the random sequence polymers synthesized by the Spanish-born New York biochemist and Nobel Laureate Severo Ochoa.

Arriving in the States in early 1965, Crick telephoned Nirenberg to get his latest code allocations. Later that Spring he met with Khorana and was brought up to date with his defined sequence polynucleotides. He also visited George Streisinger in Oregon and was updated on the work there on the amino acid sequences of mutant phage lysozyme. "By March 1965," recalled Crick, "the great majority of triplets [in the genetic code] had been unambiguously identified and just a few remained unallocated [Fig. 15.2]. It was a most exciting occasion for me, traveling about the country and seeing how the various lines of evidence fitted together."[26]

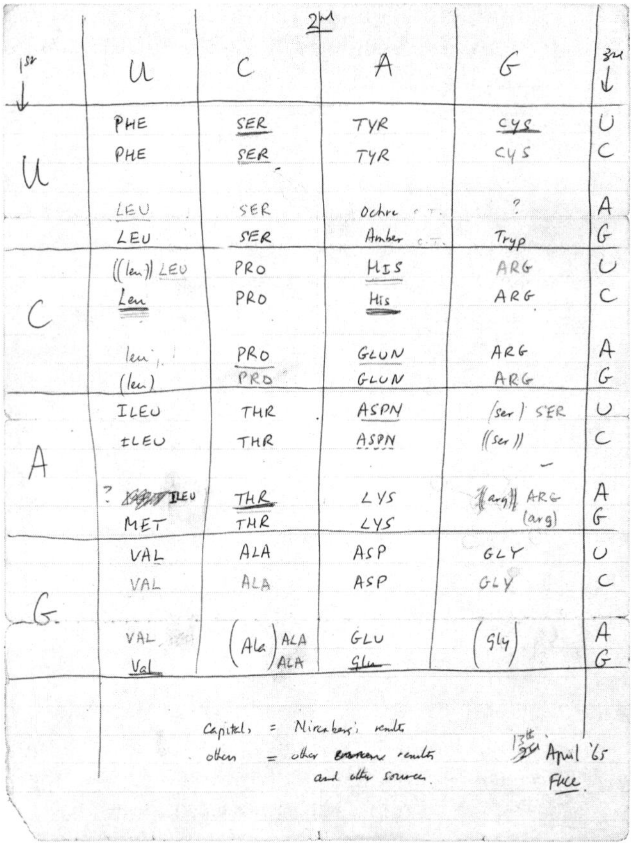

| 1st ↓ | U | C | A | G | 3rd ↓ |
|---|---|---|---|---|---|
| U | PHE | SER | TYR | CYS | U |
| | PHE | SER | TYR | CYS | C |
| | LEU | SER | Ochre | ? | A |
| | LEU | SER | Amber | Tryp | G |
| C | ((leu)) LEU | PRO | HIS | ARG | U |
| | Leu | PRO | His | ARG | C |
| | leu | PRO | GLUN | ARG | A |
| | (leu) | PRO | GLUN | ARG | G |
| A | ILEU | THR | ASPN | (ser) SER | U |
| | ILEU | THR | ASPN | ((ser)) | C |
| | ? ILEU | THR | LYS | ((arg)) ARG | A |
| | MET | THR | LYS | (arg) | G |
| G | VAL | ALA | ASP | GLY | U |
| | VAL | ALA | ASP | GLY | C |
| | VAL | (Ala) ALA | GLU | (gly) | A |
| | Val | ALA | Glu | | G |



*Figure 15.2* Crick's rough sketch of his checkerboard showing the stage reached in solving the genetic code in April 1965.

In November 1965, Khorana had updated Crick by letter with his latest news, which brought the response from Crick that he could "hardly wait to come to Madison to hear more about it all."[27] Arriving two weeks later, Crick lectured on the code and discussed with Khorana's group their biochemical confirmation of his genetic evidence for its triplet character. As he traveled he evaluated and collated the most recent data. These were soon deployed in filling up the checkerboard table that he had introduced to mark progress of the code breakers' work. Breaking the code was an open and collaborative effort.

On the 5th of May 1966, Crick was in London, where he had the great honor of giving the Croonian Lecture to the Royal Society.[28] Here, as in his Nobel lecture, Crick emphasized that the field had moved

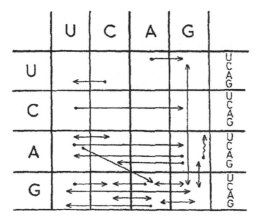

*Figure 15.3* The table of the genetic code with arrows to signify the base changes caused by mutation in the A protein of the tryptophan synthetase of *Escherichia coli*. Note one diagonal arrow indicating a change in two bases.

beyond discussion of overlapping codes and frequent nonsense-coding base sequences. Now the strength of the subject was manifest by the number of predictions from theory that were being confirmed; the data from the direct methods of the biochemists and the indirect methods of the geneticists and the virologists were proving mutually supportive. Thus he could match the data in the coding table with the reports of mutations in tobacco mosaic virus, the colon bacillus, and human hemoglobin. Representing the base changes of the mutations by arrows, he stressed that all of them are either vertical or horizontal, indicating a single base change in each case (Fig. 15.3). Only the bacillus *Escherichia coli* gave a diagonal result, a rare example of two base changes that might have an alternative explanation.

Finally in 1966, George Streisinger's work on mutant phage lysozyme that he had started in Cambridge in 1958 fulfilled a prediction of the frameshift theory—namely, that in the space on the genetic map between a plus mutation and a minus mutation the sequence of bases is read differently from the wild type.[29] By June 1966, the code breaking was complete, except for one triplet, UGA. UAA (the ochre mutant) and UAG (the amber mutant[30]) were already implicated in the mystery of how to start (chain initiation) and stop (chain termination) the translation of a stretch of a polynucleotide chain. Cambridge chemist Brian Clark and Fred Sanger's Danish postdoctoral student Kjeld Marcker succeeded in identifying the start codon—AUG,[31] and Brenner's group had by 1965 found the stop codons, Mark Bretscher by biochemical methods[32] and Anand Sarabhai by genetic methods working on the amber and ochre mutants.[33] Truly, Crick's Croonian Lecture had come at a most propitious time. Of course there were continuing concerns, in particular just how "the rate at which genes act" is controlled,[34] and questions about the origin of the code—has its structure "some stereochemical basis or is it mainly the result of historic accident?" These concerns having been aired, Crick ended on a very optimistic note and made foundational predictions for the future molecular basis of more of biology:

> The importance of the detailed work on the genetic code described here is not merely that it has uncovered the most important and central biochemical mechanisms in biology. The very existence of this exact knowledge establishes the general theoretical framework which has guided investigators for the last dozen years, and shows clearly the very different roles played in living things by nucleic acid and protein. It demonstrates clearly how natural selection can operate at the molecular level, and illuminates such concepts as the absence of the inheritance of acquired characteristics. We can now confidently look forward to placing increasing areas of biology on a molecular basis.[35]

What a climax to a Croonian lecture! "The most important and central biochemical mechanisms in biology"—have any of the other 98 Croonian lecturers of the 20th century made as great a claim?

Next came preparation for speaking at Cold Spring Harbor Laboratory in New York. The occasion was the annual Symposium, on the 2nd to the 5th of June, which in 1966 focused on the genetic code. Crick gave the opening talk, choosing as his title "The Code, Yesterday, Today and Tomorrow." This time, surrounded by informed supporters, he chose to reminisce on the past, reflect on the present, and look to the future. As the first important meeting to be held since the entire code became known, it was in Crick's words "an historic occasion." He surveyed the efforts of Oxford University's Cyril Hinshelwood and his doctoral student P.C. Caldwell in 1950,[36] those of Rochester University's biochemist Alexander Dounce in 1952,[37] and those of University of California Berkeley's George Gamow in 1954[38] to formulate a code between nucleic acids and proteins. With the advantage of hindsight he could point out their mistakes, but at the same time he could be positive about their contributions. He and others had tried to come up with solutions to the code before the advent of direct methods. In retrospect, however, it was now clear, he said, that to have arrived at the correct answer then, without the aid of the right experimental data, "would have been almost impossible."[39] And these data came with Nirenberg and Matthaei in 1961, Leder and Nirenberg in 1964, and

Khorana in 1965. What had been needed to jump-start these "direct methods" was the concept of messenger RNA.

Crick turned now to discuss the advantages and disadvantages of the several methods used in the search for the code. The first direct method (Nirenberg and Matthaei) was synthesis of a polypeptide in the cell-free system. It was limited by the kinds of polynucleotide messengers that could be prepared and the need to ignore any evidence of synthesis less than three times the background. The second direct method, triplet binding, had the advantage of permitting the study of one triplet at a time. Yet Crick warned that binding is not protein synthesis. For that, Khorana's RNA polymer technique gave "probably the best evidence" to date and could be used to confirm the results of the binding method. Yet even this, Crick noted, was open to the objection that artifacts may lurk in the cell-free system itself.

Therefore he turned to the indirect method of phage mutagenesis in whole cells—the phage-infected bacterial cell or the cells of a virus-infected tobacco plant. But he warned that this method "never tells us the identity of a particular triplet but merely the relationships *between* triplets."[40] Last for comment came the search for the sequences found on the anticodons of tRNA molecules, an idea Crick had first considered when working on tRNA with Mahlon Hoagland in 1958 (see Chapter 14). Now, with the results of Cornell biochemist Robert Holley's research, it was time to revisit tRNA. As we shall see, Holley's results played a part in Crick's formulation of ideas on the third base of the codon.

This critical discussion in Crick's Cold Spring Harbor lecture offers the reader a fine example of his concern about experimental methods and the limits to the interpretations that can be put on them. Finally, putting aside these doubts and reservations, Crick judged the elucidation of the genetic code "a great achievement":

> It is in a sense the key to molecular biology because it shows how the two great polymer languages, the nucleic acid language and the protein language, are linked together. It is not only important to know the details for their own sake, but by knowing these details we become quite confident that our general ideas, such as the sequence hypothesis, are indeed correct. It will be difficult, after this, for doubters not to accept the fundamental assumptions of molecular biology which we have been trying to prove for so many years.[41]

With the scientific program finished, participants gathered on the lawn below Blackford Hall. It was June, and Watson, aware that Crick's 50th birthday fell on the 8th, had planned a public celebration at the final banquet party. There was the usual chatter of participants and the clatter of glasses. Suddenly from a large, make-believe birthday cake on the patio beneath the Blackford porch stepped a scantily clad "Fifi" to greet the birthday boy.[42] The event caused much mirth, and Crick appeared not to mind. But privately he had thoughts of revenge, and while thanking his host, the Director of the Laboratory, John Cairns, for his hospitality, Crick invited Cairns' assistance in plotting some future mischief for Watson. A decade in the future Crick would be 60, and would be attending the chromatin symposium that June. Long before then, he advised Watson: "I scarcely celebrate my birthday myself, so why should other people."[43]

### The Wobble Hypothesis Origin of the Genetic Code

In 1965, before these public events took place, Crick had been ruminating on a subtext to his table for the genetic code. Out of these ruminations came two papers. The first, on what he called the "Wobble Hypothesis," was published that summer, and the second, on the origin of the code, was described at a meeting of the British Biophysical Society in December 1966 and published in 1968.

Looking at Crick's table for the code, one soon becomes aware that the contents of each of 8 out of the 16 boxes code for the same amino acid within that box, regardless of which base forms the third letter of the codon (Fig. 15.2); that is, there is substantial *degeneracy* in the code. This was suggestive of the existence in the code today of vestiges of an ancient duplex code (with each codon constituted of only two bases). Here Crick was cautious, because he perceived the extreme difficulty of primitive life subsequently making the change from a duplex to a triplet code. Instead, it occurred to him that "if the first two bases in the primitive codon pair in the standard way, the pairing in the third position might be *close* to the standard ones," but sufficiently different to relax the adenine/uracil[44] and guanine/cytosine pairing limitations. Then much of the code would be determined by the first two bases, but what he called the "principle of continuity" would be preserved; that is, there would be no abrupt changes in the code's evolution.

Now Crick, steeped in model building, was aware of the importance of the geometry of orientation of bases on their sugar–phosphate backbones as in the glycosidic bond between sugar and base. He distinguished four positions of that bond out of seven that should be possible in stereochemical terms. These four were sufficient to permit a degree of "wobble" in the pairing between the third base of these codons in a messenger RNA and the anticodon of a tRNA molecule. Supporting his analysis was the discovery of the nucleoside inosine (adenine minus its amino group) in several tRNA molecules. Crick learned about inosine's presence first from Holley in 1965.[45] He reckoned that the base of this nucleoside could pair in the third position with any of the four bases, as follows:

Yeast tRNA

| Reading Direction | $\rightarrow$ | $\leftarrow$ |
|---|---|---|
| | tRNA | mRNA |
| Amino Acid | Anticodon | Codon |
| Alanine | I G C | G C ? |
| Serine | I G A | U C ? |
| Valine | I A C | G U ? |

?, any one of the four bases in the third position.

True to form, Crick signed off this paper with the optimistic: "it seems to me that the preliminary evidence seems rather favorable to the theory. I shall not be surprised if it proves correct."[46] And what of the paper's light-hearted title "Codon–Anticodon Pairing: The Wobble Hypothesis" with its double entendre? It inspired Rose Feiner to write the following lines:

On a ribosome unit a messenger sat

Singing "wobble, O

wobble, O wobble";

And I said to a codon, "O why do you sit

Singing wobble, O wobble, O wobble?


Is it weakness of Watson your little inside

Or a Crick in your intercistronic divide?"

With a flop of a hydrogen bond it replied

"O wobble, O wobble, O wobble!"[47]

The wobble hypothesis also prompted discussion of the manner in which the code we know today had evolved. Naturally Crick had been ruminating on the subject for some time, as had others. Already the American microbiologist Carl Woese had discussed how an initial high error rate of translation might have been reduced to give rise to the codons we know today.[48] He had also been attracted to suggestions of a stereochemical fit between amino acids and their tRNA molecules as a factor in shaping the character of the code. He returned to this possibility in 1966. Thus the amino acid phenylalanine must somehow fit neatly into a kind of cage formed by the coding triplets—UUU and UUC—in the tRNA molecule. Woese's Cold Spring Harbor Symposium paper includes a photo of an unnamed example that he constructed with space-filling model-building equipment.[49]

Crick was opposed to this hypothesis. He had already crossed swords with researchers S.R. Pelc and M.G.E. Welton at the MRC Unit at King's College London about their claim for a fit between amino acids and the matching tRNA anticodons.[50] When shown the results of their model building, Crick found a number of errors, including the construction of all their polynucleotide sequences backward! He demanded a public retraction by the authors. When they refused to use the strong wording he had suggested, he wrote a short note to *Nature*, listing their errors and ended with the conclusion that "the models of Pelc and Welton do not support their hypothesis."[51] Privately he wrote to Maurice Wilkins suggesting that Pelc and Welton should not continue working for the MRC.[52]

In short, Crick was nearly as opposed to reliance on a *specific* stereochemical fit between amino acids and their tRNAs as he had earlier been over the fit between amino acids and the "holes in the helix" suggestions of the 1950s. This debate over amino acid–nucleic acid "recognition" goes back to that era when Crick introduced his adaptor molecules. The subsequent discovery of tRNA won the argument for him and the "impossibility of amino acid-nucleic acid recognition became an unquestioned dogma," complained Woese. He was very doubtful that Pelc and Welton's models were correct, but he judged they had strengthened the idea of some such specific interaction.[53]

Disposing of these stereochemical suggestions made the task of speculating on the origin of the code more difficult. Crick took the opportunity to discuss the subject when he addressed the British Biophysical Society in December 1966. The talk was not published at the time, but an inaccurate report of the occasion appeared in *Nature*,[54] to which Crick felt bound to reply. In his letter published shortly thereafter, he explained that he was not, as the report suggested, in favor of, but rather opposed to, the idea of "a stereochemical relationship between all amino acids and their anticodons." True, evidence against the idea he admitted was no more than suggestive, rather than conclusive. However, the main point of his talk had been "an attempt to show that a plausible theory could be constructed without necessarily assuming any stereochemical interaction of amino acids with either codons or anti-codons." He imagined the code to pass through three phases:

1. The Primitive Code, in which a small number of amino-acids were coded by a small number of triplets.

2. The Intermediate Code, in which these primitive amino-acids took over most of the triplets of the code in order to reduce nonsense triplets to a minimum. The codons produced by this process were likely to be related.

3. The Final Code, as we have it today.[55]

The crucial idea, he explained, was the transition from stage 2 to stage 3. Here, "evolutionary theory suggested that a new amino acid was incorporated into the developing code only if its introduction at the time gave a selective advantage to the primitive organism." And further, with the increasing sophistication of proteins, "no possible new amino acid could, on balance, be an advantage and the code would be frozen."[56] This would mean all life on earth derived "from the same single organism, or single cross-breeding population." Hence the name "Frozen Accident Theory."

This alternative had advantages.

1. It accounted for the degree of redundancy in the code by tracing it to an initial primitive code with very few amino acids.

2. The range in the numbers of different codons for a given amino acid would support the suggestion that those amino acids with many codons were the originals—glycine, leucine, alanine, and valine—whereas methionine and tryptophan, having only one codon each, were latecomers.

3. Similar amino acids tended to be coded by similar codons—the hydrophobic ones in the top left of the table, the hydrophilic ones in the bottom right. And the reason? When a new amino acid appeared in the primitive environment, it could more easily be "adopted" by a member of a group of codons that currently served an existing amino acid of similar nature.

4. Its universality has been due to its origin from one source and the havoc caused by any departure from the existing code once it became fully established. Thus it is the product of a series of chance events. As he later wrote, "both the number 20 and the actual amino acids in the code are at least in part due to historical accident."[57]

The original intention had been for a joint communication from Crick and Leslie Orgel to be presented by Orgel at the meeting in London. When Orgel was unable to attend, Crick went instead. Subsequently he wrote up a first draft of the talk he gave. Then he found that Orgel had also written a draft of the paper he was to have given. They therefore collated their two papers, and the *Journal of Molecular Biology* published them back-to-back in 1968.[58] Crick, reflecting on his paper, judged the theory he had presented to be "overly accommodating. In a loose sort of way it can explain anything," he wrote. If only one could get at the facts experimentally. The stereochemical theory, in contrast, could easily be checked by well-designed experiments on codon–amino acid interactions. But he cautioned that "vague models of such interactions are of little use."[59]

Reflecting on the code naturally drew Crick to discussion of protein synthesis. Here, like Woese, he was struck by the extent of the involvement in the cell of noninformational nucleic acid, the ribosomes, and the transfer RNAs. Although he saw efficiency as a possible cause, he could not "help feeling that the more significant reason for the presence of rRNA and tRNA was that *they were part of the primitive machinery for protein synthesis.*" And he repeated the suggestion he had made in 1966: "tRNA looks like Nature's attempt to make RNA do the job of a protein."[60] Two years later he went on to visualize primitive machinery for protein synthesis consisting entirely of RNA. "Possibly," he declared, "the first 'enzyme' was an RNA molecule with replicase properties."[61] Little did he or anyone else realize then that such a species of RNA exists in our own cells and plays an enzymatic role in DNA duplication! This paper gives one a sense of the speculative, theorizing world of Francis Crick's mind— imaginative, probing the unknown terrain of distant events in life's history, and yet constrained by the constantly felt need for sound evidence.

### Defending the Claims of Molecular Biology

Despite these successes, the new foundations laid by the molecular biologists continued to be challenged. Thus in 1976 an alternative to the Watson–Crick model for DNA was put forward by workers in New Zealand,[62] followed by a group in India in 1976[63] and supported by mathematician William Pohl in 1978.[64] These critics of the intertwined double helix of Watson and Crick constructed "side-by-side" or "zipper models," the chains of which could be separated with ease. Pohl favored such structures because in the case of the circular DNA found in bacteria and in the mitochondria of higher organisms, the result of its replication should be two interlocked daughter circles. But Nature clearly does not operate that way. The daughter chains are quite separate.

Reflecting on this debate, Crick recalled how these critics "feared that the Establishment would not listen to them. Quite the contrary was the case because everyone, including the editor of *Nature*, was bending over backward to give them a fair hearing." Crick spent time with Pohl and advised him that "if Nature did occasionally produce two interlocking circles, a special mechanism would have been evolved to unlink them. I believe he thought this an outrageous example of special pleading and was not at all convinced by it."[65]

Fortunately an enzyme that unlinks such circles was discovered, and determination of what is called the "linking number" (see Chapter 17) for circular DNA gave the figure predicted from an intertwined helix, not that from a side-by-side one. Crick had by this time devoted so much time to the subject that in 1979, together with two biophysicists, he wrote the paper, "Is DNA Really a Double Helix?" Their hard-hitting critique of the side-by-side (SBS) model concludes with the words: "The SBS structure is thus incorrect." Following this comes an appreciation:

> The SBS model was ingenious because it incorporated the well-established features while altering the less certain ones. It has undoubtedly made us sharpen the arguments for the double-helix. It has raised the question of how far a structure can depart from a double helix and still give the very striking absences seen in the diffraction pattern. . . . Above all it has underlined a need that has been apparent now for some time. . . . This is the solution, to high resolution, of single crystal structures of *short* lengths of the DNA double-helix having a defined base-sequence.[66]

The first success in establishing the character of base pairing from single-crystal X-ray diffraction studies came in 1959. Using mononucleotides, Karst Hoogsteen found the pairing to be the like-with-like kind, not the Watson–Crick complementary kind.[67] Fourteen years later, Alexander Rich's group at MIT found Watson–Crick pairing when they studied single crystal dinucleotides.[68] But it was not until 1979 that the same result was achieved that met Crick's request for "*short* lengths of the DNA double helix" with defined sequence.[69] Crick was naturally well aware that different kinds of base-pairing are possible outside the confines of a double-helical structure, and that both Rich's group in Boston and Richard Dickerson's in Los Angeles had discovered a new form of DNA called Z-DNA that, instead of being right-handed like the double helix, is left-handed. Needless to say, Z-DNA has a very different diffraction pattern from that produced by the A and B forms.[70]

Thus it took more than a quarter of a century, Crick remarked, for the Watson–Crick model "to go from being only rather plausible, to being *very plausible* (as a result of the detailed work on DNA fibers by Wilkins' group), and from there to being virtually certainly correct."[71] This achievement resulted from the deployment of the chemist's skills in synthesis and the crystallographer's experience with the method of isomorphous replacement.

### Criticism of the Central Dogma

Another foundation stone of the molecular biology of the 1960s was Crick's Central Dogma, described in Chapter 13. A determined critic was the cell biologist from Washington University in St. Louis, Barry Commoner. It was in 1960 in his address as retiring vice president of the American Association for the Advancement of Science that he first attacked the DNA-based molecular approach to biology, likening it to a "rootless hybrid" instead of "a true alliance between real sciences."[72] As the molecular biologists came up against a problem for their simple picture, Commoner would launch another broadside (see Fig. 15.4), culminating with his "Failure of the Watson–Crick Theory as a Chemical Explanation of Inheritance," that occupied more than six pages of *Nature* in 1968. By this time he could call on evidence from several quarters to show that all the steps in gene expression involve enzymes that not only facilitate those steps but also influence the nature of the product. The same goes for the replication of DNA itself. To Commoner, it seemed, "Inheritance is determined by a multi-molecular system vastly more complex than a DNA molecule."

> Biologists have confronted successfully—like a nest of Chinese boxes—levels of complexity. . . . The last box has now been opened. According to the Watson–Crick theory, it should have contained the single source of all the inherited specificity of living organisms— DNA. It is my view that we now know that the last box is empty and that the inherited specificity of life is derived from nothing less than life itself.[73]

Nothing in the box? Surely this was a rhetorical gesture? Crick responded neither to this challenge nor to Commoner's critique in *Harper's Magazine* 34 years later.[74] Sending a prepublication copy to Crick, Commoner wrote: "As you can see, my original critique of the central dogma is only reinforced by the more recent data." At the same time, he acknowledged "a renewed appreciation" of Crick's 1958 essay "for the clarity with which your ideas are developed."[75] Thanking him

TRANSFER OF DNA SPECIFICITY

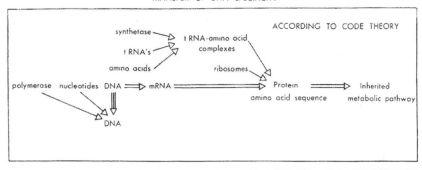

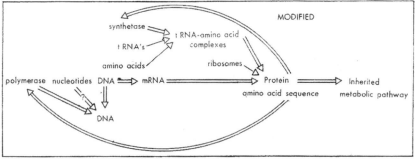

*Figure 15.4* Commoner's representation of the transfer of specificity according to the code theory and his modification. All arrows drawn with double lines represent contributors to the determination of specificity.

for writing, Crick added: "As to the Central Dogma, I see you are up to your old game of persistently misrepresenting what I wrote in 1958 (and in 1970)."[76]

In the 1950s, Commoner, Erwin Chargaff, and others had objected that gene duplication and expression must surely involve highly complex systems of interaction between proteins and nucleic acids within the whole cell. This criticism, as Crick admitted years later, was partly true.[77] But what of evidence that the direction of information flow can be reversed? In 1970 Howard Temin and David Baltimore, working independently, reported their isolation of an enzyme that supervises the transcription of the RNA sequence of a virus into a DNA sequence, thus making it an "RNA-dependent DNA polymerase," now known as "reverse transcriptase." Their papers, submitted to *Nature* in June 1970, were published rapidly and were highlighted by the News and Views column entitled "Central Dogma Reversed." It reads:

The central dogma, enunciated by Crick in 1958 and the keystone of molecular biology ever since, is likely to prove a considerable over-simplification. That is the heretical and inescapable conclusion stemming from experiments done in the past few months announced in two laboratories in the United States. For the past twenty years the cardinal tenet of molecular biology has been that the flow of transcription of genetic information from DNA to messenger RNA and its translation to protein is strictly one way.[78]

How did Crick feel when he opened his copy of *Nature* for the 27th of June 1970? This time he did not keep quiet. He was very soon penning a response that reached the editor, John Maddox, 11 days later. "This is not the first time," Crick began, "that the idea of the central dogma has been misunderstood, in one way or another." Recalling the 1950s, he explained, "all we had to work on were certain fragmentary results, themselves often rather uncertain and confused, and a boundless optimism that the basic concepts involved were rather simple, and probably much the same in all living things."[79]

Part of the misunderstanding, he implied, may have arisen because his lecture when published did not carry the explanatory diagram that he had used to represent all the possible information transfers that might exist between DNA, RNA, and protein (Fig. 15.5A). Permitting all these transfers would have made it "almost impossible to construct useful theories," he wrote. Yet such theories "were part of our everyday discussions." That was possible "because it was being tacitly assumed that certain transfers could not occur." So Crick "thought it wise to state these preconceptions explicitly." They are represented in his second figure (Fig. 15.5B), where the solid arrows represent probable transfers and the dotted ones possible transfers. Note that the RNA to DNA flow is a dotted line, thus allowing for the possibility that the sequence in RNA might be transferred to DNA as Baltimore and Temin now claimed.

It is clear from Crick's 1956 manuscript referred to earlier and illustrated in Chapter 13 that this was no afterthought. Apart from being linear rather than triangular, these diagrams carry the same message. This document supports Crick's claim that he had not envisaged the transfer RNA→DNA as impossible; moreover RNA to DNA does not figure in the second diagram, showing transfers that *never* occur. Two years later, in the publication of these claims, he stated: "the transfer of information from nucleic acid to nucleic acid ... may be possible" (i.e., in either direction). Although no evidence existed then for the transfer, RNA→DNA, no considerations of molecular structure were opposed to it (see Chapter 13). Therefore in his response to the discovery of reverse transcriptase, he recalled his thinking in the 1950s and grouped the pos-

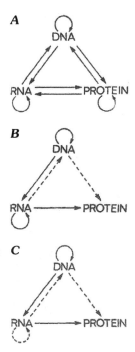

*Figure 15.5* (*A*) Crick's diagram showing all conceivable simple transfers of information between the three families of polymers. (*B*) The situation as it seemed in 1958. Solid arrows indicate probable transfers, and dotted arrows indicate possible transfers. The arrows present in *A* but missing from *B* represent the transfers judged impossible in accordance with the Central Dogma. (*C*) Tentative classification of the situation in 1970. Solid arrows indicate general transfers; dotted arrows indicate special transfers (i.e., transfers that do not occur in most cells but may occur in special circumstances); and arrows present in *A* but missing from *C* indicate transfers that the Central Dogma predicts never occur.

sible information transfers into three classes. First came the group of "general transfers," DNA→DNA, DNA→RNA, RNA→Protein, for which there was evidence in all cells. The second group, the "special transfers," do not occur in all cells, "but may occur in special circumstances" such as in virus-infected cells: RNA→RNA, RNA→DNA, DNA→Protein. The third, "unknown transfers," comprised Protein→Protein, Protein→DNA, Protein→RNA (those in Fig. 15.5A that are absent from Fig. 15.5C). In 1973 Crick recalled that one of Temin's collaborators had asked him about RNA to DNA "and I had said that I thought there was no strong reason why it should not occur. I need hardly say," he added, "that I had not forgotten that I said that."[80]

After Crick mounted this defense of the Central Dogma, he cautioned his readers about the incompleteness of our knowledge of molecular biology, mentioning the problem of the strange disease of sheep, scrapie, to which Brenner had drawn his attention. But although any of the unknown transfers of the second group "could be accommodated into our thinking without undue strain," he declared that

> the discovery of just one type of present day cell which could carry out any of the three unknown transfers [in the third group] would

shake the whole intellectual basis of molecular biology, and it is for this reason that the central dogma is as important today as when it was first proposed.[81]

He had thrown down the gauntlet in characteristic fashion. Such was his conviction that he declared he "would become a theologian" if translation from protein to nucleic acid were ever discovered.[82] Stating that he did not know of any colleague of his "who has ever suggested that the information transfer RNA→DNA cannot occur," he was brought up sharply by a correspondent from Jena, who took "much pleasure reporting that he 'found such a colleague of yours. It is no less a person than Watson,' in *The Molecular Biology of the Gene,* p. 298."[83]

"I had quite forgotten," responded Crick. Watson "must take a large responsibility for the term 'central dogma' being used incorrectly."[84] Crick enjoyed telling Watson: "you were partly to blame," but he had to admit: "I did not notice this point when I originally read your book, probably because I did not realize it would become such a catch-phrase and therefore did not worry that you would mix it up with the sequence hypothesis."[85] By this time Watson's textbook had been in use for 5 years, during which students had been reading the sentence "RNA never acts as a template for DNA." (In truth there had been times when Crick wrote out the familiar sequence DNA→RNA→Protein without mention of the possibility of the transfer RNA→DNA.)

From the beginning Crick had placed limits on the Sequence Hypothesis and suggested possible exceptions. Contrasting with this was his confidence that exceptions to the Central Dogma will not be found. Thus he wrote to Temin:

> I must tell you that I quite disagree with you about the unknown transfers. Of course, one cannot produce any argument to show that it is impossible that they should exist. It is just my opinion that they won't be found, basically because the mechanisms to make them would have to be so elaborate. However, time may show that I am wrong.

> Two other points. I do not subscribe to the view that all "information" is necessarily located in nucleic acid. The central dogma applies only to residue-by-residue sequence information. In fact, I suspect that the cell cortex holds "information" in the broad sense. However, there are philosophical (or, if you don't like that word, logical) difficulties in defining just what we mean by information of this sort. For example, the activating enzymes, the transfer RNA and the ribosomes are necessary for protein synthesis, and also define the genetic code, but they are not the sequence information itself, which resides in the mRNA. In addition, this machinery of protein synthesis is also (given the code) specified by DNA sequences. I think "information" should thus

only be used when there are at least two alternative (efficient) choices before a system. Otherwise the components, even though essential, and containing essential instructions (as the activating enzymes plus tRNA) should be classed as machinery. Now whether, say, the "instructions" in the cortex of the egg of *Drosophila* should be classed as "machinery" or "information" I really don't know. Moreover, the real state of affairs may not necessarily fall into the categories I have sketched above.[86]

Had Crick included some discussion of this enlarged conception of information in his published response in 1970, it would not have been possible for Commoner to write such an "I told you so" essay in 2002. As it was, Crick was again provoked by another *Nature* editorial in 2003. He duly complained: "It is widely assumed that DNA→RNA→Protein is the Central Dogma. This is quite incorrect." It could, he added, "usefully be called 'Watson's Dogma.'"[87] He therefore requested that subeditors should in future be advised to use the Central Dogma correctly. It refers to the *translation* of a sequence of residues in a nucleic acid into a sequence of residues in a protein. Even today, conscious of the numerous ways in which proteins alter DNA (e.g., methylation, correcting errors in DNA replication, switching genes on and off, aiding the folding of the polypeptide chains), we might still ask, as Crick did half a century ago, "Where is the machinery to effect *translation* from the amino acid sequences of a protein into the equivalent nucleic acid sequence?" The degeneracy of the genetic code poses a problem from the very start. And where in the cell should we expect to find a globular protein unfolding to reveal its amino acid sequence all accessible for copying backward into a nucleic acid? Clearly, the kind of "information" that feeds back from proteins to DNA is of a different nature and depends on the remarkable dynamic three-dimensional architecture of proteins.

## CHAPTER 16

# Speaking Out on Controversial Subjects

*What everyone believed yesterday, and you believe today, only cranks will believe tomorrow.*[1]

In the fall of 1961 the Secretary of the Medical Research Council, Harold Himsworth, received the following letter:

> I feel I should tell you that I have resigned my Fellowship (Title E) of Churchill College owing to the decision of the trustees to build a chapel. It is a comfort to me that the Medical Research Council is never likely to be as foolish as the Trustees in this respect.[2]

Crick had been appointed one of the founding Fellows of the projected new Cambridge College that was to bear Sir Winston Churchill's name. Churchill had envisaged a college on the lines of the Massachusetts Institute of Technology. Like MIT, the emphasis at Churchill would be on the natural sciences, mathematics, and engineering. Nothing was said about a chapel. When in 1958 a report on the plan and costs for the College buildings appeared in the student newspaper, *Varsity,* a number of students expressed their concern that no provision for a chapel had been made.[3] They were joined by the Dean of Gonville and Caius College, Hugh Montefiore, and the Dean of Jesus College, Barry Till. The former led a spirited campaign among the students to raise funds toward a chapel, priming the fund himself with a gift of ten guineas (£10 10s). As Mark Goldie, in his delightful essay on the subject, remarked, "The Sunday chapel boxes rattled to save the souls of the future members of Churchill College."[4] Deans Montefiore and Till also went to see Noel Annan, chairman of the Trustees' Education Committee. When asked about a chapel for Churchill, Annan, an agnostic, replied, "You won't believe it, but we forgot!"[5] They were, of course, intent on persuading the trustees to think again. Think again they did, and by the time building of the College began a central location had been assigned for a freestanding chapel, and soon piles were being driven into the ground for it. At this, members of the Governing Body, Crick among them, were stirred to action. The battle that ensued has gone down in the history of the College. Offering a glimpse of the strength of Crick's antireligious feelings, and the intimate relation in which he held his science and his agnosticism, this college battle deserves attention.

The fuse had been lit in 1960 when Montefiore persuaded the Liberal politician, the Reverend Timothy Beaumont, to aid the cause. Early in 1961 he offered to donate £28,000. This act altered the situation radically. Now a chapel could be built. Crick recalled the impact of this news on him:

> My recollection is that when I was first asked if I would be interested in becoming a Fellow I declined because of the stated intention to build a chapel. Then Teddy Bullard [Sir Edward Bullard], whom I already knew, came round to persuade me to change my mind. There's only ten pounds in the chapel fund, he said, so they'll never build it. . . . As a result I rather reluctantly let my name go forward.

> This explains to some extent why I resigned. I felt that I had been somewhat misled. Later on I did ask myself whether I would have done better to have stayed on, and argued the case. . . . If I had stayed on I think I would have suggested that the College attached itself to the church (is it St. Clements?) at the top of Portugal Place. . . .[6]

Eleven Fellows who were opposed to the concept of an Anglican chapel did not resign but stayed on to fight ultimately for a nondenominational building. Their cause was joined by a Research Fellow, John Killen, who suggested distancing the chapel from the main College buildings on a portion of College land that should be made over along with Beaumont's gift to an independent trust. His suggestion was accepted, and the chapel, although a few minutes' walk from the center of the campus, was built, but it is not formally owned by the College. The foundations for the centrally planned chapel remain covered by grass.

Crick resigned his Fellowship in September writing to Sir Winston Churchill to that effect.[7] Churchill replied expressing himself "puzzled at your reason. . . ." "A chapel," he added, "is an amenity which many of those who will live in the College may enjoy, and none need enter it unless they wish."[8] This prompted Crick to respond with a devilish suggestion:

It was kind of you to write. I am sorry you do not understand why I resigned.

To make my position a little clearer I enclose a cheque for ten guineas to open the Churchill College Hetairae fund. My hope is that eventually it will be possible to build permanent accommodation within the College, to house a carefully chosen selection of young ladies in the charge of a suitable Madam who, once the institution has become traditional, will doubtless be provided, without offence, with dining rights at the High Table.

Such a building will, I feel confident, be an amenity which many in the college will enjoy very much, and yet the institution need not be compulsory and none need enter it unless they wish. Moreover it would be open (conscience permitting) not merely to members of the Church of England, but also to Catholics, Non-Conformists, Jews, Muslims, Hindus, Zen Buddhists and even to atheists and agnostics such as myself.

And yet I cannot help feeling that when you pass on my offer to the other Trustees—as I hope you will—they may not share my enthusiasm for such a truly educational project. . . . They may even feel my offer of ten guineas to be a joke in bad taste.

But that is exactly my view of the proposal of the trustees to build a chapel, after the middle of the 20th century, in a new College and in particular in one with a special emphasis on science. Naturally some members of the College will be Christian, at least for the next decade or so, but I do not see why the College should tacitly endorse their beliefs by providing them with special facilities. The churches in the town, it has been said, are half empty. Let them go there. It will be no further than they have to go to their lectures.

Even a joke in poor taste can be enjoyed, but I regret that my enjoyment of it has entailed my resignation from the College which bears your illustrious name.[9]

Evidently Churchill did not reply but returned Crick's check "with comps" (Fig. 16.1).

His old Navy boss, Edward Collingwood, must have visited Crick just after he had posted this letter, for on the 11th of October Collingwood wrote to Crick to say "I can see you enjoyed writing it! No harm in that. Indeed, it may be a valuable outlet—an emotional release," but "I can't help being sorry you were in such a hurry to get this into the letter-box. It could not be said of you as Lord Exmouth, the famous Captain, said of himself 'No second thought of mine was ever worth the experience.'"[10]

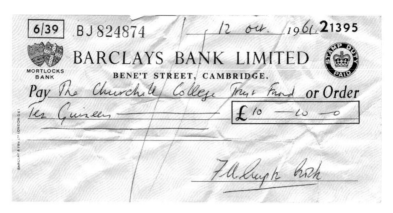

*Figure 16.1* Canceled check returned to Crick by Sir Winston Churchill.

Earlier Richard Keynes, the physiologist and great-grandson of Charles Darwin, had been trying to bring about a compromise solution to the chapel issue and had been alarmed to hear in August a rumor that Crick was threatening resignation of his Fellowship. He warned Crick that if Beaumont's money were accepted by the Trustees, the Fellows will be "powerless to reverse the decision. Hence your resignation would be futile, if a noble gesture! Will the Chapel do any really positive harm? Surely you do not rate the proselytizing powers of the Church of England very high?" Keynes did not want Churchill College to "rapidly degenerate into just another College. Resignation of persons like yourself will only accelerate this process, so please don't."[11]

Keynes followed up his letter with a telephone call about the proposal to request that all religious ornaments be movable ones. Would that make a difference for Crick? He replied that his moral conscience "is not as finely tuned as that." The Master of the College, Sir John Cockcroft, a devout Christian, knew by this time what to expect from Crick. In a draft of a long letter that Crick kept along with his other family correspondence, he had explained his position to the Master:

> I am sure there are many people in Cambridge, especially scientists, who believe that now days Christianity is a firm superstition, and who feel that this will become increasingly obvious to most educated people in the future. Apparently this does not prevent unbelievers from accepting College fellowships. Indeed some of them go so far as to have their children christened, which seems to me ridiculous. One can only conclude that they regard Christianity as a completely harmless set of beliefs and that they tolerate it as one might humour a somewhat eccentric aunt.

I cannot agree with this point of view. It seems to me shameful that a great university should be used for the furtherance of superstitions and although Christianity is no longer virulent, it is surely not yet harmless.[12]

It bothered him that Christianity in Cambridge was not "withering away," rather "there has been a minor religious revival" that he judged would not have happened "had not the colleges provided facilities for Christian worship and propaganda and if undergraduates realized just how many distinguished men do not believe in it." It riled him that "Too many unbelievers hide their faith under a bottle of port," and lend their support to religious institutions. The Nobel Laureate, Lord Adrian, is a case in point. On Ash Wednesday, as the Vicar later recalled with admiration, Adrian as Vice Chancellor, read the litany: "I believe in God the Father . . ." so "beautifully, agnostic though he was."[13]

Hence Crick found it "particularly regrettable that a new institution like Churchill which has been founded to try to shake this University out of its lethargy, should meekly follow this bankrupt tradition." Yes, he could tolerate Christians, but "the official support of Christianity by a College," no. Here we see the forceful down-to-earth expression of Crick's conviction that was so typical of him: no pretense, no ifs and buts, no punches pulled.

Recall that Crick had become "a skeptic, an agnostic with a strong inclination toward atheism,"[14] before he reached puberty. It was not a matter of rebelling against fundamentalist versions of Christianity, but rather "I just knew," he explained, "that what I was being told didn't seem to me very sensible . . . that general feeling that, you know, the Bible was true." But this skepticism at first he carried lightly, not objecting to attending chapel at Mill Hill School. Had he felt strongly on the matter he would have claimed it was against his principles to go to chapel, "and it never occurred to me I think." Indeed his fellow Mill-hillian Harold Fost was quite unaware of Crick's agnosticism either when he was at Mill Hill or at University College London. As a freshman there, Crick was even assuming that myths of the Old Testament had rational origins. Thus the myth of Eve being fashioned from one of Adam's ribs must have arisen, he thought, from contemplation of the fact that women have one more rib than men! A medical student informed by Crick of this medical treasure "almost fell off his chair with laughter."[15] Crick was learning "the hard way that in dealing with myths one should not try to be too rational."[16] Evidently he had not checked this datum out on a member of the opposite sex. But then, he came to university from Mill Hill, a boys' boarding school.

It was not until he came to work in Cambridge in 1947 that religion literally confronted him—"impinged" on his life. This was due to the fact that unlike University College London, founded deliberately as a secular institution of learning, the Cambridge Colleges had been gestated in the womb of the Church and have retained their religious traditions. They have their chaplains or deans, religious services every day, and grace is said at formal meals. The chapel bells toll for matins and evensong. In some of the Colleges "newly elected Fellows still vow in the College chapel, to uphold the triad of 'religion, learning, and research.'"[17] Although Crick found these features of Cambridge University life obnoxious, they did not prevent him from accepting dining rights at Gonville and Caius College. Churchill College was different. Here he was an "extraordinary" Fellow of a new Cambridge College, founded in 1958[18] and planned to be science- and technology-based.

So much for the Cambridge Colleges, but what about society at large? Asked, "Do you favor secularizing society more than it is in Britain at the moment?" Crick replied:

*Crick.* Well I certainly object to American Fundamentalists. But going further than that is a matter of time.

*Q.* But you object to them because they appeal over the radio to people?

*Crick.* Well yes, and it's patently untrue. The other things you can say—well, they're a matter of opinion. I happen to have a different opinion and one must be tolerant until the scientific knowledge has got to a stage where you can say other things that [show] they're indefensible. So it wasn't that . . . if I could dictate what people believe, that I would have wanted them to believe that. I think one must have a certain tolerance, but I think there's a limit to my tolerance and that comes with fundamentalists.

*Q.* Yes, so outside of fundamentalists you've got a respect for people that have a different point of view?

*Crick.* Well I wouldn't go that far as to say a respect. [*laughter*] I can't imagine—I think it'd be so silly. But that's my reaction to the majority of people.

*Q.* One could say that scientific knowledge which would throw out certain specific dogmas, such as special creation, resurrection etc., might at the same time [allow one] to preserve others. . .

*Crick.* You certainly could. But on the other hand since the foundations of the two are the same, once the result of one has been shown to be false, you might worry about the other. I think I'd worry about the method by which the things were established, like the existence of an afterlife for example, which certainly hasn't been proved or disproved. But the sort of thinking that went into it was the sort of thinking which went into special creation, and not only special creation, if you take the fundamentalist point of view, the actual [birth] date of the earth.[19]

In 1966 Crick had a series of public exchanges with the experimental ethologist and Quaker W.H. Thorpe in *Varsity*. In subsequent private correspondence between them, Thorpe sought to promote "closer understanding between Christians and Humanists." The interchange was friendly but Crick found his Quaker friend's "sophisticated interpretations" of Christian beliefs "just as unacceptable as the simple ones." On one point he agreed with Thorpe,

> that where Humanists and Christians want to achieve some end they should try to work together. You may have noticed that I was a signatory to the recent appeal to the Pope about birth control. I wish I knew what to do about disarmament but I find the problem baffling. If you have any practical suggestions I should certainly be interested. Apart from this sort of thing I really think that trying to find agreement between such totally different points of view is unlikely to be rewarding.[20]

### The Cambridge Humanists

The "faith" that Crick accused his fellow academics of hiding under a bottle of port was based on the foundations of scientific knowledge, very much in line with the teaching of the Humanists whom he supported. The British Humanist Association defined their position as follows:

> The fundamental assumption of Humanism is that man is on his own and is responsible for creating the conditions of a life worthy to be called human without the security or guidance of an absolute. Man must not look for divine guidance or expect to survive in any way after death. He must therefore choose his own values and create his own morality. These are derived from purely human experience and require no other justification.[21]

When the Chairman of the Cambridge Humanists, Victor Purcell, came to interview Crick, he was "half expecting a fiery-eyed Savonarola of Atheism full of *saeva indignation*[22] and polemical zeal." Instead he received a friendly welcome and he perceived Crick had a jovial attitude to the "folly of religion." Crick told him "he had long remained tolerant, and for years had attended wedding and funeral services as social usage demanded." When an agnostic friend pointed out to him that this was incompatible with his convictions and supportive of institutions that he opposed, he restricted himself to attending the receptions only. He also emphasized his belief that organizations like the Humanists are essential if the propaganda from the Churches is to be countered "in the interests of the moral and intellectual emancipation of the country." As for Cambridge, Crick was confident that the "disestablishment" of the Church there was "inevitable in the long run."[23] Hence his (anonymous) offer of a £100 prize for the best essay on what to do with the College chapels.[24] A rumor followed that the College Chaplains "intend to offer a £100 prize for the best essay on 'What can be done with Dr. Crick.'" So John Kendrew wrote asking Watson if he would be submitting an entry.[25]

Crick's concern about the influence of the Church of England was due in part to what he described as "a minor religious revival" in Cambridge. No doubt he had in mind the church at the center of the old town beside the market square—the University Church of Great St Mary's. In 1955 the charismatic and enterprising Mervyn Stockwood was appointed vicar there. In the five years of his ministry, the congregation grew from 200 in his first year to 2000 in his fifth, and there were weeks when between 2000 and 3000 undergraduates entered Great St Mary's. Remembrance Day in 1956 was especially noteworthy because it came in the midst of the Suez crisis. The Mayor and Council were due to attend the Remembrance service. Fearing Stockwood would mix politics with religion, the leader of the Conservatives urged loyal citizens to stay away. On the day, it was the Church that had to turn people away. The lines were still long when the limit of 2000 had been reached inside.[26]

Stockwood had in truth transformed this sleepy church. Its parish had been composed mostly of shopkeepers, and its merit to the University was as the center for its involvement in the formal events of the national and religious calendar. Under the new Vicar, the church became a center both for worship and for discussion and debate. Stockwood instituted an annual lecture series inviting prominent politicians one year and ambassadors the next. The young David Owen, a future Foreign Secretary, sharpened his debating skills at the question-and-answer time that followed the series of "informal services" Stockwood arranged. Then there was the lecture series—in 1957 it was on "Religion and the Scientists." Crick's department head, Nevill Mott, gave the first lecture on the theme of the separate nature of science and religion—just the kind of split mentality that Crick rejected. Unlike science, Mott pointed out, there is no certainty in religion "except through an act of faith."[27] Fred Hoyle, the cosmologist and friend of Crick's, contributed to the series. He explained how the evolution of the elements seemed to depend on a "whole series of quirks." The same, he argued, is true also of the origin of life. "As biochemists and biophysicists discover more and more of the detailed properties of living matter more and more random quirks will be found—apparent accidents without which life would not be possible. . . ." These events could be interpreted in two contrasting ways: either as evidence of "a deep laid scheme [theological view] or a monstrous sequence of accidents [agnostic view]." In this way he set out the alternatives of what is now called "intelligent design" and a world of chance.[28]

What Crick thought of his departmental head for inviting Stockwood in 1955 to address the physics undergraduates at the Cavendish Laboratory is not recorded. Montefiore and Stockwood sought in their own ways to fashion the gospel for a scientific and secular age; Montefiore was earnest to reclaim for religion "the high ground" of intellectual debate (i.e., God revealed in nature—the tradition of Natural Theology). The issues fought over were not just doctrinal or evidential, but underlying them was concern for the Church's loss of power and influence in an increasingly secular age. Belief in God had become almost "a kind of hobby," and religion "a minority interest, rather than the publicly accepted basis of life."[29] Crick, on the contrary, was equally concerned that in England, and especially in the Cambridge of the late 1950s, the Church still had too much power. His opposition to the chapel at Churchill College has to be seen in this context.

### The Ciba Meeting

The growing impact of science on society in the 1950s encouraged the American hero of the contraceptive pill, Dr. Gregory Pincus, to suggest to the august Ciba Foundation that the time was ripe for an unusual topic—"Man and His Future"—for their small international conferences. The "Man and His Future" conference was held in November 1962, and one of the sessions was given to eugenics and genetics. It comprised two papers, each written by a geneticist and Nobel Laureate—Hermann J. Muller, prevented from attending by illness, and Joshua Lederberg. Crick opened the ensuing discussion. Muller had devoted his paper to germinal choice, a voluntary system in which women would be able to receive sperm from a sperm bank compiled from males of chosen merit. Lederberg advocated what he called "euphenics" or improvement by medical intervention in embryological development coupled with advances in immunology enabling tissue transplants.[30]

Crick surprised the meeting by passing abruptly from stating his general agreement with these two papers to introducing his own agenda, namely, the ethical criteria behind eugenic programs. He explained:

> I think that we would all agree that on a long-term basis we have to do
> something . . . and because public opinion is so far behind, we should
> start to do something about that now. . . .
>
> I want to concentrate on one particular issue: do people have the right
> to have children at all? It would not be very difficult, as we gathered
> from Dr. Pincus, for a government to put something into our food so
> that nobody could have children. Then possibly—and this is hypo-
> thetical—they could provide another chemical that would reverse the
> effect of the first, and only people licensed to bear children would be
> given this second chemical. This isn't so wild that we need not dis-
> cuss it. Is it the general feeling that people do have the right to have
> children? This is taken for granted because it is part of Christian
> ethics, but in terms of humanist ethics I do not see why. . . .[31]

As he continued, it became apparent that, like his Cambridge friend,
Ronald Fisher, Arthur Balfour Professor of Genetics, Crick advocated
using financial means to encourage "those people who are more socially
desirable to have more children. . . ." And he reckoned that "the obvious
way to do this is to tax children," although he admitted, "This seems
dreadful to a good liberal because it is exactly the opposite of everything
that he has been brought up to believe. But at least it is logical."[32] He
agreed that money is not "an exact measure of social desirability," but at
the same time they are "fairly positively correlated." Charitably, Jacob
Bronowski suggested Crick's remarks were in the nature of "a *reductio ad
absurdum* of the method of direct control of the gene frequencies." More
importantly, confessed Bronowski, "I do not really understand what problem you [Crick] are trying to solve." If it was a question of raising the intelligence of the population, one should remember that "no one who has
known the children of accepted geniuses would suppose that a population would greatly benefit by there being several hundred of them."[33]

Crick remained quiet until one speaker, Colin Clark, the agricul-
turist, introduced the subject of the ultimate purpose of man on earth
and went on to answer that it is "to love God and obey his command-
ments." He ended by castigating "some brilliant and misguided scientists," who are supporting "a second cycle of eugenic doctrines." Crick
responded forcefully:

> I disagree strongly with Dr. Clark's remarks and with the standpoint
> from which he made them. It is clear that if we take the broad ethical
> question of ultimate ends we shall never reach any agreement. More-
> over, those of us who are humanists have a great difficulty in that we
> are unable to formulate our ends as clearly as is possible for those of us
> who are Christians. Nevertheless there are some ends that we can all
> share, even though we have these differences. It is surely clear that

good health, high intelligence, general benevolence—the qualities Muller listed—are desirable qualities which we would all agree on. We would agree also that these qualities are not uniformly distributed. There are people who are deficient in intelligence for example. . . . Surely it is a very reasonable aim to try to increase that . . . we are likely to achieve a considerable improvement . . . by using a very primitive knowledge of eugenics; that is, by simply taking the people with the qualities we like, and letting them have more children. Nobody is suggesting that we should have *enormous* numbers of people all with one father. . . . The difficulty I see concerns the techniques that are socially possible, in the present social context and in the social context of the next twenty or thirty years—a context which will change and which to some extent our views will help to change. . . .[34]

Crick's friend Peter Medawar, Nobel Laureate, immunologist, and Director of the National Institute for Medical Research, agreed with many of Crick's remarks but added

we ought to be warned by the very diversity of opinion in this room . . . my feeling at the moment is that human beings are simply not to be trusted to formulate long-term objectives—least of all Roman Catholics. What frightens me about Muller and to some extent Huxley is their extreme self-confidence, their complete conviction not only that they know what ends are desirable but also that they know how to achieve them.[35]

Medawar was later to have cause to feel some concern about his friend Crick, too.

### The John Danz Lectures

One of the talks Crick gave to the Cambridge Humanists was on "The Molecular Basis of Life" (1963). Here his purpose was to attack recent updates of vitalism, according to which any attempt to explain life calls for more than the resources of physics and chemistry, as then known. Particularly annoying to him were the writings of two physicists, Walter Elsasser[36] and Eugene Wigner.[37] When invited to give the John Danz lectures in 1966 Crick therefore chose as his theme, "Is Vitalism Dead?"

This was not to be a scholarly history of vitalism. Rather, taking vitalism as his theme, he wrote down "the sort of thing[s] that I find my friends and colleagues are saying about our present and future knowledge of biology."[38] Vitalism, he explained, "implies that there is some special force directing the growth or the behavior of living systems which cannot be understood by our ordinary notions of physics and chemistry."[39] His task, then, was to examine the most noteworthy areas of biology in which such difficulties were being encountered—the borderline between the living and the nonliving, the origin of life, and the brain and consciousness. Tackling the first of these occupied the major part of the book—a tour of molecular genetics and viral replication. The second he dispatched quickly before moving on to the third.

Here, after discussing the potential of computers to simulate humans, he turned to the notion of a soul. When did it originate in evolution? Dog-loving philosophers, he noted, "are more inclined to attribute souls to them than those who are not animal lovers. And if a dog, why not a worm, and so on?" Soul-like properties seem to emerge gradually. Then what about a baby? Does it have a soul "before birth, and at what moment does it get it, since it seems hardly likely that the unfertilized egg has a soul in the sense of which we are talking." What about the standard answers from religion? These he dismissed as "arbitrary nonsense."[40] As for the soul's survival of death and the power to communicate with a soul that is outside the body (extrasensory perception, ESP) he was understandably dismissive. He added that a Cambridge clergyman had implicated DNA in such a process—it was Montefiore—but then, added Crick: "*he* thought ectoplasm was good evidence in support of the Christian faith."[41]

The lectures concluded with an expression of his views on general education—the scandal of statutes forbidding the teaching of evolution in some southern states of the United States, compulsory religious instruction in the United Kingdom, and Cambridge University's "tremendous institutional support given to religion." He favored an alternative agenda that would provide some basic understanding of the scientific view of man and society that bridged the "two cultures" divide of C.P. Snow. All university students should take a course that he called "The Map of Science." It would cover "the broad nature of all the various sciences" and "encourage students to consider questions to which we do *not* yet know the answer. . . ." Here "the working of the brain certainly ranks high":

> It can be confidently stated that our present knowledge of the brain is as primitive—approximately at the stage of the four humors in medicine or of bleeding in therapy (what is psychoanalysis but mental bleeding?) that when we do have fuller knowledge our whole picture of ourselves is bound to change radically. People with training in the arts still feel that in spite of the alterations made in their lives by technology . . . modern science has little to do with what concerns them most deeply. As far as today's science is concerned this is partly true, but tomorrow's science is going to knock their culture right out from under them.[42]

All students should also study one science in more depth, and he agreed with Bronowski, Assistant Director of the Salk Institute, that the science of animal behavior was an excellent candidate for this role. "It teaches one of the great lessons of science in a rather personal form, namely that the familiar is not always what it seems."

In conclusion he admitted that vitalism is not dead, especially in that subject he considered backward—the science of the brain. But given adequate support of research he reckoned vitalism will be dead "but its ghost will remain. . . . A lunatic fringe always remains. There are still people today who believe that the earth is flat. . . . And so to those of you who may be vitalists I would make this prophecy: what everyone believed yesterday, and you believe today, only cranks will believe tomorrow."[43] The planned venue for these lectures was the University Presbyterian Church. When he learned this, Crick wrote warning the organizers that his lectures although not *militantly* anti-Christian, will nevertheless

> be directed against the sort of ideas at present held by many religious people. You may not know that I am an atheist and a few years ago resigned my Fellowship at Churchill College because they threatened to build a College Chapel. I myself have not the slightest objection to lecturing in the Presbyterian Church, but I think the Church Authorities might conceivably not be too keen that their building should be used for what they might regard as Anti-Christian propaganda.[44]

In 1966 Crick's book *Of Molecules and Men* was not received very kindly. Sir John Eccles, the eminent neurophysiologist and Nobel Laureate, praised the second chapter, describing it as a "Masterpiece in the exposition of one of the most important discoveries of our time." On the other hand, Eccles decried Crick's stalwart defense of chemistry as the ultimate explanatory resource for explaining biology, nor did he accept that the opponents of such reductionism are vitalists. Still less did he view with favor Crick's association of vitalism with Christians, especially Catholics, and anti-vitalism with agnostics and atheists. Eccles's most vehement criticisms were leveled at the third lecture of the book, "The Prospect Before Us." Here he judged the text to be much more superficial, "at times almost at the gossip level. And when Crick writes about the soul," complained Eccles "he cannot resist using flippancies and jibes." His conclusions about the transforming impact of modern scientific knowledge seemed to Eccles to have "the same status as dogmatic religious assertions."[45]

If Eccles showed some restraint, James Murray, writing in the *Virginia Quarterly*, did not. Focusing on Crick's discussion of the two cultures—the arts and the sciences—in his third lecture, Murray called it "conquest by violence." Take the statement that "tomorrow's science is going to knock their [the arts'] culture right out from under them." Murray called it "monumental in its arrogance and so crass in its insensitivity that in order to treat it seriously one must remind oneself that a Nobel Prize winner is speaking." Crick's vision of the death of the old literary culture and its supplanting by science in general and natural selection in particular left Murray speechless. "What can one say about such a frightening mixture of naiveté and bigotry?"[46] Murray did not consider it worth ridding ourselves of vitalism if that task requires the sacrifice of "the whole non-scientific culture." To think otherwise was a conceit.

> The social program that is supposed to carry out the cultural revolution is pure milk and water. It consists of a course of study in general science and animal behavior for all university students. The message of the apocalyptic angel turns out to be a report from the Committee on the Revision of the Undergraduate Curriculum.[47]

Only Crick's old friend C.H. Waddington in Edinburgh treated him more kindly, if somewhat flippantly, asking whether in attacking vitalism he was not "flogging a dead horse?" Apparently "the horse is still kicking, if faintly." He told Crick that his original proposal for the title of the review was "QUARKS, QUIRKS and the one and only CRICK in biology." When *Nature* refused this Waddington substituted QUARKS AND CRICK'S QUIRKS. The first title had been too long, the second suffered, the editor remarked, from "euphony." They used "NO VITALISM FOR CRICK."[48] Crick had earlier told "Wad," as he was known, that the original lectures were entitled "'Is Vitalism Dead?' ... The University of Washington Press, however, assured him that the term 'vitalism' was not understood in the States," so the vague title "Of Molecules and Men," was substituted.[49] Whether this play on Steinbeck's title *Of Mice and Men* was generally appreciated is not recorded.

### Implications of Modern Biology for Society

Another lecture Crick gave to the Cambridge Humanists had been on the topic "Is Biology Dangerous?" The opportunity to develop this theme in a lecture to a wider audience came when his old friend Annan, now Provost of Crick's old College, University College London, persuaded Crick to give the Rickman Godlee Lecture in 1968. "My dear Francis," wrote Annan, "Do do this for us! It would be marvelous to see you. [I] don't have a right to fag fellows of the College, but it would give added pleasure, if, as a Fellow, you were to accept."[50] Crick, who had recently accepted the invitation to become an honorary Fellow, complied, but three weeks before the lecture date he wrote to Annan: "I wish I could say that I am looking forward to the lecture but the nearer it gets the more I view it with apprehension! How much better it would be if you were giving it!"[51] Evidently he had some anxiety about expressing his views on this topic at a very public venue.

Under the title "The Social Impact of Biology," Crick sketched a future for society that called for rethinking our ethical and political foundations, as well as our religious ones. The event attracted one of the largest audiences the College had seen in recent years for a public lecture. After filling the Collegiate Theatre, those who came later had to be turned away. Although Crick had received unfavorable press comments on other occasions for the views he expressed, this lecture produced outrage, causing him to avoid public airings of subjects on which there clearly was so much sensitivity.[52] Interestingly, the Rickman Godlee lecture is not listed in his entry in *Who's Who*.

In the transcript of this lecture, Crick stressed our ignorance despite the recent advances made in biology; "it's really quite spectacular how much we don't know," so there is so much to discover. What then lies in the future? "In which direction" is biology going? Among the growing points he cited was our knowledge of the nervous system, which is the one part of our body that we associate most with ourselves:

> You can see this quite easily if I ask you, "Well all right, your stomach is, you know, getting inadequate and they're going to give you a new stomach, and so, do you mind having so and so's stomach?" Most people wouldn't mind very much, but if you were asked, "Would you mind having somebody else's brain?" I mean you would immediately object. It wouldn't be me![53]

Now developing "an idea of ourselves," our nature as humans

> will only come when we understand the nervous system. But when it does come I think the impact will be overwhelming . . . it will, for example, essentially make much of our literature unreadable, as much as alchemy is unreadable to modern chemists. I know many of you think that's unsound, I merely leave it with you as my personal opinion that when we understand in detail the nervous system we shall think so differently of ourselves . . . people in that time will hardly be able to understand much of our [conceptions of] mental processes today. But I think by that time practical problems will have made a considerable change in our outlook and in particular in our ethics . . . in particular we need a new ethical system based on modern science.[54]

What were these problems and what was the modern science he had in mind as the basis for their solution? On his list were overpopulation in developing countries, overcrowding in developed ones like the United Kingdom, a growing proportion of the population aged and senile, environmental damage, species extinctions, biological weapons, and addictive drugs. He attributed our reluctance to take sensible action to solve many of these problems to our Puritan tradition and to our upbringing. "Having had the experience of being brought up in one view of the world and changed over to another," he knew, he explained, "only too well that when you think a thing you don't always remove a lot of what you thought before." But he wanted his audience to try. His aim was to "raise the questions in your mind." He didn't wish to "provide ready-made answers of any sort. My aim will be fulfilled if I have made you think about these topics along your own lines."[55]

As to the advances in biology that could lead to problems in the future, he mentioned cloning, mosaic embryos, nuclear transplantation, species hybridization by cell fusion, and genetic engineering—the latter, he judged, a long way off for humans. But many of the problems we have right now are due to the fact that our evolution took place "over the last million years mainly under very different circumstances than those in which we find ourselves today." We should not "suppose that we are well adapted to our present environment, let alone to our future environment . . . it follows from this that the traditional rules of conduct are unlikely to be of much use." Anyway most of the foundations of these rules "are really unacceptable to a modern scientist." That is, "such things as revelation from some supposedly special person in some way. At the most they cannot be infallible. At the best they can only have fragments of insight." Take population. "We have to ask: do people have the right to have children, or have at least as many children as they please? It seems to me that we have to begin to consider that the answer to this question might be no, they don't have a right." Then he moved on to the question of controlling the "quality of a population." This, he admitted, raises even more problems:

> Let me ask you. Remember, there are too many babies already. You can produce babies easily—the mother can have another child—the cost is not very great. But [suppose] the child is handicapped. Wouldn't it be better to let that child die and have another one? . . . What about the child that is born incurably blind? . . . Is there any reason for keeping that child alive? And then, if that's so, if you ask those questions, you have to ask what defects are allowed? And you must know that in fact the medical profession already takes decisions on this thing. Quite a number of deformed children are allowed to die.[56]

He went on to offer some suggestions: "Wouldn't it be perhaps not better that babies were only deemed to be legally born when, say, they were two days old? . . .a short period for various reasons, including the feelings of the mother—in which the baby is examined to see if it's an acceptable member of human society. . . .we all feel uncomfortable by such a suggestion." Why is that? After all you "expect it to be done for motor cars—why should you not do it for people?"[57]

Then he turned to the aged and considered what we might call "legal death, . . . analogous to legal coming of age. It would be a formal thing at say eighty or eighty-five." Hastily, he added: "I don't mean to say you wish to kill off all people at eighty or eighty-five. A person might be deemed to be legally dead, and certain very expensive sorts of medical treatment might then be forbidden to such a person, . . . "[58] In the 1970s this problem was, of course, already present, and the issue of whether or not to turn off the oxygen "and let that vegetable, which is all one can call it, die," was with us. That for Crick was "perhaps a clear case," but what about borderline cases? The increase in the elderly population will make such problems acute. "It seems to me that we cannot continue to regard all human life as sacred." The principle should be to "look at the quality of the person involved. We must not allow the idea of the person, [as] one soul, which is really the basic theological idea behind this, to guide us."[59]

He ended the lecture with two warnings, one on the subject of the legal system. If the law on an issue like drugs or divorce is consistently flouted it "brings the law into disrespect." Speaking of pot, he opined:

> It is widely used among young people. It is quite easy to go and get some pot by just asking one or two people, and you will soon be able to get it and you will find people are taking it. It is not just something of a few flower children. The present position is in itself a bad one and I would suggest it is better to legalise it now and regulate the quality of what is provided and to whom it is sold and when, than to leave it in the present position.[60]

The other warning he made concerns education. He wanted to see our puritanical attitudes changed, and this came down to a matter of education. What are children being taught at school and by the BBC? The Christian religion of course. Provocatively, he added: Now you know, Christianity may be okay between consenting adults in private, but I really do feel the time has come to say somewhat firmly that it should not be taught to young children.[61]

What children need instead, he explained, is to understand something about the evolution of humankind and the way in which natural selection has shaped us. Where, for instance, does a humanist find foundations for his code of ethics? And from whence does he imagine our system of justice originated? What, moreover, can the humanist make of all this God worship from ancient times to the present day? It all goes back to adaptive function. All such systems from the worship of ancestors to faith in the God of Christianity have served to bind the members of a society together in the struggle for life. The same goes for our ethical systems. They provide a set of rules to govern conduct, and as such they function to preserve the cohesion of a society. A sense of guilt, a sense of right and wrong, and a conscience about our wrong-doing are imparted in education, but we may suspect that there is also an inborn element that selection has favored.

Annan subsequently offered Crick his own response to one point in the lecture. His eyebrows had shot up, he wrote, "when at one point in your lecture you said that all that was required to alter current practices in regard to a number of social problems was for us to discuss matters thoroughly and change our opinions." One hundred and ten years ago, Annan explained, John Stuart Mill took a similar line. Discuss the issues freely and truth inevitably comes to the surface. "Of course there is something in this, but no one to-day much believes in this rather simple positivist explanation of how society changes."

> I am entirely with you in thinking that unless we begin discussing the kind of topics you raised seriously and soon, we shall be unprepared for the situation which is going to arise owing to the population explosion. But at the same time I am afraid that however sensible the conclusions which may arise from such rational discussion, they will not be <u>accepted</u> as conclusions unless considerable changes have also taken place in our social structure and relationships.[62]

Crick did not repeat such performances in the latter years of his life, remarking that one needed to know more and he was "rather rash then."[63] But with the publication of his friend Jacques Monod's book, *Chance and Necessity* in the U.K. by Collins in 1972, he agreed once again to address the Cambridge Humanists, describing this talk as "A sermon with his book [*Chance and Necessity*] as the text." This was the last of a series that he had given over the years: "The Molecular Basis of Life" (1963), "Life in other Worlds" (1966), and "Is Biology Dangerous?" (1966).

*Chance and Necessity* had its origin in Monod's Robbins Lectures at Pomona College, California in February 1969. The following winter Monod gave a course on the same theme at the Collège de France in Paris. The French edition of the book, *Le Hazard et la Nécessité* (1970), met with a particularly hostile reception among French intellectuals.

Nor was there a lack of hostility to the American edition (1971), and Alfred Knopf could soon record a quarter of a million copies of that edition sold. The message of the book was the duty of scientists "to apprehend their discipline within the larger frame of modern culture." That exercise led Monod to call for the clean separation of objective knowledge (i.e., science) from ethics, but for the formulation of ethical principles in the light of current knowledge of our *biological* nature.

Dear to Crick's heart was the theme of chance and necessity, instead of intention and design. He admired, too, the powerful and logical development of Monod's argument. Although not sharing his friend's socialist inclinations, and having, he admitted, "more extreme" views on eugenics than Monod, he was keen to address the Cambridge Humanists on the book. After summarizing Monod's argument, he gave his own "gloss" on it, and in his notes wrote:

> What sets of values are <u>incompatible</u> [with our knowledge of biology] <u>or</u> how can we alter things (or nature, or ways of life so that such incompatibles be made compatible) . . . we strictly need to know much more about man, especially his higher nervous system. This then emerges as the major scientific knowledge we require (and not urban renewal, or pollution or even population control). Moreover, when this knowledge is obtained it is virtually certain that man's whole view of himself will be radically changed. It is this point of view which modern culture almost totally rejects . . .
>
> Because we live in another culture. We are surrounded by Barbarians. Sophisticated, erudite and passionate but basically barbarians because they cling to the old knowledge (which as Monod has shown) imposes a particular purpose to man or to the universe. They cannot stomach the idea of Nature evolving in an open-ended way <u>without</u> any foreseeable target. . . .[64]

Following the lecture, the chairman John Gilmour wrote thanking him "again for your time and thought and histrionic ability for us all last night. It was a great success, and we have got many forms filled up—which may lead to a rebirth of the Cambridge Humanists."[65] Quite a revivalist event! Was Crick recalling his religious upbringing when he described this talk as a sermon with Monod's book as a text?

In October 1972, Crick wrote a short essay for the press entitled "Why I Am a Humanist." It begins by stressing that there are many shades of meaning to the term. "I use it," he wrote, "in the sense defined by the Cambridge Humanists who believe 'that human problems can and must be faced in terms of human moral and intellectual resources without invoking supernatural authority.' They do not pray, and most of them disbelieve in any form of life after death. They feel we are on our own, and wonder what we should do about it." He went on:

> We have a deep need to know why we are here. What is the world made of? More important, what are *we* made of? In the past[,] religion answered many questions, often in considerable detail. Now we know that almost all these answers are highly likely to be nonsense, having sprung from man's ignorance and his enormous capacity for self-deception.[66]

Then he focused on the advances in scientific knowledge that had been achieved in the last hundred years, from astronomy, the physics of matter, nuclear physics, but especially the "intellectual revolution initiated by Darwin and Mendel," that has "shown in principle how all living things on the earth could have evolved from very simple organisms, by the subtle and beautiful mechanism of Natural Selection." Recently molecular biology has "practically obliterated the distinction between the living and the non-living":

> Against this solid background of knowledge the simple fables of the religions of the world have come to seem like tales told to children. Even understood symbolically they are often perverse, if not rather unpleasant (the story of Adam and Eve, for example). Thus modern science has given humanists wide areas of fundamental knowledge on which to base their faith. . . .
>
> Humanists, then, live in a mysterious, exciting and intellectually expanding world, which, once glimpsed, makes the old worlds of the religions seem fake-cosy and stale. It is to me a matter of great surprise and regret that so few poets have tried to live in this new world.

After describing some of the practical problems of society that humanists have been concerned with—abortion, homosexuality, birth control, and divorce—he turned to Cambridge:

> One peculiar feature of the Cambridge scene is the number of dons who hold humanist beliefs but are not prepared to do anything else about it. They are indifferent to the Cambridge Humanists and will not take action at the College level or at the University level to promote humanism or to reduce the influence of Christianity in Cambridge, even though they believe the latter to be unfortunate. The reasons given for this apathy are usually so trivial that only in Cambridge could they be accepted.

The essay ends with a typical piece of straight talk from Crick: "So now you know why I am a humanist. Are you one? What are you going to do about it?"

## The IQ Debate

In 1969 the educational psychologist Arthur Jensen published a paper on intelligence quotient (IQ) differences between blacks and whites and criticized the effectiveness of the "Head Start" program.[67] William Shockley, co-inventor of the transistor and enthusiast for eugenic policies, then pressed the National Academy of Sciences to fund a major research project on the subject. When several members of the Academy objected, Crick signed a letter of support for both Jensen and Shockley. Subsequently he wrote to Medawar describing the kind of project he favored. It would be a voluntary program to promote differential fertility and favoring those "generally considered valuable to society":

> A fairly moderate shift in fertility would make for a much more attractive and acceptable set of people. . . . Such a policy <u>in the long run</u> (within the next 100 years) is virtually certainly to be tried . . . the real issue is not race but "class" again, very broadly, between the rich and the poor. Upper and upper-middle class be encouraged to have say 3 or 4 [children] on average and manual laborers and obviously dim and disturbed people to have 0 or 1.
>
> Any reasonable selection of social virtues would produce significant and possibly massive results.[68]

Medawar replied that

> I hadn't realized you were christened "Francis" after Francis Galton, my opinions of whom are embodied in the enclosed piece from T.L.S. [*The Times Literary Supplement*],

and he explained:

> There is no means of encouraging a favoured class to reproduce without offering inducements that will not have to be withheld from a less favoured class, and politically this is simply not on. Shrill cries of "elitism" would prohibit any rational discourse.
>
> In any case your project is an example of just that kind of Utopia or holistic social engineering Sir Karl Popper, FRS, exposed and confuted in *The Open Society and its Enemies*.[69]

All Crick could write in response was that one "must distinguish sharply between what is at present politically impossible . . . and what is, in the long run, socially desirable."[70]

A stimulant to Crick's eugenic inclinations at this time was an article written by the American geneticist Richard Lewontin in 1977. Lewontin had conducted a searing critique of twin studies, pointing out the failure to break "the correlation between genetic and environmental similarities." The kind of adoption studies that would adhere to clean experimental design, Lewontin argued, would be very difficult to achieve, and the failure to do so "renders all work [to date] uninterpretable."[71] And in 1974 there were additional grounds for his skepticism. It was then that the data of Sir Cyril Burt, the authority on the heritability of intelligence, had been judged fraudulent.

Needless to say, Crick considered Lewontin's critique to be politically motivated. But he did not go public on the subject. Following the Rickman Godlee lecture of 1968, Crick had withdrawn from the debate. And when Shockley sought Crick's public endorsement for his Foundation for Research and Education on Eugenics and Dysgenics, Crick replied, "I must stick to my decision not to get involved with topics of this sort for the next few years. I am in the process of changing my scientific interests, and I find there is a limit to the number of things I can concentrate on at any one time."

Crick rarely lent his name to campaigns for public support of causes. True, at the time of the IQ controversy over the heritability of intelligence, he signed a joint letter with other eminent scientists requesting freedom to research the subject in whatever direction seemed fruitful. He pressed Max Perutz hard to persuade him to sign as well. In 2003 Crick did sign a Nobel Laureates' statement on the population problem, and one supporting the establishment of a "Darwin Day" to remind the general public of Charles Darwin's contribution. But such actions were rare. The population issue, however, did continue to concern him, and although he never directly addressed the issue of abortion publicly, he did in 1980 respond privately to the request for advice on the subject. It reads:

Dear Duval, September 18th 1980

I am not a Catholic—if anything I am an agnostic with atheistic leanings—so I certainly do not regard a human fetus as being sufficiently alive that destroying it would rank with taking the life of a mature human being. In addition such a fetus can have no knowledge that its life was taken and I do not believe it has a soul (in the sense that the soul is something that survives after death). [It cannot reason on the cause] why it should now be destroyed.

However, there is another [factor]. The fetus hardly knows what is going on, but its parents do. So while it is foolish to worry about the feelings of the fetus, or its soul (what kind of an afterlife would it have anyway, if there were a life after death) it is not unreasonable to consider the parents' feelings. This presents no problems if the parents agree but, if they differ, clearly some compromise must be sought. This will depend not only on the circumstances (will the baby be welcome at this time, etc.) but also on the beliefs of the people concerned. So I feel you did the right thing in imparting your wife's feelings but you may well wish to make her change her view (which I personally regard as irrational) in case the problem arises again. Yours sincerely Francis Crick[72]

Shortly before she died (October 1955), Crick's mother, Annie, had written to her firstborn son:

I hope the second half of your life will be full of interesting work, and possibly great success—though I can hardly hope to see it. But in spite of what you said on Sunday, I am still a little afraid that your attitude to people may be detrimental to you and I would like you to keep the words I quoted in mind:

"Know that pride,
However disguised in his own majesty
Is littleness: that he who feels contempt
For any living thing has faculties
Which he has never used: That thought with him
Is in its infancy . . ."[73]

When near the end of his life Crick was asked about the meaning of this letter he was puzzled and commented that he did not know to what she was referring.[74] Could it have been her son's rejection of the Christian religion or his disapproval of those with a scientific education who remained believers? Was it to her, pride, to be so convinced of his understanding of the matter? Yet he got on famously with the lively but conventional British broadcaster Robert Dougall, a devout Anglican, and later with the Catholic neuroscientist Christof Koch. Nevertheless he was always very clear about his position, as in his response to a correspondent in 2004:

Dear Dr. Hsu, I am agnostic with inclinations to atheism. I think the God hypothesis is bankrupt, and that we have evolved from non-living matter by Natural Selection. Our DNA model only confirmed me in these beliefs.[75]

Two years earlier he had told a reporter from *The Daily Telegraph:*

I went into science because of these religious reasons, there's no doubt about that. I asked myself what were the two things that appear inexplicable and are used to support religious beliefs: the difference between living and nonliving things, and the phenomenon of consciousness.[76]

Why did Crick feel so strongly the need to speak out on these contentious issues? It was surely his convictions as to the future of humanity and of the globe that sustains us, a conviction that Bernard Levin, columnist for the British tabloid, *The Daily Mail*, did not share. Levin felt that Crick offered a prize example of how clever people can also be silly—first over the Chapel issue, and next in 1966, when Crick complained about his friend, Noel Annan. As Provost of King's College, Cambridge, and a Humanist, Annan, by this time Lord Annan, had been "wont to read the Lesson at services in the College chapel." But Crick expected Annan as a humanist to draw the line at such public support of religious events. The tone of scorn in Levin's column at Crick's "silly" criticism caused Crick's old naval friend, Ronald Williams, to respond:

> Dr. Crick is a genuine humanist; he is concerned with the future of mankind, and is afraid for it. He feels that humanity no longer has time for comfortable mumbo-jumbo and irrational, superstitious taboos and escape-beliefs. . . . Possessed of the devil Ignorance, humanity is stampeding toward the brink of the cliff like the herd of Gaderene swine. This is why Crick is aggressive, Mr. Levin. Perhaps he is right. Perhaps you are wrong. Perhaps you should think a little more deeply before insulting a great human being whose intelligence and general knowledge are beyond your limited understanding.[77]

# *Biological Complexity*

*First molecular biology made a splash because of the apparent, astonishing simplicity. Now it continues to be exciting for the opposite reason. It couldn't be better if it had all been planned.[1]*

*When facts come in the door, vitalism flies out of the window.[2]*

At his opening John Danz lecture on 8 March 1966, Crick made the following statement: "The ultimate aim of the modern movement in biology is in fact to explain all biology in terms of physics and chemistry."[3] This surely sounded like reductionism raw and bold—eliminative reductionism. His reputation as a hard-nosed reductionist, it seemed, was well justified. Biology was to be dissolved using the exact sciences of physics and chemistry. Such a claim is a statement of the nature of being; that is, being alive is basically just physics and chemistry. As philosophers would say, this is an ontological statement.

The term "reductionism," however, is used in a variety of ways with different meanings. Often rhetorical impact overrides any literal intent. Take first the ontological meaning of the term—this represents the most far-reaching claim on behalf of the exact sciences over biology. There is more than a whiff of disciplinary imperialism about it. But are we being deceived? Was it Crick's flamboyant and provocative nature getting the better of him? Often when such claims are made they boil down to recipes for the level at which to attack and the appropriate methods to be used. Thus, although one might use as *methods* of research those that come from the exact sciences, and begin at the molecular level, working one's way up, such preferences do not necessarily entail support for *ontological* reductionism. It was characteristic of Crick to make a provocative statement, and then qualify it. He meant, he added, that the exact sciences of physics and chemistry provide "a 'foundation of certainty' on which to build biology just as Newtonian mechanics, even though we know that it is only a first approximation, provides a foundation for, say, mechanical engineering."[4]

Crick chose to speak on vitalism because of the recent "resurgence" of vitalistic ideas. He picked on the "distinguished physicist" Walter Elsasser and proceeded to excoriate him for his book *The Physical Foundations of Biology*. Elsasser's sin was that he introduced "biotonic laws" to guide mechanistic processes in living things, because there are, he claimed, "biotonic phenomena in the organism that cannot be explained in terms of mechanistic functions."[5] Conceived in the premolecular period of biology, "it is," wrote Crick, "a beautiful example of the confusion that can be brought about by ignorance":

> For example, Dr. Elsasser is surprised to find that genetic information stored in a lobster is not in its hard external parts, but in its "soft" tissue, as he calls it. He seems to think that information so stored will very easily be disturbed by thermal noise . . . it is in fact stored on a polymer [DNA]. . . . Thus the simple facts of chemistry, which as a physicist he appears to ignore, allow the cell to store an immense amount of information, very stably, at the molecular level. . . . His whole argument is based on the apparent dilemma that the large amount of information needed to construct the adult organism cannot conceivably be stored in the germ cells.[6]

Yet the amount of information needed to make a hand, he admitted, was not yet known. But conceivable mechanisms "are well within the information capacity available to us." Thus did Crick defend the explanatory potential of molecular biology in the field of embryological development, surely rendering Elsasser's "biotonic laws" redundant. Ten years later Crick was asked to advise on a paper sent to *Nature* by Elsasser on the nonreductive nature of "organismic" biology. Responding, he complained that the term organismic "could hardly be more unfortunate." To Crick it suggested "those things which are due to many complicated (so far unexplained) interactions, producing a remarkable reproducible process." He concluded, "I am getting rather tired of all this talk about reductionism, because much of it is shooting at straw men. It's getting almost as tiresome as all the nonsense written about the central dogma. I'm going to be at the Salk for eight to nine months, and if I get the time I'll write an article or a short book about it."[7]

He never wrote such a book, but when interviewed in the 1990s, he again rejected the naïve view of reductionism. "No reductionist," he claimed, "would want to explain anything as 'merely the nuts and bolts.'

It's the nuts and bolts of how a number of things interact together, and, of course, in our brain, it's a very complex interaction."[8] The neurophysiologist, John Eccles, however, complained that reductionism had become "enthroned in biology as some sort of religious dogma." He favored instead "Emergentism"—"there are new emergent properties of matter not predictable from chemistry, just as chemistry is not predictable from physics."[9] The whole, as is often said, is more than the sum of its parts. On this point Crick did not disagree. But physics and chemistry, Crick urged, have "wholes" and these are not simply the sums of their parts. Benzene, for instance, "is certainly not an arithmetical sum of its component parts. Nevertheless a theoretical chemist can deduce its behavior by doing the sum the right way; that is by using the methods of quantum mechanics."[10] This does not mean that the features of higher levels can be predicted from those of lower ones, for there is the question of the manner in which parts at a lower level are put together.

Here molecular biologists were discovering examples of the remarkable phenomenon of self-assembly. The separate protein and nucleic acid components of the tobacco mosaic virus can self-assemble spontaneously, as can the head and tail proteins and the nucleic acid of a phage particle, in both cases yielding particles capable of infection.[11] Thus, the physical forces operating between molecules can bring about organized structures. One can picture a history of successive levels of organization being built up and changes in body plan introduced that, as Crick remarked, would depend on "rare events (mutations) and what may be chance factors in the environment. . . . Consequently there is real doubt whether the actual process of evolution is predictable. It may be history rather than science. That is, chance may produce effects which basically alter the historical process. I think this distinction—between the behavior of an organism and its evolution—is of crucial importance, even if some borderline cases cause difficulty."[12] Hence, biological explanation is of two types: One explains "how, from a knowledge of its parts, one can predict its behavior," and the other type seeks to discover how it evolved. Here Crick would have sided with the philosopher Karl Popper on the distinction between the biological and the physical sciences, the former not possessing laws like those of physics that apply at all times and everywhere.

It was not, Crick explained, that all investigation of organisms should proceed from the molecular level—the "bottom-up" approach. "A biological system," he added, has long been regarded as "a hierarchy of levels of organization, the 'wholes' of one level being the parts of the next." At what point one attacks "is always a matter of tactics." A "simultaneous attack at more than one level," he believed, "will in

Chapter 17

the long run, pay off better than an attack at a single level, even though in the short run one may concentrate on one level at a time."[13] Thus the "spectacular recent advances" of the genetic code came from using a combination of approaches, studying both the whole organism and its parts. The latter were extracted chemical substances but the "whole" was the bacterial cell, albeit not a complex multicellular organism. Moving up the ladder of life, however, was not to prove easy.

### Complexity

Over the span of Crick's career in molecular biology, the simple suggestions as to how DNA unwinds and duplicates, how genetic information is expressed in protein synthesis, and what determines the fidelity of gene duplication were transformed by the discovery of complex systems of cellular control. Back in 1954, Max Delbrück had bet Crick $5 that the two strands of the double helix of DNA would not be shown to separate when duplication occurs. This was a bet Crick was happy to accept, for as the saying went, if Delbrück said an idea or experimental result was wrong—"I don't believe a word of it," was his favorite response—it stood a good chance of proving right. Granted the unwinding required was substantial, but confronted with such a problem Crick's tendency was to assume Nature has found a trick to deal with it. This would introduce complexity into the very simple idea of duplication suggested in 1953.

As reported in Chapter 13, Crick had been present at the Baltimore conference when the St. Louis biochemist Arthur Kornberg announced the success of his group in synthesizing DNA. Here the star performer was the enzyme fraction Kornberg was later to call "DNA polymerase." He had found the juice with this enzymatic activity back in 1955, but at a time when his thoughts were far removed from the Watson–Crick structure for DNA. Indeed, he later told Crick that he had not believed in their proposed mechanism of DNA replication when he started this work. He recalled,

> Watson and Crick naively proposed that the replication of DNA would proceed *spontaneously* were the building blocks to align themselves along the parental chain serving as templates. There is little likelihood that any cellular reaction, let alone one so intricate and vital as the rapid and faithful replication of DNA, could proceed without the catalysis, direction, and refined regulation that enzymes provide.[14]

In fact, Watson and Crick had left open the question whether enzymes are or are not needed for DNA duplication,[15] but to Kornberg it was unthinkable that such a process would not need them. Hence, to the biochemist the saying: "DNA makes DNA, DNA makes RNA, RNA makes Protein, and Proteins do everything else," was inadequate because proteins act to make *all* of these syntheses possible.

By the end of 1958, Kornberg had succeeded in demonstrating that the new DNA synthesized in each case mirrored in its constitution that of the five different DNA "primers" he provided. The latter, he then realized, were actually acting as *templates* for the pattern of bases on the synthesized DNA chains. There had, in short, been a specific synthesis of more of the same in each case. Not only this, but after devising a method for determining the distribution of nearest neighbors among the bases (a method for sequencing had yet to be invented), he was able to show that the two chains of the duplex do indeed run in opposite directions. This provided the most wonderful support to the Watson–Crick model, for it came not from X-ray crystallography but from biochemistry.

DNA polymerase was to have a wider future, as suspicions grew that it was one of a family of polymerases with a variety of functions. Crick recalled that changes in the bases of DNA can scarcely fall below 1 in 10,000 based on observed mutation rates of genes. But human DNA contains three billion base pairs per germ cell, so the potential for alterations in the genome leading to mutations must surely be high. How do organisms maintain such low mutation rates? Crick and Leslie Orgel decided there must be an error-correcting mechanism and they wrote to Kornberg predicting "that the enzyme he was studying that replicates DNA in the test tube . . . should contain within itself an error-correcting device. . . ."[16]

This prediction proved to be the case. But in the early 1950s no one, including Kornberg himself, had any idea just how complicated DNA replication would turn out to be. The tool kit comprised some 20 proteins, among them the helicase, a kind of biochemical tweezers, to prize open the two strands, and the primase to synthesize the "primer" (made of RNA) needed to initiate chain formation by one of the DNA polymerases. This machinery has to deal also with the opposing direction of the two parental chains, meaning that the polymerization process creating the two daughter chains has to proceed in opposite directions simultaneously (Fig. 17.1), deploying several members of the family of DNA polymerases. In truth Nature has created a remarkable miniature sewing machine that undoes the helices (using a helicase), stitches simultaneously in two directions, corrects errors by excision and patching (using the family of DNA polymerases), and sews the helices together (using a ligase), all of this going on at the same time and at breakneck speed! The extraordinary fidelity of the process lay not in "other laws" but in a complex physicochemical mechanism involving a whole regiment of enzymes. Reflecting on this incredible process, Crick remarked,

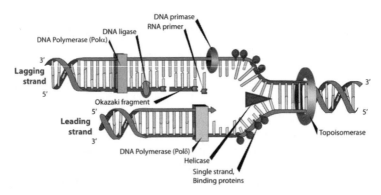

**Figure 17.1** DNA chain replication. The antiparallel nature of the chains in DNA prevents synthesis of both the daughter strands proceeding in the same direction (toward the replication fork). The leading strand shows continuous synthesis of DNA toward the fork (5′ → 3′). The laggard strand shows discontinuous synthesis (also 5′ → 3′), but here it is going away from the fork. Short stretches of RNA are first synthesized, and then they are replaced with DNA and any gaps filled. (Crick had at one time suggested that perhaps synthesis proceeds from opposite ends of the chromosome.)

> The basic idea could hardly be simpler. . . . The underlying mechanism may be simple, but if the process is biologically important, then, in the long course of evolution, natural selection will have improved it and embroidered it, so that it can work both faster and more accurately. It is because of this baroque elaboration that biological mechanisms are often so difficult to unravel.[17]

Looking back over 21 years, Crick recalled how in 1953 he and Watson "totally missed the possible role of enzymes in repair." Indeed, it was not until Crick read about (C.) Stanley Rupert's "early very elegant work on photoreactivation" that he "came to realize that DNA is so precious that probably many distinct repair mechanisms would exist."[18] The first report on this early work of Rupert's had, in fact, been squeezed into the program of the Baltimore Conference of 1956 attended by Crick.[19]

DNA repair and reverse transcriptase were not surprises to Crick, but he was taken aback by reports in the 1970s of the "editing" of messenger RNA. He first encountered this phenomenon at the chromatin symposium at Cold Spring Harbor Laboratory in 1977. The following year Walter Gilbert, Harvard University's Professor of Molecular Biology, wrote a short report for *Nature,* entitled "Why genes in pieces?" He began: "Our picture of the organisation of genes in higher organisms has recently undergone a revolution."[20] Like the film editor who cuts out unwanted stretches of film, so the protein machines of the

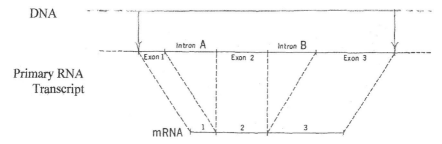

*Figure 17.2* DNA splicing. Top line, a portion of DNA of the chromosome; below, the resulting primary RNA transcript. Note excision of portions of the latter (introns) yielding the messenger RNA composed of the three exons.

cell excise unwanted stretches of the primary RNA transcript from DNA—the *introns*, leaving intact the sequences that constitute the messenger RNA—the *exons*. This editing process, known as "splicing," revealed a further level of complexity in gene expression (Fig. 17.2), thus introducing a qualification to Crick's sequence hypothesis (see Chapter 13). Considering the much greater size of the introns compared with the exons meant that there must be much RNA in the cell for which no function had yet been found, unless it was all just "junk" DNA.

In seven pages of *Science*, Crick explored the ramifications and possible significance of the discovery. He explained how, when he came to California in September 1976, he "had no idea that a typical gene might be split into several pieces." What is more, he doubted that anybody else had. And he concluded,

> There can be no denying that the discovery of splicing has given our ideas a good shake. It was of course already surmised that the primary RNA transcript would be processed in some way, but I do not share the view that has often been expressed that splicing is only a trivial extension of previous ideas. I think that splicing will not only open up the whole topic of RNA processing, . . .but in addition will lead to new insights both in embryology and in evolution. What is remarkable is that the possibility of splicing had not at any time been seriously considered before it was forced upon us by the experimental facts. . . . Lacking evidence we had become overconfident in the generality of some of our basic ideas.[21]

This paper marks the end of his discussions of contemporary developments in molecular biology and genetics. It is also the year of his first publication on neuroscience (see Chapter 19).

## New Agendas

The move of the MRC Unit to the new building on the Addenbrooke site in 1962 (see Chapter 13) had involved an expansion, bringing Frederick Sanger's biochemical group and Aaron Klug's virus group to join the staff of the former Unit. In the process the *Unit* of Molecular Biology became the *Laboratory* of Molecular Biology (LMB). The constitution drafted by the members of the Unit for the Laboratory established a set of divisions, each one independent in the choice of research agendas, appointments, etc., the members being responsible to their respective Head—Protein Chemistry (Sanger), Protein Crystallography (John Kendrew), and Molecular Genetics (Crick and Sydney Brenner). A central administrative group was then to coordinate the research programs of the several divisions. Its chairman (Max Perutz) would deal directly with Head Office in London. The degree of independence and minimal bureaucracy of this plan proved to be a factor in the continuing success of the group.

Although the first five years (1962–1966) in the new building had for Crick and Brenner been dominated by their work on the genetic code, plans for the extension of the laboratories were already under discussion in 1963.[22] This meant considering future agendas. Brenner and Crick agreed that the new knowledge should be put to work at a higher level of organization. Signaling this change was the alteration of the title of their Division of Molecular Genetics to the Division of Cell Biology. The plan was to search for a small multicellular organism and to use both molecular genetic and classical embryological methods in the study of its development. It was an early example of a plan to explore a traditional field of biology armed with the confidence, the tools, and the know-how acquired in molecular biology. From this decision came Brenner's choice of the worm, *Caenorhabditis elegans* (*C. elegans*), as a model organism for the study of development and much else.

Crick meanwhile envisaged a large-scale research program on the colon bacillus, treating it as a self-reproducing factory. The aim would be to explore the local and the global controls that operate in the cell. Known as "Project K," it was not pursued. Instead, he was attracted by the attempts made using gradients to account for organization in embryological development. Tradition had it that features of the embryo, such as the polarity between head and tail, are due to a gradient either of activity, such as metabolism, or the transfer of a form-determining substance—a "morphogen" to use the mathematician Alan Turing's term.[23] The gradients championed in the 1940s by the Chicago zoologist, C.M. Child, Crick judged, "were the results of development rather than its cause," and he commented that "an outsider to embryology has the impression that in recent years gradients have become a dirty word . . . and a feeling has grown up that diffusion is not a fast enough mechanism for establishing gradients. . . ."[24] But then he read a paper by the developmental biologist Lewis Wolpert stating that developmental fields only involve distances of a hundred cells or less. Over those distances, Crick reckoned, a morphogen traveling from "source" to "sink" could serve by its concentration to indicate positional information. Then cells along the pathway of the morphogen would "know" where they were in the organism because of the concentration of the morphogen at that point, and would develop accordingly. Encouraged by this news, Crick set to work and aided by discussions with the mathematician Mary Munro and postdoc zoologist Peter Lawrence, he explored diffusion mathematically, treating it as a random walk process (taking successive steps in random directions). He was excited to find that the result of their calculations showed diffusion to be adequate over the timescale of about 3 hours and for distances of around 70 cells—well within the range suggested by Wolpert. After mentioning some objections, Crick concluded in a characteristically confident manner that

> It is my belief that mechanisms based on diffusion are not only plausible but rather probable. Nature usually has such difficulty evolving elaborate biochemical mechanisms (for example those used in protein synthesis) that the underlying processes are often rather simple. If this serves to make the idea of diffusion gradients respectable to embryologists it will have served its purpose.[25]

Two weeks after publication of this paper in *Nature,* the editor published a letter from Elizabeth Deuchar. It reads:

> As an embryologist who started work during the heyday of "fields" and "gradients," I suppose I ought to be grateful to Dr. Francis Crick for allowing me a nostalgic look back at these long discredited concepts which he has now resurrected—or should I say canonized—with the double reputation of his own halo and some elegant mathematics. There is, however, one point that he appears to overlook: the extreme rarity with which sheer diffusion processes occur in living systems. Twenty years ago my better-informed colleagues told me about active transport and permeases. Ever since then, if materials have diffused in and out of my experimental embryos, I have regarded it as a sign that they are dying or dead. A sheet of frozen-dried tissue, extended between source and sink, might fit Dr. Crick's formulae, but—alas—it would not differentiate.[26]

Crick then wrote to Deuchar. She replied pointing to research showing not passive diffusion in embryos but *active transport*, i.e., the cells doing work to move the form-determining substances. She also expressed impatience with him for addressing only the one-dimensional model of a sheet of cells whereas "all the fundamental early differentiation processes concern balls of cells."[27] Other embryologists also wrote to him or conversed with him in a similar vein, causing Crick distress "that the main idea I wanted to put over has been completely missed."[28]

Knowing that he was to attend the forthcoming symposium of the Society for Experimental Biology in September 1970, he wrote to the organizer suggesting that he could present a paper on "A Basic Theory of the Scale of Pattern Formation" that would explain his idea once more. This was, he explained, that "all ordered patterns in biological structures originate when the region is small. For most cases the word small implies the distance of one millimeter or less. I am naturally rather hopeful that this rather rash generalization will apply over the whole of biology and, in particular, to both botany and zoology."[29] It was supported by Munro's detailed mathematical treatment.[30]

Another contributor to the meeting was Lawrence. He had an Agricultural Research Council fellowship in Cambridge University's zoology department investigating the mechanism of pattern formation in the cuticle of the milkweed bug. It was in 1969 that Crick and Brenner had first encountered him when they heard about a talk he was to give in the Department of Genetics on his work. Arriving late, these two famous figures caused a bit of a stir, but the confident speaker was not fazed. The talk excited Crick and, when they met personally, Crick suggested Lawrence might like to come and work with them at the LMB. Although taken aback, Lawrence accepted and from the ensuing collaboration came evidence of a gradient of positional information in the insect epidermis that was supportive of Crick's claims for a diffusion model.[31]

Subsequently they wrote a paper for *Science* on the work of the Spanish developmental biologist Antonio García-Bellido. This was an example of Crick's concern when an important piece of work was not being adequately appreciated and understood. Crick and Lawrence wrote their own account of how they understood García-Bellido's results and published it in *Science*, where it would be widely read. Again location in the embryo was all important. García-Bellido had identified borders that form over the insect epithelium, dividing the surface into what he called *compartments*. After a certain time in development, cells cannot cross these boundaries and their fate is then determined by the compartment in which they lie. Here was further evidence for determination of cell type by position. This work excited

Crick and Lawrence, and they remarked, "For the first time there is the real prospect of understanding the logic behind gene deployment in pattern formation."[32] History, it seems has borne them out.[33]

### Chromosome Structure

Although immersed in these developmental studies, Crick was not unwatchful for news about DNA. Two questions were proving baffling. Why do some species have so much more DNA than others and with no relation to the degree of their complexity? And how do the long DNA helices that constitute the chromosomes fit into a body the size of the cell nucleus? Even more surprising, when the cell enters the division process, the chromosomes contract to anything from a 700th to a 10,000th of their unfolded length and become clearly visible in the optical microscope as "metaphase" chromosomes, the principal actors in nuclear division. What kind of structure is it that can be folded so small?

The substance of chromosomes, known as "chromatin," is composed of DNA and protein, the latter of a particular group known as the "histones." These were believed to wind around the outside of the DNA helix, where it was assumed they control access to the genetic sequences of the bases inside the double helix. As Watson and Crick wrote in 1953: "There is room between the pair of polynucleotide chains for a polypeptide chain to wind around the same helical axis. We think it probable that in the sperm head . . . the polypeptide chain occupies this position. . . . The function of the protein might well be to control the coiling and uncoiling, to assist in holding a single polypeptide chain in a helical configuration, or some other non-specific function."[34]

But if the DNA molecule is subject to multiple folding to reduce its length so drastically, what becomes of the histones, and will they still be able to exercise control over the genes in the DNA? Curiosity was thus aroused over two related questions: First, how is DNA compacted? And, second, will discovering how it is compacted help us to answer how gene expression is controlled? For it is great to learn where the code is located and discover its messages, but what controls when this or that part of it is expressed or suppressed?

Very relevant, it seemed, to the latter question was the discovery by Roy Britten of the Carnegie Institution of Washington, that the majority of our DNA is composed of very repetitive base sequences,[35] a finding not to be expected of the carriers of highly specific genetic information. What does the organism do with such a sequence as GGTTAGGTTA... repeated several thousand times? When he collaborated with Rockefeller University biochemist Eric Davidson to compare the chromosomes of bacteria and higher organisms, they were surprised at just how much more DNA there is in the latter than in the former, most of it, as Britten had found, in repetitive sequences of the bases.[36] At this rate the *structural genes*—so-called because they code for proteins—cannot add up to more than a small fraction of all the DNA, probably about 10%. The rest, they suggested, must be there to regulate gene expression and possibly are involved in the folding of the chromosome thread.

So they set about formulating a theoretical model of the constitution of the chromosomes of higher organisms that envisaged such a system. In addition to the "structural" genes, they envisaged "receptor," "activator," "sensor," and "integrator" genes—in this clearly inspired by the "operon" model for gene expression of François Jacob and Jacques Monod. But Jacob and Monod's famous quip "Anything found to be true of *E. coli* must also be true of elephants"[37] was clearly going to need revising. Controlling gene expression in higher forms and packing the needed information into the chromosomes and the chromosomes into the cell nucleus presented puzzles worthy of Crick's attention.

In May 1971, Crick received an invitation to attend a meeting at Port Cros, an island on the French Riviera. This was not an invitation to turn down. About 50 scientists were to attend, and each day 6 hours were allotted for "refreshment." Dr. Bernardi and his colleagues had found the perfect recipe for a successful five-day event devoted to "Les Acides Desoxyribonucleiques des Eukaryotes." Crick had no hesitation in accepting.

The sessions given to DNA proved highly controversial, for the specter of so much DNA in higher organisms being composed of seemingly boring sequences of bases that are repeated again and again cried out for explanation. Britten was there to describe the work that led him to discover these "repetitive sequences." But Crick's Harvard friend, Charles Thomas, was also there seeking with "brilliant advocacy" to show that "much of DNA comprises exact multiple copies of genes." Few were convinced and many thought the repetitions were of related rather than identical sequences. In the face of divided opinion, *Nature* reported, it was Crick who "accomplished the near miracle of presiding over a final discussion" that was both probing and stimulating.[38]

Returning to Cambridge, Crick brought with him news on repetitive sequences in DNA and news about the histones. It had earlier been widely assumed that the different kinds of nuclear histone were legion. Now he learned that there are only five kinds associated with DNA in chromatin. Not news hot off the press, to be sure, but this was news to those in Cambridge to whom he reported it. Maybe chromatin would be a good research topic, because the revelation of its structure might well suggest ways in which the control of gene expression in embryological development is achieved. Could Crick engineer another double-helix triumph, this time fathoming the secret of development in higher organisms? Gradients and lineages had been interesting but not riveting. Could he make more headway by turning to the structure of the chromosomes of higher organisms (which he declared to be "probably the major unsolved problem in biology today")?[39]

Seven weeks later, Crick was due at the Summer School on "Molecular and Developmental Biology" at Erice, in Sicily. At the last minute he found the organizers were expecting him to give a talk on gradients to the students. This called for help from Lawrence in Cambridge,[40] thanks to whom all went well. But the allure of chromosome structure to Crick was growing. It was Wolfgang Hennig's discussion at Erice of work on chromosome structure and gene expression that excited him. It was the program at the Max Planck Institute for Biology in Tübingen under Wolfgang Beerman's direction. Soon Crick was planning to visit with Beerman in the fall, and a meeting was arranged for November. Crick's roving spirit had now been ensnared. Furthermore, he could once again play the game he knew so well—search for structure to discover function.

### Crick's Model

Crick described the Davidson–Britten contribution on the chromosome as a "very stimulating theoretical paper,"[41] but it was not "cashed out" in terms of molecular structure. So he set to work to formulate the outlines of such a structure and to relate his ideas to the bands long ago observed on the giant (polytene) chromosomes in the salivary glands of certain insects. These bands offer "markers" for locating genes along the length of a chromosome and distinguishing one chromosome from another. Crick wrote up his speculations on the subject in a paper he sent to *Nature* that September.

It starts in a grand way: "I wish to propose a general model for the structure of the chromosomes of higher organisms. This model is derived from ideas and data from many sources. Because I have found it impossible to set out my ideas and the supporting evidence in a short space, I merely summarize here my conclusions."[42] It ends with caution, almost apologetically: "Although the model is speculative and not fully detailed, and raises at least as many questions as it attempts to answer, I hope it may serve as a focus for discussion and for the design of experiments."[43]

Crick was sure he was on to something big, as we can see from his letter to John Maddox, editor of *Nature*.

Dear John                                   3rd September 1971

I enclose a paper for your consideration entitled "The general structure of the chromosome of higher organisms." This is the paper I mentioned on the phone. I have tried desperately to write the fuller paper,
but it is clearly going to be far too long for *Nature* and I need much
more time to write it properly.

Although I said nothing to anyone (except Sydney [Brenner]) till 18th
August the rumour that I am working on this problem has spread
rapidly. Several people have already sent me their reprints for consideration and [Guido "Ponte"] Pontecorvo (whom I have not seen recently) has invited me to a meeting next year on this subject! I am therefore
anxious to get something into print rapidly. Hence the present paper.

I apologize for the form of the paper, especially that the legends to the
figures are as long as the main body of the paper, but I am convinced
this is the appropriate way to present it. I comforted myself with the
thought that you usually put the legends in small print, but Sydney has
suggested to me that in this case you break your rule and print them
the size used in the text to make for easier reading, since so many
important points are contained in them. I leave this to your discretion.

May I ask that sub-editors do not tinker too much with the paper. . .
Many details of the paper are designed to avoid trying to offend people, though I fear that this will not be wholly successful. Of course
modifications to improve clarity are always welcome provided they
do not distort the sense. Sydney and I have spent a lot of time on the
exact wording in order to convey precisely what I mean and no more.

Could I make a special plea that I be allowed to see the proofs and to
return the corrected version to you, including those of both the figures
and the legends to the figures. Please be careful to print the Figure 3
so that the superhelix is left-handed. I would prefer to see the page
proofs as well as the galleys, but I expect this is asking too much.
(Your predecessor almost made a nonsense in 1953 of our DNA paper
by attempting to print a figure horizontally rather than vertically, but
fortunately Maurice Wilkins spotted this—I was away in Scotland. As
it was, DNA was printed as D.N.A.). . .

I have put in a real footnote at the beginning about reprint requests.
Apparently in 1961 we had 3,000 requests for our paper on the General Nature of the Genetic Code. I would estimate that without this
footnote we are likely to get between 5,000 and 10,000 requests. . . .
We should be only too delighted if you would undertake the chore of
sending these out, as you suggested.

Excuse this excitement,                          F.H.C. Crick[44]

Indeed, he had been excited, for he told Lawrence that "from the moment I picked up my pen, to the stage when the chromosome paper was ready for xeroxing was only a period of four days."[45] To *Nature* readers he explained that "a much fuller account is in preparation and will be submitted for publication in the future."[46] It was unusual for him to make such a promise in so public a location and then not to deliver. What happened?

Like Davidson and Britten, Crick assumed that most of the DNA in the chromosomes of higher organisms functions not in determining the amino acid sequence of proteins but to control gene expression. Such DNA, he suggested, is in a globular form, portions of its double-helical chain opened up to give single-stranded DNA. These stretches of exposed bases would be much easier than double-stranded DNA for the proteins regulating gene expression to "recognize." These globular patches in his model—like balls of wool—would make up the banding seen on chromosomes under the microscope (Fig. 17.3). DNA between these regions—the interbands—would be composed of double-helical DNA. It is these regions between the bands, he suggested, that code for proteins; they, he contended, contain the "structural" genes. The accepted view, on the contrary, was the reverse of this: The bands were judged to contain the genes, not the interbands.

Later he pictured his scheme in the form of a triangular diagram representing the limitations on the *recognition* of a given nucleic acid

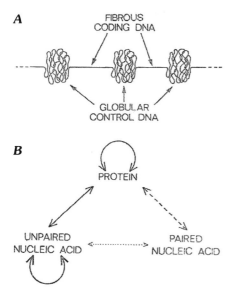

*Figure 17.3* (*A*) Crick's proposal for the functional organization of the chromosome. The straight lines are for coding and are the interband regions of the giant polytene chromosomes studied so actively by cytologists. The intricately folded globular portions are for control of gene expression. They correspond to the bands observed in polytene chromosomes. (*B*) Protein–nucleic acid interactions. (Solid lines) specific interaction between macromolecules; (dashed line) recognition of a nucleic acid sequence by a protein; (dotted line) formation of a triple helix from a paired and unpaired nucleic acid molecule.

sequence of the genetic material by a regulatory protein (Fig. 17.3) (thus reminding us of his famous 1970 diagram for his Central Dogma; see Chapter 14). Governing these relations was his "Unpairing Postulate"; namely, "the recognition sites, needed for control purposes in higher organisms, are mainly *unpaired single stranded stretches* of double helical DNA."[47]

In 1971 the focus of the structural studies in the Laboratory was tRNA, led by Klug. Although Crick was working on chromosome structure, there was no experimental program in the laboratory under way. Not until a visitor from Harvard, Roger Kornberg, came to work with Klug was such a program begun. A biochemist like his famous father, the Nobel Laureate Arthur Kornberg, Roger arrived on the last weekend in May 1972, as a visiting researcher from Harvard. His dissertation at Stanford University had been on nuclear magnetic resonance (NMR). For his postdoctoral research, he wanted to try his hand at X-ray diffraction. The faculty advised him to seek a position with Klug in Cambridge.

When Kornberg arrived Klug was away, so Kornberg questioned Klug's group about their work. It was all in the highly competitive field of tRNA and the very detailed structural analysis of the structure and self-assembly of the tobacco mosaic virus (TMV). Deciding that he did not want to work in these areas, he fell to talking with Mark Bretscher, the biochemist:

> Mark said, "Why don't you read Francis' paper on the structure of higher organism chromosomes?" I went away and read that paper and couldn't make head nor tail of it. There are certain things you have to understand about that paper, it was very instrumental in what followed. The paper had only been published a few months before I arrived and it precipitated Francis' disappearance from the lab and he was not around at all when I got there—after that he only came in occasionally for a short while and I was told he was at home, and eventually I learned that he had fallen into a really deep depression.[48]

Actually in 1971 Crick had begun to suffer from the intensity of his globe-trotting to conferences. No sooner did he return from one than he was preparing to leave for the next: June, Edinburgh; July, Sicily; September, Byurakan, Russia and Nice, France; November, Tübingen and London. In London he left the meeting unwell and was advised by his physician to rest for several weeks. But he must soon have begun preparing for the string of public and private events in 1972. The 25th of January saw him on his way to the Salk Institute, to attend the Fellows annual meeting. In March he was in Scotland. In April back in the

United States, he lectured at Harvard and MIT. Chromosome structure and gene expression were on his lecturing agenda.

First Crick, like Maurice Wilkins and Vittorio Luzzati, still accepted, as did everyone else, that the histone is on the outside of the DNA and wraps around it. Second, Crick still treasured the idea of looped out segments of nucleic acid duplexes that he had first used to account for his genetic data in 1960 (see Chapter 14). Here portions of the double helix were supposed to uncoil to create single-chain loops—his "Unpairing Postulate."[49] Crick used the loop structure in his chromosome model because he assumed, wrongly as it turned out, that proteins have difficulty recognizing long sequences of DNA when they are held within a duplex.

Reading Crick's *Nature* paper, Kornberg found the only parts that he "could get a grip on" were the diagram and the long caption beside it referring to all the authorities with whom he had discussed the subject. But Kornberg was suspicious for even "as a teenager" working in his father's lab he had known "there was a bad odor" about the histones. They had a "reputation for being very murky," making him very leery of them:

> Then Aaron [Klug] came back from holiday. I sat down and the first thing he said to me was: "Well, you know. This place is entirely free and open in regard to what you work on and with whom you do it, . . . I mentioned I had met Mark Bretscher and we had talked about Crick's chromosome paper. "Oh," responded Klug, "we will have to get Francis' permission for that, it is his topic."

They visited Francis and he was as usual welcoming and agreeable:

> Klug said: "It is a messy problem." He explained to me that it was an X-ray problem which I didn't know. He explained everything to me. He knew all about it and with his remarkable memory he could recite chapter and verse from the literature. And then he went to this huge mess in his office, piles and piles of papers, delved into it and pulled out the relevant papers. He showed me this work that was done by Luzzati. And he said: "Look you know there was this diffraction pattern, you know 110Å, 55Å, 37Å and 27Å reflections—broad reflections. It would be great for you to do a low angle x-ray problem." And I thought that sounded great and gee what a simple diffraction pattern. You know, four lines that's something I can do in a year. But contrary to what the rest of the world believed the problem had not been solved.[50]

With Roger Kornberg taking on this problem, Klug formed a chromosome group at the Lab and from the beginning Crick was a "very keen" member; he and Klug had regular meetings with Kornberg as the work progressed. Klug and Crick also had meetings on their own in which, Kornberg gathered, they discussed how to interpret his results. Indeed, it seemed that they were going to figure out the solution themselves. One entire night Crick spent most of the time pondering chromatin and at 6 a.m. telephoned Klug to discuss an idea he had. He would write Klug postcards about chromatin while on holiday. One contained a model that was "a kind of network with histone bridges."[51]

One day Francis came in to Kornberg's lab in a state of excitement and said, "We've solved it!" Kornberg then asked, "What is the solution?" Crick paced around a little and then added, "I can't tell you now, but we've done it!" and he left. Kornberg was a bit down at this, but reflected, "Well, someone has still to prove it." At this point he realized "Crick and Klug were not my supporters but my competitors." They had conceived a model of chromatin in which the DNA was wound around parallel straight chains of the histones.[52]

Young Kornberg, who Klug described as "a chip off the old block," had the skills and the patience to face the challenge presented by the histones. He read the literature, including the report from two biochemists in Capetown,[53] describing their extraction of histone from sea urchin nuclei. They had not used the harsh extractive procedures generally used for the histones but a milder one that yielded the five main types of histone, called at that time F1, F2A1, F2A2, F2B, and F3. The sizes of the resulting extracts suggested to Kornberg the presence of dimers, each made of two histone types. Surprisingly, although looking for dimers, he found a tetramer $(F2A1)_2(F3)_2$ and the smaller "oligomeres" F2A2 and F2B. The tetramer is a four-part molecule not unlike hemoglobin. Kornberg proposed that it forms a repeating unit along the chromatin fiber, whereas the oligomers define the path connecting successive such units. The DNA "would form some path on the [surface of the] tetramer and the remainder of the DNA would connect tetramers ... making a flexibly jointed chain of repeating units ... rather like beads on a string"[54] (Fig. 17.4).

Reflecting on this in 2004, Klug judged it "was the psychological breakthrough. I'll tell you why. Because most people thought that the bulk of the genes were sequestered by the protein which wrapped around them. If you look at the models that Wilkins had done, and so on, the protein protected the DNA. But here it showed if you had a ... hemoglobin-type structure ... this would be a globular protein and it couldn't wrap around a DNA helix."[55]

Where, then, could these globular histones be in the chromosome? Not on the outside of the DNA, but, suggested Kornberg, in the form of

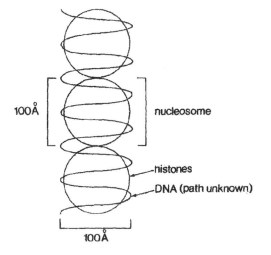

**100Å** nucleosome

histones
DNA (path unknown)

**100Å**

*Figure 17.4* Highly schematic drawing of the structure of a chromatin fiber. Some features, such as the path followed by DNA, have been drawn in an arbitrary way and should not be taken literally.

"beads" around which the DNA thread is wound, 200 base pairs per bead. The chromatin consists of a succession of these beads forming as a whole "a flexible jointed chain." While writing up the discovery for *Science*, Kornberg was excited by the electron micrographs of chromatin fibers published in the 25 January 1974 issue of *Science*.[56] The authors, Ada and Donald Olins, described their pictures as "beads on a string." To Kornberg the string had to be DNA and the beads were histones around which in his model he had wound the DNA. Klug recalled,

> We proved it by electron-microscopy and x-rayed the fraction and all that. That was a psychological breakthrough and it meant a whole new way of looking at it, because it meant DNA was on the outside and it was accessible to other factors which would switch genes on and do/undo the chromatins and so on. You see, DNA is on the outside, it's not sequestered.[57]

Kornberg described his views on the arrangement of histones and DNA as "a working hypothesis." Its appeal lay in the manner it brought together several strands of research on chromatin structure. For the first time Wilkins' X-ray diffraction data on chromatin could be explained effectively, and the "beads on a string" had the degree of flexibility that chromosomes possess, but which a continuous super-helix lacks. Keen to tell Crick about it, Kornberg ventured to his office. "He was sitting in his chair some distance from the board. I stood at the board and wrote down what I had found. His eyes were transfixed on the board. He was intent, perceiving every single word without any loss, for he is the world's best listener."[58]

Following that occasion Crick and Klug embraced Kornberg's model. Indeed, "Francis adopted it as if it were his own. This was a lesson to me, for there are those who resist new knowledge simply because it involves losing face, so they continue to object on non-rational grounds. Not so Francis and Aaron. They were too smart for that. Others may steal your thunder, but your reaction should not be to find fault with it, but to latch on to it—add it to your vocabulary and take it to the next stage."[59] Kornberg decided to tell Brenner about his model. He found him sitting at that little table beside the door where he often worked. "'What are you doing?' Sydney asked. So I told him about my model of the nucleosome. He was dumb-founded. 'That's fantastic. You've got it!' he declared. Brenner then asked me about my future plans, and I said I was getting out of the LMB. I had a position at Harvard awaiting me, and nothing for the future at the LMB. 'I'm going to talk to Francis,' Brenner responded. A position was soon offered to me after that, although how much it was due to Sydney's intervention is not clear."[60]

In truth Crick acted in this matter as one whose ego had been punctured, for Kornberg's nucleosome model ruled out any model for chromatin of the kind Crick had with such enthusiasm constructed. On the other hand, he knew how to bounce back and champion Kornberg's model. The work of the chromosome group continued in Cambridge throughout the 1970s. Although Crick left for a sabbatical at the Salk Institute in La Jolla, California in 1976, and relocated permanently there in 1977, he kept in frequent contact with Klug by letter, was a co-author for two of the group's papers, and spoke on the subject at five international meetings. By this time, Klug had suggested a further level of coiling, this time involving Kornberg's "beads," later called "nucleosomes." Not only was the DNA double helix wound around the nucleosome, but the nucleosomes, he ventured, were in their turn wound into a wider helix making a structure he called the "solenoid" (like a hose wound round a drum).

Now two Danish scientists, Arne Leth Bak and Jesper Zeuthen, in Aarhus, Denmark, together with Crick were introducing yet one more level of coiling that made what they called a "super-solenoid." These successive levels of coiling would give all told a 10,000-fold compaction of the DNA helix, making it so fat and short that it should be clearly visible under the optical microscope as a mitotic chromosome. Such spirals and superspirals—a hierarchy of helices—is not, they admitted, "a new idea," but they were able to suggest "the various levels of the hierarchy,"[61] and offer evidence for such a super-super-super coiled double helix from their electron microscope pictures. The paper was submitted in January 1977 to the National Academy of Sciences *Proceedings.*[62]

Come the summer, Crick left California for Europe to be with his former colleague Dan Brown and with Bak and Zeuthen at Aarhus University as Visiting Professor. Meeting them for the first time, he was able to examine their experimental work. To his horror, he discovered that the structures they had isolated were distortions of the native state—they were sausage-like artifacts and it was on these that the paper had been based! The lesson to Crick was clear—do not accept research on trust until you have met the authors and visited with them![63]

With the advent of the nucleosome, Crick's model of 1971 was now only of historical interest. As recently as 1973, 12 authors had cited it at the Cold Spring Harbor Symposium on Chromosome Structure and Function, but their comments, if any, were not upbeat. The next year came the Olins' "v bodies" and Kornberg's "beads" (nucleosomes). But here, as in the case of Crick's first attempt at the structure of DNA and that of collagen, he returned to join the fray. Enthusiastically he followed the field and collaborated with Klug and others on the implications of Kornberg's structure. Hence came his discussions with Klug on whether DNA could develop "kinks" at regular intervals to overcome the stresses of being wound into a superhelix[64] and his lucid explanation of the mathematical treatment of supercoiling that the mathematician F. Brock Fuller had presented elsewhere in a highly technical manner.[65] Multiple-supercoiling models, however, have not enjoyed universal assent.[66] But Kornberg's nucleosomes have found a lasting place in our modern knowledge of chromatin structure. Crick's paper had provided the stimulus for Kornberg to turn to this subject.

### The Origin of Life

Accepting organization as the crucial feature distinguishing living things from the nonliving raises the question of life's origin. For the Darwinian evolutionist, this question necessitated an appeal to an historical process that over long periods of time and under the influence of natural selection has fashioned such organization in primitive systems of organic matter. As we have seen in Chapter 15, Crick had collaborated with Orgel on the origin of the genetic code. Now he was to discuss the broader question of the origin of life with Orgel and with his friend, the cosmologist Tommy Gold.

In 1971 Orgel and Crick were invited to a meeting on "Communication with Extraterrestrial Intelligence," held at Byurakan Observatory, in Soviet Armenia. The substance of their contributions they later published in *Icarus,* the journal devoted to extraterrestrial science. Entitled "Directed Panspermia," the paper describes the "theory that organisms were deliberately transmitted to Earth by intelligent beings on another planet." The approach is characteristic of Crick—speculative, critical, and thought-provoking. Crick did not really believe in the idea, but wanted to stimulate discussion and promote further research.[67] Orgel, as Director of the Salk Institute's Chemical Evolution Laboratory from its establishment in 1964, was like-minded. In this way they introduced the subject that interested them into a meeting given to extraterrestrial intelligences. Subsequently, Crick expanded their paper into the book *Life Itself* (1981). When Peter Medawar read their *Icarus* essay, he suggested Crick should recast it for the fashion magazine *Vogue*. Crick replied that they had originally considered *Playboy* magazine. Admitting in 1977 that they lacked any scientific support for this "way-out" idea, the lesson Crick learnt from this "exercise in imagination" was that the subject was too backward to permit its refutation.

The obvious question to ask about the origin of life is: Where did it happen? If life had originated here from a chemical soup in the oceans of the primitive Earth, why has it not been possible to simulate the process in the lab? True, the Chicago graduate student Stanley Miller had shown the spontaneous synthesis of a number of simple organic compounds, including the amino acids, glycine, and alanine, in 1953, but only if oxygen was scarce. Was there so little oxygen in the atmosphere of the primitive Earth? That was and is debatable.[68] In the absence of living organisms, small carbon compounds like those Miller synthesized would have accumulated, but only if the Earth's atmosphere was reducing (i.e., no appreciable amount of oxygen). But Crick and Orgel drew attention to other problems with the "life-from-soup-here" idea. One would expect several different pathways to have been followed from simple carbon compounds to life, and if life originated here, these would have left their traces in the enormous variety of living forms today. Yet what do we find? A single genetic code, barring a few trivial exceptions that can be accommodated. Moreover, the products of this code, the proteins, are all constituted of amino acids drawn from the very same vocabulary of just 20. This "set of 20 is so universal," wrote Crick, "that its choice would appear to date back to the very beginning of all living things."[69] There are, nevertheless, several thousand different proteins in organisms. Yet consider a polypeptide chain 200 amino acids long; how many different chains of that length could there be, assuming all possible sequences of the 20 different amino acids along the chain are realized? The answer is approximately $10^{26}$, an inconceivably large number. Somewhere there has been a "bottleneck." Could it be that life evolved elsewhere, and that only a very limited sample of that life reached this earth?

By the time Crick and Orgel wrote their paper, it had become clear that the 19th century ideas about life driven here by the pressure of light or carried by meteors from another solar system, the doctrine of panspermia, were ruled out. Carl Sagan, the astronomer and science popularizer, reckoned that exposure to so much radiation on their journey through space would kill them. As for meteors carrying live objects, he told Orgel and Crick, the probability of one coming from outside our solar system and landing on Earth was indeed slender. But a chance remark of Gold's gave Orgel and Crick an idea. Gold had remarked that if astronauts reached another planet "some of the bacteria they carried would reach the primitive ocean, there to survive and multiply long after the death of the astronauts." Now suppose, suggested Crick and Orgel, "a primitive form of life was deliberately planted on the Earth by a technologically advanced society on another planet . . . could life have started on Earth as a result of infection by microorganisms sent here . . . by means of a special unmanned spacecraft?"[70]

To explore the feasibility of this scenario, called by them "directed panspermia," they asked whether our own technological society will in the foreseeable future be capable of infecting a planet outside our solar system. If we can imagine how that could be done, then the reverse scenario is possible. Life in the form of microorganisms could have been sent here by an advanced community on another planet. The form of the genetic code, the alphabet of the amino acids, and much else would already have been set in stone in these microorganisms and the uniformities we observe would have been the result. Could mice and men be sent in a spaceship, asked Crick in his popular book, *Life Itself?* Unfortunately the journey would outlast a human lifetime, not to mention the life span of a mouse. The astronauts could be frozen to preserve them, he suggested, or they "must breed in the spaceship; not," he added, "my idea of the Good Life."[71] Better to send microorganisms that can be preserved for long periods and that behave robustly under extreme situations.

Directed panspermia shifted the problem of the origin of life away from this planet, but the difficulties remained of imagining just how the basic machinery of life originated. As we have seen, protein synthesis is a complicated process, and it seems as if many of the "nuts and bolts" needed to be in place before it would function correctly. When we strip down life to its fundamental processes, remarked Crick, "we cannot help being struck by the very high degree of *organized complexity* we find at every level, and especially at the molecular level."[72]

The proteins he called the "machine tools of the cell," whereas the nucleic acids are "the dumb blonds of the biomolecular world, fit mainly for reproduction (with a little help from the proteins) but of little use for much of the really demanding work."[73] But to synthesize the proteins involved creating a translation process from the language of the nucleic acids to that of the proteins. Nature achieved this by inventing "the assembly line some billions of years before Henry Ford."[74] Such a mechanism Crick judged to be "far too complex to have arisen in one blow. . . . Indeed, the major problem in understanding the origin of life is trying to guess what the simple system might have been."[75] Yet the scientist "is possessed of an almost boundless optimism concerning his ability to forge a wholly new set of beliefs, solidly based on both theory and experiment." He ranked the origin of life with other major origin questions, plus the "the nature of consciousness and the 'soul.' To show no interest in these topics" he declared, "is to be truly uneducated, especially as we have a very real hope of answering them in ways which would have been regarded as miraculous even as recently as Shakespeare's time."[76]

The message in *Life Itself* was clear. Speaking from the intensity of his conviction and the independence of his mind, he looked ahead. He had no truck with complacency or pessimism. "Negative prophecies," he pointed out, have been upset time and again. The problem of the origin of life "and the related problem of life on other worlds, is so important to us that in the long run, it will be bad luck if we fail to find the answer."[77] Crick's Caius College acquaintance Hugh Montefiore took home a different message from *Life Itself.* Citing Crick's admittance that the origin of life "appears to be almost a miracle," Montefiore was ready to count it "yet another example of matter with a bias to assemble itself in ways that lead to the emergence of intelligent life,"[78] thus making it a goal-directed process. Crick had been frank but, as the passage in question shows, he had expressed himself with characteristic precision:

> An honest man, armed with all the knowledge available to us now, could only state that in some sense, the origin of life appears at the moment to be almost a miracle, so many are the conditions which would have had to have been satisfied to get it going. But this should not be taken to imply that there are good reasons to believe that it could *not* have started on the earth by a perfectly reasonable sequence of fairly ordinary chemical reactions.[79]

### The RNA World

Retracing biological complexity to the first known forms of life brings us back to the problem that had attracted Crick's enquiring mind in 1947—"the division between the living and the non-living." Although in 1953 he judged he had found *the* secret of life, he was now admitting that other secret combinations are needed to unlock the door to the primal complexity of the cell. Discovering the mechanisms of protein synthesis and DNA replication revealed the cell's fundamental complexity. How it arose remained a field for speculation. In *Life Itself*, Crick evaluated the plausibility of several possible scenarios. Like Orgel and Carl Woese, his prejudice was that "nucleic acid (probably RNA) came first, closely followed by a simple form of protein synthesis."[80] The candidacy of RNA had first been suggested by Crick at a meeting of the British Biophysical Society in London in 1966[81] and independently by Carl Woese in 1967. To Woese, tRNA was "such an odd molecule, a kind of molecular misfit, more like a protein than a nucleic acid. Maybe it is with us now because it was present very early in the evolution of the translation process."[82]

In *Life Itself* Crick admitted RNA was "not ideal" for this pioneer role. Yes, it can form three-dimensional structures, a feature of all the protein enzymes we know. Such RNA structures, however, seemed to lack "any catalytic activity." On the other hand, if small organic molecules in the primitive soup were to combine "neatly with certain folded RNA molecules to produce a primitive 'enzyme,' some 'rather crude catalytic activity' might have resulted."[83] That same year (1981) Thomas Cech and co-workers announced their discovery of enzymatic function in an RNA molecule, to which Sidney Altman and colleagues added a further example two years later.[84] Crick's hopes for primitive RNA enzymes now took on a degree of plausibility that formally they had lacked. Such RNA molecules called "ribozymes" have become the basis of the so-called primordial "RNA world" in which life might have evolved whether here or on some distant planet.

By the time of the Cold Spring Harbor Symposium on "The RNA World," in 1994 both Crick and Orgel, like many others, had become skeptical about primordial RNA. "Protein-based life," remarked Crick, "was already in existence 3.6 billion years ago." That left so little time "to get life started,"[85] and Orgel reported the chemists' growing doubts about RNA as the "first self-replicating molecule."[86] Hence by the end of the century, a question mark still hung over naturalistic accounts of the origin of life. True, one could still retreat to Crick and Orgel's panspermia, deflecting the topic to exobiology—the study of life beyond the Earth. Meanwhile, the complexity of higher organisms was taking molecular biologists further and further into the complex hierarchies of organization. Was Crick ever attracted to a radically different approach such as topology? Here the French mathematician René Thom did not impress him. He seemed "somewhat arrogant," wrote

Crick, and it appeared he "understood very little about how science was done."[87] So although others turned to catastrophe theory, Crick's thoughts returned to the subject that had so long intrigued him—the brain.

## CHAPTER 18

# Leaving the "Old Country"

*. . .have settled in to our house, bought a car, <u>immense</u> office (by Cambridge standards)—almost two times as big as Max's—I have even managed to pass my California driving license test.[1]*

On the 27th of March 1977, the following report appeared in London's *Sunday Times*, together with a photo of Crick entitled "Crick: tax exile."

> Dr. FRANCIS CRICK, the man who shared a Nobel Prize for what has been called the most important scientific advance of the century, has left Britain because of our tax laws. Crick is planning to take up a permanent post at the Salk Institute in La Jolla, California, where he went last September as a visiting professor on what was thought to be a one year sabbatical. . . .
>
> Cambridge's hopes of seeing Crick again rest in the law which allows tax exiles to return to Britain for up to 90 days a year once they have been away one year. So the man who, in Cambridge, produced one of the key scientific advances of the century might end up as no more than a visiting professor to Britain.[2]

Contacted in California by the *Sunday Times*, Crick reminded them, "You know I never talk to the Press, what makes you optimistic this time?"[3] Those in the Cambridge lab were more forthcoming. Apparently Crick had told them it was a change in tax laws which made him decide to leave. The chief culprit was the 1974 Finance Act. It was targeted at the practice of the Lonrho Company to pay its directors salaries in the Cayman Islands, where they attracted no tax either there or in the United Kingdom. The Act would tax three-quarters of such overseas earnings whether they were brought back to the United Kingdom or left elsewhere. U.S. earnings were, of course, included in the Act. This meant a significant cut in Crick's earnings from his lecturing and awards overseas. He had hoped that such earnings (£14,000 in 1974) would help him to live off an MRC pension, which would be relatively low because of his late entry into MRC employment.

Following the *Sunday Times* report came a report in *The Times* on the tax issue. The top marginal tax rate, the report noted, is 98% on investment income and 83% on earned income. The salaries of top research scientists were frozen at £8,000–£9,000 for five years, and retirement at 65 was mandatory. Aimed at the policy of this socialist government, and its Chancellor of the Exchequer, Denis Healey, the report ended "That such a distinguished scientist in the latter stages of his career is forced to such a conclusion, because he is unable to provide a satisfactory income for the remainder of his life, should give Mr. Healey pause for thought. If, today, he can begin the process of restoring sense to the tax system, it will not be a moment too soon."[4]

Perutz told the press, "It is terribly sad for us, he had a tremendously stimulating influence here." Aaron Klug added, "Obviously we are not going to stop working and go into mourning, but it is very sad. He was such a tremendous spirit in our laboratory. He had a grasp of so many different lines. He is such a good critic, quick to see an avenue to explore or a weakness in someone else's work."[5]

Was this the whole story? It had been a big decision. After spending over a quarter of a century in Cambridge, 27 years at the Unit that had grown up around Perutz, why leave now? Was this just about money or also the result of some confrontation in Cambridge, some bitterness between him and his colleagues in the Laboratory, the MRC in London, or the wider community of Cambridge University? No, it was none of these. When in 1975, Crick was considering his future, retirement at the MRC was still mandatory at 60, making the 8th of June 1976 his last day as a salaried MRC employee, an event for which he was not ready. Nor did his colleagues consider he was. He had long been a "powerhouse" at the lab and at 60, colleagues said, "there is no sign whatever of any decline in power."[6] At this point he was approached by two Fellows of his College, Gonville and Caius, asking if he would agree to have his name put forward for the position of Master. A series of informal and frank discussions then took place "with some hilarity about what Masters really do, and what they really HAVE to do; the mood was very cordial and for all his sceptical habits of mind," Crick, it seemed, "was truly interested and wanted to learn about the inner mysteries of colleges, to see if he could fit in and do a good job."[7] Shortly thereafter he left for Sweden to attend the 75th anniversary of the Nobel Foundation (1975); he took with him the statutes of the College to study. He explained how he "analyzed the job (apart from the business of the chapel, which would have been a little problem shall we say? Although, you know, maybe we could have come to some agreement)." But the job "involved all the things I don't like doing: like sitting on committees, you know, being polite to lots of people, and administration!"[8]

On his return, he therefore withdrew his name. In La Jolla a year later, a friend from Balliol College Oxford called him to ask if he would like to become Master of Balliol. "He was very taken aback when I immediately said, 'No thank you!,'" recalled Crick. Then he had to explain why he had decided that being the head of an Oxbridge college was not for him. This decision did not lead him to look abroad for continued employment. The move, when it came, resulted from an unanticipated chain of events, not from deliberate planning.

As he explained, it arose as a result of his spending a year's sabbatical at the Salk Institute in California in 1976. Why this institute? Because he was well known there having long been one of its nonresident Fellows, meaning that he paid annual visits in an advisory capacity. Indeed, his association with Jonas Salk's enterprise goes back to its very beginnings. Even before the Institute became a legal entity (December 1960), Crick met with Jacob Bronowski, Jacques Monod, and Salk in London and in Paris to formulate the bylaws and debate "other fascinating matters!" In 1962, he was invited to become a nonresident Fellow and, with the MRC's agreement, subject to the forfeiture of 2 months' MRC salary a year, he accepted. He then combined the required annual visit to La Jolla to attend the combined meeting of resident and nonresident Fellows in February with visits he would make to other institutes and universities on the lecture trail before heading back to Cambridge. The visiting Fellows functioned at this stage rather like a visiting committee. The aim, Crick explained to Miss Brumfitt at the MRC, was

> to discuss the affairs of the Institute, and also to talk to trustees, who will also be meeting there at the same time. It is thus not a meeting in the usual sense of the word, but rather a series of private consultations. As such it does not have a title.[9]

Careful to keep the MRC informed, he listed his lecture engagements for his absence. Thus, from 9 January to 8 March, 1964, he told Miss Brumfitt his schedule: 16–23 January in Hyderabad, India; 6–11 February, Vanuxem Lectures, Princeton; 21 February, James W. Sherrill Lecture at the Scripps Clinic, La Jolla; and 5 March, North Carolina

State University, Raleigh.[10] He was to continue with such a pattern of January/February tours, starting his schedule in 1965 at Chicago, in 1966 in Denver, and in 1967 in Evanston.

The nonresident Fellowship was ideal for Crick at this stage in his life. It offered him the means to cross the Atlantic annually, visit his many friends in different parts of the United States, give lectures—often well-endowed ones—and take part in building a high-powered research institute from the ground up at La Jolla. In this way, he avoided what he called "the comparative isolation of Cambridge."[11]

By 1976 that isolation began to trouble him because, as he explained,

> In 1973 my second six years of non-resident fellow expired, and they asked me if I would do a third one. Jacques Monod said "yes" and I said "no." I thought I've done enough, and I decided . . . I'd travel less for a while. . . . But I found that after two or three years of that, that I was getting a bit out of touch with what was going on [on] this side of the Atlantic. And I must have mentioned this to Leslie Orgel [at the Salk] and he said: "Why don't you come sabbatical to [the] Salk?" That was the origin of coming here.[12]

Permanent residence was not the intention at that time. Then the President of the Salk Institute got in touch with Crick about Orgel's idea, but Crick had to tell him that he had just permitted his name to be put forward for the Mastership of Gonville and Caius. "So I had to tell Frederic de Hoffman, who was the President here then, you see. And he said, 'Oh, we were hoping to make arrangement for you,'" that is, on a long-term basis. This was the first intimation of such an idea, but at this stage it was just a possibility. "And I think we met at Claridges [Hotel] . . . and so it had been sort of explored and I would raise various objections, you see, saying 'I can't really cut myself off from England.' [Hoffman would reply] 'Oh, you can go back every summer and we'll pay your fare', and all that kind of stuff. . . . So he got me a fellowship from the [Ferkauf] Foundation for a year here." With the long-term possibility still in the air, Crick applied for immigration visas for himself and Odile just in case. Meanwhile, de Hoffman

> looked around for another foundation who would endow a chair. And that was arranged, by which time I think we [had] bought our condo because it looked to me as if it was going through. And the prices were going like this [he raised his hand toward the ceiling], which is why we did buy it. Well then there was an interim because I said to Odile, "Dear, well how do you feel about it?" because we hadn't discussed it. She said, "I'd like three months to think it over." So it was in limbo for . . . the first

three months of 1977. And then she said: "yes." And then I wrote to the MRC and everybody and said: "You know, I'm gonna stay here."[13]

The retiring age with the MRC had been a difficulty for remaining in the United Kingdom, but as he recalled, it was extended to 65 before he made his decision to move:

> I don't know really what decided me except I've always liked it here and it seemed a good opportunity. And then I thought, well then I can perhaps switch to the brain, you know. I don't think I thought of switching to the brain before that . . . I mean I vaguely thought about it but [the move] promoted it. It wasn't as if there was one over-whelming reason. I was sort of a bit cross about the way things were going politically in England, you know. Our salaries were getting very behind. There was a stage about then when the younger scientists were receiving less money than the assistants because there were two different trade unions. And the Government cheated the academic union by refusing to look at their claim before some time had run out [five years], you see. And taxation was very high, and so on.

> There were a number of subsidiary reasons but the basic reason was that it seemed so very pleasant here, and they made me a very reasonable offer, and I couldn't think of any reason why not to accept. I would say it was not something I was positively looking for, you see. . . . When I said to Leslie [Orgel] about coming on sabbatical it certainly wasn't in my mind at all. If it hadn't been for the President here wanting me to stay, I wouldn't have approached the subject. So it was one thing led to another . . . more than anything else, rather than any careful planning.[14]

Crick felt very much at home in southern California but not in other parts of the United States, especially not in New York. The prosperity, relaxed way of life, and the ocean, mountains, and desert near at hand appealed to him. Indeed, the Anza-Borrego Desert so attracted him that he was later to have a house built at Borrego Springs to which he could escape for complete peace and the enjoyment of the "subtle colors and the wide expanse of sky."[15] If there had been winter rains, the following February, the desert would burst into flower—all that arid land transformed into a sea of colors. Unlike Odile, Crick felt no strong ties to Europe, and in England, none of his generation of Cricks was still alive. His son Michael was living in Seattle. Only his daughters Gabrielle and Jacqueline remained in the Old Country, and Crick offered to bring them to the United States.

Odile's attitude was different. She had enjoyed visits to her maternal aunt in the quiet little village of Ménerbes in Provence. Ever since her childhood visits to her grandmother on the rue Lamarck in Paris, Odile felt an attachment to France. And Provence—the fragrance of lavender and olive trees, the golden fields of sunflowers, those sun-kissed tomatoes, the sound of the French language, and the sight of the blue Mediterranean—was another world, and those who visited it wanted to come back. Hence, it was predictable that Odile was cautious. Francis reported that Odile "quite likes La Jolla, but she is uncertain whether she would like to go on living here. However, she is prepared to try the winter of 1977."[16] Some 2 months later, responsive to Crick's wish, she agreed to make their future there. And Francis? He became the J.W. Kieckhefer Distinguished Research Professor, thanks to the Foundation of that name, established by J.W. Kieckhefer, who had made a fortune in timber. Before February 1977 was over, Perutz knew the decision and reported it to Sir John Gray, the MRC's Secretary:

> As you may have heard, Crick decided to take a year's leave of absence and to spend it at the Salk Institute in La Jolla. I am sorry to report that he has now decided to resign from the Council's service and to accept a post offered him by the Salk Institute. This is a great shame as we shall miss the stimulus and forceful criticism which he provided, not to speak of his bounding vitality.[17]

### Arrival on Sabbatical

No sooner was Crick installed in the office prepared for him at the Salk in the fall of 1976 than he began writing enthusiastically to Aaron Klug in Cambridge, "I have an immense office to myself, two rugs provided by Leslie [Orgel] and a large green plant. It has a lovely view over the ocean and I can sit and watch people hang gliding above the cliff."[18] Barely a fortnight later he wrote again, "settled in to our house, bought a car, immense office (by Cambridge standards)—almost two times as big as Max's—I have even managed to pass my California driving license test."[19] At 60, the boyish humor and sense of rivalry was still there. A year later, he was telling André Lwoff, of the Pasteur Institute, that even though burdened with the title "Professor" he had

> no teaching duties whatsoever. Nor am I a member of the Salk Faculty, though, I may, if I wish, attend Faculty meetings. I have always found that the important thing, if one wishes to lead a quiet life, is not to have a vote. Then nobody bothers you for your opinion unless they really want it.[20]

Could all of those advantages, though, cancel out the loss of Cambridge—the history, the immaculate green of the College quads, their closure on all sides affording privacy, the open feel of the "Backs," the glory of the architecture, the associations with great names in the sciences and the arts, but most important Odile and Francis Crick's social circle going back to the 1940s? That was the question the neuroscientist Richard Stevens put to Crick when he was considering an offer from the Salk Institute. "How would it be living in Southern California," Stevens asked. "After Cambridge how did he put up with La Jolla? Crick responded giving me a powerful sales talk—'Wonderful. It is warm. You can swim every day and enjoy better health.' The intellectual climate at the Salk he enjoyed too. No, he had no regrets."[21] No bitter winds sweeping across the fens, no streets choked with student cyclists—watch out or they'll run you down—and no College Chapels! Then Sydney Brenner wrote from Cambridge telling him, "The country is in a mess and the MRC gets worse by the day. Just today, Max and I were discussing organizing a strike of Directors! Perhaps Jim Gowans who will take over from Gray in March can change it again."[22] Crick replied hoping the change would make a difference, then added gleefully, "the weather here continues to be unusually good. Cool air but lots of sunshine."[23]

If one follows the coastline north from San Diego, passes La Jolla Cove with its seals and pelicans, goes on past the Scripps Oceanographic Institute and beyond it the nudist "Black's Beach," then ahead the hang gliders can be seen floating high above the sea. Atop the bluffs is the gliders' runway, and nearby one can spy two long concrete buildings of modest height, looking out over the ocean and home to the Salk Institute for Biological Studies, so named in honor of Jonas Salk, the leader of the team at the University of Pittsburgh Medical School that in 1955 developed the Salk vaccine against poliomyelitis. Supported by the March of Dimes appeal, Salk won the gift of land on Torrey Pines Mesa to build his dream of a research institute. Working with the renowned architect Louis Kahn, Salk wanted "to create a facility worthy of a visit by Picasso," a "crucible of creativity."[24] The result is certainly original, and judging from the regular pilgrimage of architects to visit the site, the profession of architects agrees.[25]

Here the bells do not sound for matins, and the space between the buildings—the central court—is not enclosed. Instead, standing in that space, your eyes are directed to the blue of the ocean and sky looking straight out to sea. Louis Kahn had envisioned a lush garden, or perhaps just two lines of poplar trees running midway between the buildings from end to end. But when the Mexican architect Luis Barragán visited the site with Kahn, he told him, "Don't put one leaf nor plant, not one flower, nor dirt. Absolutely nothing—and I told him, a plaza . . . will unite the two buildings, and at the end you will see the line of the sea."[26] When they met Jonas Salk, Barragán said to him, "If you make this a plaza, you will gain a façade—a façade to the sky."[27]

Salk had wanted a private courtyard, but he eventually gave in and two years after the buildings were up, in 1967, the plaza was created. The result was not an Oxbridge transplant to California but a poem to space, stone, and concrete, dominated by Kahn's cubist style. The buildings, which were planned to reduce upkeep to a minimum and allow for flexibility in laboratory space, have an air of economy. Why use stone or brick when you can pour concrete? Why build fixed, load-bearing walls between labs when you can erect movable ones? (This idea was to facilitate the changing space requirements of the labs over time and to promote communication among researchers. Crick thought the plan misguided, but many researchers have thought otherwise.)

What a contrast with the historic Cold Spring Harbor Laboratory, where Jim Watson, who had been Director since 1968, had an old whaling house torn down and faithfully reconstructed and the Wawepex cottage restored. Before Watson's time, the village fire-house had been towed across the harbor to a site on campus. So you could be forgiven for imagining that at Cold Spring Harbor Laboratory, you had been transported to an historic location in New England—quite quaint and certainly beautiful.[28] At the Salk, you cannot but know you are in America and at the frontier with a future that has no place for ornamentation and tradition. But the very austerity of the design and the blandness of the building materials inculcate a feeling of awe. Here, one could attack big problems in science and maybe solve them.

These buildings were the end result of much pondering of several designs. Odile recalled her husband's feelings of impatience in the early days of the Institute with all the time and resources going into planning the buildings and the slow progress on the side of scientific research. After a meeting with Bronowski in August 1963 to update him on progress, Crick warned Salk that ". . . it does not appear to me that the Trustees are pulling their weight." Their role is "to raise money for the Institute, and to see that it is put on a much firmer financial foundation. . . . It will damage the Institute very much if building is delayed for lack of money."[29]

In a sense, La Jolla was a home away from home for Crick. His good Cambridge friends Leslie and Alice Orgel, and Edward Bullard and his second wife, Ursula, were already there. The Orgels came to the Salk in 1964, and the Bullards came to the nearby Scripps Institute of Oceanography 10 years later. Bronowski, the well-known mathematician and presenter of television programs, had died in 1974, but his wife Rita (née Coblentz) the sculptress, continued to live in La Jolla. She was a special friend of Odile's.

### Shaping the Research Agenda

The major role of the nonresident Fellows had been to shape the character of the research program, chiefly as advisers on the choice of candidates to fill the resident Fellow positions. In establishing policy here, Crick found himself among forceful personalities with goals he did not always share. Consider the eminent scientific bureaucrat Warren Weaver, already on the Board of Trustees, who was then appointed a nonresident Fellow also. Weaver had spearheaded the Rockefeller Foundation's support of the application of the physical sciences to biology in the 1930s and 1940s, but he upheld the need to study the whole organism in its own right and opposed thoroughgoing reductionist agendas. Assured of the Trustees' support, he began to throw his weight around. He advised the need for the Institute to have a "naturalist" or "biologist's biologist" among the nonresident Fellows and suggested the paleontologist George Gaylord Simpson or the evolutionist Ernst Mayr.

Crick would have none of this and wrote Weaver expressing his opposition. He told Weaver he could not envisage either Simpson or Mayr seeing "eye-to-eye with the other nonresident Fellows," or being "able to communicate with the existing Fellows. . . . They will want to add biological workers for their own sake, and we shall have endless trouble dissuading them."[30] To his friend, Jacques Monod, he explained the reason for his frustration over the addition of Weaver to the ranks of the nonresident Fellows. He had earlier advised that the Institute be registered as a private corporation, thereby permitting a legal structure that gave the scientific staff a determining role in academic policy decisions. Hence, they could not one fine day find themselves overruled by the board of trustees.[31] Academic policy would be the responsibility of the Fellows, not the Trustees. The latter were not unilaterally to appoint Fellows, and this was just what they had done in Weaver's case. To Brenner, he mentioned the possibility of resigning.

Although Crick had expressly insisted on being free from the organizational side of the Institute's work, and to have no vote, he, in fact, was a powerful influence on the course the Institute followed, for his advice on academic matters was sought and followed. It did not escape Salk and Bronowski that Crick would take great pains to advise on the suitability of potential Fellows. He was the same conscientious Crick that we knew from earlier years, also the same forceful personality, as

Weaver must have felt as he read the last paragraph in Crick's letter to him: "I am quite unimpressed by all this talk about molecular biologists forgetting we are dealing with 'living organisms.' It is just this sort of vague nonsense that I am against. We are all agreed we should move towards whole cells, tissues, and organisms. The problem is how to do it, and who to appoint. I've started to collect names in a small way, and hope to have a list by the end of the summer."[32]

Bronowski had no more success in broadening the scope of work at the Salk than did Weaver. He wanted to turn the philosophy of science from its concentration on physics "to study the logic of biology." The Institute, he hoped,

> will be a pioneer in taking the problem of life, the problem of human life, much deeper. I hope that it will see that man's philosophy, his view of the universe, his life of the mind, are as much a part of life as are the physical processes. The Institute seems to me above all the place where science and the humanities can be linked in concrete studies. I hope I can continue to help to create such links.[33]

Several years later, Salk sent Crick a document entitled "Proposals for the Academic Development of the Institute." These would broaden the scope of the research. Crick had wanted the concentration to be on cell biology, but "a major effort in human biology," he agreed was possible. After all, he was himself increasingly interested in whole organisms, including man. His reservations lay with the Bronowski-inspired "application of biology to society." If the Institute undertakes "the underlying fundamental research," Crick had "little doubt that . . . its application will follow."[34]

Just as Crick had been influential in shaping the research agenda of the LMB in Cambridge in the years following his return from Brooklyn, so he was the major force in establishing the neurosciences at the Salk with the appointment of researchers like Terrence Sejnowski, since 2004 the holder of the Francis Crick Chair of Computational Neuroscience and director of one of the five laboratories devoted to the neurosciences. Of these, the Vision Center Laboratory was devoted to the field studied so intensively by Crick in conjunction with Christof Koch, Caltech's Professor of Computation and Neural Systems. Crick's long-time friend Leslie Orgel was still running the Chemical Evolution Laboratory where candidate pathways to the origin of life are explored. Laboratories given to genetics, gene expression, regulatory biology, and structural biology all reflected Crick's former interests. To have been so influential without formal status on matters of organization and policy betokens not only the respect his advice commanded, but also the forceful powers of expression he possessed and used if goals dear to him were endangered.

## Prospects and Retrospects

Crick often made predictions about the future in terms of how many years it would take to solve this or that problem and attain one goal or another. The talk he gave in 1978 on the topic "Contemporary Frontiers" to General Electric Company's International Symposium on "Science, Invention, and Social Change, is an example.[35] Here, he discussed the value of the lessons he had learned from his experiences in molecular biology for making forecasts about the future. At this time, he had been at the Salk Institute for six years, and, save for chromosome structure, he was now committed to the neurosciences. As he reflected on the field he was leaving, he stressed that "at all levels [of biology] our ignorance is profound." Hence, to understand biology "at all, even in outline, will take a very long time. It is all too easy to enumerate the major problems which will eventually have to be solved. It is quite another matter to try to foresee which ones will yield in our own lifetime." But we can learn from the classical period of molecular biology (1945–1966). He hoped that he might be "forgiven for saying that these developments have been the most successful part of biology in the last thirty years." Therefore, he explained, "In this attempt to forecast the future it will pay us to analyze the reasons for this success."[36] "We should start," he advised, "with a general theoretical reason. Thus, we now know that Nature is engineered, in detail, at the molecular level. Thus, results at this level, if they can be obtained, are bound to be important at this and at all higher levels." Second, methods have been developed for studying the relevant molecules in the test tube and in the cell. The latter task has been aided by the use of "highly specific chemical inhibitors (often antibiotics)," and by deploying genetics, "which can be used to make delicate alterations to a protein while it is actually in the intact cell. . . . Experience has shown that without the use of both these approaches—the purified component and the intact organism—the problem is likely to get into a mess."[37]

Added to this, molecular biology had the advantage from 1953 onward of acquiring "more by luck than by anything else, a good theoretical framework for the work of the following decade." But he judged "the major reason for molecular biology's success was the introduction of rapid, cheap, and powerful methods." Among these, "the use of microorganisms totally transformed the subject, especially on the genetic side. Many of these techniques had the following property.

They were not one-shot experiments but could be used tens or hundreds of times in one experiment. That was why it was essential that they should be both rapid and cheap."[38]

Progress since 1966, he judged, had been steady, "but in the last three or four years the pace has quickened. Again this has been due to the invention of new techniques. The most important of these is 'genetic engineering'—really DNA fragmentation, selection, and multiplication." He singled out two techniques as especially important: (1) the synthesis of a required protein, like insulin or hemoglobin, by inserting the relevant gene into a microorganism, and (2) Fred Sanger's technique for sequencing nucleic acids. The latter, he confidently forecast will result in "an immense explosion in DNA sequence data. Merely to record it usefully will involve storing it in computers. Computation will also be needed to help understand what the sequence means." He told a like story for the proteins where "a similar explosion appears to be starting. . . . We now have powerful 2-D chromatograms for displaying several thousand different proteins on an autoradiograph."[39]

Clearly, Crick had grasped the potential of the new phase in molecular biology, while at the same time, disengaging himself as far as possible from active involvement. Now established at the Salk Institute, he could hope to achieve his final goal of immersion in the neurosciences while remaining mindful of the lessons learned in molecular biology. He could enjoy the vitality of science in the United States and the wealth of its support from both the private and public sectors. As one of the Salk Institute's founding members, he could devote his energies to building its program in the neurosciences.

## CHAPTER 19

# Taking the Plunge: Neuroscience

 *I don't think there was a general expectation that I would achieve much in neurobiology. I think the general reaction was, "Who is this old guy coming into our field?"[1]*

Escaping from the past was not as easy as the move to La Jolla might suggest. Chromosome structure was a hot subject. Crick had already put quite an investment of time and effort into it. He was still to attend five chromosome meetings and be involved in the writing of eight papers on the subject. Then there was the surprise and excitement over the discovery of split genes (see Chapter 17) and the continuing irritant over so-called nonsense or "junk" DNA. Yet his keen interest in the brain dates back to the time he became a nonresident Fellow at the Salk Institute in 1963 and had spied the name David Hubel in a footnote to an article in the literary magazine *Encounter*. It referred to a paper that this Harvard physiologist wrote about his work on the physiology of vision. Imagine Hubel's surprise when one of the reprint requests came from Crick.

Eagerly and with amazement, Crick read Hubel's account of the classic experiments he and his Swedish-born Harvard colleague, the physiologist Torsten Wiesel, had carried out on what they called the "functional architecture" of the primary visual center in the brain. Inspired by this work and eager to form a group working in the same field, Crick suggested that a meeting be arranged at the Salk to introduce the Fellows to the subject. Among the speakers Crick had suggested inviting were Hubel, Roger Sperry, the latter famous for his split-brain experiments, and Rita Levi-Montalcini, the developmental neurobiologist. Hubel recalled how, early in January 1964,

> The phone rang. "This is Jonas Salk" came the voice from the other end. "Yes . . . ?" I answered, ". . . I'm sorry, who did you say was

speaking?" It was indeed Jonas Salk, calling to invite me to a meeting at the Salk Institute in La Jolla, the object of which was to inform a remarkable cast of molecular biological celebrities, each more famous than the next, about the state of the field of neurobiology.[2]

In the midst of teaching at Harvard University, Hubel decided to come for just one day. His one-hour slot was on Sunday, 23 February. When he had covered their first major discoveries in visual perception, his time was up. As he later explained:

> I was interrupted many times with questions, especially from a young man in the front row with a French accent, whom I took to be a very bright teenager, but who turned out to be Jacques Monod. When I finished, Francis Crick jumped up and said, "But you were supposed to tell us something about visual deprivation and learning." I pointed out that my hour was up, but that seemed to make no difference—so I went on for another hour. At which Crick rose up and said, "But I heard that you and Wiesel had done work on color. Are you not going to tell us about that?" Time and schedules seemed to present no problems to these molecular biologists, so I went on for still another hour. I felt like Fidel Castro. Their enthusiasm seemed boundless, and I began to think that our work was not so boring after all.[3]

The audience numbered just ten—the resident Salk Fellows Melvin Cohn, Renato Dulbecco, Edwin Lennox, and Leslie Orgel, and the nonresident Fellows Francis Crick, Salvador Luria, Jacques Monod, Leo Szilard, Seymour Benzer, and Warren Weaver—about as formidable an audience as you could imagine. So critical and intimidating were they that the last speaker coming to the stand, Crick remembered, could be seen "visibly trembling as he reached it."[4] No doubt a contributing factor to this state of anxiety was the free-for-all to interrupt speakers during their talks, including perhaps Crick's well-known question: "Why are you doing these experiments anyway?"

Writing to Crick, Hubel recalled the event nearly four decades later: "when I met such a fantastic bunch of molecular biologists (not least yourself) for the first time, and gave what was supposed to be an hour's talk but turned into three—and on subsequent many visits to the Salk and to you. The first meeting was to me a revelation."[5] Crick replied: "I still have vivid recollections of your first visit to the Salk."[6]

Hubel and Wiesel had found order in the seeming "impossible jungle of the brain." A human brain contains some $10^{11}$ neurons (the cells with their connecting fibers) and around $10^{15}$ synapses (connections)—all packed into the skull. This packing is no less dense in the brains of the cats they were studying. They were recording the "firing" of cells

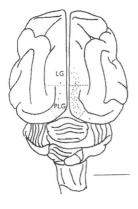

**Figure 19.1** The dorsal view of the cat brain showing the entry points of 45 microelectrode penetrations into the striate cortex or V1 LG, Lateral gyrus; PLG, postlateral gyrus.

in the striate cortex at the back of the cat's brain (the primary visual cortex[7] or V1 that receives the input from the ganglion cells of the retina) (Fig. 19.1). Here, 100,000 neurons and more lie beneath a square millimeter of the surface.

Diffuse illumination yielded no obvious electrical responses in these cells. Neither, initially, did a stationary bar of light against a dark background. But an accidental shadow *moving* across the same background caused rapid firing. The cells were responding to the motion and its direction. Thus alerted, Hubel and Wiesel persisted with their experiments and discovered a variety of cells in the brain that responded to different features of the retinal input.

Their painstaking experiments, each run lasting many hours, involved projecting a spot of light onto a screen and moving it around until it caused a cell in the striate cortex under investigation to fire. The "restricted area" of the retina thus illuminated is known as the cell's "receptive field." The light stimulus received there was being transmitted to this cortical cell making it "fire" and a minute electrode planted in the brain picked up that signal. This is called "single-cell recording." It identifies the cell in question with the stimulus, and the nature of that stimulus then serves to distinguish the kind of cell that fired. Hence the term "functional anatomy."

This technique had first been applied years before to the ganglion cells of the retina—a more convenient location than the brain but really an "outpost" of it (Fig. 19.2). These were the first such cells to reveal their secrets. Their receptive fields are divided into two regions in a concentric arrangement, one region excitatory and the other inhibitory. Some ganglion cells were fired by a tiny spot of light in the center of the receptive field surrounded by a ring of darkness and others by a dark center surrounded by a ring of light (Fig. 19.3). Because of the

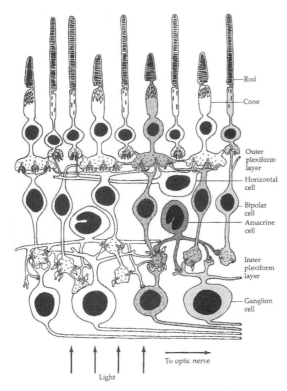

**Figure 19.2** Diagrammatic representation of the pathways in the retina through which the rod's and cone's responses to light are fed to the ganglion cells, and then to the brain.

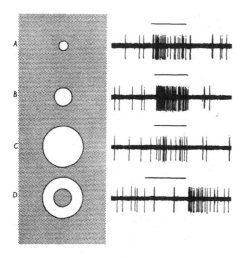

**Figure 19.3** Responses from an "on" center cell. *A*, *B* and *C*, central light spot 1°, 2°, and 14° diameter. *D*, annular light stimulus, inner diameter 2° outer diameter 14°. Horizontal time line above each response record (of the axonal spikes) indicates when 1 sec. light stimulus was given. In *C* the illumination has extended beyond the limits of the "on" center causing some inhibition of the "on" response. In *D* the response from the annular "off" region of illumination comes the moment the light is turned off.

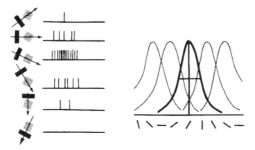

**Figure 19.4** Edge "Detectors." Different orientations of the bar of light are shown on the left. The arrows indicate direction of motion of the bar. On the right is the orientation tuning curve for this cell (the heavy line). The curves for cells with different orientation specificities are represented by the fine lines.

antagonism between the region inhibitory to light with the region excitatory to it, experiments using uniform illumination of the whole of the receptive field yielded virtually no firing from these cells.

Hubel and Wiesel applied their method to cells in V1. The receptive fields of these cells, they discovered, were also divided into light-excitatory and light-inhibitory regions, and their shapes were not concentric but very elongated, giving the appearance of an edge or border between dark and light regions (Fig. 19.4). Therefore, these cells responded maximally to a bright bar or slit of light against a dark background. They were "orientation-specific." Moreover, to Hubel's and Wiesel's delight, they discovered that the orientation of the light/dark edge of the bar or slit had to be close to a specific angle to obtain maximal firing. Another remarkable feature of these cortical cells is their organization in V1. Those cells responding to the same orientation of light were found lying one above another in a columnar arrangement. But cells adjacent to them formed columns responding to an orientation a degree or two different.

Evidently, features of the retinal image were being distributed to and processed by specific sets of neurons in a functional arrangement. They had discovered a functional architecture to the cortex. Could it be that the silhouette of an object is constructed by the brain when certain of its cells bring together a whole series of these responses to differently oriented stimuli? They illustrated this point using Van Gogh's "Self-portrait in a Hat," a work of art composed entirely of oriented short lines. It seemed to Hubel and Wiesel that the brain contains a hierarchy of centers that progressively bind together the elemental features to which the different kinds of cortical cells are "tuned."

Hubel described this revelation of the functional anatomy of the striate cortex in the first hour of his marathon lecture at the Salk Institute. The take-home message was the importance of borders, edges, and silhouettes. As Crick later put it: "The retina is processing the information coming into the eye in such a way that it partially eliminates redundant information. What is sent to the brain shows mainly the interesting parts of the visual field, where the light distribution is not uniform, largely ignoring the dull parts where it is fairly constant."[8] The adaptive significance of this visual selectivity was clear, especially the response to moving stimuli. Consider the shadow cast by a hawk flying overhead, and on the ground beneath it, a mouse. Cells that fire in the mouse's retina when the edge of the hawk's shadow moves across its field of vision offer a timely warning of danger. So, too, a cat seeking a mouse or a frog a fly for dinner will discriminate the object most readily when it moves revealing its silhouette.

## Testing the Waters

Established as the Kieckhefer Distinguished Research Professor at the Salk in the Spring of 1977, Crick could look back to his first meeting with Hubel and forward to clearing the decks in order to plunge into neuroscience. To Hubel he admitted to feeling the need for a "proper background in the nervous system. This means learning some elementary human neuroanatomy which is really the most frightful subject I have ever had to tackle. More to the point, I have lots and lots of questions . . . ."[9] Next he told Brenner that he planned "to make a real effort in the next few months to grasp the general layout of the higher nervous system of mammals. . . ."[10] For starters, he needed to master the nomenclature of the brain—the cortex, its lobes named frontal, occipital, parietal, and temporal; its organs, the amygdala, cerebellum, colliculus, corpus callosum, hippocampus, hypothalamus, thalamus, lateral geniculate nucleus, basal granules, and so on; and their locations and connections. For Crick's reading, Hubel recommended A. Brodal's *Neurological Anatomy* (1969). Fortunately, Crick and Hubel were to meet quite frequently. Hubel's mother lived in San Diego, so he often came by the Salk when visiting her. Such visits continued through the seventies. In the late 1960s, Hubel and Wiesel spent a summer at the Salk.

Not long thereafter, the editors of *Scientific American* began to plan a special issue on the brain and they asked Crick "as a newcomer to neurobiology, to make some general comments on how the subject strikes a relative outsider."[11] Called "Thinking about the Brain," Crick's contribution, his first in this field, evaluates the strengths and weaknesses of several of the approaches being used, an ideal exercise for him at this point in time.

He had, he explained, been interested in neurobiology for more than 30 years, but only in the last two had he "attempted to study it seriously." He began by separating those topics that "appear at least capable of explanation by familiar approaches of one kind or another from those for which no ready explanation, even in outline, seems available at the present time." Indeed, "[They] appear to me to defeat our present understanding. We sense there is something difficult to explain, but it seems almost impossible to state clearly and exactly what the difficulty is. This suggests that our entire way of thinking about such problems may be incorrect. In the forefront of the problems I would put perception . . . in particular visual perception."[12]

This point was brought home to him when he reflected on the attempts of those working in artificial intelligence (AI) to model the information handling involved in just seeing: "When one reflects on the computations that must have to be carried out before one can recognize even such an everyday scene as another person crossing the street, one is left with a feeling of amazement that such an extraordinary series of detailed operations can be accomplished so effortlessly in such a short space of time." Yet he warned that AI's reliance on the "analogy between a computer and the brain, although it is useful in some ways, is apt to be misleading." In the standard desktop computer, information is processed serially and with rapidity, whereas in brains, the processing is in parallel and much slower. Failure of an element in the computer can wreck its performance, but the brain can often continue to function, thanks to the element of redundancy—a feature of its parallel processing and plasticity. It does not compute mathematical sums as rapidly and reliably as a computer, but by adjusting the connections (synapses) between the nerves "in complex and subtle ways," the brain can "adapt its operation to experience." For instance, "human beings can recognize patterns in ways no contemporary computer can begin to approach."[13] This is thanks to their *connectionist* architecture, about which more anon. (He was writing this back in 1979.)

Could psychology help? Yes, pure psychology (by which Crick meant describing behavior and relating input to output) is essential, but judged from the "standards of hard science" (i.e., the physical sciences), it is "rather unsuccessful." Furthermore, the science also includes describing our thoughts and feelings—"introspection." Here, he warned, "we are deceived at every level . . . mainly because what we can report is only a minute fraction of what goes on in our head." This deception, he judged, "is why much of philosophy has been barren for more than 2000 years and is likely to remain so until philosophers learn to understand the language of information processing."[14] (Philosophy always brought out the brash side in Crick!)

Nor is introspection the only method that can mislead. Neurophysiological experiments can do as much. That is why we need parallel studies of external behavior and internal processes. Take the fallacy of the homunculus:

> I was trying to explain to an intelligent young woman the problem of understanding how it is we perceive anything at all, and I was not having any success. She could not see why there was a problem. Finally in despair I asked her how she herself thought she saw the world. She replied that she probably had somewhere in her head something like a little television set. "So who," I asked, "is looking at it?" She now saw the problem immediately.[15]

However, it is easier to state the fallacy than to avoid slipping into it. "This," he judged, is because "we have an illusion of the homunculus: the self." In addition, there is what he called the "fallacy of the overwise neuron." We tend to think of neurons in sensory physiology in terms of "color detectors, edge detectors, etc., but a single 'edge detector' does not really tell us that an edge is there. What it is detecting is, loosely speaking, 'edginess' in the visual input, that is, a particular type of nonuniformity in the retinal image that might be produced by many different objects. One of the objectives of theoretical neurobiology is to try to turn such vague concepts as 'edginess' into mathematically precise descriptions."[16]

What attracted Crick to Hubel's and Wiesel's work was that it opened the black box of the visual system, revealing the manner in which the visual data are analyzed. Their boss Stephen Kuffler had in 1953 shown how the "ganglion" cells of the retina begin this analysis. They applied Kuffler's analysis to V1 and revealed how it mapped the "landscape" of the retina. V1's geography is *retinotopic*. Neighbors among retinal cells are represented by neighboring neurons in V1. Indeed, there is a succession of mappings of visual data in a number of centers. The data are not being viewed on a kind of television screen, but their elements are being *distributed* and analyzed, and the particular cells firing in this or that center are *constructing* representations or *symbols* of the seen object at many levels through the hierarchy of the brain. As Crick explained: "It is difficult for many people to accept that what they see is a symbolic interpretation of the world—it all seems so like 'the real thing.' But in fact we have no direct knowledge of objects in the world."[17] (Well might he at this point have paid tribute to Immanuel Kant, who in the 18th century distinguished between the "noumenal" world—what is "really" out there but unknown to us—and the "phenomenal" world—the world as we perceive it. But then Kant was a philosopher.)

Cognitive psychologists and behaviorist psychologists before them had a preference for keeping their hands dry. Therefore, they treated the brain as a black box. No wet science for them. But as Crick explained, "unless the box is inherently very simple a stage is soon reached where several rival theories all explain the observed results equally well. Attempts to decide among them often prove unsuccessful because as more experiments are done more complexities are revealed. At that point there is no choice but to poke inside the box if the matter is to be settled one way or the other."[18]

This passage should remind us of the unpublished paper Crick wrote on protein structure determination in 1951 (see Chapter 5), where he surveyed the several methods and pointed out their weaknesses. In 1979, however, he could reflect on the experience of his career in molecular biology, how the deliberate search for new methods advanced the field, and how important it was to establish a "broad framework of ideas" into which to fit all of the different approaches from biochemistry and genetics. The DNA structure had provided such a framework for molecular biology. But for the brain, he contended, "a broad framework of ideas" is "conspicuously lacking."[19] As a mechanist looking for a framework, he opted for the patterns of connectivity of the parts—hence, the importance to him of learning the anatomy.

### The AI Approach to Vision

Crick widened his circle of friends in the neurosciences when he attended a meeting of the Neurosciences Research Program (NRP) in Boston in May 1978. This organization had been set up by the MIT biophysicist Francis O. Schmitt in 1962 to promote interdisciplinarity and bring together neurophysiologists, cognitive scientists, neuroanatomists, and more. Attending this meeting, Crick renewed his acquaintance with the young Italian cognitive scientist Tomaso Poggio, then working at the Max Planck Institute for Biological Cybernetics in Tübingen (see Chapter 20). It also served to renew his friendship with the British cognitive scientist David Marr. With a mathematics degree from Cambridge, Marr had turned in 1966 to theoretical neuroscience. It was Crick and Sydney Brenner who had then provided him with an office at the MRC lab in Cambridge until he "became a partial convert to the AI (artificial intelligence) approach and moved to MIT."[20]

When Crick brought Poggio and Marr to spend the month of April 1979 at the Salk, all three lunched together frequently while they worked on a theoretical paper on the visual cortex.[21] Something of the flavor of their lunchtime debates can be gleaned from the final section of Marr's classic book *Vision*. There, Marr constructed an imaginary dialog between a skeptic and a defender of the information-processing point of view. Although Marr called his discussants "imaginary," he acknowledged that the dialog was based on these lunchtime conversations. Many of the skeptic's remarks reflect Crick's point of view.[22]

These discussions were given over to the subject of the paper Marr and Poggio had come to La Jolla to write—the hyperacuity of the eye. They used the information-processing approach, and, with the analogy of the computer in mind, they constructed a mathematical model of the processes whereby this remarkable "positional accuracy" of our visual system—hyperacuity—might be achieved. Because Poggio and Marr had to return to Boston before the meeting, it was Crick who presented their paper at the NRP's Spring colloquium held at the Woods Hole Marine Biological Laboratory (29 April to 4 May). It begins with a typically Crickish sentence: "That we see the world as well as we do is something of a miracle."[23]

Crick enjoyed these two enthusiasts for the AI approach because they were also mindful of neurological details. As a doctoral student, Marr had been "caught up" in the excitement that the neurophysiologists' studies of single-cell recording had created. Turning to AI, however, he perceived the limitations of the physiological approach. It could not tell one how to construct a hand detector. "Do the single-unit recordings," he asked, "—the simple and the complex cells—tell us much about how to detect edges?"[24] The answer was no. Just how these cells did their detecting remained unclear. Crick agreed.

Marr's move to MIT had as its aim to seek the principles upon which one might build an analog to mimic the work of the eye. What *procedures* would such a machine need to perform? This question Marr and Poggio referred to as the *computational* level. Below that comes the *algorithmic* level—the set of rules or steps by which the computation might be achieved—and below that the level of *hardware implementation*—the neurons and their connections or the chips in a computer. "The levels idea is crucial," Marr explained. "Visual perception cannot be understood without it—never by thinking just about synaptic vesicles or about neurons and axons, just as flight cannot be understood by studying only feathers. Aerodynamics provides the context in which to properly understand feathers."[25]

***Figure 19.5*** This figure is described by David Marr as "a discretely sampled and quantized image of Abraham Lincoln, an imaginary primal sketch." In spite of our efforts we cannot perceive Abraham Lincoln unless we defocus the image or squint. But the zero-crossings in the larger channels have yielded "an approximate repression" of his face.

Crick had found Marr stimulating and his ideas useful but was not convinced that the computational and algorithmic levels are as independent from the implementation level as Marr claimed. Such independence justified the cognitive scientist in ignoring neurons, but it flew in the face of Crick's understanding of evolution as a process that has had to make do with whatever starting material was at its disposal, in this case the neurons. Nor was Crick (alias *Skeptic*) happy to leave out the structure—function relation, for he had built his reputation in molecular biology on clarifying function by discovering structure. Consider the following passage from Marr's imaginary dialog:

> *Skeptic*: One has a strong urge to tie explanation to structure eventually—that, of course, was the impact of molecular biology. It has to be done here, don't you think? Or do you see the endeavor as totally hopeless?[26]

Marr doubted "if it can ever be done completely. The complexity barrier is just too great." From his machine perspective in MIT's Artificial Intelligence Laboratory, however, he envisaged a series of stages in the construction of the shape and arrangement of objects in the visual field. Starting with what he called the primal sketch (Fig. 19.5), there followed the 2½-D [dimensional] sketch, and finally the 3-D model.[27] Each stage involved a further computation and added detail to the retinal representation of the image.

Crick recognized Marr's insight and also appreciated his suggestion of the need to construct a "sequence of representations" of the visual data. But he (alias *Skeptic*) was not happy with so much talk about computation:

> *Skeptic:* The brain, after all, is made of neurons, not silicon chips. But I suppose I'll get used to it. Still vision is the construction of descriptions, they must be implemented neurally, mustn't they? So couldn't

one hope to look for neurophysiological correlates of the 2½-D sketch or of a piece of a 3-D model? That I would find convincing.[28]

After all, finding a mathematical theorem that produces the correct result is not enough. *Skeptic* explained:

> I cannot really accept that the computational theory is independent of the other levels. To be precise, I can imagine that two quite distinct theories of a process might be possible. Theory 1 might be vastly superior to theory 2, which may be only a poor man's version in some way, but it could happen that the neural nets have no way of implementing theory 1 but can do theory 2 very well. Effort would thus be misplaced in an elaborate development of theory 2.[29]

It was Marr's and Poggio's belief that one can formulate mathematically the operations that ganglion cells in the retina perform upon the image and that this formulation would then at least be suggestive of how these cells carry out their function. Marr's first stage in the representation of the image—the primal sketch—he envisaged as the identification of what he called "primitives," where the intensity of the image changes abruptly from positive to negative as at the edge of a dark object against a bright background (or the reverse). Using what is called the "zero-crossing" theorem of the mathematician B.F. Logan, these primitives can be represented by a discrete line or border. In detecting these borders, the ganglion cells were envisaged as filtering the image data, thus making plausible the attempt to devise a mathematical formulation of the operation these cells appear to perform in creating the primal sketch. Armed with this algorithm, one could design a program for a computer that could create the primal sketch.

We learn from the detailed and enthusiastic 3000-word letter Crick wrote to Poggio after the NRP meeting that Crick did cover these important ideas of Marr.

> Let me say first that I didn't give a very good talk myself and I'm glad neither of you was there to hear it. It wasn't that I did not go to considerable pains to prepare it. . . . In the talk I tried to do three things. First to describe, very briefly, David's general approach to vision. Second to outline the stereopsis theory and third, to discuss acuity and our hypothesis. . . . You would have been appalled at what I had to skip over. . .

> One of the good things of the meeting was that I got to know people, specially the younger people.[30]

A popular meeting place in Woods Hole was the bar of the Captain Kidd. There Crick encountered Jennifer Lund, the cytologist who worked with Hubel. Famous neuroscientists such as Vernon Mountcastle and Herbert Jasper also met him for the first time there. But it was the younger folk who impressed him most. He judged their work to be excellent and he found that "they are much more receptive to ideas than some of the older ones. The field is crying out for someone to pull it all together or at least put the thinking about it into some sort of order. So I was very glad I went. I plan to invite some of the younger ones to La Jolla, on two-day visits, next academic year, so I can really digest what they are doing. Such a joy having you both here. I do hope you'll decide to come again."[31] Sadly, Marr's untimely death from cancer came at age 35 the following year.

Having thus immersed himself in the mathematical approach of his cognitivist friends Marr and Poggio, Crick confessed to the latter, "What I don't think anyone can say is just where all this theoretical work is heading, but the only way to answer that is to try and see."[32] He was indeed "testing the waters." A decade later, reflecting on the styles of explanations in neuroscience, the one he found closest to his own inclination was not that of Marr and Poggio. It was that of his San Diego friend, V.S. Ramachandran, for whom modeling visual perception on the structure of a rational argument had little appeal. Nor did "Rama" feel the need to solve elaborate equations by using Fourier analysis, the DOG (difference of Gaussians), and the convolution of functions as did Marr and Poggio. Instead, he believed perception "is essentially a 'bag of tricks'; that through millions of years has evolved numerous short-cuts, rules of thumb and heuristics which were adopted not for their aesthetic appeal or mathematical elegance but simply because they worked. . . ."[33]

Crick quoted this utterance with hearty approval.[34] And Rama quoted Crick as declaring "God is a hacker," an analogy that fitted Rama's "utilitarian" theory admirably. Crick had earlier asked Hubel for his opinion on the representation of the visual system as analyzing spatial frequencies using Fourier theory—the mathematicians' favored approach—rather than describing their action in the imprecise terms of feature detectors. Hubel replied that his "prejudice" was "totally against the cortex analyzing things in Fourier terms." Crick responded, "I don't believe in most of the ideas involving Fourier transforms, since they seem to me to be forced."[35] After having immersed himself in the Marr and Poggio paper, he was more forthright on the subject. He now found the zero-crossing idea "to have become inflated with repetition. The original idea is a nice one but the theoretical foundations, especially in two dimensions, are weak." As for the cells acting as filters $(\nabla^2 G)$, he thought that "the original idea was that this was the most efficient representation," but he now questioned this.[36]

## Parallel Distributed Processing

The target of Rama's criticism was the kind of computation used by Marr and Poggio. His own "utilitarian" approach, he explained, "is not inconsistent with modern connectionist approaches to vision,"[37] the mathematics of which differs from that of von Neumann style computation, based on mathematical logic.[38] In a connectionist model, the processing of the data is distributed through a network of units running in parallel and interconnected. The network then learns the correct output by adjustment of the strength of these connections. This geometry thus reflects the parallel organization of so much of our nervous system. This approach was being explored at several centers, but especially at Carnegie Mellon University in Pittsburgh and at UCSD's (University of California, San Diego) Institute of Cognitive Science just down Torrey Pines Road from the Salk. There, from January 1982 onward, a group 16 strong met weekly to discuss connectionism, and from their collaborative work came the book, *Parallel Distributed Processing* (abbreviated to PDP). More than 1100 pages and with 16 authors, this two-volume publication became an unlikely best seller. The editors were James McClelland and David Rumelhart, both at UCSD at the time of the book's inception.

Crick, who had become an Adjunct Professor in UCSD's psychology department, was a member of the group. He remembers how he used "to walk to the small informal meetings of their discussion group." But now "the land over which I strolled has been turned into enormous parking lots."[39] Characteristically, the style of Crick's contributions to these meetings was both critical and skeptical, shaped by the priority he gave to seeking models that genuinely imitate the brain. His frank and forceful style accepted at Cambridge did not desert him at UCSD. "How," I asked "Jay" McClelland, "did he react to it?" He responded:

> I just felt that he changed the tenor of our meetings by his presence. We had started to meet before he started to come and I had been very insistent that various other "silver-backed gorillas" not be allowed to attend, and I got the word that Crick had found out about this and he was planning on coming. And I said: "It's not going to be the same any more," and I was right about that. He was a very tall man, . . . so even though he sat in a very low sofa at the back of the room he still somehow dominated the whole scene. He was obviously trying not to over-dominate things because he did sit in the back and I appreciated that wasn't his intention. But at the same time the tenor of the whole PDP enterprise was to peek outside the box and not let anything overly constrain you, and Rumelhart in particular was very important in helping at least me to get to that point of thinking—yes, we just want

to understand these computations and explore the implications of these ideas, and when Crick started to attend [it was] just like we always had to talk about biological possibility. So we weren't making as much progress exploring the issues that were at the forefront of my own thinking. I didn't mind him being frank and forceful, but I did mind him redirecting the conversation.[40]

Crick was not disruptive, explained McClelland. "Interrupting would be too strong" a word, but "he was insistent, he knew what was in his mind and when the opportunity arose to ask questions he would do so and would force the issue to a specific subject of interest to him." McClelland recalled his own model of net control, which he had thought at the time was "a brilliant piece of work," but Crick strongly opposed:

It relied on using multiplicative connections and Crick just didn't like that idea. . . . Two different signals would come to the same place [on a dendritic spine] at the same time, one would gate the other. . . . Because Crick was so focused on the exact microstructure of the way the nervous system actually worked he couldn't imagine that they could be organized tightly enough for this process to occur.[41]

On this point, Crick was insistent and he was later proved right—that is, where neurological evidence rules.

In a similar vein, Crick expressed his concern about theorists who do not recognize the importance of animal models accessible to invasive experimentation. This, he reckoned, should be a consideration when choosing which sensory modality to explore. This concern surfaces in his essay for the PDP book. There, he explained that since the macaque monkey provides a model for the visual system, "the solutions of visual problems should be easier to bring down to earth than linguistic ones." In the draft read by McClelland as an editor, Crick then added that the absence of an animal model for language meant that its analysis "should be relegated to the cloud cuckoo land that is only to be found in departments of psychology and linguistics." "I told him," recalled McClelland, "that it expressed his point of view so exactly that he shouldn't delete it, but he might want to think about it." Crick deleted it and substituted the following: "This does not mean that linguistic problems may not suggest valuable ideas about the working of the brain. . . . It does mean that they may be more difficult to test at the level of neuronal organization and function."[42]

In 1979, Crick had reminded readers of *Scientific American* that the conventional computer does not offer us a good model for the brain, the chief reason being that it operates on the data sequentially, whereas most of the pathways between neurons in the brain are in par-

Output Patterns

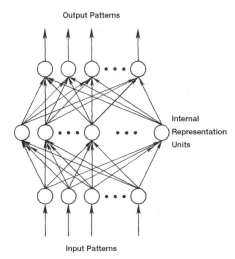

Internal
Representation
Units

Input Patterns

*Figure 19.6* A very simple "parallel distributed processor" network with "hidden" or "internal representation" units.

allel. Among these are the two million fibers that run from our retinas to the brain. This enables the visual data to be distributed for analysis in a variety of ways simultaneously. Moreover, unlike the brain and unlike a parallel processor, a conventional computer needs a central point of control, providing a routine for everything—that is, the operating system. A parallel processor, like the brain, has to be trained— that is, it can learn through experience. What it learns is a pattern of firing of its units that is then stored in particular *distributions* of stronger and weaker connections between the units of the processor. The stored pattern can be recalled subsequently—that is, it has been remembered.

In its simplest form, a parallel processor consists of three layers of units, a set of input units receiving information, a set of output units responding to that information, and an intervening layer of units, the so-called "hidden units," with connections between all members of neighboring layers (Fig. 19.6). For a crude analogy with the brain, consider the ganglion cells in the retina represented by the input units of the computer network and neurons in the striate cortex of the brain represented by the output units. The hidden units of the middle layer could represent that so-called "way station" between eye and cortex— the lateral geniculate nucleus (LGN). In a "supervised" network, the strengths of the outputs from each unit can be adjusted by a supervisor training the network until the required response is achieved.

But how are units at several levels to be adjusted? Crick explained:

> The recent excitement has sprung mainly from a neat algorithm . . .
> "the back propagation of errors" . . . back-prop for short. It can be
> applied to any number of layers. . . . The error signals [provided by
> the teacher or in the case of the brain by another part of it] are used to
> adjust the weights of all the connections to the top or output layer and
> this information is "back-propagated" to the hidden units in the mid-
> dle layer. . . . The results that can be achieved with such simple nets
> are astonishing.[43]

A striking example of such a network was demonstrated in 1987 by
computational neuroscientists Terry Sejnowski and Charles Rosenberg.
They had designed one network they called NETtalk that could be taught
to convert an input of English text into spoken English. To anyone who
has listened to NETtalk's progressive advance from jumbled sounds to
spoken English, the experience is not soon forgotten, and Crick noted
what "a heady sense of euphoria" resulted from such achievements.

Back-prop is a mathematical algorithm whose effect is to modify in
successive stages the network's approach to the correct output. Crick
had reservations and asked:

> But is this what the brain actually does? Alas, the back-prop nets are
> unrealistic in almost every respect, as indeed some of their inventors
> have admitted. They usually violate the rule that the outputs of a sin-
> gle neuron, at least in the neocortex, are either excitatory or inhibito-
> ry ones, but not both. It is also extremely difficult to see how neurons
> would implement the back-prop algorithm.[44]

"And where in the brain," asked Crick, "are the units that compare
the output of each unit with the information from the teacher?" Such
units (neurons) "should have novel properties and would be worth
looking for, but there is no sign of back-prop advocates clamouring at
the doors of neuroscientists, begging them to search for such neurons.
Nor are they concerned that they do not believe their own children
learn to speak using such a simple back-prop network inside their
heads."[45] What a curious situation! But, as we have seen, many cognitive psychologists such as McClelland and Rumelhart did not share
Crick's biological constraints. Proof of the wisdom of their vision is
today evident in the very fundamental contributions that PDP research
has made to control mechanisms, search engines, security systems, and
so much else. Crick, whose mathematical skills were by no means as
limited as he sometimes suggested, was nevertheless unhappy with
this example of the mathematization of learning that back-prop pro-

vided, and he suspected that "within most modelers a frustrated mathematician is trying to unfold his wings. It is not enough to make something that works. How much better if it can be shown to embody some powerful general principle for handling information, expressible in a deep mathematical form, if only to give an air of intellectual respectability to an otherwise low-brow enterprise."[46] When in 1994 Crick delivered the first David Marr lecture in Cambridge, a mathematician in the audience asked how mathematicians could help the neuroscientists. Crick's reply was surprisingly forthright and negative.

### Neural Nets

Although back-prop seemed to be an inappropriate algorithm for the nervous system, connectionism itself—the study of imaginary parallel processors—was attractive. It is not a matter of actually building such networks but of simulating them on a conventional (serial) computer. The aim is to simulate the performance of the brain, and thus, they are often referred to as "neural nets." In his *Scientific American* paper of 1979, Crick had explained how the brain in higher organisms is a combination of "precision wiring" determined by the genes and "associative nets," the neurons of which are all connected to one another, but their strengths are "adjusted 'by experience' on the basis of well-defined rules."[47] In this way the behavior of the system as a whole is fine-tuned and learning can occur.

When the Cambridge mathematician Graeme Mitchison visited the Salk in 1980, Crick wrote to his friend Poggio excitedly, telling him how they were having "a lot of fun scampering through neurobiology and talking about associative nets."[48] From the behavior of these nets, they were led to consider a possible mechanism at work when we dream in our sleep and to suggest its function. Three years later, their paper "The Function of Dream Sleep" appeared in *Nature*. Crick recalled in a public lecture how he had been "trying to think about the brain" when Mitchison came:

> He and I were wondering about the behavior of neurons and it's fairly obvious that you can't entirely restrict yourself to looking at a single nerve cell. You must look at how groups, or must *think* about how groups of nerve cells behave and how they interact. And people have made very simple models of such interactions . . . which are sometimes called neural nets. They are a number of units which are somewhat like neurons which connect together and we know how to play with these things and the nets are arranged so that they can remember things.[49]

Playing with these nets, Crick and Mitchison noted that as they taught the net more and more tasks, "it got overloaded and . . . confused and it didn't do the right thing. . . . Wondering how to improve it we invented a process called 'reverse learning.'" This involved isolating the network from external input and cycling or "running" it repeatedly. The result showed an improvement in its behavior. Here, they saw analogies with the dream state known as REM sleep, so named because of the rapid eye movements during dreaming. The other term for it, paradoxical sleep, refers to the remarkable brain wave activity revealed in the encephalogram, yet the person is difficult to arouse, sensory input to the cortex having been blocked. The brain, it appeared, is just as active at sleep as when awake!

In their paper, Mitchison and Crick treated the cerebral cortex

> as a network of interconnected cells which can support a great variety of modes of mutual excitation. Such a system is likely to be subject to unwanted or "parasitic" modes of behaviour, which arise as it is disturbed either by the growth of the brain or by the modifications being produced by experience. We propose that such modes are detected and suppressed by a special mechanism which operates during REM sleep and has the character of an active process which is loosely speaking, the opposite of learning. We call this "reverse learning," or "unlearning."[50]

Crick had been impressed by the fact that so many of the synapses in the neocortex originate and terminate locally. Such a dense network could well be serving to store associations—engrams—in the form of patterns of local excitation as in PDP. The presence of so many excitatory neurons, rather than inhibitory neurons, suggested to him the power of such networks, isolated from external input in REM sleep, to be very active in generating dreams. Disturbed by hallucinatory drugs, epilepsy, or migraine, however, the process could become unstable, generating wild and fearful hallucinations. Maybe dreaming plays a part in controlling such tendencies, removing the overload of multiple patterns of firing produced by the day's events. There was the parallel here with the behavior of computer networks mentioned above that become overloaded. Isolating the network and running it repeatedly achieved clarification of its content.

Of course, the subject matter of dreams is normally forgotten, unless the subject awakens during the dream. Such forgetting, they assumed, is due to the lack of transfer from short-term to long-term memory. Reverse learning, Crick and Mitchison explained, differs from this, being a "positive mechanism" that "changes synaptic strength so that the dream is not just forgotten but actively 'unlearned'. . . . Put more loosely, we suggest that in REM sleep we unlearn our unconscious dreams. We dream in order to forget."[51] This forgetting stabilizes the connections that remain to continue as memories in the associative network.

Supporting the mechanism of reverse learning, although not its association with dreaming, was the recent work of the connectionist John Hopfield's group at Caltech. Sent before publication to Crick by Hopfield, their paper gave Crick and Mitchison the opportunity to repeat and confirm Hopfield's quantitative study. It showed that neural nets do get overloaded and that their performance in learning can be improved by running the nets in the absence of external input.[52] Of course, these were idealized nets in a computer program, not live neurons. Truly, supportive evidence was hard to find.

Four days after their paper was published, Crick gave a radio broadcast on this work, and *The Guardian* included an account under the title "Dreaming is Life's Dustbin." Soon, the editorial office of *Nature* was flooded with criticisms both wild and sober. Rex Chapman wrote:

> Dear Sir Francis, . . .I no longer sleep and neither does anyone else who has passed 4th Dhyāna. Your hypothesis is therefore invalid. <u>Scientia</u>, i.e., knowledge obtained by objective research, is only the 1st of 8 grades of wisdom. A little knowledge is a dangerous thing![53]

Another correspondent gave his opinion that the paper, although suitable for a journal of philosophy or science fiction, yet "as a piece of would-be science it is thoroughly rubbishy."[54] Turning to more informed correspondents, Professor Dick Swaab, Director of the Netherlands Institute for Brain Research, wrote judging the paper "detrimental to serious sleep research. Since 'ideas have legs' as a writer once put it, it seems necessary to block the road as soon as possible." After offering friendly advice he added, "What is there to forget for neonates? The trauma of birth?"[55] He was referring to the fact that newborn babies are in REM sleep about 8 hours per day. To be fair, Crick and Mitchison had attributed this to "development" (i.e., error corrections to the wiring-up process). As Mitchison explained, "the wiring up of the cortex, which is probably going on apace at that [neonate] stage, will doubtless produce lots of misconnections and 'parasitic modes' that dreams can remove (according to our theory)."[56]

Nor were clinicians happy with the paper. Take Robert B. Daroff and Ivan Osorio at the Department of Neurology, University Hospitals of Cleveland. They focused on Crick's and Mitchison's expectation that REM sleep deprivation, because it would prevent reverse learning, should result in an accumulation of parasitic modes. These would find expression in "bizarre associations," obsessions, and hallucinations. But depriving subjects of REM sleep using monoamine oxidase inhibitors did not produce any such "psychological deficits." Daroff and Osorio then turned to clinical evidence including "two patients with spino-cerebellar degenerations who had total absence of both REM sleep and dreaming but displayed no behavioral abnormalities." Their conclusion was that Crick and Mitchison's hypothesis was not supported by the clinical data. Patients with no REM sleep whatever "are not crazy."[57]

Although a copy of their critique found its way into Crick's correspondence, it was actually a letter submitted to the editors of *Nature* for publication. After the editors declined to publish it "without providing a reason [or] a critique," Osorio and Daroff published it in *Annals of Neurology*.[58] To them, it appeared, *Nature*'s editors "wanted to protect Crick's image and could not bear to look at the facts."[59]

From Crick's old hometown, Cambridge, came one letter to the Editors of *Nature*. The author was Alan Fine in the Department of Experimental Psychology. For him, speculation is "scientific to the degree that it allows of falsification," a view with which Crick certainly agreed, and one might add, Karl Popper, philosopher and author of the doctrine of falsification. But the ad hoc dismissals of contrary evidence in the paper and the authors' admission that "a direct test" of their theory "seems extremely difficult" concerned him. Fine continued:

> I might equally, and equally persuasively, argue that dream sleep functions to provide the animal with a risk-free and non-committal mechanism by which to model consequences of alternative behaviours. I should, however, address my speculations to the ordinary forum of cranks and hobby-horse riders, the "Letters to the Editor." I suggest that the scarce and coveted columns of *Nature* be reserved for Science.[60]

Leaving aside this sarcasm from Cambridge, it comes perhaps as no surprise that some eight months passed between Crick's original submission of the paper on 8 October and its publication. Meanwhile, he had written asking advice on the literature to Jim Horne, Director of the Loughborough Sleep Research Centre. Horne replied with a detailed and very helpful response, the result of which was a revised manuscript from Crick sent to *Nature* in March 1983 and published on the 12th of May. Following the paper's publication, letters kept arriving on Editor John Maddox's desk and were sent on in batches to Crick. Freudians accused Crick of not knowing his Freud and of ignoring the *meaning* of dreams. The idea of reverse learning was not new anyway—Hippocrates had it first, and Freud had quoted him!

Unabashed, Crick's interest in the subject did not wane. He invited the veteran of dream research Michel Jouvet to come from France to spend a month at the Salk to interact with him on the subject. Jouvet later described his host to be "a very polite man, sometimes charming, sometimes extravagant," who had "elaborated a hypothesis of the function of dreaming, for no unknown frontiers can resist either molecular biological or British imperialism!"[61]

They had "very lively discussions." Searching for a test of the hypothesis, they decided to check Crick's assertion that there should be a direct relation between the size of the cortex and the amount of REM sleep. What about the dolphin? suggested Jouvet. It has nearly as large a cortex as man but is supposed to have no REM sleep. Clearly this claim should be tested. As it turned out, there were dolphins at the Naval base in nearby San Diego. They telephoned the research director, an Admiral, but "neither Crick's British passport nor Jouvet's French one was good enough: top secret research, no visiting!"[62] Crick was not downcast. "Absence of proof is not proof of absence," he concluded. More than a decade later, we find him writing to Mitchison about a paper supportive of the "unlearning" idea. "Rather encouraging," he remarked, "and makes me hope that unlearning may turn out to have some biological utility."[63]

The association between REM sleep and the consolidation of learning has not gone away. Although reverse learning may not be favored, "structural reorganization" to quote the Salk's computational neurobiologist Terry Sejnowski is in. Today, in dream research, Crick and Mitchison get mentioned here and there. Attention does get drawn to their analogy between dreaming and the behavior of neuronal networks. The philosopher Owen Flanagan, for one, likened reverse learning to "pruning," and "akin to putting things in the trash, or less drastically, of neatening file divisions" on your hard drive.[64]

How original was Crick and Mitchison's hypothesis, bearing in mind that the association between dreaming and memory actually goes back to David Hartley in the 18th century? After Freud's psychoanalytic study of dreaming, it was in the 1950s, following the discovery of REM sleep, that theorizing about the function of dreaming took off. Memory consolidation was the favorite. Yet it was Crick and Mitchison who first attributed such consolidation to a process operating in reverse. Today, the jury is still out on the subject.[65]

This episode is one more example of Crick moving from one topic to another, sampling the various research methods, and getting the feel of the field before settling on a route to his goal—the nature of perception and the neural basis of consciousness. Mindful of the crucial importance in molecular biology of novel experimental methods and of the framework that the structure of DNA provided for the development of that science, he urged the need for such resources in the neurosciences. A broad framework of ideas such as the sequence hypothesis and the Central Dogma of molecular biology, he noted, was "conspicuously lacking"[66] in the neurosciences. As for neuroanatomy, it was much in need of new methods to lift it out of its backward state.[67] Once more, Crick was to play the role of provocateur.

## CHAPTER 20

# From the Searchlight to the Soul

 *We believe that the problem of consciousness can, in the long run, be solved by explanations at the neural level.*[1]

July of 1981 found Crick in the old German university town of Tübingen in Baden-Württemberg. Tübingen, as the saying goes, "has no university—it is a university." The town is remarkably well-preserved and within it sits the five-centuries-old Eberhard Karls University. Yet, it has a modern medical clinic, a science park, and four Max Planck Institutes and can boast nine Tübingen-born Nobel Laureates, five of them from the life sciences. As for history, it was in Tübingen in 1869 that the young Swiss physiologist Johannes Friedrich ("Fritz") Miescher described his isolation of a substance from cell nuclei to which he gave the name "nuclein," now known as DNA. But Crick's destination was neither the University nor the old Castle Laboratory in which Miescher isolated DNA. It was instead the Max Planck Institute for Biological Cybernetics on Spemannstrasse, where the young Italian computational neuroscientist Tomaso Poggio worked as a Scientific Assistant to the founder of the Institute, Werner Reichardt. Crick was anxious to discuss with Poggio an idea he had about short-term memory.

Crick's first visit to the Institute had been in 1977. That was when he first met Poggio and the computational neuroscientist Valentino Braitenberg. Now, in the summer of 1981 and fresh from a meeting in Aspen, Colorado, Crick was excited about the suggestion made at Aspen by the German theoretician Christoph von der Malsburg. There might, explained von der Malsburg, exist a way neurons are able to alter the "weights" of synapses rapidly and transiently, thus forming ever-changing networks of firing that connect rapidly changing sensory experiences.[2] This possibility led Crick to consider reports of changes in the shapes of the numerous spines found on the little

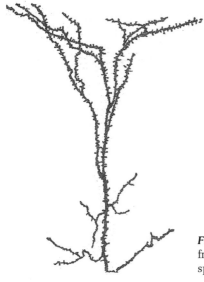

*Figure 20.1* Portion of the dendritic tree from a pyramidal cell showing numerous spines.

branches of dendritic trees that "feed" information to the neuron (Fig. 20.1). Noting their presence on nine-tenths of all cortical neurons, the idea had been advanced that "memory might in part be stored in their shape."[3] Changing the shape could conceivably alter the "weighting" of a synapse on the spine, thereby facilitating, or conversely impeding, transmission across that synapse. There was potential here, for as Crick explained, there are some $10^{14}$ neurons in the cortex, each bearing anything from 300 to 5000 spines and all the spines possessing one, and occasionally more than one, synapse.

Poggio and his research student, Christof Koch, came to Crick's aid. They, too, had been interested in the spines. They proceeded to calculate the effects of these shape changes and confirmed the possibility of such a process, with Koch writing it up in his doctoral dissertation. That winter, Crick sent a paper on the subject to *Trends in Neuroscience* in which he suggested that perhaps contractile proteins such as actin or myosin are present in the spines, there to bring about the suggested alterations of shape. "It is not outrageous," he wrote. "Both the time needed for a muscular gel of this type to contract and the time for ions or small organic molecules to diffuse over the small distances involved are short enough for the proposed mechanism to work, as is the time for the contractile gel to relax when activity ceases."[4]

Debate over the function of the spines had been ongoing since the 1970s, when microfilaments presumably composed of actin were seen in them. But debate over their extent and possible function continued.[5]

Crick was later to refer guardedly to the function of these spines in *The Astonishing Hypothesis*. There we find: "The spine is a fairly elaborate structure and we are a long way from understanding its function completely. I suspect that a spine is a key evolutionary invention that allows for far more sophisticated processing of the incoming signal [to the neuron] than would be possible without it."[6]

This second visit to Tübingen would prove to be important to Crick not only because of his contribution to the current reevaluation of the role of dendritic spines, but because it introduced him to the young man who would become his last collaborator, Christof Koch. That same year (1981), Poggio and Koch moved from Germany to the Massachusetts Institute of Technology (MIT). There as a postdoc, Koch recalled that he and Shimon Ullman "devised ways to explain visual attention on the basis of artificial neural networks. Subsequently, Shimon and I visited Francis for a stimulating and vigorous week-long exchange of ideas. The pace of our interaction intensified when [in 1986] I became a professor at the California Institute of Technology in Pasadena, a two-hour drive from La Jolla."[7]

Koch, a tall athletic figure, has about him that agile and daring spirit of a rock climber combined with the intensity and passion of a research scientist. Born in Kansas City, he was brought up in a succession of European countries, but he graduated from the Lycée Descartes in Morocco. He studied physics and philosophy in Tübingen, followed by research at the Max Planck Institute for Biological Cybernetics. Here was the making of a collaboration between an unlikely pair—Crick, debunker of religion, philosophers, and the immortal soul, and Koch, raised in a devout Catholic family. In 2004, he expressed "much sympathy for the Roman Catholic tradition of the transcendent and immortal soul"[8] and admiration for philosophers like Schopenhauer, Nietzsche, and the young Wittgenstein.[9] Indeed, it was the philosophers Nietzsche and Wittgenstein, not Crick, who first drew Koch to the subject of consciousness. But Koch, like Crick, had his focus on discovering the physical events associated with consciousness. They called them the "neural correlates of consciousness," or the NCC. Koch was soon sharing Crick's passion. An intimate friendship developed between them—Crick aged 70 and Koch less than half his age.

For Crick, collaborating with a much younger scientist worked well. There were 12 years between James Watson and Crick, 11 between Sydney Brenner and Crick, and 40 between Koch and Crick. All three were vital to the expression of Crick's creative energy. For all three, the years of collaboration were memorable. Watson's collaboration over DNA had lasted two years (October 1951–July 1953), followed by a short period in the academic year 1955–1956 working on virus structure. Brenner's extended from 1957 to 1976 and was renewed at the Salk Institute in 2001—23 years in all. Koch's close collaboration with Crick began in 1986 and continued to the end of Crick's life—18 years. Of the three, it was Crick's collaboration with Watson that at times strained their friendship—the hoax letter from Linus Pauling and the refusal to permit a BBC talk in 1953, followed by the revealing personal account of the discovery of DNA in 1968. But Crick forgave and subsequently enjoyed Watson's hospitality at Cold Spring Harbor Laboratory, and Watson, the Crick's hospitality in Cambridge. As for rivals, Crick did not bear grudges, nor did Gerald Edelman, the Nobel Laureate, whose book *Neural Darwinism* Crick had attacked and its claims dissected in a nine-page review in 1989.[10] In the sunset of his life, they became good friends.[11]

### Visual Attention: The Searchlight

Crick's first step toward tackling the subject of consciousness is to be found in a paper he wrote on visual attention published in 1984. Attention, understood as the *act* of attending, could be studied by psychologists with little or no mention of the word "consciousness." Behaviorist psychologists in the first half of the 20th century had outlawed consciousness and left the brain to the physiologists. Consciousness was no longer at the center of their discipline. Nor did the majority of their successors, the cognitive psychologists, want to remove the ban. And most of those who did were looking at consciousness from a functional standpoint and had few words for the brain's machinery. They were concerned with consciousness as a feature of the *mind*, rather than of the brain.

Crick's initial aim was to relate the work of the psychologists on attention to recent work on the anatomy and physiology of the brain. He had been excited by the research of the cognitive psychologist at Princeton, Anne Treisman, on visual attention. He then had "extensive discussions" with his long-time British friends, the mathematicians Hugh Christopher Longuet-Higgins[12] and Graeme Mitchison. Also included in these discussions were Max Cowan's neuroscience group that President Frederic de Hoffman and Crick had brought to the Salk Institute in 1980.

Treisman visualized attention by the well-known analogy of the searchlight or spotlight that searches from one object to another in the visual field to locate and illuminate the subject of concern. She had been a pioneer in studies of a subject's ability to identify one small item in a pattern, for instance, a letter *S* among a dense pattern of *T*s or a

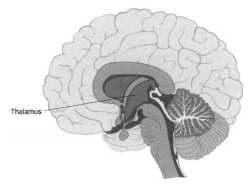

**Figure 20.2** The location of the thalamus and its connections in both directions with the cerebral cortex.

green *T* in a pattern of green *S*s. The latter required much more time than the former and it was, commented Crick, "as if the brain were searching the letters in series, as if the brain had an internal attentional searchlight that moved around from one visual object to the next, . . . In this metaphor the searchlight is not supposed to light up part of a completely dark landscape but, like a searchlight at dusk, it intensifies part of a scene that is already visible to some extent."[13]

Treisman had the idea that this process of "lighting up" might be causing the "binding" of the different sensory data of an object together, so that when we attend to an object *T*, its several features of shape, color, location, etc., although recorded in different parts of the brain, are perceived as belonging to one and the same object. This should account for the unity we experience in visual perception. How to achieve this unity is known as the "binding problem."

Crick, wanting to locate this searchlight in the brain's actual hardware, turned to a part of an organ at the base of the brain, the thalamus (Fig. 20.2). Described as the "gateway to the cortex" because most of the sensory input to the cortex passes through it, Crick's attention was drawn to that part known as the reticular complex. It has numerous connections both to and from the forebrain. He pictured the "searchlight" of attention in the form of neurons in the reticular complex that produce very rapid short bursts of firing directed here and there in the cortex. The state of attention would be associated with these local patterns of firing that last for fractions of a second. The effect of the searchlight would be to establish "*temporary* 'conjunctions' of neurons,"[14] thus "binding" together the several features of the object to which attention was being directed. "What do we require of a searchlight?" he asked.

It should be able to sample the activity in the cortex and/or the thalamus and decide "where the action is." It should then be able to intensify the thalamic input to that region of the cortex, probably by making the active thalamic neurons in that region fire more rapidly than usual. It must then be able to turn off its beam, move to the next place demanding attention, and repeat the process.[15]

In support of this idea, he drew upon von der Malsburg's idea about transient memories. Short-lived networks of firing neurons—cell assemblies—would arise and die away as the searchlight of attention roves over the visual field. Crick admitted the ideas were speculative. Yet he hoped they "might begin to form a useful bridge between certain parts of cognitive psychology, on the one hand, and the world of neuroanatomy and neurophysiology on the other."[16]

### A New Collaboration

Like Crick, Koch had been nurturing a desire to tackle the mystery of consciousness. It had fascinated him since his student days in philosophy. His doctoral thesis had been on nonlinear processing, and he used this approach to explore the computer modeling of neurons, treating them as "little computers" in their own right. Next, he and his colleague Shimon Ullman began a search for the neural circuitry underlying shifts in visual attention. Crick read their paper in 1985.[17] An invitation to the authors to visit Crick at the Salk followed. Such meetings together in La Jolla increased after Koch moved west to become Caltech's Professor of Cognitive and Behavioral Biology in 1986. Crick enjoyed these meetings immensely, for as Odile explained, he appreciated colleagues who would contradict and interrupt him and not be upset when he told them they were talking nonsense. If you can "walk over" the other person, where's the fun and what's the point? Like Crick, Koch would talk with you about any subject—he was just such fun to be with. But there was trouble—Koch is a skilled and daring rock climber. When asked, "Has he had accidents?" Crick's grave reply was "Yes he has." Odile remembers how whenever Koch was off on one of his mountaineering exploits, Crick worried until Koch returned.

Crick and Koch had a special working relationship that focused on the many papers they wrote together and the two books—Crick's *The Astonishing Hypothesis* (1994) and Koch's *Quest for Consciousness* (2004). *The Astonishing Hypothesis* carries the dedication to Koch, "without whose energy and enthusiasm this book would never have been written."[18] It was the regular meetings with Koch, several days at a time, over Koch's book that kept Crick's spirits up while his frame weakened. In the preface, Koch thanked his mentor "without whose guidance, insight, and creativity, this book simply would not have happened." He dedicated it to "Francis and his searing, uncompromising search for the truth, no matter where it takes him, and to his wisdom and his ability to gracefully accept the unavoidable. I do not know anybody else like him."[19]

What made Koch such an attractive collaborator was that although he was a convinced neural net-worker for whom the computer was his laboratory, he has always been concerned to achieve modeling that reflects the properties of real neurons. This was evident in his early papers and especially in his impressive book on the single neuron.[20] Their first collaborative paper appeared in 1990. Its grand title, "A Neurobiological Theory of Consciousness" (1990)—very Crickish—signaled a bold announcement.

From the beginning, they avoided the tiresome business of trying to define consciousness: "We suggest that the time is now ripe for an attack on the neural basis of consciousness. Moreover, we believe that the problem of consciousness can, in the long run, be solved by explanations at the neural level."[21] From his experience in genetics, Crick had come to the conclusion that worrying about arriving at a precise definition of a scientific concept before initiating an experimental investigation was not the way to go. As was evident back in the 1940s, there was no consensus about how to define the gene, and the molecular gene was to make matters worse, not better. Yet progress in the field had been remarkable. Why then follow the example of the philosophers and worry about the definition of consciousness? Clearly, there are different forms of consciousness: awareness, self-consciousness, the sensation of pain, of the locations of parts of our bodies, and those *subjective* and *private* experiences each of us has of a pain, a color, a sound that philosophers call "qualia." The best strategy would be to explore first a form of consciousness that appears the least mysterious and about which we know something—visual awareness.

Crick wrote the first draft of this paper. It begins with a "prolegomenon" that contains two assumptions. The first is that "there is something that requires a scientific explanation," because although we are not aware of the processes underlying consciousness, such processes do exist.[22] The second is that "all the different aspects of consciousness . . . employ a basic common mechanism or perhaps a few such mechanisms."[23] Then follows a list of six topics to be excluded from their agenda. This "framework" identified a focused and achievable research agenda that was intended to act as a corrective and a stimulant to the field.

One issue was how to handle the descriptions and models of consciousness supplied by the cognitive psychologists. Here, Crick and Koch pointed to a "major handicap"—"the pernicious influence of the paradigm of the von Neumann digital computer" that resulted in psychological models containing "various boxes labeled 'files,' 'CPU,' 'character buffer,' and so on." How were these to be translated "into the language of neuronal activity and interaction?" The root problem was the cognitive psychologists' reliance upon "present-day computers," and these, declared Crick and Koch, "make extensive use of precisely-detailed pulse-coded messages. There is," they declared, "no convincing evidence that the brain uses such a system and much to suggest that it does not."[24] In short, the von Neumann computer is digital but the brain is analog. The research program they next described incorporates the searchlight model of attention described above with its suggested transient memories, but this time they enlarged on the concept of synchronized firing and the kind of memory required.

Their aim was to overcome the problem created by the fact that the hardware of the brain is composed of various kinds of neurons, but all of them are either excitatory or inhibitory. When the sciatic nerve fires (excitatory), muscles in the leg flex. When the vagus nerve fires (inhibitory), the heart slows down. All either kind can do is to "fire." They are on/off switches, albeit controlled in a sophisticated, nonlinear manner. True, they can fire at different rates, thus offering a measure of the intensity of the stimulus. Then recall the variety of chemical neurotransmitters that cross the gap in the synapses between neighboring neurons. They are affected in specific and different ways by drugs. But the surest way for a researcher to establish the *meaning* of the message carried by a sensory neuron was to locate the sense organ from whence it comes and find out what kind of stimulus to that organ causes it to fire—the technique of single-cell recording as used by David Hubel and Torsten Wiesel.

So how does this hypothetical searchlight in the brain identify the correct neurons so that attention is directed to the intended object? Von der Malsburg had already suggested the idea of synchronized firing of the relevant neurons in 1986, but he had at that time no direct evidence. When three years later, Charles Gray and Wolf Singer in Frankfurt observed firing patterns among groups of neurons ranging between 35 and 75 cycles per second (Hz) but oscillating around 40 Hz, the idea caused considerable excitement. For in addition to this pattern of firing, these German workers noted that presenting a single and discrete stimulus resulted in a *synchronization* of these oscillations in the relevant neurons. Could it be that when the searchlight of attention

***Figure 20.3*** Figure/ground ambiguity—a vase or two profiles?

"shines" on a particular cortical region, these synchronized oscillations occur? Koch explained the idea by the analogy of Christmas tree lights. "Your task is to make a group of lights at the top of the tree stand out. One way to achieve this is to increase the flickering rate of that group of lights. Another is to trigger each light in the group at the same time. Simultaneous flashing this set of lights increases its saliency considerably. . . . The same logic also applies to the brain, with the 'observer looking at the tree' replaced by some neural network."[25]

Crick illustrated a change in attention from the experience of viewing the famous figure of a vase that one moment appears as a white vase against a black background, yet a moment later appears as the silhouettes of two faces in black against a white background (Fig. 20.3). Consider the brain "as a vast crowd of muttering neurons." Then "the oscillating neurons are like a group of people who suddenly start singing the same song."[26] When you see the white vase, one set of neurons is firing in a synchronized manner, when you see the black faces another set is firing.

Crick and Koch were very excited by this prospect. Writing for the Caltech magazine, *Engineering and Science,* Koch declared,

Crick and I think that there's a special pattern of electrical activity. . . . [We] think this is the crux of it. It's a bit of a leap . . . but we think that this synchronized oscillation could be the neuronal trace of consciousness . . . if you are aware of an event, *all* the nerve cells involved in the perception of that event anywhere in the brain fire at the same time. That is, they fire in a synchronized manner. Other events that you are not aware of—like the sound of traffic outside your window—excite other neurons simultaneously, but these neurons fire randomly. They may even fire at the same rate, but they don't fire in synchronization.[27]

There is no escaping the flavor of excitement here. They felt that they were really on to something big—a neural correlate or *the* neural correlate of conscious awareness! Was this to be another "double helix" moment?[28] This was the state of excitement in which Crick decided to write what was to be his last book.

### The Astonishing Hypothesis

Crick was 78 years of age when *The Astonishing Hypothesis* was published in 1994. In this, his last book, he presents a very accessible account of his approach to "the mystery of consciousness." Here, too, his missionary zeal to slay the immortal soul is much in evidence. He begins Chapter 1 by throwing down the gauntlet:

> The Astonishing Hypothesis is that "You," your joys and your sorrows, your memories and your ambitions, your sense of personal identity and free will, are in fact no more than the behavior of a vast assembly of nerve cells and their associated molecules. As Lewis Carroll's Alice might have phrased it: "You're nothing but a pack of neurons." This hypothesis is so alien to the ideas of most people alive today that it can truly be called astonishing.[29]

A number of reviewers did not find it astonishing at all. Nor did his editor, but she had the wit to add the subtitle: "The Scientific Search for the Soul." Some readers apparently failed to appreciate the irony intended. Crick went one better with the title he gave the last chapter: "Dr. Crick's Sunday Morning Service." Now he could respond to all those sermons he had listened to in Northampton and Mill Hill School as a boy. He urged that if the scientific evidence supports the Astonishing Hypothesis, one will be able to argue that for man "a disembodied soul is as unnecessary as the old idea that there was a Life Force." This, he realized would contradict "the religious beliefs of billions of human beings alive today."[30] Thinking of the millions of Fundamentalists in the United States who accept the literal interpretation of the Bible, he was no longer as confident as he had been in the 1960s that this new knowledge will in the foreseeable future change public opinion. This prevalence of religious beliefs he attributed to the early history of humankind. "It is more than likely," he wrote, "that the need for them was built into our brains by evolution. Our highly developed brains, after all, were not evolved under the pressure of discovering scientific truths but only to enable us to be clever enough to survive and leave descendants."[31]

Evolution also saw to it that our brains have been fashioned to guess the most plausible interpretation of our environment from what in Paleolithic times was very limited evidence. Therefore, we possess an "almost limitless capacity for self-deception." Unless disciplined by the practice of scientific research, "we shall often jump to wrong conclusions, especially about rather abstract matters." How it all began, where we go when our life ends here, and other such questions are not safely answered by common sense alone or by pure imagination and certainly not by appeal to tradition.

Crick argued persuasively on several fronts for his "amazing hypothesis." First, he pointed to the presence of so much unconscious activity in the brain. He cited examples of responses to visual stimuli by those with damaged vision. Such people swear they cannot see the stimulus, yet they succeed in pointing to it. Evidently, some unconscious visual processing (blind sight) is guiding their response.

He was also stimulated by the cognitive psychologist Ray Jackendoff's suggestion that we are only aware of a small portion of the work our brains do in processing sensory data. Surprisingly, Jackendoff claimed it is an intermediate level in the processing of sensory data that constitutes the substance of consciousness, not the highest level. In speech, he suggested, it is neither the level of syntax nor that of conceptual structure but the spoken sounds. In music, it is the succession of notes, rather than their metrical structure or grouping. Turning to vision, he found in David Marr's theory the intermediate level he sought—the 2½-D sketch (see Chapter 19). We can infer the 3-D sketch, he explained, but we are not directly conscious of it. So much computation is going on, in short, without our being aware of it. Our conscious thoughts represent only the tip of the iceberg beneath which lies a world of unconscious processes. And those very reasonings that we attribute to our conscious efforts are the results of unconscious computations.

Crick took this idea a stage further when he wrote a postscript to *The Astonishing Hypothesis.* Following a suggestion from his old Argentinean friend Luis Rinaldini in 1986, he set down his own ideas on the subject of free will and put them into this postscript. He visualized the brain making plans for future actions by a process of computation. In the same way, he suggested, the brain arrives at "decisions" on accepting or rejecting such plans. One is aware of the plans and the decisions but not the computations that led to them. "Then," he concluded, "such a machine . . . will appear to itself to have Free Will, provided it can personify its behavior—that is, it has an image of 'itself.'"[32] True, we think that by introspection or by confabulation we can discover the steps we took to arrive at these plans and make these decisions, thus preserving our sense of free will, but we are mistaken. After writing this out, he came across an account of a patient who seemed to have lost her will; her brain damage was located in a region called the anterior cingulate cortex. This suggested to him that here was the center of the will. Without more ado he marched into Sejnowski's informal tea group at the Salk to announce that the "seat of the Will" had been discovered! Crick was being provocative again.

But so far, all that has been claimed is, first, that consciousness is involved in far less of what we identify as "mental" activity than we customarily assume and, second, that so many of the "thoughts" that underlie the content of consciousness are in fact computations inaccessible to consciousness. But what about the truly conscious residue? Take the sense of awareness. How can we account for its *intentionality*, that is to say, the sense we have of our minds being directed to any object of our attention? Crick did not attempt to offer a material account of this defining feature of mental activity. Intentionality does not appear in the index of *The Astonishing Hypothesis,* or in that of Koch's *Quest for Consciousness.* Would not the 19th century German philosopher Franz Brentano, who coined the term, turn in his grave if he knew this!

Then there is the problem of *qualia,* meaning the blueness of blue, the bitterness of lemon, or the painfulness of pain. How, asked Jackendoff, does what your neurons are doing give rise to "blueness *as you or I experience it?*"[33] In 1990, Crick and Koch recognized that there was a problem here, but what they expected of an *adequate* theory of consciousness is that it "should explain *how* we see color."[34] Many critics would surely not consider that statement to constitute an adequate "explanation" of the mystery of consciousness for it leaves out the qualia. The favorite example used by philosophers to discuss qualia is the sensation of pain. Wittgenstein did so in his *Philosophical Investigations*, exploring the question by appealing to our use of language as follows:

> But isn't it absurd to say that a *body* has pain?—And why does one feel an absurdity in that? In what sense is it true to say that my hand does not feel pain, but I in my hand?

> What sort of issue is: Is it the *body* that feels pain? How is it to be decided? What validates saying that it is not the body?—Well, something like this: if someone has a pain in his hand, then the hand does not say so . . . and one does not comfort the hand but the sufferer: one looks into his face.

How am I filled with pity *for this man*? How does it come out what the object of my pity is? (Pity, one may say, is a form of conviction that someone else is in pain.)[35]

Crick and Koch preferred to concentrate on the neural *correlates* of consciousness—those factors that are invariably present when consciousness occurs, but not present when it is absent. These would be the "sufficient conditions" for consciousness. Any discussion of a *causal* hypothesis was deferred. The structure of Crick's theoretical approach was built on two principles: the localization of function and the interaction of the parts. The former refers to the numerous centers in the brain that process particular features of the sensory data, as seen in studies of the visual centers (V1, V2, V3, MT, etc.). The interaction is made possible through the numerous reciprocal connections between these centers. Any idea of the progressive analysis and construction of the image by a *unidirectional* flow of information up through the hierarchy was ruled out by the evidence of extensive feedback down from "higher" centers to "lower" ones. This explanatory formula rules out any simplistic notions of feed-forward processing leading to a "grandmother cell" at the apex of a pyramid.

Appealing to the principle of localization, Crick turned to the several parts of the thalamus and introduced his *Processing Postulate*. It states that "each level of visual processing is coordinated by a single thalamic region."[36] He suggested that, for instance, the three main regions of the thalamus may coordinate David Marr's three stages of visual modeling: primal sketch, 2½-D, and 3-D (see Chapter 19). In addition, Crick was attracted to the idea of there being among a hierarchy of processing units certain ones that "may exert some sort of global control over the others. Sets of neurons that project widely over the cortex, such as those from the claustrum and the intralaminar nuclei of the thalamus, could well play such a role."[37] Thus, the thalamus, which the eminent neurologist Wilder Penfield had associated with consciousness in 1975, was being singled out again. The very word "thalamus," from the Greek for "inner chamber," signifies its location wrapped around by the cortex, thus admirably placed to interact with any part of it.

At this point, Crick's confidence gave way to pessimism. There just was for him no

one set of ideas that clicked together in a convincing way to make a detailed neural hypothesis that has the smell of being correct. If you think I appear to be groping my way through the jungle you are quite right. Work at the research front is often like that. I do feel, however, that I now have a better understanding of what the key problems are than I did ten years ago. At times I even persuade myself that I can glimpse some of the answers, but this is a common delusion experienced by anyone who dwells too long on a single problem.[38]

Having expressed his disappointment at this uncertain state of the subject, he threw caution to the winds and sketched the outlines of a possible tentative scheme. It involved his processing postulate for the thalamus, two-way connections between thalamus and cortex sufficient to permit them to act as reverberatory circuits for short-term memory, the "searchlight" attentional mechanism of the thalamus, and the 40-Hz frequency of firing to achieve "binding." He hoped nobody would call it the Crick (or the Crick–Koch) theory of consciousness:

> While writing it down my mind was constantly assailed by reservations and qualifications. If anyone else produced it I would unhesitatingly condemn it as a house of cards. Touch it and it collapses. This is because it has been carpentered together, with not enough crucial experimental evidence to support its various parts. Its only virtue is that it may prod scientists and philosophers to think about these problems in neural terms, and so accelerate the experimental attack on consciousness.[39]

### The Response

The response to *The Astonishing Hypothesis* was very diverse, as one might expect on such a subject. Take the correspondents who wrote to the Yahoo website. Several took exception to the publisher's subtitle, "The Scientific Search for the Soul," and expressed surprise and sympathy for Crick. A New York physician exclaimed: "My God! (if you'll pardon the expression). Does Crick need the money that bad?" And how come there is nothing about the soul in the book? Any scientist or theologian, "knows that a 'Scientific Search' for the soul is as meaningless an effort as describing the sound blue makes."[40] Another wished he had got a nickel every time Crick wrote about the need for further experiments, or judged the facts on the subject were unclear. "Unfortunately academic humility doesn't make a good book."[41] The most interesting aspect of the book to a Florida correspondent was the "unconscious account of a Nobel Prize winner's frustration in trying to untie the Gordian knot of consciousness." The result was an "unarticulated pessimism" about science's ability to succeed. "It happens in the best of families," he explained, "Nobel Laureates Nirenberg, Crick and Edelman being just the most recent examples. We don't expect physicists turning to biology and philosophy to be successful any more than we expect retiring baseball star Cal Ripken to be as good playing chess."[42]

When we turn to the professional reviewers, there is more engagement over fundamental issues. Crick's friend, the philosopher John Searle, writing in the *New York Review of Books*, discussed six books on consciousness in 1994. For a simple and direct account of the workings of the brain he rated *The Astonishing Hypothesis* the best. He also admired Crick's ability to summarize the neuroscience and to integrate it with work in other fields. His eagerness to admit the poor state of knowledge and his enthusiasm for speculation were both applauded. On philosophical issues, however, Searle found Crick wanting. The price for his "contempt for philosophy" was that he made philosophical mistakes. Thus, his account is strangely out of tune with the reductionist claims he made. "Nothing-butery" talk (like 'You're nothing but a pack of neurons') should be classed as *eliminative* reduction, because "we get rid of the reduced phenomenon by showing it is really something else." Consciousness, the sense of self, qualia, are just neurons firing, just as lightning is a mighty spark, and that is just an electrical discharge. Yet Crick clearly admitted consciousness exists. When neurons for pain fire, they cause the feeling of pain, but to Searle, "they are not the same thing as the feeling." Crick "gives away the game," wrote Searle, when he judged consciousness to be "an emergent property of the brain,"[43] that is to say it *emerges* from the complex *organization* of neurons in the brain.

The result is that we find claims for reduction on one page and for emergentism four pages later. But these are traditionally regarded as alternatives! Crick defined the term "emergent" applied to the brain as behavior that "does not exist in its separate parts, such as the individual neurons. An individual neuron is in fact rather dumb. It is the intricate interaction of many of them together that can do such marvelous things."[44] Distancing himself from any mystical overtones to the term "emergent," Crick explained what he understood by it. "While the whole may not be the simple sum of the separate parts, its behavior can, at least in principle, be understood from the nature and behavior of its parts *plus* the knowledge of how all these parts interact."[45]

To Searle, the most interesting feature of consciousness was the private and subjective experience it involves—the qualia. Assuming that the physiological events leading to these subjective feelings will be discovered, this still leaves the problem of getting "over the hump from electrochemistry to feeling." This is the "hard part of the mind–body problem that is left over after we see consciousness must be caused by brain processes and is itself a feature of the brain."[46] Setting the problem of qualia on one side, as Crick did in *The Astonishing Hypothesis*, was unacceptable to Searle, for he judged it is not just one aspect of consciousness, "it *is* the problem of consciousness." Strangely, Crick seemed to feel that qualia are a problem only because we cannot communicate to another person precisely our sensations of redness, pain, etc. If we could wire ourselves up with another person, and both our past experiences were on record, we could overcome the problem. Our qualia would then be public. But for Searle, communicating our qualia is only part of the larger problem, which is that "you can give a complete causal account of why we feel pain, but that does not show that pains do not really exist."[47] Jackendoff expressed the point very clearly when he wrote: "It is one thing to provide neurological *distinctions* among qualia—to say that one bunch of neurons is activated for blue, another for red, another for saltiness—but quite another to explain how blueness *as you or I experience it* arises from what our brains are doing."[48]

### The Sequel: A Framework for Consciousness

Crick expected some readers to be disappointed with *The Astonishing Hypothesis*, for as he explained, "It deliberately leaves out many aspects of consciousness they would dearly love to hear discussed." But he was dealing with a subject "about which there is little consensus, even as to what the problem is. Without a few initial prejudices," he explained, "one cannot get anywhere."[49] In 1994 he had reported that the significance of oscillation and synchronized firing of neurons (40 Hz) was still a subject of debate. "On balance," he confessed, "it is hard to believe that our vivid picture of the world really depends entirely on the activities of neurons that are so 'noisy' and so difficult to observe."[50] Six years later, Crick and Koch reported their position as "agnostic," and in 2003 they were definitely negative: "We no longer think that synchronized firing, the so-called 40 *Hz* oscillations, is a sufficient condition for NCC."[51] This position is repeated in Koch's book, *The Quest for Consciousness.*[52] Nonetheless, they left open the possibility that it might still be *one* of the *necessary* conditions.

In 2001, they had been bold enough to confront the "hard problem" (i.e., the neural source of the subjective features of consciousness), and they echoed their philosopher-critics, admitting

> It is not clear how any physical process, such as neural activity, can give rise to a subjective phenomenon such as awareness. The search for the neural correlates of consciousness is therefore an empirical investigation that remains initially neutral on issues of causality, seeking instead to identify and characterize patterns of neural activity that specifically correlate with conscious experience, rather than with unconscious perception or action.[53]

Their paper, "A Framework for Consciousness," of 2003 looks more promising. It marks their retreat from "theory" and "hypothesis" of consciousness in favor of "framework" defined as "a suggested point of view for an attack on a scientific problem," one that is often suggestive of testable hypotheses. They explained

> Biological frameworks differ from frameworks in physics and chemistry because of the nature of evolution. Biological systems do not have rigid laws, as physics has. Evolution produces mechanisms and often sub-mechanisms, so that there are few rules in biology that do not have occasional exceptions.

> An example from molecular biology might be helpful. The double helical structure immediately suggested, in a novel way, the general nature of gene composition, gene replication and gene action. This framework turned out to be broadly correct, but it did not foresee, for example, either introns or RNA editing. And who would have guessed that DNA usually starts with the synthesis of a short stretch of RNA, which is then removed and replaced by DNA? The broad framework acted as a guide, but careful experimentation was needed for the true details to be discovered. This lesson is broadly applicable throughout biology.[54]

Unfortunately, their framework for consciousness introduces a veritable cornucopia of ideas that defy summary. The framework itself, they claimed, "knits all these ideas together, so that for the first time we have a coherent scheme for the NCC in philosophical, psychological and neural terms."[55]

The central idea is that of coalitions of neurons that are in competition. These could be called cell assemblies, but they are transient and in competition with one another. Selective attention can bias their competition. There are coalitions at the back of the brain and at the front. They interact "extensively, but not exactly reciprocally." "Consciousness," they wrote, "depends on certain coalitions that rest on the properties of very elaborate neural networks." The actual NCC, they speculated "may be expressed by only a small set of neurons, in particular those that project from the back of cortex to those parts at the front. . . ."[56] That meant that *while numerous areas of the cortex would at one time or another be involved, only a limited set of neurons make all the difference between conscious and unconscious mental activity.* In truth, the framework was primarily a focused empirical agenda. But Crick was not downcast because, as he recalled, "exploratory research is really like working in a fog. You don't know where you're going. You're just groping. Then people learn about it afterwards and think how straightforward it was."[57]

In summarizing the contributions of others to the problem of consciousness, Crick and Koch could not resist having a swipe at physicists who try "to apply exotic physics to the brain, about which they seem to know very little, and even less about consciousness." Roger Penrose was their target. Author of *The Emperor's New Mind* (1989) and *Shadows of the Mind* (1994), Crick pursued him relentlessly in correspondence. Georg Kreisel tried to convince Crick not to waste time on the matter, but it was characteristic of Crick to feel the need to attack such "nonsense," especially when it came from so revered a physicist. Nor did philosophers escape criticism. "Listen to their questions," he advised, "but don't listen to their answers."[58]

Working in a fog was surely the name of the game. Yet Crick could argue that the brain is deploying mechanisms in the form of computations. Where you assume that you are thinking thoughts, more than 90% is unconscious neural computation. Where you imagine that you are consciously making decisions, rival coalitions of neurons are forming and competing for domination. These processes occupy many different and changing locations in the brain. When you are conscious of the activity, however, a special set of neurons will also be firing, and we need to find where they are and what is special about them.

This approach to the mystery of consciousness was informed by two other guiding ideas. The first was evolution. What is the evolutionary significance of consciousness? As organisms evolved to become more complex and the sensory information they receive greater, a situation of overload would have arisen had not some kind of summarizing of the data been introduced. Crick and Koch likened the process to former President Ronald Reagan's demand of his aides to limit their memoranda to him on any topic to a single page. Consciousness, they suggested in 1995, likewise provides the brain with an "executive summary" of the data. Thus, the function of visual awareness

> is to produce the best current interpretation of the visual scene, in the light of past experience either of ourselves or of our ancestors (embodied in our genes), and to make it available, for a sufficient time, to the parts of the brain that contemplate, plan and execute voluntary motor outputs (of one sort or another).[59]

Thus equipped, an animal can make decisions rapidly, as when faced with a predator or coming upon prey.

Second, Crick stressed on many occasions that our commonsense notion of what goes on in our brains is probably radically mistaken. Time was when it was assumed that living things are such by virtue of a vital force within them that is absent from stones and rocks. We do not believe that anymore because the intricate machinery of living organisms has been exposed to view. Now we should be prepared to question our assumptions about consciousness and the soul—that which thinks, desires, fears, loves, and wishes—that so many believe resides in them.

Maybe many educated people do accept the idea that consciousness is at bottom just neurons at work. But, objected Crick, "such people have often not seen the full implications of the hypothesis." He acknowledged that he found it "at times difficult to avoid the idea of a homunculus. One slips into it so easily. It will not do to explain all the various complex stages of visual processing in terms of neurons and then carelessly assume that some aspect of the act of seeing does not need an explanation because it is what 'I' do naturally."[60]

Returning at the end of *The Astonishing Hypothesis* to the subject of the soul in his "Sunday Morning Sermon," Crick had argued, "If the scientific facts are sufficiently striking, and if they support the Astonishing Hypothesis, then it will be possible to argue that the idea that man has a disembodied soul is as unnecessary as the old idea that there was a Life Force."[61]

Would the religious accept such evidence? He doubted it. Millions of American Fundamentalists in America still do not accept the scientific evidence on the age of the Earth or the evolution of species. Why should we expect them to accept the evidence opposed to the very existence of an immortal soul?

He mused on this "obstinate clinging to outmoded ideas." Moral ideas, he suggested, "impressed on us at an early age often become deeply embedded in our brains" so that altering them can be difficult. "But how," he asked, "did such ideas originate in the first place?" To answer this question, he turned to the early history of humankind when they were hunter–gatherers and tribe fought neighboring tribe for territory. The need for a strong bond between members of a tribe was thus at a premium, and "a shared set of overall beliefs" would strengthen it. Crick judged it "more than likely" that the need for them was built into our brains by evolution.[62]

Let us look back, now, on this last phase of Crick's scientific career. It lasted for a quarter of a century. It offered unrivaled opportunities that Crick, despite growing infirmity and terminal illness, exploited to the full. He had no lab to run and no administrative responsibilities or teaching duties. His Professorship was endowed by the J.W. Kieckhefer Foundation. Applying to the Systems Development Foundation, he obtained funds that enabled him to invite experts to visit the Salk and interact with him. Hubel, Benzer, Mitchison, Jouvet, Poggio, and Koch were among those who spent time at the Salk in this way.

As the record shows, Crick's publication rate did not decline as the years went by. In the quarter century that he devoted to the brain and consciousness, he as sole author and together with Koch published 23 papers and short articles on consciousness, a number of which have been reprinted. *Scientific American* published his first paper in the field, "Thinking about the Brain," in 1979 and "The Problem of Consciousness" with Koch in 1992, the latter reprinted in a special issue of *Scientific American* in 1997. The aim of many of them was to attract more researchers to the study of consciousness. He wanted to rescue the subject from the philosophers, many of whom, he found, did not want to know about neurons and were not inclined to do experiments. There is Daniel Dennett who loves thought experiments and considers he has not only explained consciousness, but explained it away. He seems, judged Crick and Koch, "not to believe in the existence of consciousness in the same way as we do," and worse, he told them, neurons "are not my department."[63] The Australian philosopher David Chalmers considers the mind a separate realm with its own laws. We may discover all the details of the mechanism of attention that enable one to be aware. This Chalmers called the *easy problem*. The *hard problem* is to explain scientifically the experience when you are aware. How is the leap made from neural mechanisms, however intricate, to the *experience* of being aware? Crick applauded Chalmers for coining the hard/soft problem distinction,[64] but he did not like the kind of solution Chalmers sought. Then there were the cognitive psychologists, most of whom did not want to be troubled any more than did the philosophers with the hardware of the brain. Yet he was convinced that we will never be able to explain away the mysteries of consciousness until we learn more about the neurophysiology involved. In 2001, Crick and Koch asked:

> Why, then, is consciousness so mysterious? A striking feature of our visual awareness (and of consciousness in general) is that it is very rich in information, even if much of it is retained for only a rather brief time. Not only can the system switch rapidly from one object to another, but in addition it can handle a very large amount of information in a coherent way at *a single moment*. We believe it is mainly these two abilities, combined with the very transient memory systems involved, that has made it appear so strange. We have no experience (apart from the very limited view provided by our introspection) of machines having complex, rapidly changing and highly parallel activity of this type. When we can both construct such machines and understand their detailed behavior, much of the mystery of consciousness may disappear.[65]

From this, it should follow that in the future, given sufficient understanding of the brain, a conscious machine could be built. Our brains are mechanisms, our thoughts are computations. But what about our feelings?, asked the interviewer in the imagined dialog with Koch in his book, *The Quest for Consciousness.* Koch answers: "Aye, there is the rub. Right now, there are no set answers. . . ." Roundworms and snails may be just "bundles of zombie agents" and might not have any feelings whatever, but for higher forms of life, evolution took a different course. It evolved "a powerful and flexible system whose primary responsibility is to deal with the unexpected and to plan for the future."[66] These responses could equally have come from Crick, but not the following response on the last page of *The Quest for Consciousness*:

*Question:* What about religion? . . .

*Response:* . . .every conscious act or intention has some physical correlate. With the end of life, consciousness ceases, for without brain there is no mind. Still, these irrevocable facts do not exclude some beliefs about the soul, resurrection, and God.[67]

# Eighty-Eight Years

*. . .all our lives in England start with Francis, "larger
than life," a titan on a street corner, a shining flame—
even to those of us who'd never heard of DNA.[1]*

*A few months ago I drove Francis to the pharmacy at
Thornton Hospital. He pointed to a spot and said, "You
park here; I'll only be a minute." He was frail and I
wanted to offer a hand. But I did as I was told and
watched this slow, halting figure walk about 20 yards to
the entrance, through the door, and into the building. He
paused twice to rest before going on. It was painful to
watch. Then, a little later, the same thing as he returned
to the car. He got in, slowly adjusted his long legs and
his cane, closed the door, then turned to me, and with a
wee smile, pointed straight ahead and said, "Home
James, and don't spare the horses!"[2]*

Many a sufferer from the terminal stages of colon cancer would
find it difficult to show such good humor and fortitude. When
Crick received the news of his diagnosis in 2001, Christof Koch was
present in the room. Crick put the phone down, reported the news, and
for a few moments stared into space. Then, with no further comment
on the matter, he went back to his reading. Koch was amazed at Crick's
composure. "How would it be," he mused, "to have as a father someone with such detachment?"[3]

As Crick's administrative assistant for the last eight years of his life,
Kathleen Murray watched anxiously while he weakened. Too frail to
travel to the celebratory events that marked the jubilee of the DNA discovery, he sent videos to welcome the participants. In these videos, he
seemed very much his old self. The legendary enthusiasm and intensity of expression were still there, along with his gracious appreciation

of what others had done. The video shown in Bilbao, Spain, in which he paid so warm a tribute to the principal speaker his long-time friend, Arthur Kornberg, was particularly moving.[4]

Murray, however, knew what an ordeal preparing these videos had been, and how drained he was following their completion. Seeing his life slipping away, the indomitable spirit within Crick pressed on with determination. Murray wanted to say to him, "Just let go!" But his pursuit of the neural correlates of consciousness had taken possession of him. In this obsession, he became more and more withdrawn from the life outside his office door. Gone were his regular 4 o'clock tea breaks in Terry Sejnowski's laboratory. Gone were the lunches in the sunshine on the patio with colleagues. Now, Murray would order his lunch to be brought to his office. And while Crick read through paper after paper, the search for the numerous items on his extensive reading list went on. That remarkable stamina for concentration had not deserted him even in his last year, although he did complain that he used to be able to concentrate for eight hours, now it was only six! Kathleen was at her desk in the foyer outside, Bernice Walker at the far end at the foyer entrance, and, in the library below, the Librarian, Carol Bodas. Between the three of them the hunt went on, searching the Web, the Institute Library, the UCSD (University of California, San Diego) Library, and further away for the sources Crick wanted to see. Walker and Bodas "spent hour after hour, researching, printing, photocopying, and logging articles that enabled Francis to keep on going."[5]

Aaron Klug visited Crick in April 2004 when he was in hospital and was amazed at the way in which he

> seemed to take charge of his own case from his doctors, it was hard to believe that that powerful personality, that I had witnessed in so many other happier situations, could ever be extinguished.

> I could feel even then his impatience that he had little time left and [I could] sense the underlying "rage against the dying light." I hear from Kathleen Murray how this emerged so forcefully in the last few hours, when he managed to finish writing his paper with Christof.[6]

The source of Crick's obsession and the subject of the paper was the possible role in consciousness of a mysterious structure in the brain known as the claustrum. Twenty years before, he had pondered the role of the reticular complex of the thalamus. Now in 2004, it was the turn of the structure lying on top of that complex. Crick and Koch explained, "The word *claustrum* means 'hidden away.' It is a thin, irregular sheet of grey matter, one sheet on each side of the head. . . .

Viewed face on, it has an irregular outline, for primates not unlike that of the contiguous United States."[7]

This was not the first time that Crick had wondered about the involvement of the claustrum in consciousness. Before him the eminent neurosurgeon Wilder Penfield had had such thoughts. Now, however, with new techniques available to explore its structure, what could be learned? Its location for a start was suggestive. Surrounded on its upper side by the cerebral cortex, and lying over the thalamus, the claustrum makes numerous contacts back and forth with the frontal region of the cortex. The dendritic trees of its neurons have numerous spines, and those that lack spines are connected by what are known as gap junctions, permitting electrical transmission in both directions across the synapses between neurons. Such junctions, he reasoned, might be aiding the synchronized firing of "far flung populations of neurons." That suggested to Crick and Koch that an appropriate analogy for the claustrum is that of "a conductor coordinating a group of players in the orchestra," they representing "the various cortical regions," some of the players at the back of the brain, others at the front. "Without the conductor, the players can still play but they fall increasingly out of synchrony with each other. The result is a cacophony of sounds." Accepting that the neuroanatomy of the claustrum "is compatible with a global role in integrating information," they urged further investigation, especially if it "plays a key role in consciousness. What could be more important?" they asked. "So why wait?"[8]

This last sentence is vintage Crick, looking forward to the future and doing so when others in his situation might well have been looking back. He had every reason to be confident about the future. Consider what had been achieved in his latter years. Consciousness was no longer a shibboleth for murky science either among the 35,000 members of the Society for Neuroscience or for the research-grant-awarding National Institutes of Health and National Science Foundation.

Fourteen years ago, his first paper with Christof Koch, "Towards a Theory of Consciousness," ushered in 23 further co-authored papers and two books. Ten years had passed since the publication of *The Astonishing Hypothesis* and since the establishment of the Center for Consciousness Studies at the University of Arizona in Tucson. The first of its biennial conferences: "Toward a Science of Consciousness," was held in 1994, and the same year saw the first issue of the *Journal of Consciousness Studies*. In truth, the quarter of a century that had passed since Crick's first publication on the brain had witnessed a transformation in which not only was consciousness put on the map, but the black box approach of traditional cognitive science and computational modeling had begun to feel the *need for relevance to the results of invasive neurophysiological approaches.* It was surely Crick's influence that played a major role in bringing about this change of attitude.

Heartening though these advances were, what practical outcomes were there? If it was early days for neuroscience, what of molecular biology? What hope that all the resources put into the new science by the MRC had been justified by successful developments in therapeutics? In the 1960s, Crick had admitted the absence of any dramatic cases where "the new knowledge has led to a revolution in Medicine."[9] Indeed, so eminent a figure as Sir Ernst Chain of penicillin fame had been deeply skeptical as late as 1973. Molecular biology, he predicted, will not enable us to produce "tailor-made drugs," or cure hereditary diseases "by genetic engineering. . . . This," he declared, "is science fiction which does harm rather than good to the image of science."[10] Three years later, it was the microbiologist and Nobel Laureate Frank Macfarlane Burnet who judged molecular biology to be "an evil thing" for the advance of medicine. Its "practical applications," he opined, might even prove "sinister."[11] During the decade of the 1970s, however, the foundations had been laid for a recombinant DNA technology that was to transform the situation. Crick's long-held view "pursue the fundamental science, and the medical benefits will follow" was evidently being justified. One could add that the social/legal benefits were also, for consider how effective DNA fingerprinting has proved to be in freeing innocent prisoners who had been languishing in jail for years following wrongful conviction or in apprehending suspects based on DNA evidence left at the crime scene.

Then, too, understanding the mechanisms of heredity and variation at the molecular level not only removed the mystique associated with heredity, but was also having repercussions in evolutionary theory, especially phylogeny and classification. The study of DNA sequences offered molecular data for reconstructing the evolutionary tree of life that have upset traditional schemes of classification and confronted phylogenists with evidence of extensive horizontal (lateral) gene transfer. As astronomers today view events in the past thanks to the vast distances light must travel to reach us, biologists can now trace lineages in the DNA sequences of species that go back through the history of life.

Such devotion to science, such single-minded pursuit of Nature's mysteries—surely Crick must have been guarding every precious moment of his time for science? True, there was his card listing all the requests he would not undertake, from accepting honorary degrees, to appearing on television, and more. But was he, as the introducer of the American version of the DNA film "Life's Secret" enquired, living the life of a hermit in the Californian desert?[12] Far from it. At that time, he was courageously fighting cancer, but in earlier years, as mentioned in Chapter 14, Seymour Benzer observed that Crick "was into all kinds of adventures on the personal level. . . . No shrinking violet, that's for sure." What were these "adventures"? Benzer, working at the Unit in 1957–1958 had seen not only the quantity, but also the variety of Crick's mail as it arrived at the MRC Unit's Hut, some of it marked PERSONAL. He realized that the enthusiasm with which Crick immersed himself in scientific adventures was being directed elsewhere as well. These adventures included liaisons with members of the opposite sex. For there are times, he explained, in a long marriage when there can come the want of novelty and excitement and the challenge that seduction offers. But Crick set his ground rules—such relationships were always to be private and temporary in nature, for his marriage to Odile was not to be undermined. He treated sexual adventures as he did intellectual ones. Both were fired by his thirst for life in all its fullness. And when he met a pretty young woman who was also feisty and sharp, he could be very attentive, expressive of his feelings, and receptive to hers. It was not just in science that he was the world's best listener. Indeed, Georg Kreisel once complained to Odile about the times when Crick's attention would be totally devoted to the young woman in their company. Odile responded with characteristic insight. "Her very presence," she explained, "brings out in Francis a show of verbal fireworks and the compulsion to make an impression takes over completely. Anything which gets in the way is either brushed aside or totally ignored."[13]

The passion that Crick felt for science he could also feel for pursuits that seemed out of character. Thus, in 1963, his friends were greatly surprised when he took up gardening, the hobby so beloved by many a Brit. The opportunity arose when he acquired the 18th century thatched "Well Cottage" located midway between the villages of Kedington and Sturmer in Suffolk. The Cricks had chanced upon it while Odile was teaching Francis to drive in her Austin Mini. Spying two semidetached cottages next to a beautiful old farm house, they approached the owner about buying them. He, it turned out, was in need of cash, and happy to sell. Although declared "unfit for human habitation," Crick bought the two cottages in 1963. He restored them, converting them into a single dwelling.

Next, he bought more land and with customary thoroughness embraced the art of gardening. He pored over catalogs and ordered vast quantities of daffodils and narcissi—earlys, mid-season, and lates, trumpets, small- and large-cupped, red- and yellow-cupped. Odile did most of the planting under the venerable apple trees. It was then the rose's turn. The subject was researched, the catalogs were explored for prickly old-style "hybrid-teas" to modern "floribundas," and the new but old-style fragrant varieties created by David Austin. He then found the red flowered dwarf form of *Rosa moyesii* v. "Geranium" that Peter Lawrence so admired.

Digging the holes and preparing the dirt was undertaken by wiry old Bill Sephton, a neighbor. Francis and Odile then performed the final act of planting. Crick's vision did not stop here. He liked the idea of creating a grand estate, so he added rhododendrons and unusual trees. He was the visionary gardener; Odile and Bill Sephton were the working gardeners. The result was of course magnificent, for gardening had become his hobby, and Odile and Mr. Sephton were there to execute his wishes. Friends and family visited him there, secret liaisons could hide there, pot could be smoked, and, very occasionally, LSD experienced. All this happening not far from the village church of St. Peter and St. Paul, its interior considered the "Westminister Abbey" of Suffolk! American tourists were often seen making brass rubbings there, and Odile and Francis, like them, admired the fine sculptures on the tombs. Although there are two pubs in the village, the White Horse and the Barnardiston Arms,[14] neither was frequented by Francis and Odile, for as Odile explained, Well Cottage was for them an escape from the busy social life in Cambridge, a place where quiet and anonymity were preserved. They did not sell this treasure in the country, since 1974 a Grade-2 listed building, until they ceased their regular summer visits to England in 1995.

### *Friends*

Over the years, many of Crick's friends from earlier days lost touch with him, but when they heard him broadcasting on the radio, several of them wrote to remind him of old times and congratulate him on his success, telling him that he sounded just the same as he had some 20 or 30 years before. There were those hints of a laugh or a chuckle when he made an amusing aside, which those who knew him well did not miss. When the Queen awarded him the Order of Merit in 1991, he saw his school friend Harold Fost in London and received a letter from Harold's sister Pamela. Such good friends had young Tony and Francis been as schoolboys with Pam that the Fosts quite expected one of them to marry her some day. In 1991, Pam reminded Francis how he had taught her to do crossword puzzles while she was in the tub, by calling out the clues (from the other side of the bathroom door). Nor was she the only one to learn the art from Francis. Joan Wolfenden wrote to him after reading the article about him in the *Sunday Times Magazine* in 1994. She reminded him of the magical 2 weeks in Aunt Alice's house in the Buckinghamshire village of Whiteleaf. "Not many boys had your outsize brains and I think of you every day when I have a go at *The Times* crossword. You were the first to show me how! I shall never forget your mother thanking my mother for taking you away for two weeks. She said it had restored her sanity and she gave mother a beautiful set of coffee spoons."[15]

Crick's first collaborator, Jim Watson, came to La Jolla knowing it would be the last time he would see the friend he had first met 53 years before. Watson was then an uncouth postdoc aged 25 and Crick became like an elder brother to him, advising him on etiquette as well as science. Watson did his best to cheer the family when Crick, suffering a bad spell, was unable to enjoy the occasion in the manner he would have liked. Crick's second collaborator, Sydney Brenner, had been keeping in touch from the time he received a part-time appointment at the Scripps Institute close by the Salk Institute in 1995. When not in the Far East or in Ely near Cambridge, he would be in La Jolla, and in 2001 as Distinguished Professor, he once again became Crick's colleague at the Salk.

For his third collaborator, Christof Koch, Crick had become a father figure. Although Koch wrote the entire text of his big book *Quest for Consciousness,* it represented the outcome of a collaboration that had become a daily matter through e-mail, fax, letter, and telephone. Then there were the visits from Koch. He would drive down from Pasadena one day, stay two nights in the Crick's guest room, and drive back the following morning. This collaboration was very precious to Crick. Indeed, they became so familiar with each other that when they were "sitting in comfortable wicker chairs in his study at home," recalled Koch, "we could infer each other's thoughts from a few words or a paper we were looking at."[16]

Among his Cambridge friends who visited him in his last years were two who had become very special friends: his first secretary Alison Auld and his third, Pauline Finbow. Alison had been hired to help with the move of the lab to the Hills Road site in 1962, at a time when the work involved was very varied. One day she would pop off to auction sales, buying items for conversion of a greenhouse into a lab. Another day, she would go to the London docks to collect the baggage of a visiting researcher. Such work provided an income, but it was archaeology that fascinated her. She was into the mood of the swinging 60s, and often stayed up to party all night. Typing out Crick's handwritten letters the next morning could often be difficult. Crick complained: "Alison, you've missed three lines. . . . Why don't you put a ruler on it?" "What a good idea Francis!" she replied. When she wanted time off to go on an archaeological dig, he would reorganize his routine when possible. Had he any enthusiasm for archaeological ruins? Oh no. She recalled a holiday with the Cricks on their yacht *Kiwi II* in 1965. They were sailing down the Italian coast. When Alison spied a historic site coming into view and asked to stop, Crick would refuse to put ashore to view it—"only an old pile of stones," he declared. As each historic site came into view, the order was given: "Sail on!"[17]

In 2003, Alison visited the Cricks in their beautiful house in La Jolla. Notwithstanding his illness, Crick wanted her to see the sights. "No, no," she objected. "I'm definitely happy slopping around the pool." But Crick would say: "I think Alison would like to do this," or "Why don't you show her that?" Then off they went "on a huge adventure," said Alison. Their destination was the Anza-Borrego Desert and the Cricks' house at Borrego Springs. Built in 1995 by a local builder to Francis and Odile's design, it is modest in appearance, blending nicely into the arid and rocky scenery where barrel cactus, creosote bush, and desert agave grow. Entrance is through a handsome hardwood front door designed by Odile. Floor-to-ceiling glass provides a wonderful view of the surrounding desert from within. As in every project he undertook, the details were thoroughly mastered, and alternative plans were researched. Deciding on the floor plans called for model building in plasticine on cardboard bases.[18] Evidently, as Michael Crick pointed out, it was not only in the lab that Crick appreciated the value of model building.

Once again, Crick researched the horticultural literature, this time for plants suitable to grow in the desert, such as rock daisies, Arizona lupins, fishhook cacti, and Bigelow's monkey flowers. To support his acquisitions, he set up a complex system of watering in which the garden was divided into sectors and the very limited water supply was directed to one sector after another in a cycle.

For the trip to Borrego Springs, Odile was up at dawn, getting all of the food ready, keeping a list of all the items needed to take to the desert house. Although 80 years of age, she did all the driving, as well as all the packing up. "Can you believe it?" exclaimed Alison. They went on a safari with other visitors to see desert life by night. Odile remembered the giant desert scorpion and many other nocturnal animals. Crick stayed behind at the desert house for he did not relish the bumpy ride over rough terrain in a jeep-like vehicle.[19] Yet to Crick it was immensely pleasing to be visited by his first personal secretary (shared with Brenner) and one who became a family friend and more, also modeling for Odile.

Pauline Finbow, a secretary for four years (1967–1971), also came out to La Jolla to see the Cricks. Francis, she recalled, was the only person she had worked for "who would sit down with you, discuss what you are doing—'How is your life sorting out, is there a problem, can I help?'" He had attended her wedding in 1967. Mischievous as ever, he gave her a little book entitled, *How to Undress in Front of your Husband*.[20] And when she became pregnant, Odile asked Pauline to model for her. "I was much too embarrassed at the time," she recalled, so she declined. But she did not forget how, after she "broke the news to Francis," that she was pregnant, she received a telephone call from a local Family Planning Clinic "asking me to make an appointment to go and see them?!!!!! I think Francis was quite alarmed at the fact that I would be leaving! I don't know what he thought I'd do but I was quite delighted about being pregnant and I certainly wanted the baby and said as much to the girl who called me from the Clinic. Francis had obviously got in touch with the FPC and asked them to telephone me."[21]

As for working at the LMB, those "were wonderful days." As for Crick's sexual proclivity, "Believe you me," replied Pauline laughing, "I was aware of it. I suppose there were times when he would chase me round the desk, but he never caught me. I was shy and newly married, and had not come across anyone quite like this before."[22] When Crick moved to the Salk Institute, he asked Pauline to look after the letting of *The Golden Helix* and his other rental properties in Cambridge. Number 4 Croft Lodge—the Cambridge house on Barton Road, in which his mother lived in her latter years—he had taken down and built 20 flats in its place. By this time, all but six of the 20 had been sold, and these were rented out. There was also an apartment at 1 Poynton Place and one on Quainton Close off the Newmarket Road.

Crick had surely forgotten the occasion when he had been unwell and Pauline was summoned to the house to take dictation. Directed upstairs to the bathroom by Odile, she found him in the bath, no soap bubbles obscuring his physique, ready to dictate as if they were in the office. Today, the faucet on the side of the bath is in the form of an elegant sculpture of a woman. Turning on the faucet, water issues from between her legs. This erotic piece was typical of the work of their friend, the erudite sculptor John Gayer-Anderson. Along the length of the bath panel is a fire-breathing dragon painted by Odile's friend, the Italian artist, Rodolfo De Sanctis.

Even without these distractions—they were added later—Pauline had to concentrate hard at her work. Crick kept in touch with Pauline right to the end. In 2003, he thanked her for her reports on the Jubilee week in Cambridge and expressed surprise at the number of attendees. She must have mentioned life after death for Crick went on to caution her about mediums.

There is no good evidence that mediums can "connect to the other side." Its all done by well known tricks, which can easily deceive people. When tested carefully, preferably by so-called magicians, mediums have always been shown to be frauds. So the scientific evidence supports the hypothesis that there is no life of any kind after death. If we could understand the neural basis of consciousness we would be in a better position to decide the question. In time this should happen.[23]

One special friend who could not pay such a visit was Jacques Monod, for he had died in 1976. Crick retained a vivid memory of his exploits sailing with Monod in the Mediterranean and the Aegean Seas. Crick's pleasure in sailing went back to summer holidays in the 1940s with friends boating on the Norfolk Broads, but in 1964, with Nobel Prize money, he bought a share in a 47-foot racing cruiser, described by the sales agent as "a very nice ketch" (a two-master).[24] Called *Kiwi II*, it was kept in Naples. Standing on it wearing the cap from the Naval officer's uniform he had worn to go to Russia in 1945, and sporting a British flag, he must have looked the part of a grand old salt. But at the wheel, he admitted, he was "always a rather bumbling amateur." When in the summer of 1969, he accompanied Monod on his 37-foot sloop (single-master) sailing to Corsica, he had an experience he never forgot. It was on the return journey to St. Tropez in the night when a storm hit them. Crick remembered how

> As the waves got higher and the wind blew more strongly I became mildly apprehensive, although trusting in Jacques to get us through, even with the handicap of having me to help him. He was clipping himself on, as he moved about the boat, lit dramatically from above, with darkness all around us. Finally I said to him, "Jacques, exactly what do I do if you fall overboard?"[25]

This was not the only time when rough weather was experienced. Gabrielle Crick recalled the time when they were warned to move *Kiwi II* to a safer mooring, and the boat tipped to such an angle in the process that she was very frightened. On another occasion, John Kendrew was on board when Crick's mariner skills were put to a test in the Bay of Naples. "It is clear," remarked Kendrew, "that when Jim Watson made his famous remark about Francis' modest mood he had never seen him on a boat."[26]

While a Mediterranean storm made him apprehensive, hapless errors by others could make him furious. His secretary Sue Foulkes recalled the one time when he was "really angry." It was in the spring of 1972 involving a flight to Glasgow and on by bus to Aviemore. The travel agent had sent him his airline ticket made out to Dr. Prick! Sue's diary entry reads: "Francis refuses to use it and wants a new one issued. Then a woman in the Medical School damaged his car backing out [of the parking lot]. He nearly shouted when I told him."[27] It was his beautiful white Lotus Elan Sprint sports car—0–60 mph in 6.6 seconds, top speed 121 mph, "superb road holding, and lightning-quick steering response."[28] It survived this minor damage, but subsequently a serious crash while returning from Crick's weekend cottage led to its demise. The Lotus' body was fiberglass.

Although these expensive indulgences only became possible for Crick in the 1960s, it was the social life of the 1950s in Cambridge that old friends remember with particular nostalgia. Freddie Gutfreund admitted that "Portugal Place may have had a reputation for lively parties that almost turned to orgies. However," he added, "it was also and more important, a comfortable meeting place around the kitchen and dining table. This is where friends met and learned from each other, entertained each other, got good scientific and personal advice and found people with complementary interests."[29] They became part of the Crick's extended family. Bachelors among them, like Watson and Gutfreund, were taken shopping by Odile. Members of this extended family all lived within walking distance of Portugal Place. They did not own cars and they did not need them. Some took turns cooking for dinner parties, seeking to rival Odile's exquisite culinary skills.[30] After living in La Jolla for many years, Odile remarked on the absence of "interesting cooks"! And she added, "Life is very different, but as Sydney Brenner said on a recent visit rather nostalgically 'those days are past.'"[31]

In the 1960s, Crick was often away when their party-going friends got together. Odile would then write to him describing the event. In February 1967, for instance, Crick was in La Jolla for the nonresident Fellows' meeting, and Odile reported:

> The last weekend was rather hectic. Rudolpho [De Sanctis] arrived for Jan Ellison's party which turned out to be quite a success. He went as a novice and we arrived as a pair as sacred & profane love. I arrived in my housecoat and shed that to reveal black fancy bra top and black net tights, gold shoes—long black gloves, piled up hair-do, fancy gold necklace—Much admired, especially by young girls (Toi & John's Israeli girl Abigail). We got back about 3 a.m. Jan was splendid, really in his element and bullied his guests into coming in fancy dress. Result was very gratifying—Ricky absolute scream as a dark-haired, buxom dancing girl! Had everyone fooled for quite a while with his undulating movements & long fingernails! Priss very reptilian with glittering green headdress covered in snakes. Theresa [Popham?] gorgeous as

Danté's Beatrice with marvelous blonde wig and demure long dress! Unfortunately no suitable men for her—Tristram came dressed in black with "portcullis" tunic and hatchet, all from Mermaid Theatre.[32]

## Honors, Degrees, and Medals

Although Crick could look back on the many honors he had received, he took pomp and ceremony, medals, and titles with a pinch of salt. If they carried a handsome sum of money, naturally his attitude was different. All his medals, awards, and a statue formed a "shrine" kept first by his mother and after her death by Aunt Ethel. He routinely refused honorary degrees. When offered one he would reply—"I could not now accept your offer without offending those universities I have refused in the past."[33] Why take time to go to a University in which you had never worked and go through the charade of accepting a degree and give a speech? Becoming a Fellow of a prestigious scientific institution, however, was a very different matter. He was delighted to accept the Fellowship of the Royal Society in 1959, the Society's Royal Medal in 1972, and the Copley Medal in 1976. In 1960, his old college, University College London, made him an Honorary Fellow.[34] Other honorary fellowships followed, as described in Chapter 13.

Crick was away when the government offered him a CBE (Commander of the British Empire) in 1963. Kendrew had to telegram Crick to discover his response and pass the information on. Predictably, it was negative. The original invitation had been sent to The Green Door. When a second letter reached him, he replied that he had left the Green Door 14 years ago, but he was "flattered that the Prime Minister (Harold Macmillan) has me in mind, but as I hope my secretary has already told you, I would prefer to decline."[35] In so doing, Crick was in good company. There are 30 others who are known to have refused the offer of a CBE, among them Alan Bennett, Albert Finney, John Cleese, and Evelyn Waugh.[36]

Crick also declined a knighthood. It was enough that he was assumed by some to be Sir Francis—even by the 1971 edition of *Encyclopedia Britannica* no less[37] and adopted frivolously among the scientific community in Tübingen.[38] He claimed that it was the University of Washington that started the error when he lectured there in 1966. It made sense for Kendrew to become Sir John, an acknowledgment of his considerable government work, but not for Crick who had done no such work. Besides, having a "Sir" in front of your name distances you from others by conferring a special status and that he did not want. Nor did he wish to appear supportive of a government that he judged funded scientific research inadequately. When in 1962, the Queen came to open the Laboratory of Molecular Biology on the new site, Crick was among those who did not attend. This, he explained, was not because of antiroyalty sentiments but because Cambridgeshire's High Sheriff had instigated the choice instead of leaving it to the scientists to settle on someone with interest and knowledge of science, such as Prince Philip.[39] As the Queen's representative in the County, the High Sheriff did have the right to act. However, the absence of Crick and Sydney Brenner enabled Max Perutz to invite Jim Watson from America instead. Perutz subsequently remarked pointedly that the Queen took a lively interest in the subject of their work.

The Order of Merit in 1991 was a different matter, because it came from the Queen, not the government. Moreover, there are only 24 British recipients at any one time, most of them coming from the arts, theater, literature, and science. It is awarded for "exceptional service to the Crown or for the advancement of arts, learning, law and literature." When Perutz learnt that Crick was accepting, he faxed him to express his delight; "doubly so, because the secretary asked me several months ago to explain to her what DNA is." Hearing no more on the subject, Perutz had wondered if the Queen "had changed her mind, or worse, whether you had turned it down."[40]

At a lunch for Order of Merit recipients, given by the Queen in Buckingham Palace, Francis and Odile met his former colleagues Fred Sanger and Max Perutz, violinist Yehudi Menuhin, philosopher Isaiah Berlin, opera singer Joan Sutherland, aviation pioneer George Edwards, Australian painter Sidney Nolan, and others. Seated next to Odile at lunch was the art historian Ernst Gombrich, whom she found to be a fascinating conversationalist. But she found tedious Yehudi Menuhin's harangue at the change in the public's taste in music to the detriment of classical music. Little did he know that Odile owned a Hohner piano accordion, Tango II, on which she played Ian Lendler's drinking songs and other hearty Austrian marching songs. She had learned them in Vienna in 1936. Perhaps, too, Menuhin was not aware of Francis's penchant for musicals and plays rather than J.S. Bach and opera.[41]

Crick enjoyed his conversation with the legendary Sutherland. When he had been making waves in science, she had been winning international acclaim in the opera world where her nickname was "La Stupenda." Although not an opera fan himself, Crick loved to learn about the theatre,[42] and Sutherland surely could recall many a dramatic role. As for conversation with the Queen, that was simple, joked Crick. They were instructed not to introduce a topic but wait for the Queen to do so. This rule ensured she would not be embarrassed by a subject unfamiliar to her. So one waited for her to speak. Winston Churchill knew that the safe subject for royal conversation was horse breeding and horse racing, but despite having enjoyed horse riding, Crick had little interest in these topics. Horses, he once remarked "are so stupid. They don't run to be first, they only run not to be last."[43]

The significance of the Order of Merit to Crick was that it acknowledged the importance of the science about which he felt so passionately. It was deeply gratifying that his work as a scientist was honored alongside the achievements of those in the arts who then and in previous years had been so honored—sculptor Henry Moore, painters Graham Sutherland and Ben Nicholson, composers Benjamin Britten and Michael Tippett, novelists E.M. Forster and Graham Greene, and the great Shakespearian actor Laurence Olivier. Just as these recipients of the Order of Merit were key figures in the achievements of the Arts in Britain and the Commonwealth, so Crick had been at the heart of the molecular revolution in the life sciences. Justly, he now joined a line of scientists who were recipients of the Order, famous names in the history of science stretching back to the beginning of the 20th century: Lord Rayleigh (in 1902), J.J. Thomson, Ernest Rutherford, William Bragg, F.G. Hopkins, Arthur Eddington, James Jeans, John Cockcroft, Dorothy Hodgkin, Howard Florey, Paul Dirac, Alexander Todd, Andrew Huxley, Fred Sanger, and Max Perutz.

He could recall the events surrounding the discovery of the structure of DNA with the utmost pleasure. "Am I glad that it happened as it did?" he asked. "I can only answer that I enjoyed every moment of it, the downs as well as the ups."[44] Not only did it provide excitement and drama, it also proved to be the turning point in his career. As he recalled, "the DNA structure made Watson and Crick. After all, I was almost totally unknown at the time, and Watson was regarded, in most circles, as too bright to be really sound."[45] Crick might have added, too, that before the discovery, Bragg had been looking forward to the departure of Mr. Francis Crick from the Cavendish Laboratory.

Yet how free and uncluttered had been their lives in the 1950s. Did Crick worry then about future employment? No, he always felt that something would turn up. Funds were available, and competition for them was minimal. Recall that these years saw the expansion of the U.K. Atomic Energy Authority and an unprecedented demand for physical scientists at Harwell, Windscale, Winfrith, and Dounreay. X-ray crystallography was only just establishing its institutional identity, and biophysics was receiving a new breath of life. Indeed, "It was an event," Crick recalled, "when a letter arrived at the laboratory for me." So paperwork was at a minimum and as for making transatlantic telephone calls, that was out of the question. Should that rare event—good weather in Cambridge—happen, "one took the afternoon off to enjoy it."[46]

Why, then, did Watson's manuscript, *Honest Jim*, upset him so much? First, because he had long guarded his privacy, shied away from cameras, and refused requests to permit the publication of photos of him. When Watson was seeking photos for *The Double Helix*, Crick refused him, and Watson failed to persuade Odile instead to oblige. Even two decades later, when Crick was writing his autobiography, he revealed rather little about his personal life, his theme being the lessons we can learn about how to do science from the mistakes we have made. Unlike François Jacob, in his autobiography *The Statue Within,* Crick did not bare his soul. Second, his passionate commitment to science led him to expect to be insulated from the prying eyes and ears of the journalists and paparazzi who plague politicians, pop singers, and film stars.

By 1967 Crick realized Watson was determined to publish, and he wrote to Kendrew:

> The only thing which might persuade Jim to alter his mind would be Bragg's withdrawal of his Foreword. I must tell you that I take a pretty dim view of Bragg's not having done this already, in view of my sustained objection to this book and I hesitate to tell him directly as we have had quite enough unpleasantness already. If there had been as much about Bragg in the book as there is about me (in Jim's girlish style) he would never have agreed to write the foreword in the first place.[47]

As was mentioned in Chapter 1, Crick did change his mind about the book, so that when the suggestion of making a film came in the 1970s, he entered into the project with enthusiasm, and—no surprise— it was at Portugal Place that a lively party for the film crew was held. So lively, in fact, that the director regretted having left his filming equipment behind. Crick was by this time well aware of the attractive features of the book to a general reader. As he explained in 1979:

> No doubt it is fascinating to read just how a scientific discovery is made; the misleading experimental data, the false starts, the long hours spent chewing the cud, the darkest hour before the dawn, and then the moment of illumination, followed by the final run down the home straight to the winning post.

> And what a cast of characters! The Brash Young Man from the Middle West, the Englishman who talks too much (and therefore must be a genius since geniuses either talk all the time or say nothing at all), the older generation, replete with Nobel Prizes, and best of all, a Liberated

Woman who appears to be unfairly treated. And in addition, what bliss, some of the characters quarrel, in fact almost come to blows. . . . Surely the script must have been written, not in heaven, but in Hollywood.[48]

In the 1980s, Crick cooperated with the BBC when they produced the docudrama *Life Story*. Watson was played by Jeff Goldblum, Crick by Tim Pigott-Smith, Maurice Wilkins by Alan Howard, and Rosalind Franklin by Juliet Stevenson. Crick considered Stevenson's portrayal of Franklin "the really key performance. . . . She is not only the true center of the film," Crick judged, "she is almost the only person who really appears to be *doing* science."[49] But this was the aspect of Watson's book that had particularly concerned him. He was not strolling on the "Backs," partying, or drinking at The Eagle all the time, and he had long since given up playing tennis. His intensive study of protein crystallography, however, had brought him to the forefront of the interpretive side of the subject. Thus equipped, he, if anyone, could read the message in Franklin's diffraction patterns of DNA.

It was not until 2002, however, when Brenda Maddox's *Rosalind Franklin: The dark lady of DNA* was published[50] that widespread public concern was expressed about the manner in which Franklin's data had been used. Consulted by Maddox in 2001, Crick wrote her lengthy letters and persuaded her to receive instruction from the experts on X-ray crystallography. But he was not happy with the end result. He was adamant that Patterson analysis had not been the way to go with DNA, and those who had condoned or encouraged Franklin in that choice had been wrong. This included the approval of Franklin's strategy from Dorothy Hodgkin. But she, noted Crick, had also advised Franklin that "there were enough pieces of general information available . . . to make model building a reasonable course to pursue separately."[51] Earlier, the arrival of Pauline Cowan (see Chapter 11) at King's College London had brought the suggestion to Franklin that Cowan, an outstanding young X-ray crystallographer, should help with the DNA work by model building. Franklin refused.[52]

After all the warm and heartfelt appreciation of Crick's achievements, did he have regrets as resentment grew not so much over the use of her data as over the manner in which he, along with Watson and Wilkins, had obscured Franklin's contribution? First one needs to ask: "Was there an expectation in X-ray crystallography that those who obtain the diffraction patterns should be left in peace to solve them? Or was it general practice to publish the data so that others might attempt solutions?" When asked this question, Crick replied that they should be given reasonable time, but if the data proved unyielding one would expect their publication.[53] Clearly, by the end of 1952, he felt that Franklin had had more than enough time and should by then have published her data. And what of the accusation that they had stolen Franklin's data? To her biographer, Brenda Maddox, he responded, "As to the 'theft' issue, I will make no comments, or you will accuse me of feeling guilty about it—which I don't."[54]

Franklin's MRC report of 1952 was not a publication, but an internally circulated report. Working in another MRC Unit, Watson and Crick had every right to see it. Was there really a problem about citing it? Well, the gentlemanly approach would have been at least to advise her of their using it. But behind their reluctance to specify the report in their acknowledgments was the cold war between Wilkins and Franklin, and to cite her report would have placed her contribution above Wilkins. The Crick–Wilkins friendship thus played a decisive role here (see Chapter 10) as it did over Crick's subsequent opposition to Watson's book *The Double Helix* (see Chapter 1).

Vittorio Luzzati, who knew Crick very well, described the obscuring of Franklin's contribution "unfortunate," for he insisted, Crick was a very *decent* person, a word, he remarked, that does not have an equivalent in French, but is defined in English as "considerate of others, having a sense of what may be fitly expected of one."[55] It was certainly not out of character for Crick to move in on the work of others, but omitting citation of the specific source of crucial data was uncharacteristic. Crick did have regrets on this matter, and he acknowledged that specific mention of the report should have been made in the full paper in 1954 describing how they arrived at the structure.[56] (This was in the *Proceedings of the Royal Society*, a scientific periodical that in 1988 he half jokingly had called "an obscure journal.") As has already been mentioned, Watson wrote this paper while Crick was busy writing up his doctoral dissertation.[57]

Did Crick have other regrets? In his latter years, he reflected on the manner in which on some occasions his enthusiasm or impatience would break through and he would be hurtful. It was more often the established and the pompous speakers who became the targets of his sharp words than the young and inexperienced. But as we have seen (Chapter 17), Henry Harris, although in 1962 the head of a research institute, found defending his critique of messenger RNA from Crick's interrogation "a matter of life and death." Nor had Bernard Davis, Harvard's professor of microbiology, forgotten some two decades after the event how Crick had stopped him as he explained a possible solution to a problem concerning the genetic code. Crick had suddenly stood up and requested that "all those who have not thought of this solution before, please put up your hands." No hand went up, and Davis had to sit down. Years later, reviewing Crick's book, *What Mad Pursuit*, Davis noted its modest tone. But he had not forgotten their encounter and commented, "For those of us who have been crushed by Crick for making a silly suggestion at a scientific meeting this tone may seem a mellowing with age, but his autobiography gives a deeper insight. There Crick wrote: 'Unfortunately I have sometimes been carried away by my impatience and expressed myself too briskly and in too devastating a manner.'"[58] Had he, perhaps, remembered an incident at the chromatin symposium at Cold Spring Harbor in 1977? Charles Weissmann recalled how

> in the discussion following a presentation by a young scientist, Crick remarked: "You amateurs should keep out of this field." In response to this I prepared the T shirt with "Amateur" written on the back, and then, facing the audience, I introduced my talk saying that this field (transcriptional control) was rather new to me and I hoped the audience would bear with me, whereupon I turned to the blackboard and revealed the logo. It was a big success (not with Francis).[59]

## The Last Year

During 2003, members of Crick's family visited him: Michael and his family came from Seattle, Gabrielle from London, and Jacqueline and her family from Sussex. They did not expect him to see the year out. Close to the end, Watson visited, but was saddened to find his long-time friend so ill.

Crick kept rigidly to the regimen advocated by his oncologist, Dr. Barbara Parker, in whom he had placed his confidence. A clinical professor of oncology at UCSD's Department of Medicine and an affiliate of UCSD's Thornton Hospital in La Jolla, she has a special concern for quality of life issues under chemotherapy. Odile judged that "Dr. Parker did everything that she possibly could, taking over Francis' entire medical treatment."[60] She followed up the leads that Crick obtained from his colleagues as to new possibilities. The traditional bursts of chemotherapy followed by rest periods meant a very fluctuating state of well being. The cure was then worse than the disease. Steady low-level treatment proved to be more manageable. When the chemotherapy did not work, a spell of radiation was substituted, creating fewer adverse effects, but no check to the disease. By the end of 2003, all treatment was brought to a close. When the cancer had been diagnosed in 2001, it had been too late for the surgical removal of all the metastasized tissue.

Through these last years, Odile was constantly at hand, devising bland yet inviting meals that alleviated the unpleasant effects of the chemotherapy on the digestive system. Helping him dress and be ready for the day were no mere formalities. As Auld observed: "She has to feed exactly the right foods, do this, do that—and he is often agitated and bad-tempered at times and she has to calm him and keep the show on the road all the time. And I think she is finding that very stressful."[61] Except during the aftermath of a course of chemotherapy, Crick was usually able to recover his spirits, especially when faced with a lively visitor. When John and Brenda Maddox came over to La Jolla, John challenged Crick over his claims about consciousness. That was enough to set him off on a powerful hour-long response.[62]

There were long-time friends who did not make it to La Jolla. Harold Fost, a close school friend living in the Channel Island of Jersey, had last seen him in London on the occasion of Crick's Award of the Order of Merit in 1991. Traveling to California in 2003 was definitely ruled out. And Georg Kreisel, still obsessed by fear of health problems, could not face seeing his old friend in so weakened a state.

### The Crick Jacobs Center

Outside that intimate world. Crick could marvel at the expansion of the neuroscience community from approximately 3000 members in the early 1980s to more than 35,000 today. Reflecting on those early unsure years of the Salk Institute and comparing them with its condition at the turn of the century must have been very gratifying. Especially rewarding for him was that it had become a center of excellence in neuroscience.

In 2001, together with Terry Sejnowski, Crick had suggested the need for computational biology at the Salk, for as he urged "in the right hands, computers could be used to convert vast amounts of information about the brain into models of brain function that could be tested in our laboratories."[63] Clearly, his study of the brain had caused him to move up the hierarchy, some way from the molecular level of his former triumphs. Computer modeling now seemed to be very relevant.

He was deeply moved when a group of his colleagues came to his house to give him the news that they had obtained the resources to establish the Center he envisioned. They wished to name it the "Crick Jacobs Center" in honor of him and the benefactor, Dr. Irwin Jacobs, chairman of the board of San Diego–based Qualcomm Inc. It would be interdisciplinary, aimed at integrating the different approaches to achieve an "understanding of the organization of signaling systems and the functional neuroanatomy of the brain, from the molecular to the systems levels."[64] The Crick–Koch paper on the claustrum was Crick's attempt to prepare a review of a subject for the future research of his colleagues in the Center.

He was still revising the paper in the last week of July, 2004, when he was taken in a weakened state to hospital, but he assured Koch he was going to be alright. On Wednesday, July 28, as Koch recalled, Crick "was editing our last manuscript, dictating corrections to Kathleen Murray . . . and asking that she return the next day with the edited text."[65] But there was not to be a next day. Crick died that evening. The manuscript he left betrayed the troubled state in which he had been working. Uncharacteristic was the tendency to ramble. Koch now had to edit and in the process trim the text of his much-loved mentor and co-researcher.

### Two Memorials

Asked by Michael how his father would like his funeral to be, Crick with his ready wit as sharp as ever, responded, "Surprise me!" The family knew that as a convinced humanist, he wanted no grave and headstone. He was to be cremated, his ashes to be scattered on the ocean waves. Unlike Einstein, no one would be able to study his brain or like Jeremy Bentham have his skeleton mounted at University College London, or like Darwin have his corpse redirected from the local funeral home to be placed in a tomb in Westminster Abbey.[66] Nor would Alcor Life Extension Foundation "preserve his brain and body pattern down to the molecular level" using its "cryonics" procedure. Although the company had offered to waive the $120,000 fee for "world famous Dr. Crick,"[67] he ignored their invitation.

One week after Crick's death, a private memorial was held at the Salk Institute in memory of his life, followed on the 27th of September by a larger public memorial held on the plaza of the Salk. President and CEO of the Salk Richard Murphy opened the proceedings by declaring how privileged they were to have known him. "We can describe him, but we can never fully capture him. He was beloved, and he gave an enormous amount to science, to humanity, and to the Salk Institute. We will miss him deeply." Then Watson described his time at the Cavendish in the 1950s. There he "witnessed Crick's irrepressible brain in action . . . like watching Fourth of July fireworks. Never before had I witnessed such disciplined intelligence." Aaron Klug, who had joined Crick at the LMB in 1959, noted how Crick "became, as it were, the conscience of the new subject of molecular biology. . . . He led, he inspired, he conducted an extensive correspondence, he corrected mistakes, settled disputes, he cajoled, he joked." Sydney Brenner, who had worked with Crick from 1957 until 1976 agreed. "He was the creative influence in the early days of molecular biology, catalyzing both theory and experiment. . . . He was not interested in adding another brick to the edifice of science, but in building new edifices." Tommy Poggio asked, "Was Crick arrogant, aloof?" No, quite the reverse: "He dedicated an incredible amount of his time and energy to junior researchers such as David [Marr] and me—and to the problem of how the brain works. He took us skeptically but seriously."[68] From the family came endearing tributes from Michael and from Jacqueline. Michael asked what made his father tick? Was it to become famous, popular, or wealthy? Not so, although he did enjoy the fine things of life. It was his love of problem solving and his opposition to vitalism and religion. Then there was that compulsion to win the argument that drove him on. His two popular books, *Of Molecules and Men* and *Life Itself*, express these features clearly. Jacqueline spoke frankly about her father as a family man

> not a hands-on, bedtime-story, teach-you-to-ride-a-bike kind of dad. But our house was full of life, laughter and exciting people, and Francis, with his enormous vitality, was at the center of it. Family life mostly revolved around the dining table where Francis tried to impart his vast wealth of knowledge to his sometimes reluctant daughters. The fruit bowl was the center of our education, as we learned how the apple revolves around the orange once a year while spinning around its core once a day.[69]

At this memorial, one had the feeling that Crick's spirit was indeed living on, as Brenner remarked. This impression was heightened by the musical intermissions when the pianist Jim Guerin played English folk songs, Gershwin tunes, and a collection of Beatles songs, including of course "Sgt. Pepper's Lonely Hearts Club Band," "Yellow Submarine," and "Kansas City," all Crick's favorites, so often played at Portugal Place.

Little over a year before, Crick had made all those video recordings of good will messages to the many celebratory events of the Golden Jubilee of the Watson–Crick structure for DNA. One such video was seen at the Cold Spring Harbor meeting that spring. Crick, who in 1953 had been looking for simple solutions, now half a century later stressed the complexity of contemporary molecular biology and neuroscience. To him it seemed that there is "no limit to the problems that now confront us. I shall not live to see their solution, but many of you should survive long enough to see many radically new techniques and striking discoveries. Good luck to you."[70]

# Epilogue

Francis Crick is survived by his first wife Doreen, now 96 years of age, his son Michael, and his daughters Gabrielle and Jacqueline. His second wife Odile survived Francis for three years, passing away on the 5th of July 2007. Events to celebrate her life were held at La Jolla and Cambridge and, on the 12th of October 2007, the Odile Crick Memorial Exhibition of her art was held at the Salk Institute. Crick is survived by his four grandchildren Alex, Camberley, Francis, and Kindra, the offspring of his son Michael and his daughter-in-law, Barbara (née Davis); and by two grandchildren, Mark and Nicholas, adopted by his younger daughter Jacqueline and her husband Christopher Nichols. His elder daughter Gabrielle is single; she accompanied her mother on her last European holiday in 2006.

Crick's friend Maurice Wilkins died on the 5th of October 2004. James Watson is now Emeritus Chancellor of Cold Spring Harbor Laboratory having retired in October of 2007. Sydney Brenner continues his scientific work at the Salk Institute. Christof Koch continues as the Lois and Victor Troendle Professor of Cognitive and Behavioral Biology, and Professor of Computation and Neural Systems at the California Institute of Technology. Crick's long time friend Georg Kreisel lives in retirement in Salzburg, Austria.

A stained glass window depicting a double helix and commemorating Crick was installed in the dining hall of Gonville and Caius College, Cambridge, in 1992–1993. Memorials to Crick's memory include a sculpture "Discovery" unveiled in the Northampton town center in December, 2005, The Francis Crick Prize Lectures (established 2003 by Sydney Brenner) at The Royal Society, London, and The Francis Crick Graduate Lectures at the University of Cambridge begun in 2005. The Westminster City Council unveiled a green plaque to Crick on the front façade of 56 St George's Square, Pimlico, London SW1, the apartment rented by Crick while working for the Admiralty (1946–1949). Both The Eagle pub in Cambridge and the Austin Wing of the Cavendish Laboratory display plaques recording the discovery of the structure of DNA. The "Golden Helix" symbol (still painted yellow) remains firmly in place above his former home in Portugal Place, Cambridge.

In November 2005, James Watson donated the DNA sculpture that is to be found outside Thirkill Court, Clare College, Cambridge. On the helices is written: "The structure of DNA was discovered in 1953 by Francis Crick and James Watson while Watson lived here at Clare." Mention is also made of Rosalind Franklin and Maurice Wilkins.

MARTIN PACKER AND ROBERT OLBY
*May 2009*

# Sources and Notes[*]

## Books and Journals

| | |
|---|---|
| *AH* | Crick F. 1994. *The astonishing hypothesis.* Simon and Schuster, New York, London. |
| *Chromosome* | Crick F. 1971. General Model for the chromosomes of higher organisms. *Nature* **234:** 25–27. |
| *CIBA* | Ciba Foundation, 1957. *Symposium on the nature of viruses.* Little, Brown, Boston 1957. |
| *CSHSQB66* | Crick FHC. 1966. The genetic code–yesterday, today, and tomorrow. *Cold Spring Harbor Symp Quant Biol* **31:** 3–9. |
| *Daedalus* | Olby R. 1972. Francis Crick, DNA, and the central dogma. *Daedalus* **99:** 938–937. Reprinted in Holton G. ed., 1972. *The twentieth century sciences: Studies in the biography of ideas.* W.W. Norton, New York. |
| *DFL* | de Chadarevian S. 2002. *Designs for life: Molecular biology after World War II.* Cambridge University Press, Cambridge. |
| *DHn* | Watson JD. 1980. *The double helix. A personal account of the discovery of the structure of DNA* (ed. G. Stent), A Norton Critical edition. Norton and Co., New York. |
| *Eighth Day* | Judson H. 1996. *The eighth day of creation: Makers of the revolution in biology*, expanded edition. Cold Spring Harbor Laboratory Press, Cold Spring Harbor, New York. |
| *Future of Man* | Wolstenholme G. 1963. *Man and his future: A Ciba foundation volume* Little, Brown, Boston. |
| *Gen Code* | Crick FHC, Barnett L, Brenner S, Watts-Tobin RJ. 1961 General nature of the genetic code for proteins. *Nature* **192:** 1227–1232. |
| *GGG* | Watson J. 2002. *Genes, girls, and gamow: After the double helix.* Knopf, New York. |
| *Haemoglobin* | In Roughton FJW, Kendrew JC. eds., 1949. *Haemoglobin: A symposium based on a conference held at Cambridge in June 1948 in memory of Sir Joseph Barcroft.* Butterworths Scientific Publications, London; Interscience Publishers, New York. |

---

[*]Page references followed by ff denote "and the following pages"; Ibid denotes that the reference is the same as the preceding reference; Op cit denotes that the reference is the same as the previously cited reference as indicated in the parentheses.

| | |
|---|---|
| *Kreiseliana* | Odiffreddi P. ed., 1996. *Kreiseliana: About and around Georg Kreisel.* A.K. Peters, Wellesley, Massachusetts. |
| *Life Itself* | Crick F. 1981. *Life itself: Its origin and nature.* Simon and Schuster, New York. |
| *McElroy/Glass* | McElroy W, Glass B. 1957. *A symposium on the chemical basis of heredity.* The Johns Hopkins Press, Baltimore. |
| *Memorial* | *Remembering Francis Crick.* From presentations given at a memorial service for Francis Crick, on 3 August 2004, and from speakers at the celebration of his life on 27 September 2004. Salk Institute, La Jolla, California. |
| *Molecules* | Crick F. 1966. *Of molecules and men.* University of Washington Press, Seattle and London. |
| *My Life* | Brenner S. 2001. *My life in science.* Biomed Central Ltd., London. |
| *Passion* | Watson JD. 2000. *A passion for DNA: Genes, genomes, and society.* Cold Spring Harbor Laboratory Press, Cold Spring Harbor, New York. |
| *PDH* | Olby R. 1974. *The path to the double helix: The discovery of DNA.* Macmillan, London; revised edition (1994), Dover, New York. |
| *Phage Origins* | Cairns J. et al. eds., 1966. *Phage and the origins of molecular biology.* Cold Spring Harbor Laboratory Press, Cold Spring Harbor, New York. |
| *Quest* | Koch C. 2004. *The quest for consciousness: A neurobiological approach.* Roberts & Co., Englewood, Colorado. |
| *Rickmann* | Crick F. Rickman-Godlee Lecture, University College London, 21 September 1968, PP/CRI/E/1/16/13. |
| *SEB58* | Crick FHC. 1958. On protein synthesis. *Symp Soc Exp Biol* **12:** 138–163. |
| *The Box* | Dougall R. 1973. *In and out of the box: An autobiography.* Collins, Harvill Press, London, p. 144. |
| *Third Man* | Wilkins M. 2003. *The third man of the double helix: The autobiography of Maurice Wilkins.* Oxford University Press USA, New York. |
| *WMP* | Crick F. 1988. *What mad pursuit: A personal view of scientific discovery.* Basic Books, New York. |

## Unpublished

| | |
|---|---|
| *DTAH* | Crick FHC. 1995. On degenerate templates and the adaptor hypothesis. *PP/CRIH/1/38.* |
| *MsWMP* | Unpublished portion of Crick's manuscript for *WMP*, 1987, *UCSD, MSS600/10/A1.* |
| *Recollections* | Crick F. 1999. Recollections, sent with cover letter to Mary Woolnough, 22 December 1999, UCSD, MSS600, Box 90, Folder: Correspondence 2000–2001. |

## Manuscript Collections

| | |
|---|---|
| *CaltechArchives* | *Oral Histories Online; Delbrück Papers.* |
| *CCAR* | Churchill College Archives: Churchill papers. |
| *Churchill/AVHL* | The Papers of Archibald Vivian Hill, Churchill Archives Centre, Cambridge, UK. |

| | |
|---|---|
| *CSHLA,W* | Cold Spring Harbor Laboratory Archives: Watson Papers. |
| *CSHLA,B* | Cold Spring Harbor Laboratory Archives: Brenner Papers. |
| *Kendrew Papers* | Modern manuscripts, Bodleian Library, Oxford, UK, J.C. Kendrew Papers: GB 016. |
| *FD21/13–14* | Medical Research Council, Francis Crick, Personal File. Now deposited at the National Archive, Kew, Richmond, Surrey, UK. |
| *FRKN* | Churchill College Archives: Churchill papers. |
| *OSUSC/PP* | Oregon State University, Special Collections, Pauling Papers. |
| *PP/CRI* | Wellcome Library for the History and Understanding of Medicine: The Francis Crick Papers. |
| *PP/FGS* | Wellcome Library for the History and Understanding of Medicine: The Frederick Gordon Spear Papers. |
| *UCSD* | University of California, San Diego, Mandeville Special Collections Library |
| | *MSS600,* Francis Crick Scientific Papers. |
| | *MSS660,* Francis Crick Family Personal Papers. |

## Preface

1. Oscar Wilde, cited in Holroyd M. *Works on paper: The craft of biography and autobiography,* p. 4. Little & Brown, London.

2. Oliver Sacks, letter to Crick, 28 June 2003, *UCSD, MSS600:92/S.*

3. Miescher F. 1871. Ueber die chemische Zusammensetzung des Eiters *Med Chem Unters* **4:** 441–460. Submitted 1869.

4. Francis Crick, Cherwell-Simon Lecture, 17 May 1966, *PP/CRI/E/1/14/8.*

5. Capote T. 1965. *In cold blood: A true account of a multiple murder and its consequences.* Random House, New York. Capote called it a "non-fiction novel."

6. *Phage Origins*

7. Crick, letter to the author, 14 November 1966, author's archive.

8. Crick, letter to Watson, 13 April 1967, PP/CRI/I/3/8/4.

9. *PDH.*

10. *Daedalus.* Reprinted in 1972 with revisions following comments from Dr. Crick and from his aunt, Mrs. Arnold Dickens, and a postscript added to include further comments from Dr. Crick.

11. Watson J. 1968. *The double helix: A personal account of the discovery of the structure of DNA.* Atheneum, New York.

12. Ridley M. 2006. *Francis Crick. Discoverer of the Genetic Code.* Harper Collins, New York.

## Chapter 1

1. Edward Collingwood to Georg Kreisel, reported in Kreisel to Crick, 15 May 1954, Collingwood to Crick, 27 September 1959, and Edward Andrade to Crick, 11 October 1962. (For further details, see Note 12 below.)

2. Although forming a terrace, these houses are best described as town houses.

3. The account of the party is taken from interviews with Odile Crick, Michael Fuller, Frederick Sanger, and Mark Bretscher.

4. FRS, 31 July 1968. Notes of a not-Watson, *Encounter*, p. 64. Reprinted under the authorial FXS, in *DHn*, p. 180.

5. Crick, letter to Watson, 30 October 1962, *CSHLA,W:9/36*.

6. Crick explained that they never got around to giving it a coat of gold paint (personal communication). As a single helix, it was to symbolize helices in the proteins, especially his first success—the Fourier transform of a helix.

7. Fuller M, 2005. Recollections of life as a technician. . . , CD of a talk at the Cavendish Laboratory, *Cavendish Laboratory Archives*, and interview with the author, 3 March 2004.

8. Hildegard Lamfrom's family emigrated to Oregon from Germany in 1937 and established the very successful Columbia Sportswear Company. Hildegard became a widely respected biochemist and she knew how to throw a good party.

9. The style and the literary allusions point in the direction of the literary critic George Steiner. When the original essay in *Encounter* was reprinted in the Norton edition of Jim Watson's *The double helix*, the "FRS" was altered to "FXS" because Steiner was and is not an FRS (i.e., Fellow of the Royal Society). More appropriately he is a Fellow of the British Academy. Like Crick he is a founder Fellow of Churchill College.

10. FRS, op cit (Note 4), p. 178.

11. John Steinbeck, quoted in *The New York Times,* 26 October 1962, pp. 1 and 12.

12. *WMP,* p. 81. The remark was made by Dr. Frank Putnam. Actually, in 1954 Crick had been informed of a comment made by his former Naval boss during World War II, Edward Collingwood, on the subject. Learning of the birth of Crick's third child in 1954, Collingwood commented "Crick will need a Nobel Prize which, rumour has it, he will one day get for his helices." Kreisel, letter to Crick, 15 May 1954, *UCSD, MSS660:5/3*. Then, in 1959, congratulating Crick on his election to Fellowship of the Royal Society, Collingwood wrote "We now await the Nobel Prize and Royal Medal." Collingwood, letter to Crick, 27 September 1959. *UCSD, MSS660:2/6*. A week before the telegram arrived from Stockholm, Crick's physics professor Edward Andrade wrote to congratulate him on the award of the Gairdner Prize and added "Now for the Nobel Prize!" Andrade, letter to Crick, 11 October 1962, *PP/CRI/A/3/4*.

13. Crick, letter to Jacques Monod, 31 September 1961, *PP/CRI/D/1/2/8*. Also cited in Zallen DT. 2003. Despite Franklin's work, Wilkins earned his Nobel. *Nature* **435:** 15.

14. Crick, quoted in *Cambridge Evening News*, 19 October 1962.

15. *Who's who, 1962. An annual biographical dictionary.* A & C Black, London. (The item "Recreations" was omitted from his entry in more recent editions.)

16. Deryck Harvey, Dr. Crick moves on to mice and men, *Cambridge Evening Reporter*, 20 October 1972.

17. FRS, op cit (Note 4), p. 64; *DHn*, p. 180.

18. The description of the Nobel ceremony is based on the manuscript account by Max Perutz and the full account of the 1950 celebrations by Philip Hench, "Reminiscences of the Nobel Festival, 1950," see Nobelprize.org. Also based on conversation with Francis and Odile Crick. For a view of Crick receiving his award in Stockholm, see the clip from Sveriges Television, Arkiv Forskåjningen, *CSHLA*.

19. Because of injuries sustained in a serious automobile accident, the physics laureate Lev Landau was unable to attend.

20. At the time of the discovery of the structure of DNA, Crick was a member of the Medical Research Council Unit housed in the University's Cavendish (physics) Laboratory under the direction of Professor Sir Lawrence Bragg. Crick's position was an MRC appointment in the unit that was to become the "MRC Unit for Molecular Biology" in 1957. He would move in 1962 to a new building outside of the town, this time associated with the University's Postgraduate Medical School and called "The MRC Laboratory of Molecular Biology." From 1950 to 1953, Crick was also a doctoral student at Cambridge University and a member of Gonville & Caius College. Bragg moved to the Royal Institution in London in January 1954 and was succeeded as Cavendish Professor by Nevill Mott.

21. Engstrom A. 1964. Presentation speech, *Nobel lectures including presentation speeches and laureates' biographies 1942–1962*, p. 753. Elsevier, Amsterdam, London, New York.

22. For Watson's banquet speech, *Les Prix Nobel en 1962*. The Nobel Foundation, Stockholm, 1963. For manuscript versions, see *CSHLA,W.*

23. Crick, letter to Stanley, 6 November 1962, *University of California at Berkeley, Bancroft Library, Wendell M. Stanley Papers, 7/121.*

24. Crick, letter to Pauling, 24 October 1962, *OSUSC/PP, "The race for DNA."* See http://osulibrary.orst.edu/special collections/coll/pauling/dna/corr/index.html/.

25. Crick, letter to J.D. Bernal, 1 November 1962, *Royal Institution London, J.D. Bernal Papers, J 178.*

26. For the Santa Lucia Festival of Lights, see http://www.serve.com/shea/germusa/lucia.html/.

27. *The New York Times*, 2 February 1962, p. 14.

28. *WMP*, p. 76.

29. Op cit (Note 27), p. 14.

30. Crick, personal communication.

31. Susan Foulkes (neé Barnes), e-mail to the author, 30 July 2005.

32. Nan Robertson, Love and work now Watson's double helix, *The New York Times*, 26 December 1980, p. A24.

33. Crick, letter to Watson, 13 April 1967, *PP/CRI/1/3/8/4*, p. 6.

34. Ibid, p.4.

35. Ibid.

36. Lwoff A. 1968. Truth, truth, what is truth?, *Sci Amer* **219:** 133–138: reprinted in *DHn,* p. 230.

37. Sinsheimer R. 1968. The double helix, *Science and Engineering,* p. 4. California Institute of Technology; reprinted in *DHn,* p. 192.

38. Friedberg E. 2005. *The writing life of James D. Watson,* p. 58. Cold Spring Harbor Laboratory Press, Cold Spring Harbor, New York.

39. Alan Schwartz, letter to Watson, 1 June 1967. *CSHLA,W:* Cited by Friedberg E. 2005. *The writing life of James D. Watson,* p. 60. Cold Spring Harbor Laboratory Press, Cold Spring Harbor, New York.

40. Sir [William] Lawrence Bragg, letter to Watson, 19 April 1967, *CSHLA,W:23/2.*

41. See Watson JD. 1990. Bragg's foreword to *The double helix.* In *Selections and reflections: The legacy of Sir Lawrence Bragg* (ed. John Thomas and Sir David L. Phillips), p. 112. Royal Institution, London. Also see Hunter G. 2004. *Light is a messenger. The life and science of William Lawrence Bragg,* p. 242. Oxford University Press, Oxford.

42. Kendrew, letter to Watson, 23 April 1967, *CSHLA,W:23/2.*

43. Crick, op cit (Note 33), p. 6.

44. Kendrew, op cit (Note 42).

45. Watson, letter to Kendrew, 19 June 1967, *CSHLA,W:23/2.*

46. Leonard D. Hamilton, letter to Crick, 16 May 1967, reporting content of letter from Robert Montgomery to President Nathan Pusey, Harvard University, *UCSD, MSS660: 2/25.*

47. *WMP*, p. 84.

48. *WMP*, p. 81.

49. *WMP*, p. 80.

50. *DHn*, pp. 1–3.

51. Bernal, letter to Kendrew, 20 December 1966, *CSHLA,W:23/2;* published in Friedberg, op cit (Note 38), pp. 50–51.

52. Dubos RJ. 1950. *Louis Pasteur: Free lance of science*, p. 389, Charles Scribner's, New York.

## Chapter 2

1. *Recollections*, p. 6.

2. Crick, 1993. Mother's ambition for me, interview, *People's Archive.*

3. Winifred Dickens (née Crick), postcard to the author, undated (1970), author's archive.

4. Michael Crick, e-mail to the author, 9 March 2007, author's archive.

5. William Latimer was the son of Richard Latimer, who was married to Mary Crick. She and Walter Drawbridge had the same grandfather but different grandmothers.

6. Editor of the "Agents' Guide," 1890. *Where to buy in Northampton, an illustrated local review*, p. 31. Robinson Son & Pike, Brighton.

7. Nonconformist, that is, dissenting from the practices and governance of the established Church of England (the Anglican Church).

8. Recent Wills, *Northampton Daily Echo*, 28 December 2003. (In terms of today's purchasing power, this figure would translate to approximately £1,918,000.00 according to www.measuringworth.com.)

9. Dickens, letter to the author, 18 December 1970, author's archive.

10. Crick, Jr. W, Soddy F. 1939. *Abolish private money or drown in debt. Two amended addresses to our bosses.* Joseph Sault, London.

11. *Northampton Daily Echo,* 19 March 1925. Walter's outbursts remind us of claims made in Henry Ford's *Dearborn Independent.* See Neil Baldwin, 2001. *Henry Ford and the Jews. The mass production of hate.* Public Affairs, New York.

12. Dickens, letter to the author, undated (February 1970), author's archive.

13. Dickens, op cit (Note 9).

14. Dickens, letter to Crick, 5 November 1985, *USDC, MSS660:3/13.* ("It was quite a climb to this place and I thought it was a sort of museum [and he was allowed to live there presumably as a sort of caretaker]. . . .")

15. Ibid.

16. Obituary: Mr. Harry Crick, *Northampton Independent,* 6 February 1948, p. 8.

17. *WMP*, p. 10.

18. Susan Phillips, e-mail to Martin Packer, 26 September 2005, reporting recollection of 98-year-old Mrs. Nutt (née Jarman). This encounter with the Crick family probably occurred in 1926. Author's archive.

19. Frederick Followell, e-mails to the author, 14 February 2006 and 23 February 2007. Apparently, Annie recommended the Montessori Method to Ivy Parker when she would have children to educate. Author's archive.

20. *Recollections*, p. 6.

21. William Crick. See http://www.northampton.org.uk/history/history.htm; Matt Ridley, 2006. *Francis Crick. Discoverer of the genetic code*, p. 7. Harper Collins, London.

22. Crick, 1993. Parents: Dealing with a scientific child, *People's Archive*; "Frank Dickens," 2005. *Oxford dictionary of national biography*, vol. 16, pp. 78–79. Oxford University Press, Oxford; interview with Michael Crick, 11 July 2007.

23. *Recollections*, p. 4.

24. Obituary: Mr. F.F. Wilkins, *Northampton Independent,* 5 November 1913, p. 21.

25. Crick, personal communication. As a boy, Francis sometimes stayed at his grandmother Sarah Cleaver Crick's house in Northampton. His memory of her was of a "rather grumpy old lady." *Recollections,* p. 3.

26. Annie's grandson Michael was very definite regarding this point. Interview with the author, 11 July 2007, La Jolla.

27. Michael Wittet, letter to Ms. Wilkins, 14 January 1956, *MSS660, 10/39.*

28. Clive Binfield, letter to the author, 31 August 2005, author's archive. Professor Binfield added, ". . .even edging into philistinism."

29. The word "crick" has many meanings and there is no consensus as to why the village was so named.

30. Marquess is a male title of nobility, one step below that of duke. It is also spelled Marquis. The female equivalent is Marchioness.

31. For the Compton family, see 2003. *Burke's peerage baronetage & knightage clan chiefs Scottish feudal barons,* 107th ed., vol. 2, pp. 2926–2929. Burke's Peerage & Gentry, Stokesley, UK.

32. Michael Crick, op cit (Note 4).

33. For Weston Favell, see 1937. *The Victoria history of the counties of England. A history of Northamptonshire*, vol. 4, pp. 107–111. Oxford University Press, Oxford.

34. *Recollections*, p. 6.

35. Ibid.

36. Sacks O. 2001. *Uncle Tungsten*: *Memories of a chemical boyhood.* Alfred Knopf, New York.

37. When a 5th grade class of students from Pacific View School visited Crick in his office at the Salk Institute in 1999, he told them about these experiments. One of the letters of thanks from members of the class came from Cocinne Oshima, 21 June 1999. She commented, "My mum would never let me blow up anything." *PP/CRI/J/1/4/15/1.*

38. *Recollections*, p. 6.

39. This was the repertory company that employed Errol Flynn during 1933–1934. He appeared in 22 plays between December 1933 and May 1934. Until the summer of 1934, however, Crick was at Mill Hill during term time. McNulty T. 2004. *Errol Flynn.*

*The life and career*, p. 26. MacFarland and Co., Jefferson, North Carolina and London. Flynn's own account of his time in Northampton is amusing. See Flynn E. 2003. *My wicked, wicked ways. With an introduction by Jeffrey Meyers*, p. 181. Cooper Square Press, New York.

40. *WMP*, pp. 9–10.

41. Thompson B. 1903. The late Mr. W.D. Crick. Personal reminiscences by Beeby Thompson. *J. Northamptonshire Nat. His. Soc.* **12**: 134–144; Thompson B. 1903. Obituary: Mr. W.D. Crick, Fellow of the Geological Society, *Quart J Geol Soc* **60**: 1xxx.

42. Walter D. Crick, letter to Charles Darwin, 18 February 1882, February No. 9049. Henry E. Huntington Library and Art Gallery, San Marino, California. Darwin, letter to Crick, 26 March 1882, DAR 205.3/8225, Darwin Papers, Cambridge University Library.

43. Darwin C. 1882. On the dispersal of freshwater bivalves. *Nature* **25**: 529–530.

44. Crick WD. 1902. Obituary: The Rev. George Nicholson, B.A. *J Northampton Nat Hist Soc* **11**: 185–191.

45. Charles Larkman, 23 December 1903. Address at the funeral of W.D. Crick. In "Funeral Today," *Northampton Daily Echo*.

46. *WMP*, p. 10.

47. Obituary, 1937: Charles Larkman, *Congregational Yearbook* pp. 664–665; Binfield, op cit (Note 28).

48. For George Russell, see Binfield, op cit (Note 28).

49. *WMP*, p. 10.

50. When the statue needed to be relocated to make way for a garden of remembrance in 1935, the Vicar of St. Andrews Church wanted it taken right from his parish.

51. Construction continued for many years, and it was finally completed in 1911. The total cost was £9,000 (in present U.S. dollars, the equivalent would be approximately $800,000). Mr. Sharman was the builder.

52. Binfield, op cit (Note 28).

53. Odile Crick, personal communication.

54. Lunch conversation with Crick, undated (2002), La Jolla, California.

55. Mee A. ed. 1909–1910. *The children's encyclopedia*, vol. 1, p. 3. Amalgamated Publishers, London.

56. Ibid, vol. vii, p. 4120.

57. Ibid, vol. v, p. 2866.

58. *WMP*, p. 9.

59. Arthur Mee was a Methodist and literary editor of the *Daily Mail*. His motive for launching and editing *The children's encyclopedia* was "to encourage the raising of a generation of patriotic and moral citizens." See http://en.wikipedia.org/wiki/Arthur_Mee; Sir John Hammerton, 1946. *Child of wonder: An intimate biography of Arthur Mee*. Hodder & Stoughton, London.

## Chapter 3

1. Michael Hart, letter to the author, 14 May 1969, author's archive.

2. Editorial and School Notes, 1925. *The Northamptonian* **6**: 1–2.

3. Crick, letter to Woolnough, 22 December 1999, *Recollections; UCSD, MSS600:90*, p. 2.

4. *A short history of Northampton Grammar School* (1947). Reprinted 1991, in *A short history of Northampton Grammar School for Boys*, p. 84. The Guildhall Press, Northampton.

5. The term "Oxbridge" refers to both the Universities of Oxford and Cambridge. Crick could not remember to which colleges of which of the two universities he had applied; therefore, the term is used frequently throughout the chapter. Application to the universities was made through their colleges.

6. Crick, typescript for the Sloan autobiography *MsWMP* (some portions were excluded from the book), *PP/CRI/L/5/5/1/1.*

7. For Special Operations Executive, see http://www.spartacus.schoolnet.co.uk/OSE-cammaerts.htm.

8 Crick, interview number 5. "Relationship with religion," http://www.peoplesarchive.com.

9. Kathleen Murray, personal communication.

10. A barrister is a lawyer who can plead in the higher courts of England.

11. Colinvaux R. 1971. *Carver's carriage by sea,* 12th ed. Stevens & Son, London. Colinvaux subsequently embezzled his father-in-law's estate and shortly thereafter committed suicide; Joan Wolfenden, letter to Crick, 3 May 1994, *UCSD, MSS660:10/46.*

12. Harold Fost, telephone conversation with the author, 25 January 2006.

13. John Shilston, letter to Packer, 10 August 2008, author's archive.

14. In many famous fee-paying boarding schools, boys appointed as *prefects* and *monitors* had a role in maintaining discipline (outside the classroom). Serious misdemeanors might result in a caning from the head prefect, but not at Mill Hill.

15. Shilston, letter to the author, 6 September 2005, author's archive.

16. There was no grade between distinction and pass, but a numerical score could be cited.

17. Crick, interview number 8. "School," http://www.peoplesarchive.com.

18. *MsWMP,* p. 11.

19. Shilston, op cit (Note 15), p. 2.

20. Hart, op cit (Note 1).

21. Shilston's contribution to Roderick Braithwaite, Obituary, Professor Francis Crick OM and OM (1930–1934), *Old Millhillian*, 2005, p.174.

22. Charles Priestley chaired the International Commission on Atmospheric Turbulence meeting in Switzerland in 1958.

23. Braithwaite R. 2006. *The history of the Mill Hill Foundation 1807–2007*, p. 159. Phillimore & Co Ltd., Chichester, UK.

24. Shilston, op cit (Note 15), p. 174.

25. Crick, letter to headmaster of Mill Hill School, 25 October 2000; Braithwaite, op cit (Note 23), p. 160.

26. *WMP*, p. 12.

27. Braithwaite, op cit (Note 23), p. 159. The Walter Knox Prize was shared between Crick and Troughton. See Mill Hill School, Foundation Day Friday, 7 July 1933, Program, author's archive.

28. Ibid.

29. Kreisel, interview with the author, 20 February 2002.

30. Hart N, North J. 2004. *The world of UCL 1828–2004,* 3rd ed., p. 47. UCL Press, London.

31. Ibid, p. 30.

32. Crick, personal communication.

33. Shilston, op cit (Note 15), p. 1.

34. Ibid.

35. Williams S. 2003. *The people's King. The true story of the abdication*, p. 257. Penguin Books, London.

36. Shilston, op cit (Note 15), p. 2.

37. Andrade EN da C. 1923. *The structure of the atom,* 3rd ed. Bell, London.

38. University College London Physics Finals Papers, 1937, University College London Archives.

39. *WMP*, p. 13.

40. Andrade EN da C. 1962. Some personal reminiscences, in *Fifty years of X-ray crystallography* (ed. P.P. Ewald), p. 512. International Union of Crystallography, Utrecht, The Netherlands.

41. Crick, personal communication.

42. Gribbin J, Gribbin M. 1997. *Richard Feynman: A life in science*, p. 51. Dutton, New York.

43. Odile Crick, interview with the author, 20 June 2003.

44. Packer, telephone conversation with Joan Barker, 26 September 2005.

45. Crick, interview with the author, 8 March 1968.

46. Andrade EN da C. 1934. A theory of the viscosity of liquids. *Philosophical Magazine* **17:** 497 and 698.

47. *WMP*, p. 13.

48. See Hunter G. 2004. *Light is a messenger. The life and science of William Lawrence Bragg*, p. 203. Oxford University Press, Oxford.

49. Andrade, letter to the Admiralty, 28 March 1946, author's archive.

50. This was named for David Anderson, president of the Institute of Civil Engineers. In the first year of the war, 2.3 million Anderson shelters were distributed.

51. Panter-Downes M. 1947. Letter from London, *The New Yorker book of war pieces: Anthology*, p. 3. Schocken Books, New York.

## Chapter 4

**Note:** *The book-length accounts that deal with the British mining program in World War II are referred to in Notes 5, 14, 20, and 24 below. In these sources, there is little mention of civilian scientists who played so decisive a part in reorganizing the program and thereby establishing a successful strategy. None of these sources refers to Massey or to Crick. (In Note 14, Lieutenant Commander T.G.P. Crick mentioned by Poland on p. 26 was of no relation to Francis Crick. Poland does mention Bullard and Collingwood.)*

1. Harrie Massey, interview with the author, 10 March 1967, author's archive.

2. Cited in Snell JL. 1967. *Illusion and necessity; The diplomacy of global war, 1939–1945*, p. 38. Oxford University Press, New York. See also http://www.guardian.co.uk/theguardian/2008/sep/04/from.the.archive.

3. Panter-Downes M. 1947. Letter from London, *The New Yorker book of war pieces: Anthology*, pp. 3–5. Schocken Books, New York; Whiting C. 1999. *Britain under fire. The bombing of Britain's cities, 1940–45*. Pen & Sword Books, Barnsley and South Yorkshire, UK.

4. The Sea Lords are naval members of the Admiralty, i.e., the Board administering the Royal Navy.

5. Commander Webb ED. 1956. *H.M.S. Vernon. A short history from 1930 to 1955*, p. 20 ff. The Wardroom Mess Committee HMS Vernon, Portsmouth.

6. *DFL*, p. 20.

7. Alexander C. 1966. Obituary: Anthony Foster Crick. *New Zealand Medical Journal* **65:** 404; Thomas TF. 1966. Obituary: Anthony Foster Crick. An Appreciation. *Aust Radiologist* **10:** 168.

8. The term "boffin" was used during World War II to refer to scientists working on weapons, radar, and the like. It was popularized by Nigel Balchin in his 1943 book *The small back room*. Collins, London.

9. *MsWMP*, p. 10.

10. Mott N, Massey H. 1933. *The theory of atomic collisions*. Clarendon Press, Oxford.

11. *MsWMP*, p. 12.

12. Michael Gunn, letter to the author, 21 January 2008, author's archive.

13. *MsWMP*, p. 8.

14. Rear Admiral Poland EN, 1993. *The torpedomen: HMS Vernon's story 1872–1986*, p. 145. Kenneth Mason Publications Ltd, London.

15. *MsWMP*, pp. 9–10.

16. Report on A/S experimental work being carried out by the Admiralty, 25 July 1941. Papers on the Admiralty Advisory Panel on Scientific Research, 1941–1943. *National Archive, Kew, Richmond, Surrey, ADMI/15197 Paper No. 11*.

17. Massey H. 1976. D.R. Bates—A sixtieth birthday tribute. In *Atomic processes and applications: In honor of David R. Bates' 60th birthday* (ed. P.G. Burke and B.L. Moiseiwitsch), p. 6. North Holland Publishing Co., Amsterdam and New York.

18. *MsWMP*, p. 14.

19. *MsWMP*, p. 15.

20. Americans called the products "tailored assemblies." Captain Cowie JS. 1949. *Mines, minelayers and minelaying*, p. 158. Oxford University Press, London.

21. *MsWMP*, p. 16.

22. *MsWMP*, p. 17.

23. *MsWMP*, pp. 17–18. See also Cowie, op cit (Note 20), p. 159.

24. Lincoln FA. 1961. *Secret naval investigator*, pp. 144–145. William Kimber, London.

25. *MsWMP*, p. 19.

26. *MsWMP*, p. 20.

27. Lincoln, op cit (Note 24), p. 146; damaged 77, sunk 31. Cowie, op cit. (Note 20), p. 153.

28. *MsWMP*, p. 21.

29. Massey, interview with the author, 10 March 1968.

30. *MsWMP*, p. 24.

31. *MsWMP,* p. 25.

32. Poland, op cit (Note 14), p. 222; Cowie, op cit (Note 20), p. 155.

33. Cowie, op cit (Note 20), pp. 160–161.

34. Ibid, p. 158. For details on the technical advantages of the horizontal magnetic sensors that the British used, as opposed to the vertical sensors used by their opponents, see pp. 99–101 and 152 ff.

35. Poland, op cit (Note 14), p. 227. Information from University College London Archives.

36. Ibid, p. 224.

37. *WMP,* p. 15.

38. The Science Museum, 1946. *Naval mining and degaussing. Catalogue of an exhibition of British and German material used in 1939–1945,* p. iv. H.M. Stationary Office, London.

39. Information from University College London Archives.

40. Marriage certificate, dated 18 February 1940, General Registry Office, Division of St. Pancras.

41. Michael Crick, e-mail to the author, 5 May 2005.

42. Fost, telephone conversation with the author, undated.

43. Mary Clark, *WW2 People's War—Mary Clark's War Memories,* BBC.

44. Massey, op cit (Note 1), p. 6.

45. Cuthbert Collingwood (1748–1810), First Baron. He took over from Nelson at Trafalgar and fired the first shot; see also Obituary: Sir E. Collingwood, *The Times,* London, 26 October 1970.

46. Ibid, Obituary: Sir E. Collingwood.

47. *MsWMP,* pp. 10–11.

48. Subsequently, Sir Edward Bullard (1907–1980). He had a major role in the retrofitting of degaussing coils to protect ships from magnetic mines. He and his second wife later became close friends of the Cricks in La Jolla, California.

49. Crick, personal communication.

50. David Bates, letter to Crick, 23 November 1972, *USCD, MSS660:1/11.*

51. Flannery R. 2003. David Bates, 1916–1994, in *Physicists of Ireland. Passion and precision* (ed. M. McCartney, A. Whitaker), p. 264. Institute of Physics Publishing, Bristol and Philadelphia.

52. *Kreiseliana,* p. 25.

53. *Kreiseliana,* p. 27.

54. Kreisel, interview with the author, 21 January 2004.

55. Kreisel, interview with the author, 20 February 2002.

56. *Kreiseliana,* p. 32.

57. *The Box,* p. 144.

58. Poland, op cit (Note 14), p. 196.

59. *The Box,* p. 146.

60. *The Box,* pp. 146–147.

61. Crick, personal communication.

62. *WMP,* p. 15.

63. Crick, Intelligence memorandum, *UCSD, MSS660/20/13*.

64. Crick, letter to Jones. In Jones RV. 1978. *Wizard war: British scientific intelligence, 1939–1945*, pp. 524–525. Coward, McCann & Geoghegan, New York.

65. Ibid.

66. Crick, personal communication.

67. *WMP*, p. 16.

68. Dyson FJ. 2007. *A many-colored glass. Reflections on the place of life in the universe*, pp. 58–59. University of Virginia Press, Charlottesville, Virginia.

69. See http://www.trooping-the-colour.co.uk/.

70. Crick's conversation with Philip Low, postdoctoral fellow with Terence Sejnowski at the Salk Institute.

71. This is the first floor, in UK terminology.

72. *The Box*, pp. 163–164.

73. *The Box*, p. 165.

74. Pauling L. 1946. Analogies between antibodies and simpler chemical substances. *Chem Eng News* **24:** 1064.

75. *WMP*, p. 17. Note the author's italics.

76. Schrödinger E. 1944. *What is life? The physical aspect of the living cell*, pp. 46–47. Cambridge University Press, Cambridge.

77. Ibid, p. 61.

78. Ibid, pp. 47–48.

79. Crick, in *PDH*, p. 247.

80. Dyson, op cit (Note 68), p. 59.

### Chapter 5

1. Collingwood, letter to Crick, 9 September 1947, *UCSD, MSS660:2/6*.

2. Crick diary entry, *UCSD, MSS660/16/5*.

3. The Strangeways Laboratory, founded and directed by Dr. Thomas Strangeways, began its life in 1905 as a hospital for the clinical study of rheumatoid arthritis. Twenty-three years later, it became the Strangeways Research Laboratory, directed by Honor Fell. See *History of the Strangeways Research Laboratory (Formerly Cambridge Research Hospital)* 1912–1962 (E. Dorothy Strangeways, "1905–1926"; Frederick Gordon Spear, "1927–1962"; Honor B. Fell, "Cell Biology"). This booklet was published by the Laboratory, undated (1962).

4. Muriel Wigby, telephone interview with the author, 12 March 2004.

5. *WMP*, p. 21.

6. Brown A. 2005. *J.D. Bernal. The sage of science*, p. 115. Oxford University Press, Oxford.

7. Crick, letter to Archibald Vivian Hill, 1 May 1947, *Churchill/AVHL ll 4/1/18*.

8. Crick, letter to Hill, 2 July 1947, *Churchill/AVHL ll 4/1/18;* Mellanby memorandum, 7 July 1947, *FD21/13*.

9. *WMP*, pp. 18–19.

10. Hill, letter to Crick, 4 July 1947, *Churchill/AVHL II 4/1/18*.

11. Edward Mellanby, in a memorandum on meeting with Crick, 7 July 1947, *FD21/13*.

12. Crick's application for Studentship for Training in Research Methods, 7 July 1947, *FD21/13*.

13. Britain's largest protected wetland, the Broads, has long been a popular location for boating holidays. Crick probably began his sailing here.

14. Crick, letter to Hill, 31 July 1947, *Churchill/AVHL II 4/1/18*.

15. Mellanby, letter to Fell, 31 July 1947, *FD21/13*.

16. *WMP*, p. 21.

17. Douglas Lea was a former Fellow of Trinity College Cambridge and author of the 1946 classic on radiation genetics entitled *Actions of radiations on living cells*. Cambridge University Press, Cambridge.

18. Crick, personal communication. Crick told the author that Lea's death followed his obtaining a life insurance policy and occurred the day before the family was due to go on holiday.

19. Crick, letter to Mellanby, 12 August 1947, *FD21/13*.

20. Fell, letter to Mellanby, 14 August 1947, *FD21/13*.

21. *WMP*, p. 22.

22. Crick, personal communication.

23. *Third Man*, p. 109. Crick and Maurice Wilkins did not first meet during the War. See Maddox B. 2002. *Rosalind Franklin. The dark lady of DNA*, p. 157. Harper Collins, Glasgow; Hunter G. 2004. *Light is a messenger. The life and science of William Lawrence Bragg*, p. 187. Oxford University Press, Oxford. It was Harrie Massey and Wilkins who became acquainted in Berkeley, California while working on the Manhattan Project. See *Third Man*, p. 79; *WMP*, p. 20.

24. Glauert A. 1987. Fell, Dame Honor Bridget, *Biogr Mem R Soc* **19**: 257–258.

25. Hill, letter to Mellanby, 9 July 1947, *FD21/13*.

26. Collingwood, letter to Crick, 1 September 1949, *UCSD, MSS:660:2/6*.

27. Collingwood, op cit (Note 1).

28. Crick FHC, Hughes AFW. 1950. The physical properties of cytoplasm: A study by means of the magnetic particle method. I. Experimental, *Exp Cell Res* **1**: 73.

29. Ibid, p. 74.

30. Ibid, p. 77.

31. Ibid.

32. Ibid, p. 78.

33. Peter Lawrence, personal communication. Lawrence remarked that "instead of looking down the microscope, Francis would ask me to draw what he would see if he did!"

34. Fell, letter to F.J.C. Herrald (MRC), 12 July 1948, *FD21/13*.

35. Hughes A. 1952. *The mitotic cycle; the cytoplasm and nucleus during interphase and mitosis*, p. 209. Butterworths Scientific Publications, London.

36. Ibid, p. 89.

37. See Bradbury S. 1967. *Evolution of the microscope*, p. 76. Pergamon Press, Oxford and New York; Barer R. 1942. Cytological techniques, in *Cytology and cell physiology* (ed. G. Bourne), 1st ed., p. 76. Clarendon Press, Oxford.

38. Hughes, op cit (Note 35), p. 17.

39. *WMP*, p. 22. See also *Strangeways Laboratory, Annual Report for 1949*, pp. 7–8.

40. *Strangeways Laboratory, Annual Report for 1949*, pp. 7–8; Jacobson W, Webb M. 1951. The two types of nucleic acid during mitosis. *J Physiol* **112**: 2P–4P (*Proceedings of the Physiological Society*, May 1950).

41. *Third Man,* p. 123.

42. Crick, letter to Hill, 7 March 1949, *Churchill/AVHL II 4/1/18.*

43. Hill, letter to Crick, 11 March 1949, *Churchill/AVHL II 4/1/18.*

44. Crick, letter to Hill, 18 March 1959, *Churchill/AVHL II 4/1/18.*

45. Odile Speed, letter to Crick, undated (winter, 1948–1949), *UCSD, MSS660:2/25.*

46. Margaret (surname unknown; Crick did not recall) letter to Crick, 2 January 1949. *UCSD, MSS660:11/15.*

47. Crick, personal communication.

48. Edie Hammond, letter to Crick, undated (summer, 1945), *UCSD, MSS660:4/13.*

49. Ibid.

50. Hammond, letter to Crick, undated (December 1948), *UCSD, MSS660:4/13.*

51. Hammond, letter to Crick, undated (January or February 1949), *UCSD, MSS660:2/25.*

52. The Women's Royal Naval Service (WRNS) is the British equivalent of America's Women Accepted for Volunteer Emergency Service (WAVES).

53. Lunch conversation with Odile and Francis Crick, La Jolla, 5 February 2002.

54. Edith Weisz, telephone interview with the author, 24 August 2007.

55. *WMP*, p. 43.

56. Freddie Gutfreund, interview with the author, 6 March 2007.

57. Marguerite Speed, telephone interview with the author, 29 August 2007.

58. Southampton Row is in a part of London called Holborn. The College is now the Central St. Martins College of Art and Design and is part of the University of the Arts London.

59. Odile Speed, letter to Crick, undated (September 1947), *UCSD, MSS660:2/25.*

60. The divorce decree was dated 25 March 1947; it became final 8 May 1947.

61. Odile Crick, interview with the author, 11 July 1968.

62. Tony Crick, letter to Francis and Odile, 5 August 1949, *UCSD, MSS660:2/13.*

63. Annie Crick, letter to Crick, undated (January 1949), *UCSD, MSS660:2/12.*

64. Odile Speed, letter to Crick, 1 January 1948, *UCSD, MSS660:2/25.*

65. Odile Speed, letter to Crick, undated (December 1948), *UCSD, MSS660:2/25.*

66. Ibid.

67. Odile Speed, letter to Crick, undated (early 1949), *UCSD, MSS660:2/25.* (Maman is the French equivalent of "mummy" or "mommy.")

68. Odile Crick, interview with the author, 20 June 2003.

69. Odile Speed, letter to Crick, undated (early summer, 1949), *UCSD, MSS660:2/25.*

70. *WMP*, p. 44.

## Chapter 6

1. Crick, letter to Hill, 7 March 1949, *Churchill/AVHL.*

2. Crick, personal communication.

3. Crick, letter to Edward Mellanby, 22 February 1949, *FD21/13.*

4. Mellanby, memorandum, 1 March 1949 regarding his interview with Crick, *FD21/13.*

5. Crick, op cit (Note 1).

6. Perutz M. 1998. *I wish I'd made you angry earlier. Essays on science, scientists, and humanity,* p. 313. Cold Spring Harbor Laboratory Press, Cold Spring Harbor, New York.

7. Crick, 1990. W.L. Bragg: A few personal recollections. In *Selections and reflections: The legacy of Sir Lawrence Bragg* (ed. J. Thomas and Sir D. Phillips), p. 109. Royal Institution of Great Britain, London.

8. Bragg, letter to Mellanby, 21 March 1949, *FD21/13.*

9. Perutz, letter to Arthur Landsborough Thomson, 22 March 1949, *FD21/13.*

10. Crick's memorandum gave his reasons for wishing to work on protein structure and transfer to the Cavendish Laboratory, *FD21/13.*

11. The office of the Medical Research Council (MRC, at the Laboratory of Molecular Biology, Cambridge), letter to Crick, 2 May 1949, *FD21/13.*

12. Conditions of service on the staff of the Medical Research Council, *MRC Archives, PF540.*

13. Dina Fankuchen, interview with the author, 3 June 1970, *PDH,* p. 258.

14. Bragg, interview with the author, 6 December 1967, *PDH,* p. 264.

15. Bragg WL. 1912. The diffraction of short electromagnetic waves by a crystal. *Proc Cambridge Philos Soc* **17:** 43–57. The formulation in use today was introduced in 1913; see Bragg WH, Bragg WL. 1913. The reflection of X-rays by crystals. *Proc Roy Soc A* **88:** 428–438.

16. Perutz M. 1993. Co-chairman's remarks. *Gene* **135:** 9–13. Reprinted as How the secret of life was discovered. In Perutz, *I wish I'd made you angry earlier,* p. 204 (see Note 6).

17. Crick wrote "It is not true that I twice flooded his [Bragg's] office, since in fact his office was on the opposite side of the lab to the room in which I worked with water; although it is true that I did twice cause a flood it was not due to the rubber tubing around the condenser but the rubber tubing around a suction pump." See Rough notes on your manuscript, Crick, letter to Watson, 31 March 1966, *CSHLA,W:9/38,* p. 1. Watson then qualified his text to read "the corridor outside his office," *DHn,* p. 10.

18. *WMP,* p. 57.

19. Crick, interview with the author, 8 March 1968.

20. *WMP,* p. 58.

21. See Bragg WL. 1962. The growing power of X-ray analysis. In *Fifty years of X-ray diffraction* (ed. P. Ewald), p. 130. The International Union of Crystallography, Utrecht.

22. Robertson JM. 1962. Problems of organic structure. In P. Ewald, ed., op cit (Note 21), p. 154.

23. Ibid, p. 153.

24. Kendrew JC, Perutz MF. 1949. The application of X-ray crystallography to the study of biological macromolecules. *Haemoglobin,* p. 171.

25. Perutz MF. 1949. Recent developments in the X-ray study of haemoglobin. *Haemoglobin,* p. 136.

26. Perutz, letter to Landsborough Thomson, 6 March 1950, *FD21/13*.

27. *WMP*, p. 49.

28. *WMP*, p. 50.

29. Perutz, interview with the author, 12 January 2002.

30. David Davies, interview with Georgina Ferry, 2007. In Ferry G. 2008. *Max Perutz and the secret of life*, p. 148. Chatto & Windus, London.

31. Crick. 1951. The determination of the structure of proteins by X-ray crystallography. Prospects and methods, *UCSD, MSS600:10A-2*, pp. 3–4.

32. Ibid, p. 15.

33. Crick FHC. 1952. The height of the vector rods in the three-dimensional Patterson of haemoglobin. *Acta Crystallogr* **5**: 386.

34. *WMP*, p. 51.

35. Perutz, letter to the MRC, 9 October 1950, *FD21/13*.

36. Crick, letter to the MRC, received 24 March 1949, *FD21/13*.

37. Crick's 1950 application to enter the Cambridge PhD program, from his curriculum vitae, ca. 1940s, *UCSD, MSS660*, pp. 12 and 18.

38. Gutfreund, e-mail to the author, 30 September 2006.

39. Crick FHC. 1953. The unit cells of four proteins. *Acta Crystallogr* **6**: 221.

40. Crick, personal communication. It was this claim, not the alleged accusation of stealing (that Crick denied he made), that had infuriated Bragg, according to Crick. See also Crick, op cit (Note 17), p. 2.

41. Watson, interview with the author, at Cold Spring Harbor Laboratory undated (2005); Perutz, letter to Crick, 7 October 1994, *UCSD, MSS660:11/15*.

42. Crick F. 1954. The structure of the synthetic α-polypeptides. *Sci Progress* **42**: 211.

43. Perutz, letter to Crick, 7 October 1994, *UCSD, MSS660:11/15*.

44. *WMP*, p. 45.

45. *WMP*, p. 46.

46. Crick, interview with the author, 8 March 1968.

47. William Cochran, letter to the author, 19 July 1968.

48. Cochran W, Crick FHC. 1952. Evidence for the Pauling–Corey α-helix in synthetic polypeptides. *Nature* **169**: 235. (This was published before the paper cited in Note 51 below.)

49. Ibid.

50. Crick, interview with the author, 7 August 1972, cited in *PDH*, p. 313.

51. Cochran W, Crick FH, Vand V. 1952. The structure of synthetic polypeptides. I. The transform of atoms on a helix. *Acta Crystallogr* **5**: 581–586.

52. Crick, op cit (Note 33), pp. 384 and 386.

53. *WMP*, p. 78.

## Chapter 7

1. Crick, interview with the author, 8 March 1968.

2. Ridley M. 2006. *Francis Crick. Discoverer of the genetic code*, p. 45. Harper Collins, New York.

3. *WMP*, p. 64.

4. McElheny V. 2003. *Watson and DNA. Making a scientific revolution*, p. 1. Perseus Publications, Cambridge, Massachusetts.

5. *DHn*, p. 31.

6. *WMP*, p. 64.

7. Crick, interview with the author, 8 March 1968.

8. Avery OT, MacLeod CM, McCarthy M. 1944. Studies on the chemical nature of the substance inducing transformation of pneumococcal types. *J Exp Med* **79**: 137–158.

9. Crick did not accept the alternative view that phage multiplication is a case of an enzyme-like protein catalyzing the formation of more of its kind.

10. Astbury WT, Bell FO. 1938. Some recent developments in the X-ray study of proteins and related structures. *Cold Spring Harbor Symp Quant Biol* **6**: 109–118; Bell, 1939. "X-ray and related studies of the structure of the proteins and nucleic acids." PhD thesis, Leeds University, UK.

11. *Third Man*, p. 123.

12. Geoffrey Brown, cited in *PDH*, p. 354.

13. Odile Crick, personal communication.

14. *Third Man*, p. 178.

15. Schrödinger E. 1944. *What is life?*, p. 61. Cambridge University Press, Cambridge.

16. Crick, *WMP*, p. 75.

17. *Passion*, p. 4.

18. Perutz M. 1998. *I wish I'd made you angry earlier. Essays on science, scientists, and humanity*, p. 188. Cold Spring Harbor Laboratory Press, Cold Spring Harbor, New York.

19. *Passion*, p. 24.

20. Crick, interview with author, 8 March 1968.

21. Astbury WT. 1947. X-ray studies of nucleic acids. *Symp Soc Exp Biol* **1**: 66–76.

22. Watson, personal communication.

23. *Third Man*, p. 128; Sir John Randall, letter to Rosalind Franklin, 4 December 1950, cited in *PDH*, p. 346.

24. This letter was first published in the 1974 edition of *PDH*, p. 346.

25. *DHn*, p. 37.

26. Notes for Franklin's talk at the DNA seminar of 21 November 1951, *FRKN:1/1*.

27. *DHn*, p. 49.

28. Manuscript transcribed in *PDH*, p. 357 ff.

29. Raymond Gosling, in the 1987 British Broadcasting Company (BBC) film *Double Helix*, a transcript of the American version of *Life Story*.

30. *Double helix: The DNA story*, 2003 BBC film transcript.

31. Ibid.

32. *DHn*, p. 59.

33. Cochran W, Crick FHC. 1952. Evidence for the Pauling–Corey α-helix in synthetic polypeptides. *Nature* **169**: 234–235.

34. Bragg, letter to Hill, 18 January 1952. *Churchill/AVHL*.

35. Perutz, letter to JG Duncan, 15 January 1952, *FD21/13*.

36. Landsborough Thomson, letter to Perutz, 15 January 1952, *FD21/13*.

**Chapter 8**

1. Crick, interview with the author, 8 March 1968, cited in *PDH*, p. 388.

2. Chargaff E. 1968. A quick climb up Mount Olympus. *Science* **172**: 637.

3. For a brief biography, see http://www.amphilsoc.org/library/mole/c/chargaff.html.

4. Chargaff E. 1950. Chemical specificity of nucleic acids and the mechanism of their enzymatic degradation. *Experientia* **6**: 201–209.

5. *DHn*, p. 78.

6. Chargaff E. 1978. *Heraclitean fire. Sketches from a life before nature*, p. 101. Rockefeller University Press, New York.

7. Crick, op cit (Note 1).

8. *PDH*, p. 387.

9. Subsequently, John Griffith repeated the calculations and found that his first results were incorrect. Crick, personal communication.

10. *PDH*, p. 388.

11. Perutz, letter to JG Duncan, 19 May 1962.

12. Duncan, letter to Perutz, 27 May 1952, *FD21/13*.

13. Perutz, letter to Duncan, 16 June 1952, *FD21/13*.

14. Crick FHC. 1952. Is α-keratin a coiled coil? *Nature* **170**: 883.

15. Crick FHC. 1953. The Fourier transform of a coiled-coil. *Acta Crystallogr* **6**: 685–689.

16. Crick FHC. 1953. The packing of α-helices: Simple coiled-coils. *Acta Crystallogr* **6**: 689–697.

17. Pauling's diary for 15 September 1952, *OSUSC/PP*.

18. Crick, op cit (Note 14).

19. Pauling L, Corey RB. 1953. Compound helical configurations of polypeptide chains: Structure of proteins of the α-keratin type. *Nature* **171**: 59–61.

20. Crick, personal communication.

21. Pauling, letter to Jeffreys Wyman, 17 September 1952, *OSUSC/PP:438/15*.

22. Ibid.

23. Jerry Donohue, letter to Pauling, 9 November 1952, *OSUSC/PP:9.001/1.12*

24. Pauling, letter to Donohue, 19 November 1952, *OSUSC/PP:9.001/1.12*

25. Crick, letter to Pauling, undated (February 1953), *OSUSC/PP:68/11*.

26. Kreisel, letter to Crick, undated (early 1951), *UCSD, MSS660/5/3*.

27. *WMP*, p. 47; *PP/CRI/H/1.10*.

28. Roger Herriott, letter to Al Hershey, 16 November 1951. Part of this letter appeared in *Phage and the origins of molecular biology*, 1966. (ed. J. Cairns, G. Stent, and J. Watson), p. 102. Cold Spring Harbor Laboratory Press, Cold Spring Harbor, New York.

29. Linus Pauling, letter to Peter Pauling, 21 January 1953, *OSUSC/PP:68/11*.

30. *DHn*, p. 91.

31. Crick, interview with the author, 8 March 1968.

32. Peter Pauling, letter to Linus Pauling, 13 January 1953. Cited in Maddox B. 2002. *Rosalind Franklin. The dark lady of DNA*, p. 191. Harper Collins, London. This letter was not among the Pauling papers given to Oregon State University.

33. Pauling L, Corey RB. 1953. A proposed structure for the nucleic acids. *Proc Natl Acad Sci* **39:** 84–97.

34. *DHn*, pp. 93–94.

35. Crick, letter to Watson, 31 March 1966, *CSHLA,W:9/35*, p. 2.

36. *DHn*, p. 96.

37. *DHn*, p.98.

### Chapter 9

1. *WMP*, p. 74.

2. *DHn*, p. 101.

3. Ibid.

4. Crick, interview with the author, 8 March 1968.

5. *DHn*, p. 100.

6. Linus Pauling, letter to Henry Allen Moe, 19 December 1952, *OSUSC/PP* Science, Box 14/014, Correspondence, John Simon Guggenheim Memorial Foundation, 1952, folder 14/7.

7. Bragg to Watson, Recollections in 1966/1967, *CSHLA,W:5/38*.

8. Wilkins, letter to Crick, Thursday, undated (February 1953); *PDH*, p. 402; also see *Third Man*, p. 204.

9. Crick FHC, Watson JD. 1954. The complementary structure of deoxyribonucleic acid. *Proc R Soc A* **233:** 85; *DHn*, p. 280.

10. Crick, op cit (Note 4).

11. *Third Man*, p. 201.

12. *Third Man*, p. 205.

13. *Third Man*, p. 206; Crick, personal communication.

14. Crick, in *PDH*, p. 405.

15. Crick, in *PDH*, pp. 403–405; see also *DHn*, p. 103; Bretscher, letter to Crick, 28 April 2003; Crick, letter to Bretscher, 13 May 2003, author's archive.

16. Bretscher, letter to Crick, April 28, 2003; Crick, letter to Bretscher, 13 May 2003.

17. *PDH*, p. 403.

18. Jordon DO. 1952. Physicochemical properties of the nucleic acids. *Prog Biophys* **2:** 51–89; Gulland JM, Jordan DO, Taylor HFW. 1947. Deoxypentose nucleic acids. II. Electrometric titration of the acidic and basic groups of the deoxypentose nucleic acid of calf thymus. *J Chem Soc* **1947:** 1131–1141.

19. Broomhead J. 1950. "An X-ray investigation of certain sulphonates and purines." PhD thesis. Cambridge University, UK. Note that this was work performed in the Subdepartment of Crystallography, Cavendish Laboratory.

20. Watson, letter to Max Delbrück, 26 February 1953; *PDH*, p. 409, *CaltechArchives*.

21. *PDH*, pp. 411–412. Watson has frankly admitted that he had not understood the crystallographic argument about the C2 symmetry; Watson, interview with the author, 12 October 2007; Watson, 2007. *Avoiding boring people. Lessons from a life in science*, p. 106. Random House, New York. But the symmetry argument had been important not only for the antiparallel chains, but for another reason. It had led Crick to insist on 36° of angular separation between the residues on the polynucleotide chain. Hence, this crystallographic datum was important to Watson as he was trying to build the backbones, even though he did not build them in an antiparallel manner. It is true that only later did the symmetry relationship of the two chains become obvious to him, namely, when he discovered base pairing.

22. Crick, letter to Klug, 2003, *UCSD, MSS600:109*. Crick explained: "From my point of view the base pairs confirmed the anti-parallel chains deduced from the C2 symmetry, but Jim's view was different. He thought the C2 symmetry confirmed the anti-[parallel] symmetry deduced from the base pairs (which he himself had not noticed)." See also Watson, *DHn*, p. 112. There Watson mentions his "lukewarm response to Chargaff's data."

23. *DHn*, p. 114.

24. *PDH*, p. 412.

25. Chargaff E. 1950. Chemical specificity of nucleic acids and mechanism of their enzymatic degradation. *Experientia* **6**: 209.

26. *DHn*, p. 115. Crick did not recall acting in this way; Crick, personal communication.

27. *DHn*, p. 116.

28. Crick in *PDH*, p. 414.

29. Ibid.

30. *DHn*, p. 120

31. Wilkins, letter to Crick, 7 March 1953, in *PDH*, p. 414.

32. This quote not identified, but supported by Wilkins. See: *Third Man*, p. 211.

33. *Third Man*, p. 211.

34. *DHn*, p. 120.

35. *Third Man*, p. 212.

36. Wilkins, op cit (Note 31).

37. *Third Man*, pp. 212–213.

38. *Third Man*, p. 213.

39. *Third Man*, p. 214.

40. Wilkins, letter to Crick, 18 March 1953, in *PDH*, p. 417.

41. Maddox B. 2002. *Rosalind Franklin. The dark lady of DNA*, p. 209. Harper Collins, London; PDH, p. 418.

42. Wilkins, letter to Crick, 12 March 1953, in *PDH*, p. 418.

43. For Franklin's research notebook, see Klug A. 2004. The discovery of the DNA double helix. *J Mol Biol* **335**: 15; Elkin LO. 2003. Rosalind Franklin and the double helix. *Physics Today* **56**: 42–49.

44. Crick, personal communication.

45. Maddox, op cit (Note 41), p. 211.

46. Crick, personal communication.

47. Odile Crick, interview with the author, 20 March 1970.

48. Harold Fost to the author, enclosed with letter from his second wife, Doreen Schofield-Fost, February 2006.

49. Crick, letter to Michael Crick, 19 March 1953. Copy in author's archive.

### Chapter 10

1. Witkowski JA. 2002. Mad hatters at the DNA tea party. *Nature* **415:** 473–474.

2. Ibid. Bragg had seen the model earlier but they had worked through the night to build this 6 ft. model.

3. *DHn*, p. 118.

4. *DHn*, p. 120.

5. Hager T. 1995. *Force of nature: The life of Linus Pauling*, p. 427. Simon & Schuster, New York.

6. Ibid, p. 428.

7. Ibid, p. 427.

8. Delbrück, letter to Watson, 5 March 1953, *CaltechArchives*.

9. Robert Corey, letter to the author, 2 November 1967.

10. Confirmed by Jack Dunitz, letter to Packer, 4 November 2007.

11. Brenner S. 2001. *My life in science as told to Lewis Wolpert*, p. 26. BioMed Central Ltd., London.

12. Dunitz, op cit (Note 10).

13. John Maddox, remarks cited in *The New York Times*, 25 February 2003.

14. *GGG*, p.19.

15. Here, Crick and Watson followed the example set by the physicists Nevill Mott and Harrie Massey. Crick, letter to Asimov, 27 September 1976, *CRI/D/1/3/1*.

16. *DHn*, p. 240.

17. Kreisel, letter to Crick, 30 April 1953, *UCSD, MSS/660:5/3*.

18. Crick F. 1974. The double helix: A personal view. *Nature* **248:** 766.

19. Crick F. 1953. The molecular structure of the gene. Draft text (holograph, in pencil), *PP/CRI/H/1/12/1*, p. 5.

20. Jacob F. 1988. *The statue within. An autobiography*, p. 269. Basic Books, New York. Translated from Philip F. 1987. *La statue intérieure*. Éditions Odile Jacob.

21. Watson JD, Crick FHC. 1953. Genetical implications of the structure of deoxyribonucleic acid. *Nature* **171:** 964–965; *DHn*, p. 241.

22. Crick, op cit (Note 19).

23. Watson and Crick, op cit (Note 21), p. 966. Watson and Crick wrote, "Whether a special enzyme is required to carry out the polymerization, or whether the single helical chain already formed acts as an enzyme, remains to be seen." This was the first time that Crick suggested that a nucleic acid might itself act as an enzyme; *DHn*, p. 245.

24. *DHn*, p. 245.

25. *DHn*, p. 246.

26. Ibid.

27. *WMP*, p. 66.

28. Crick FHC, Watson JD. 1954. The complementary structure of deoxyribonucleic acid. *Proc R Soc A* **223:** 80–96; reprinted in *DHn*, pp. 282–283.

29. Pauling L. 1956. Specific hydrogen-bond formation between pyrimidines and purines in deoxyribonucleic acids. *Arch Biochem Biophys* **65:** 164–181.

30. Furberg S. 1950. An X-ray study of the stereochemistry of the nucleosides, *Acta chem Scand* **4:** 751–761; also 1949, An X-ray study of some nucleosides and nucleotides. PhD Thesis, London.

31. *DHn*, p. 291.

32. *DHn*, p. 289.

33. *GGG*, p. 30.

34. Crick, letter draft to Randall, undated (March 1953).

35. Watson's 23 March 1953 typed draft of *A structure for D.N.A.*, PP/CRI/H/11.

36. *DHn*, p. 240.

37. *DHn*, p. 241.

38. Wilkins, letter to Crick, 18 March 1953; *PDH*, p. 418.

39. Crick, seven-page holograph draft of an undated letter to Wilkins, p. 2.

40. Ibid, pp. 2–5.

41. Crick, op cit (Note 39), pp. 6–7.

42. Perutz M. 1969. Letter to the editor. *Science* **164:** 1537; *DHn*, p. 207.

43. Ibid, p. 208.

44. *DHn*, p. 240.

45. Crick, personal communication, undated (2003).

46. Crick and Watson, op cit (Note 28), p. 82; *DHn*, p. 277, footnote 1.

47. Crick, first meeting with the author to discuss the 1953 discovery, following Crick's Mendel Lecture at University College London, 19 November 1966.

48. Delbrück to Linus Pauling, text written on Delbrück's 17 April 1953 letter to Warren Weaver, *Caltech Archives, Delbrück Papers.*

49. Pauling, letter to Delbrück, 20 April 1953, *Pauling Papers, LP Science, 9.001/1.39.*

50. Hager op cit (Note 5), pp. 429 and 667.

51. Broomhead JM. 1950. The structure of pyrimidines and purines. II. A determination of the structure of adenine hydrochloride by X-ray methods. *Acta Crystallogr* **1:** 324–329; Broomhead JM. 1951. The structure of purines and pyrimidines. IV. The crystal structure of guanine hydrochloride and its relation to that of adenine hydrochloride. *Acta Crystallogr* **4:** 92–100.

52. Clews CJB, Cochran W. 1949. The structure of pyrimidines and purines. III. An X-ray investigation of hydrogen bonding in aminopyrimidines. *Acta Crystallogr* **2:** 46–57.

53. *DFL*, p. 238.

54. Calder R. Nearer the secret of life, *News Chronicle*, 15 May 1953, p. 1.

55. Form of "life unit" in cell is scanned, *The New York Times,* 16 May 1953; Clue to chemistry of heredity found, *The New York Times*, 13 May 1953, p. 17.

56. X-ray discovery, *Varsity*, 30 May 1953, p. 1.

57. Olby R. 2003. Why celebrate the golden jubilee of the double helix? *Endeavour* **27:** 80–84; Olby R. 2003. A quiet debut for the double helix. *Nature* (Special Jubilee Supplement) **421:** 402–405.

58. *DFL*, Chapter 6.

59. Crick and Watson op cit (Note 28), pp. 93–94; *DHn*, pp. 289–290.

60. Collingwood, letter to Crick, 23 May 1953, *UCSD, MSS660:2/6*.

61. Watson, letter to Kendrew, 22 January 1954, *Kendrew Papers*.

62. Crick F. 1953. Polypeptides and proteins: X-ray studies. PhD thesis, p. 1. Cambridge University, Cambridge, UK.

63. Ibid, p. 2.

64. Sanger F. 1949. The terminal peptides of insulin. *Biochem J* **45:** 563–574; Sanger, 1988. Sequences, sequences, and sequences. *Ann Rev Biochem* **57:** 1–28; Soraya de Chadarevian, 1996. Sequences, conformation, information: Biochemists and molecular biologists in the 1950s. *J Hist Biol* **29:** 361–386.

65. Ryle AP, Sanger F, Smith LF, Kitai R. 1955. The disulfide bonds of insulin. *Biochem J* **60:** 541–556.

66. Crick, op cit (Note 62), p. 145.

67. Crick, op cit (Note 62), p. 3.

68. *WMP*, p. 47.

69. Crick, op cit (Note 62), p. v.

## Chapter 11

1. Crick, interview with Horace Judson, 19 November 1975, *Eighth Day*, p. 264.

2. Clifford Mead, letter to the author, 27 November 2006. This letter turned out to be a hoax written to Crick by Jerry Donohue, Peter Pauling, and James Watson and "signed" by Linus Pauling, dated probably July 1953. Mead suggested that this letter was among those that Peter Pauling had retrieved and sold some years ago; see also *GGG*, pp. 33–34.

3. Crick, personal communication, undated (May 2004).

4. Bragg, letter to Harold Himsworth, 22 November 1956, *FD21/13*.

5. Bragg to Watson, undated, *CSHLA,W:5/38*.

6. Crick, letter to Felix Haurowitz, 13 July 1953. *Manuscripts department, Haurowitz manuscripts, Lily Library, Indiana University.* As it transpired the contract from the MRC did not arrive to Crick in Brooklyn until February 1954. It was a 7-year appointment; Crick was not yet eligible for "membership (unlimited)." At this point Francis, Odile, his son Michael, and their daughter Gabrielle had been in America for 6 months, all the while uncertain about their future. An MRC note, probably dated 25 July 1953, read "Before we can take any action . . . it will be necessary to decide how long this Unit at Cambridge is likely to go on." Crick, letter to the MRC, accepting the 7-year appointment, 4 February 1954, *FD21/13*.

7. Perutz, letter to Dorothy Hodgkin, undated (July 1950), *Bodleian Library, Hodgkin Papers, MS.Eng.c, 5707, folder H183*; see also Berol D. 2000. "Living materials and the structural ideal: The development of the protein crystallography community in the 20th century." PhD thesis, p. 165 ff, Princeton University. Ferry G. 2007. *Max*

*Perutz and the secret of life*, p. 139. Chatto & Windus, London. (This outburst from Perutz suggests that he must have felt the threat of a competitor.)

8. Crick, letter to David Harker, 21 January 1953, *PP/CRI/D/1/1/8*. Sent in response to Harker's letter of 7 January 1953.

9. Tatiana Yates, e-mail to the author, 7 October 2005, author's archive.

10. Michael Crick, e-mail to the author, 5 January 2009, author's archive.

11, Yates, op cit (Note 9).

12. Odile Crick, interview with the author, 20 March 1970.

13. *Eighth Day*, p. 264.

14. Crick, letter to Watson, 7 October 1953, *CSHLA,W:9/35*.

15. Crick, letter to Watson, undated (February 1954), *CSHLA,W:9/35* (the "+" sign is used because Odile was pregnant with their second child, Jacqueline, born in King's Lynn, UK, 12 March 1954); Crick, letter to Watson, 3 February 1954, *CSHLA,W:9/35*, in which he also said, "We live <u>very</u> quietly here mainly because we are so broke."

16. Michael Crick, e-mail to the author, 1 July 2007, author's archive.

17. Hauptman HA. 1998. David Harker, *Biographical memoirs of Fellows of the National Academy of Sciences*, pp. 127 and 139; Tulinsky A. 1999. The protein structure project, 1950–1959: First effort of a protein structure determination in the U.S. *Rigaku J* **16**: 8–15; http://www.nap.edu/readingroom/books/biomems/dharker.html.

18. Victor Luzzati, e-mail to the author, 30 August 2004; http://www.psc.edu/science/Hauptman/Hauptman.html.

19. Bragg WL. 1962. The growing power of X-ray analysis. In *Fifty years of X-ray diffraction* (ed. P.P. Ewald), p. 132. The International Union of Crystallography, Utrecht.

20. Crick, letter to Kendrew, undated (December 1953), pp. 3–5, *Kendrew Papers; NCUACS78.7.98/R.85*.

21. Ibid, p. 4.

22. Luzzati, e-mail to the author, 30 August 2004, p. 2.

23. Crick, op cit (Note 20), p. 6. One month earlier, Kendrew had reported that "Francis is trying to make up his mind whether to go to Pauling in 1954/5 but my feeling is that he will decide <u>not</u> to do so and to return here in September." Kendrew, letter to Watson, 8 November 1953, *CSHLA,W:23/2*.

24. Magdoff BS, Crick FHC. 1955. Ribonuclease II. Accuracy of measurement and shrinkage. *Acta Crystallogr* **8**: 464–465.

25. Magdoff BS, Crick FHC, Luzzati V. 1956. The three-dimensional Patterson function of ribonuclease II. *Acta Crystallogr* **9**: 161.

26. Crick, letter to Watson, 7 March 1954, *CSHLA,W:8/35*.

27. Crick, letter to Kendrew, 8 February 1954, *Kendrew Papers*.

28. Yates, e-mail to the author, 7 October 2005.

29. Luzzati, e-mail to the author, 30 August 2004.

30. Crick, letter to Kendrew, 18 May 1956, *Kendrew Papers*.

31. Louise Johnson, e-mail to Gutfreund, 9 January 2007.

32. Tulinsky, op cit (Note 17), p. 13.

33. Odile Crick, telephone conversations with the author, 11 October 2004.

34. Crick, telephone conversation with the author, 16 June 2003.

35. Crick F. 1954. The structure of the synthetic α-polypeptides. *Sci Prog* **166:** 217.

36. Perutz, letter fragment to Hodgkin, undated (1950) *Bodleian Library, Hodgkin Papers, MS.Eng.c, 5707, folder H183;* David Berol, 2000. "Living materials and the structural ideal: The development of the protein crystallography community in the 20th century." PhD thesis, Princeton University, p. 165 ff.

37. Wrinch D. 1937. The cyclol hypothesis and the "globular" proteins. *Proc R Soc A* **161:** 505–524; Wrinch D. 1953. The hypothesis of parallel rod-like chains in horse haemoglobin. *Acta Crystallogr* **6:** 638–646.

38. Harker D. 1957. Contribution to discussion following Barbara Low's paper entitled Importance of helices in molecules of biological origin. In *A symposium on the chemical basis of heredity,* p. 575. Johns Hopkins University Press, Baltimore.

39. Todd AR. 1954. Chemical structure of the nucleic acids. *Proc Natl Acad Sci* **40:** 749.

40. Crick FHC. 1954. The complementary structure of DNA. *Proc Natl Acad Sci* **40:** 756–757.

41. Luzzati, personal communication, 19 September 2003.

42. Luzzati, e-mail to the author, 30 August 2004.

43. Crick and Luzzati, personal communications.

44. Crick, letter to Himsworth, 15 June 1953, *FD21/13.*

45. Luzzati, op cit (Note 42).

46. Crick, op cit (Note 27), p. 10.

47. Crick FHC. 1953. General discussion. In *Nature and structure of collagen* (ed. J.T. Randall and S.F. Jackson), p. 249. Butterworths, London; Academic Press, New York; Cohen C, Bar R. 1953. Helical polypeptide chain configuration in collagen. *Nature* **175:** 2783–2784.

48. John Randall, Notes on Crick's structure for collagen, 4 January 1954, *Churchill College Archives, Randall Papers, RNDL2/1/3.*

49. Crick, letter to Watson, 3 February 1954, *CSHLA,W:9/35.*

50. Kendrew, letter to Watson, 23 January 1954, *CSHLA,W:23/2.*

51. Crick, letter to Kendrew, undated (December 1953), *Kendrew Papers.*

52. Tony North, letter to the author, 13 May 2005.

53. Linus Pauling, letter to Peter Pauling, 14 January 1954, *OSUSC/PP.*

54. Rich A, Crick F. 1955. Structure of polyglycine II. *Nature* **176:** 780–781.

55. Ramachandran GN, Kartha G. 1955. Structure of collagen. *Nature* **176:** 593–595.

56. Rich A, Crick F. 1955. The structure of collagen. *Nature* **176:** 915–916.

57. Cowan PM, McGavin S. 1955. Structure of poly-L-proline. *Nature* **176:** 501–503. From receipt to publication, this paper waited for 3 months; Crick and Rich's paper waited for only 1 month.

58. *WMP,* p. 67. But Crick goes on to explain why collagen is very important from a medical aspect.

59. North, op cit (Note 52).

60. Gutfreund, e-mail to the author, 7 January 2007.

61. Crick, report on the year in Brooklyn, 30 September 1954, *PP/CRI/E/1/2/2.*

## Chapter 12

1. Crick, 1955. On degenerate templates and the adapter hypothesis. Circulated to members of the RNA Tie Club, January, *DTAH, PP/CRI/H/I/38*, p. 4.

2. Luzzati, personal communication, undated (April 2003).

3. *WMP*, p. 91.

4. George Gamow, letter to Watson and Crick, 8 July 1953, *GGG*, p. 262.

5. Gamow G. 1953. *Mr. Tompkins learns the facts of life*, p. 30. Cambridge University Press, Cambridge.

6. Ibid, p. 264.

7. *WMP*, p. 91.

8. Gamow G. 1954. Possible relations between deoxyribnucleic acid and protein structures. *Nature* **173**: 318.

9. Crick, letter to Watson, 3 February 1954, *CSHLA,W:9/35*.

10. Crick, letter to Watson, 22 November 1953, *CSHLA,W:9/53*.

11. Crick, letter to Watson, 4 December 1953, *CSHLA,W:9/35*.

12. Crick FHC. 1958. Protein synthesis. *Symp Soc Exp Biol* **12**: 140.

13. *WMP*, p. 94.

14. Alex Rich, source not located. Note similarity of his account to Gamow's in note 15.

15. Gamow, letter to Crick, 8 March 1954, *PDH*, p. 431.

16. Crosbie GW, Smellie RMS, Davidson JN. 1953. Phosphorus compounds in the cell. 5. The composition of the cytoplasmic and nuclear ribonucleic acids of the liver cell. *Biochem J* **54**: 287–291.

17. Watson, letter to Crick, 13 February 1954, *PP/CRI/H/1/42/3*.

18. Crick, letter to Watson, 21 February 1954, *CSHLA,W:9/35*.

19. Ibid.

20. Watson, letter to Crick, 28 March 1954, *PP/CRI/D/2/45/88*.

21. Watson, op cit (Note 17), p. 7.

22. For an excellent account of the history of this approach, see Kay L. 2000. *Who wrote the book of life? A history of the genetic code.* Stanford University Press, Stanford.

23. *DTAH*, p. 4.

24. Ibid.

25. *DTAH*, p. 7.

26. *DTAH*, p. 7–8.

27. *DTAH*, p. 6.

28. *DTAH*, p. 9.

29. *DTAH*, p. 17.

30. Watson, letter to Crick, 10 October 1955, *PP/CRI/D/2/45*. Sidney Brenner recalled that Crick's adaptor molecule idea "was just pooh-poohed by everybody" at that time. See McElheny VK. 2003. *Watson and DNA. Making a scientific revolution*, p. 89. Perseus Publishers, Cambridge, Massachusetts.

31. Crick FHC, Griffith JS, Orgel LE 1956. Commaless codes. A note for the RNA Tie Club, p.1, *UCSD, MSS600:A1/1454; PP/CRI/H/2/3*.

32. *WMP*, p. 99.

33. Crick FHC, Griffith JS, Orgel LE 1957. Codes without commas. *Proc Natl Acad Sci* **43:** 417, Fig. 1.

34. Ibid, pp. 417–418.

35. Ibid, pp. 416 and 418.

36. Ibid, p. 420.

37. Delbrück, letter to Rich, 9 November 1955, *PDH*, p. 431.

38. Rich A. 2003. Does RNA form a double helix? In *Inspiring science. Jim Watson and the age of DNA* (ed. J. Inglis, et al.), p. 155. Cold Spring Harbor Laboratory Press, Cold Spring Harbor, New York.

39. Watson JD. 1954. The structure of tobacco mosaic virus. *Biochem Biophys Acta* **13:** 10–19.

40. Crick FH, Watson JD. 1956. Structure of small viruses. *Nature* **177:** 475. For insightful analysis of the changing role of a plant virus as a model organism, see A. Creager, 2002. *The life of a virus. Tobacco mosaic virus as an experimental model, 1930–1965.* University of Chicago Press, Chicago and London; Morgan GJ. 2003. Historical review: Viruses, crystals, and geodesic domes. *Trends Biochem Sci* **28:** 86–90.

## Chapter 13

1. Crick, letter to Cy Levinthal, 1 February 1957, *PP/CRI/D/1/12.*

2. Fraenkel-Conrat H. 1956. The role of the nucleic acid in the reconstitution of active tobacco mosaic virus, *J Am Chem Soc* **78:** 882–883.

3. Crick, More about the chemistry of heredity, *PP/CRI/H/2/8,* p. 1.

4. As Franklin explained at the Ciba Foundation meeting, the RNA is embedded in the protein toward the hollow center but the whole particle is in fact a hollow cylinder. However, one refers loosely to the core composed of RNA and to the coat of protein.

5. Crick, op cit (Note 3), p. 6.

6. Crick, letter to Sydney Brenner, 4 November 1955, *CSHLA,B.*

7. Crick, letter to Watson, 13 April 1955, *CSHLA,W:9/35.*

8. Crick, in the discussion following Norman Pirie's paper "Material in virus preparations not necessary for the manifestation of characteristic virus properties," *CIBA*, p. 68.

9. Crick and Watson, "Virus structure: General principles," *CIBA*, p. 5.

10. Ibid, p. 11.

11. Chargaff E. 1957. Base composition of deoxypentose and pentose nucleic acids in various species. In *McElroy/Glass*, p. 521.

12. *DTAH*, p. 17.

13. Chargaff E. 1957, Discussion. In *McElroy/Glass*, p. 528.

14. Delbrück M, Stent G. 1957. On the mechanism of DNA replication. In *McElroy/ Glass*, pp. 699–736.

15. Crick, Discussion. In *McElroy/Glass*, pp. 746–747.

16. Kornberg A. 1989. *For the love of enzymes. The odyssey of a biochemist*, p. 121. "The significance of the double helix did not intrude into my work until 1956. . . ." Harvard University Press, Cambridge, Massachusetts.

17. Kornberg A. 1957. Pathways of enzymatic synthesis of nucleotides and polynucleotides. In *McElroy/Glass,* pp. 579–608.

18. Kornberg, personal communication, 7 February 2003, Bilbao, Spain.

19. Spiegelman S. 1957 Nucleic acids and the synthesis of proteins. In *McElroy/Glass,* p. 232. The author's italics. (Spiegelman used the term "polyenzyme" instead of the more common term "multienzyme" model [p. 233].)

20. Ibid, p. 265.

21. Matthews EF. 1957. Discussion. In *McElroy/Glass*, p. 527.

22. Volkin E, Astrachan L. RNA metabolism in T2-infected *Escherichia coli*. In *McElroy/Glass*, pp. 686–694.

23. Crick, letter to Brenner, 17 July 1956, *CSHL,B.*

24. Crick, "Ann Arbor visit," *PP/CRI/H/2/6.*

25. *WMP*, p. 108. (The "earlier" is probably a reference to July 1956.)

26. Crick, Ideas on protein synthesis (Oct. 1956), *PP/CRI/H//2/6.*

27. Spiegelman, op cit (Note 19), p. 237.

28. Glass B. 1957. A summary of the Symposium on the Chemical Basis of Heredity. In *McElroy/Glass,* p. 780.

29. Crick, op cit (Note 26).

30. Crick, op cit (Note 1).

31. *Eighth Day*, p. 339.

32. *SEB58,* p. 138.

33. The first symposium was held in Cambridge. Participants had to "bring their own towels, soap and ration books"—a sign of postwar shortages. *PP/FGS, Box C28.*

34. *SEB58,* pp. 138–139.

35. *SEB58,* p. 139.

36. *SEB58,* p. 143.

37. Bretscher, interview with the author, 8 March 2004.

38. *SEB58,* p. 144.

39. Ibid.

40. Ibid.

41. Dounce AL. 1952. Duplicating mechanism for peptide chain and nucleic acid synthesis, *Enzymologia* **15:** 251–258.

42. *SEB58,* p. 152.

43. Ibid.

44. Benzer S. 1957. *McElroy/Glass*, p. 93; see also Holmes F. 2006. *Reconceiving the gene. Seymour Benzer's adventures in phage genetics.* Yale University Press, New Haven, Connecticut and London.

45. *WMP*, p. 108.

46. *SEB58*, p.153.

47. *WMP*, p. 109.

48. 1983. *Chambers 20th century dictionary,* p. 369. W. & R. Chamber Ltd, Edinburgh; 1983. *Webster's encyclopedia unabridged*, p. 423. Gramercy, New York; 1952. *Nouveau petit Larousse illustré dictionnaire encyclopédique*, p. 313. Librairie Larousse, Paris.

49. *Eighth Day*, p. 337.

50. Cyril Darlington, personal communication, undated (1970).

51. *WMP*, p. 110.

52. Klug, interview with the author, 9 March 2004.

53. Jacob F. 1988. *The statue within. An autobiography*, pp. 287–288. Basic Books, New York. Originally published in 1987 as *La statue intérieure*. Éditions Odile Jacob, Paris.

54. Brenner, interview with the author, 9 March 2006.

55. Horace Barlow, letter to Crick, 17 April 1956, *UCSD, MSS660:4/42*.

56. An *Extraordinary Fellowship* carries no teaching or administrative duties.

57. Brenner, op cit (Note 54).

58. Michael Ashburner, letter to the author, 12 September 2007; Ridley M. 2006. *Francis Crick, discoverer of the genetic code*, p. 110, Harper Collins, New York.

59. Darlington CD. *Recent advances in cytology,* 2nd ed., p. 549. J. & A. Churchill, London; see also *PDH*, p. 449.

60. Sir James Gray, Cambridge University's Professor of Zoology (1937–1954).

61. Guido Pontecorvo, letter to Crick, 8 October 1958, *UCSD, MSS660:8/33*.

62. Crick, personal communication.

63. Sanger, interview with the author, 10 March 2004.

64. Kendrew, letter to Watson, 6 June 1954, *CSHLA,W:23/2*.

65. *DFL*, p. 209 ff.

66. Crick, personal communication.

## Chapter 14

1. Crick FHC. 2002. Cambridge and the code, *The Scientist* **16:** 18.

2. Crick, letter to Brenner, 20 October 1955, *CSHLA,B*; cited in *DFL*, p. 191.

3. *WMP*, p. 103.

4. *WMP*, p. 105.

5. Crick, Experimental Notes headed "Biochemical Work on Lysozyme," *PP/CRI/H/1/1*.

6. Crick, letter to Brenner, 6 July 1955, *CSHLA,B*: also in *Eighth Day*, p. 309.

7. *WMP,* p. 105.

8. Crick, op cit (Note 6), p. 308.

9. Pauling L, Itano HA, Singer SJ, Wells IC. 1949. Sickle cell anemia, a molecular disease. *Science* **110:** 543–548.

10. *WMP*, p. 105.

11. *WMP*, p. 103; Beadle GW. 1945. Genetics and metabolism in *Neurospora Physiol Rev* **25:** 660; *PDH*, p. 148.

12. Hoagland MB, Zamecnik PC, Stephenson ML. 1957. Intermediate reactions in protein biosynthesis. *Biochem Biophys Acta* **24:** 215–216.

13. Hoagland MB. 1996. Biochemistry or molecular biology? The discovery of "soluble RNA" *Trends Biochem Sci* **21:** 79; reprinted in Jan Witkowski (ed.), 2005. *The inside story: DNA to RNA to protein. Readings from* Trends in Biochemical Sciences, p. 219. Cold Spring Harbor Laboratory Press, Cold Spring Harbor, New York.

14. Mahlon Hoagland, letter to the author, 1 April 1970; cited in *Daedalus*, p. 256.

15. Hoagland M. 1990. *Toward the habit of truth: A life in science*, p. 104. Norton, New York and London.

16. Ibid, p. 116.

17. Crick, letter to Watson, 14 November 1957, *PP/CRI/D/2/45*.

18. Brenner, letter to Joseph Needham, 23 February 1949, *CSHLA,B*.

19. Brenner, letter to Crick, undated (December 1954), *CSHLA,B*.

20. Brenner, interview with the author, 19 March 1970; also *Daedalus,* p. 265.

21. *My Life,* p. 63.

22. Bretscher, interview with the author, 8 March 2004.

23. Crick, op cit (Note 1), pp. 18–19.

24. *WMP,* p. 128.

25. *My Life,* p. 180.

26. Crick, op cit (Note 1), p. 19.

27. Crick, letter to Brenner, 21 August 1956, *CSHLA,B.*

28. Seymour Benzer's term for a gene defined in functional terms by the *cis–trans* test. The introduction of this term signified the subdivision of the classical gene.

29. Benzer S. Adventures in the rII region, *Phage Origins*, p. 162.

30. For an excellent account of the development of the program for phage research in Cambridge, see *DFL,* p. 195 ff. For further information on the phage experiments, see Brock TD. 1990. *The emergence of bacterial genetics.* Cold Spring Harbor Laboratory Press, Cold Spring Harbor, New York., p. 137 ff.

31. Benzer, *Caltech Archives: Oral histories on line;* transcript of interview 19 October 1990, p. 60.

32. Ibid, p. 62.

33. *WMP*, p. 127.

34. Benzer, letter to Delbrück, 3 February 1955, *CaltechArchives*.

35. Brenner S, Benzer S, Barnett L. 1958. Distribution of proflavin-induced mutations in the genetic fine structure, *Nature* **182**: 984.

36. Brenner S, Barnett L. 1959. Genetic and chemical studies on the head protein of bacteriophages T2 and T4, in *Structure and function of genetic elements, Report of Symposium held June 1–3, 1959*, pp. 86–93. Brookhaven National Laboratory, Upton, New York.

37. Crick, letter to Jacques Fresco, 7 January 1960, *Daedalus,* p. 261; *My Life,* p. 92.

38. *WMP,* p. 125.

39. Crick, interview with the author, 8 March 1968.

40. Lerman LS. 1961. Structural considerations in the interaction of DNA and acridines, *J Mol Biol* **3**: 18–30.

41. Brenner S, Barnett L, Crick FHC, Orgel A. 1961. The theory of mutagenesis, *J Mol Biol* **3**: 123 [submitted 16 December 1960].

42. *WMP,* p. 99.

43. *WMP,* p. 128.

44. *Gen Code,* p. 1229.

45. Crick, laboratory notes, cited in *Eighth Day*, p. 442.

46. Crick, letter to Delbrück, 19 June 1961, *CaltechArchives*, fldr.41.

47. Latarjet R. 1961. Introduction, and Deoxyribonucleic acid. Structure, synthesis and function, *Proceedings of the 11th Annual Reunion of the Société de Chimie Physique*, p. xi.

48. Crick, 1961. Contribution to discussion, in *Proceedings of the 11th Annual Reunion of the Société de Chimie Physique*, p. 188.

49. *WMP*, p. 129.

50. *WMP*, p. 130; *Eighth Day*, pp. 463–464.

51. Barnett L, Brenner S, Crick FHC, Shulman RG, Watts-Tobin RJ. 1967. Phase-shift and other mutants in the first part of the rIIB cistron of bacteriophage T4, *Philos Trans R Soc London B* **252**: 509 ff; *My Life*, p. 97.

52. *Gen Code*, p. 1229.

53. *WMP*, p. 133.

54. Ibid.

55. Crick, letter to the American Biophysical Society, 9 October 1961, *PP/CRI/E/1/10/4/3*.

56. *Gen Code*, p. 1230.

57. Crick, laboratory notes; see Mark Bretscher, Manuscripts and Correspondence, on the Web at www2.mrc-lmb.cam.ac.uk/archive/manu-corr-bretscher.html.

58. *Gen Code*, p. 1227.

59. *Gen Code*, p. 1230.

60. *Gen Code*, p. 1231.

61. *Gen Code*, p, 1232.

62. Ibid.

63. Crick FH. 1962. The genetic code. *Sci Amer* **207**: 66–74.

64. Crick, letter to Roger Jones about the BBC recording, 18 December 1975; Crick, "The rII System and the Genetic Code," *S299 Radio Programme 5*, Radiovision, Notes & Diagrams as a Basis for the Talk, p. 1, *UCSD, MSS600:1/4*.

65. *My Life*, p. 97.

66. Crick, letter to Raymond Latarjet, 16 November 1961, *PP/CRI/E/1/9/5*.

67. Crick, letter to Marshall Nirenberg, 16 November 1961, *PP/CRI/D/1/1/14*.

68. Margerison T. Scientists have cracked the code of life, *The Sunday Times,* 31 December 1961, p. 8.

69. Davy J. A major advance, *The Observer*, 31 December 1961, p. 4.

70. Crick, letter to Nirenberg, 4 January 1962, *CRI/E/1/10/4/2*.

71. 1962. Advances in biology spur hopes of solving heredity's chemical secrets this year, *New York Times,* February 2, p. 14.

72. Crick, op cit (Note 63), pp. 66–74.

73. Crick, letter to the Editor of *Scientific American,* 3 September 1962, *PP/CRI/H/3/9*; see also Chris Beckett, 2004. For the record: The Francis Crick Archive at the Wellcome Library, *Medical History* **48**: 245–260. (Nirenberg's paper was published two issues after Crick's.)

74. Op cit (Note 71).

75. Barnett et al., op cit (Note 51), p. 501.

76. Brenner S, Barnett L, Katz ER, Crick FHC. 1967. UGA: A third nonsense triplet in the genetic code. *Nature* **213:** 449–450; *WMP,* p. 135.

77. Sarabhai AS, Stretton AOW, Brenner S, Bolle A. 1964. Colinearity of the gene with the polypeptide chain. *Nature* **201:** 13–17.

78. Okada Y, Terzaghi E, Streisinger G, Emrich J, Inouye M, Tsugita A. 1966. A frame shift mutation involving the addition of two base pairs in the lysozyme gene of phage T4. *Proc Natl Acad Sci* **56:** 1692–1698.

79. *WMP,* p. 135.

80. Finch J. 2008. *A Nobel Fellow on every floor. A history of the Medical Council Laboratory of Molecular Biology,* p. 179. MRC Laboratory of Molecular Biology, Cambridge.

81. Eric Miller, letter to Audrey Glauert, 13 February 2002, author's archive.

## Chapter 15

1. Crick FH. 1967. The genetic code. The Croonian Lecture 1966, *Proc R Soc B* **167:** 346.

2. Pardee AB, Jacob F, Monod J. 1959. The genetic control of cytoplasmic expression of "inducibility" in the synthesis of β-galactosidase by *E. coli, J Mol Biol* **1:** 165–178.

3. Riley M, Pardee AB, Monod J, Jacob F. 1960. On the expression of a structural gene, *J Mol Biol* **2:** 216–225.

4. Crick, interview with the author, 8 March 1968. (For Volkin–Astrachan, see Chapter 13.)

5. Jacob F. 1988. *The statue within. An autobiography,* p. 320. Basic Books, New York (translated from *La statue intérieure* [1987]).

6. Crick, What are the properties of genetic RNA? A note for the RNA Tie Club, *CSHLA,B.* Cited in *Eighth Day,* p. 420.

7. *Eighth Day,* p. 422.

8. Crick, interview with the author, 8 March 1968.

9. Brenner S, Jacob F, Meselson M. 1961. An unstable intermediate carrying information from genes to ribosomes for protein synthesis, *J Mol Biol* **3:** 576–581.

10. Gros F, Glibert W, Hiatt H, Kurland C, Risebrough RW, Watson JD. 1961. Unstable ribonucleic acid revealed by pulse labeling of *Escherichia coli. Nature* **190:** 581–585.

11. Harris H. 1987. *The Balance of Improbabilities. A Scientific Life,* p. 148. Oxford University Press, Oxford; Brenner, interview with the author, July 2007.

12. Harris H. 2005. An early heresy in the fifties, *Trends Biochem Sci* **19:** 303–305. Reprinted in *The Inside Story. DNA to RNA to Protein* (ed. Witkowski), pp. 257–258. Cold Spring Harbor Laboratory Press, Cold Spring Harbor, New York; Sir Henry Harris, letter to the author, 11 May 2009, author's archive.

13. Crick, DNA—The molecule of heredity, broadcast 29 November 1955, BBC, European Service, General News Talk, *The Frontiers of Knowledge; UCSD, MSS600, 531 4A,* p. 1.

14. Ibid, p. 6.

15. Crick, Cracking the genetic code, broadcast 22 January 1962, BBC, European Service, *The Frontiers of Knowledge,* No. 256, *PP/CRI/H/3/8.*

16. John Maddox, 1980. One of the men who changed the course of biology, *Listener,* 10 July 1980, 2669: 43.

17. Ibid.

18. John Inglis, telephone conversation with the author, 13 July 2007; Inglis, letter to the author, 2 March 2009, author's archive.

19. Watson, letter to Crick, 6 July 1990, *CSHLA,W:9/35.*

20. Originally a wartime radio program, *The Brains Trust* was moved to television in the 1950s. An American version was aired from WTTW Channel 11 in Chicago with great success.

21. Crick, personal communication.

22. Crick FHC. 1963. On the genetic code: Deductions about the general nature of the code are drawn from results of biochemical experimentation. *Science* **139**: 462.

23. Ibid, p. 463.

24. Ibid, p. 461.

25. Ibid, p. 464.

26. *CSHSQB66,* p. 3.

27. H. Gobind Khorana, letter to Crick, 17 November 1965, and Crick, letter to Khorana, 22 November 1965, *PP/CRI/E/1/13/19.*

28. The first Croonian lecture had been delivered in 1738. It is considered the most prestigious of the Society's lectures for the biological sciences.

29. Streisinger G, Okada Y, Emrich J, Newton J, Tsugita A, Terzaghi E, Inouye M. 1966. Frameshift mutations and the genetic code, *Cold Spring Harbor Symp Quant Biol* **31**: 77–84.

30. The opal chain-stopping mutant was found by Hilliard Bernstein. In naming it after him the English translation was used—*Bernstein* means amber. When a second mutant was discovered Brenner called it "opal." See *My Life,* p. 102.

31. Marcker KA, Sanger F. 1964. *N*-Formyl-methionyl-sRNA, *J Mol Biol* **8**: 835–840.

32. Mark Bretscher, 1965. "Studies on the genetic code and protein synthesis," PhD thesis, Cambridge University; Bretscher MS, Marcker KA. 1966. Polypeptidyl-σ-ribonucleic acid and amino-acyl-σ-ribonucleic acid binding sites on ribosomes, *Nature* **211**: 380–384.

33. Sarabhai AS, Stretton AOW, Brenner S, Bolle A. 1964. Co-linearity of the gene with the polypeptide chain, *Nature* **201**: 13–17; Brenner S, Stretton AOW, Kaplan S. 1965. Genetic code: The "nonsense" triplets for chain termination and their suppression, *Nature* **206**: 994–998.

34. Crick, op cit (Note 1).

35. Ibid.

36. Caldwell PC, Hinshelwood C. 1950. Some considerations on autosynthesis in bacteria, *J Chem Soc* **4**: 3156–3159.

37. Dounce AL. 1952. Duplicating mechanism for peptide chain and nucleic acid synthesis, *Enzymologia* **15**: 251–258.

38. Gamow G. 1954. Possible relations between deoxyribonucleic acid and protein structure, *Nature* **173**: 318.

39. *CSHSQB66,* p. 6.

40. *CSHSQB66,* p. 7.

41. *CSHSQB66,* p. 9.

42. A description of this event opens Matthew Ridley's very compact biography of Crick in the Eminent Lives Series: Ridley M. 2006. *Francis Crick: Discoverer of the genetic code.* Harper Collins, New York.

43. Crick, letter to Watson, 24 July 1974, *CSHLA,W:9/38.*

44. Recall that in RNA, uracil takes the place of thymine.

45. Holley RW, Apgar J, Everett GA, Madison JT, Marquisee M, Merrill SH, Penswick JR, Zamir A. 1965. Structure of a ribonucleic acid, *Science* **147:** 1462–1465.

46. Crick F. 1966. Codon–anticodon pairing: The Wobble Hypothesis, *J Mol Biol* **19:** 548–555.

47. Rose Feiner, Wobble poem, *PP/CRI/D/1/2/1.*

48. Woese CR. 1965. The evolution of the genetic code, *Proc Natl Acad Sci* **54:** 1546–1552; Woese, 1967. *The genetic code.* Harper and Row, New York.

49. Woese CR, Dugre DH, Dugre SA, Kondo M, Saxinger WC. 1966. On the fundamental nature and evolution of the genetic code, *Cold Spring Harbor Symp Quant Biol* **31:** 723–736.

50. Pelc SR, Welton MGE.1966. Stereochemical relationship between coding triplets and amino-acids, *Nature* **209:** 868–872.

51. Crick FH. 1967. An error in model building, *Nature* **213:** 798.

52. Crick, letter to Wilkins, undated (1967).

53. Woese et al., op cit (Note 49), p. 726.

54. 1966. Editorial, *Nature* **212:** 1397; Woose et al., op cit (Note 49), pp. 732–733.

55. Crick FH. 1967. The origin of the genetic code, *Nature* **213:** 119. For Crick's lecture notes for this presentation, see *PP/CRI/E/1/14/16,* manuscript dated December 1966.

56. Ibid.

57. Crick FH. 1968. The origin of the genetic code, *J Mol Biol* **38:** 377.

58. Ibid, pp. 367–379; Orgel LE. 1968. Evolution of the genetic apparatus, *J Mol Biol* **38:** 381–393.

59. Ibid, p. 378.

60. Ibid, p. 371; *CSHSQB66,* p. 7.

61. Ibid, p. 372.

62. Rodley GA, Scobie RS, Bates RH, Lewitt RM. 1976. A possible conformation for double-stranded polynucleotides, *Proc Natl Acad Sci* **73:** 2959–2964.

63. Sasisekharan V, Pattabiraman N. 1978. Double stranded polynucleotides: Two typical alternative conformations for nucleic acids, *Curr Sci* **45:** 779–783.

64. Pohl WF, Roberts GW. 1978. Topological considerations in the theory of replication of DNA, *J Math Biol* **6:** 383–402.

65. *WMP,* p. 72.

66. Crick FH, Wang JC, Bauer WR, 1979. Is DNA really a double helix?, *J Mol Biol* **129:** 456.

67. Hoogsteen K. 1959. The structure of crystals containing a hydrogen-bonded complex of 1–methylthymine and 9-methyladenine, *Acta Crystallogr* **12:** 822–823. (Hoogsteen base-pairing has since been found in certain cellular DNAs.)

68. Rosenberg JM, Seeman NC, Kim JJP, Suddath FL, Nicholas HB, Rich A. 1973. Double helix at atomic resolution, *Nature* **243:** 150–154.

69. Wang AH-J, Quigley GJ, Kolpak FJ, Crawford JL, van Boom JH, van der Marel G, Rich A. 1979. Molecular structure of a left-handed double helical DNA fragment at atomic resolution, *Nature* **282:** 680–686.

70. *WMP*, p. 73.

71. *WMP*, pp. 73–74.

72. Commoner B. 1961. In defence of biology, *Science* **133:** 1748.

73. Commoner B. 1968. Failure of the Watson–Crick theory as a chemical explanation of inheritance, *Nature* **220:** 340.

74. Commoner B. 2002. Unravelling the DNA myth. The spurious foundation of genetic engineering, *Harper's Magazine,* February 2002, pp. 39–47.

75. Barry Commoner, letter to Crick, 9 January 2002, *UCSD, MSS600:91:C2002.*

76. Crick, letter to Commoner, 10 January 2002, *UCSD, MSS600:91:C2002.*

77. Crick, personal communication.

78. 1970. News and Views: Central dogma reversed, *Nature* **226:** 1198–1199.

79. Crick F. 1970. Central Dogma of molecular biology, *Nature* **227:** 561.

80. Crick, letter to the author, 22 August 1973, author's archive.

81. Crick, op cit (Note 79), p. 563. (For the Rous sarcoma virus, see H.M. Temin, 1964. Homology between RNA from Rous sarcoma virous and DNA from Rous sarcoma virus-infected cells, *Proc Natl Acad Sci* **52:** 328.)

82. Editorial, 1970. Two ways to protein, *Lancet* **2:** 31.

83. Horst Malke, letter to Crick, 29 September 1970, *CSHLA,B:9/38*; Watson, 1965. *Molecular biology of the gene,* p. 298. W.A. Benjamin, New York.

84. Crick, letter to Malke, 20 October 1970, *CSHLA,W:9/38.*

85. Crick, letter to Watson, 20 October 1970, *CSHLA,W:9/38.*

86. Crick, letter to Temin, 3 August 1970, *PP/CRI/D/1/2/14.*

87. Crick, letter to Donald Kennedy (Editor-in-Chief), 11 April 2003, *UCSD, MSS600:109.*

## Chapter 16

1. *Molecules*, p. 99.

2. Crick, letter to Himsworth, 8 September 1961, FD21/13.

3. *Varsity*, 17 May 1958.

4. Mark Goldie, 2004. The chapel affair, *Churchill College Magazine,* p. 31.

5. Ibid.

6. Ibid, pp. 35–36.

7. Crick, letter to Sir Winston Churchill, rough draft, 19 August 1961 (presumably not sent for several weeks), *MSS660/2/1.*

8. Churchill, letter to Crick, 9 September 1961; Goldie, op cit (Note 4), p. 28; *CCAR800/2:106/13.*

9. Crick, letter to Churchill, 12 October 1961, in Goldie, op cit (Note 4), p. 28. Hetaira is a woman employed in public or private entertainment.

10. Collingwood, letter to Crick, 11 October 1961, *UCSD, MSS660:2/6.*

11. Richard Keynes, letter to Crick, 22 August 1961, *UCSD, MSS660:2/1.*

12. Crick, letter to Sir John Cockcroft, undated, *UCSD, MSS660:2/1*.

13. Stockwood M. 1982. *Chanctonbury Ring: An autobiography*, p. 89. Hodder and Stoughton, London.

14. *WMP*, p. 10.

15. *WMP*, p. 12.

16. Ibid.

17. Goldie, op cit (Note 4), p. 30.

18. The first graduate students arrived in 1960; the first undergraduate followed in 1961.

19. Crick, interview with the author, 9 January 1991.

20. Crick, letter to W.H. Thorpe, undated, *PP/CRI/D/1/19*.

21. British Humanist Association, Gower Street, London.

22. *Saeva* means fierce.

22. Victor Purcell, Dr. Crick and humanism, *Varsity*, 1972.

24. Report in *Varsity*, 1963; *UCSD, MSS660/1/24*. The winner of the prize was P.J. Lewis, with his descriptions of the individual College chapels and their history plus a detailed survey of their suitability for conversion to secular use, 1964, *UCSD, MSS660/1/25*.

25. Kendrew, letter to Watson, undated, *CSHLA,W:9/41*.

26. Stockwood, op cit (Note 13).

27. Nevill Mott in: Stockwood M. (ed.). 1959. Religion and the scientists, p. 11. Student Christian Movement.

28. Fred Hoyle. In Stockwood, op cit (Note 26), p. 64.

29. Montefiore H. 1988. *Communicating the gospel in a scientific age*, p. 5. Saint Andrews Press, St. Andrews, UK. Montefiore once wrote to Crick following a discussion with him at Gonville and Caius College on telepathy: "Brain molecules give off a kind of template which can be picked up by other people who have brain protein molecules of a similar kind [in a different code from that in which memory is stored]," Montefiore, letter to Crick, 25 November 1959, *UCSD, MSS660:2*.

30. Lederberg J. 1963. Biological future of man, in *Man and his Nature. A Ciba Foundation Volume* (ed. Gordon Wolstenholme), p. 266. J. & A. Churchill, London.

31. Crick, Discussion, op cit (Note 30), p. 274–275.

32. Crick, op cit (Note 30), p. 276.

33. Jacob Bronowski, op cit (Note 30), p. 285.

34. Crick, Discussion, op cit (Note 30), pp. 294–295.

35. Peter Medawar, Discussion, op cit (Note 30), pp. 295–296.

36. Elsasser W. 1958. *The physical foundation of biology*. Pergamon Press, London.

37. Wigner E. 1961. In *The logic of personal knowledge: Essays presented to Michael Polanyi on his Seventieth Birthday 11 March 1961* (ed. Edward Shils). Free Press, Glencoe, IL.

38. *Molecules*, p. xii.

39. *Molecules,* p. 16.

40. *Molecules*, p. 85.

41. *Molecules*, p. 86; Montefiore, op cit (Note 29).

42. *Molecules*, p. 96.

43. *Molecules*, p. 99.

44. Crick, letter to Organizers of the John Danz Lectures, 14 December 1965, *PP/CRI/E/14/5*.

45. Sir John Eccles, 1967. Review of *Of molecules and men*, Zygon, **2:** 283; *PP/CRI/E/14/5*.

46. James Murray, 1967. Review of *Of molecules and men*, *Virginia Q* **42:** 514; *PP/CRI/E/14/5*.

47. Ibid, p. 517.

48. C.H. Waddington, letter to Crick, 14 October 1967, *PP/CRI/D/1/120*.

49. Crick, letter to Waddington, 9 November 1967, *PP/CRI/D/1/120*, p. 1.

50. Noel Annan, letter to Crick, 14 December 1967, *PP/CRI/E/1/16/13/2*.

51. Crick, letter to Annan, October 1, 1968. *PP/CRI/E/1/16/13/2*.

52. For a purely factual report, see 1968. News and views: Logic of biology, *Nature* **220:** 429–430: "Legal birth at two days, legal death at eighty-five and parents' dedication of one twin to genetic research were some ideas discussed by Dr. Francis Crick at University College London, last week."

53. *Rickman*, p. 9.

54. *Rickman*, p. 29.

55. *Rickman*, p. 30.

56. *Rickman*, p. 22.

57. *Rickman*, p. 24.

58. Ibid.

59. *Rickman*, p. 23.

60. *Rickman,* pp. 26–27.

61. *Rickman*, p. 27.

62. Annan, letter to Crick, 4 November 1968, *PP/CRI/D/1/1/1*.

63. Crick, personal communication.

64. Crick, Chance and Necessity, 12 October 1972, *PP/CRI/E/1/20/15*.

65. John Gilmour, letter to Crick, 26 October 1972, *UCSD, MSS660:1/31*.

66. Crick, 1966. Why I am a humanist, *Varsity, 8* October; *PP/CRI/H/4/4*.

67. Jensen AR. 1969. How much can we boost I.Q. and scholastic achievement?, *Harv Educ Rev* **33:** 1–123.

68. Crick, letter to Peter Medawar, 31 January 1977, *PP/CRI/D/2/25*.

69. Medawar, letter to Crick, 4 February 1977, *PP/CRI/D/2/25*.

70. Crick, letter to Medawar, 17 February 1977, *PP/CRI/D/2/25*.

71. Lewontin RC. 1975. Genetic aspects of intelligence, *Annu Rev Genet* **9:** 403.

72. Crick, letter to Duval, 18 September 1980, displayed for sale via the Web, Thompson, Antiquarian bookseller, San Francisco.

73. Annie Crick, letter to Crick written (it appears) shortly before she died (1955), *UCSD, MSS660:2/12*. Quote was paraphrased from Lyrical Ballads, with other poems 1800, Vol 1. by William Wordsworth.

74. Crick, personal communication.

75. Crick, letter to Howard Hsu, handwritten for typing, 2004, *UCSD, MSS600, awaiting filing.*

76. Crick, to *The Daily Telegraph,* Martin Packer to the author, 10 November 2007, author's archive.

77. Ronald Williams, letter to Bernard Levin, *Daily Mail,* 25 September 1966, with newspaper clipping: Bernard Levin, So silly and so sad, *UCSD, MSS660:10/43.*

## Chapter 17

1. Kreisel, letter to Crick, undated (1978), *UCSD, MSS660:6/7.*

2. *Molecules,* p. 22.

3. *Molecules,* p. 10.

4. Ibid.

5. Elsasser W. 1958. *The physical foundations of biology.* Pergamon, London; *Molecules,* p. 21.

6. *Molecules,* pp. 19–20.

7. Crick, unpublished letter to the editor of *Nature,* Spring 1976, *PP/CRI/D/1/2/8.* (The book mentioned was *Life Itself,* written after he moved to La Jolla. See Note 17.)

8. 1994. The consciousness of Francis Crick, *Amgen's Magazine of Biotechnology* **3**: 45.

9. Sir John Eccles, 1967. Review of *Of molecules and men, Zygon* **2**: 281; *PP/CRI/E/14/5.*

10. *Molecules,* p. 14.

11. See papers given at the Ciba Conference in London in 1965: Wostenholme GEW, O'Connor M. 1966. *Ciba Foundation Symposium: Principles of biomolecular organization.* Little Brown, Boston.

12. *Molecules,* pp. 15–16.

13. *Molecules,* pp. 12–13.

14. Kornberg A. 1989. *For the love of enzymes: The odyssey of a biochemist,* p. 121. Harvard University Press, Cambridge, Massachusetts.

15. The precise wording as published, and as found in Crick's original draft, reads "Whether a special enzyme is required to carry out the polymerization, or whether the single helical chain already formed acts effectively as an enzyme remains to be seen." *PP/CRI/H/1/12/1.*

16. *WMP,* p. 111.

17. *Life Itself,* p. 69.

18. Crick F. 1974. The double helix: A personal view, *Nature* **248**: 767.

19. Goodgal SH, Rupert CS, Herriott RM. 1957. Photoreactivation of *Hemophilus influenzae* transforming factor for streptomycin resistance by an extract of *Escherichia coli* B. *J Gen Phys* **41**: 451–471; in *McElroy/Glass,* pp. 341–343. See also Friedberg E. 1997. *Correcting the blueprint of life: An historical account of the discovery of the DNA repair mechanism,* p. 49 ff. Cold Spring Harbor Laboratory Press, Cold Spring Harbor, New York.

20. Gilbert W. 1978. Why genes in pieces?, *Nature* **271**: 501.

21. Crick F. 1979. Split genes and RNA splicing, *Science* **204**: 264–271.

22. *DFL,* p. 285 ff.

23. Turing AM. 1952. The chemical basis of morphogenesis, *Philos Trans R Soc B* **237:** 37–72.

24. Crick F. 1970. Diffusion in Embryogenesis, *Nature* **225:** 422.

25. Ibid.

26. Deuchar E. 1970. Letter to the Editor, Diffusion in Embryogenesis, *Nature* **225:** 671.

27. Elizabeth Deuchar, letter to Crick, 18 February 1970; Crick, letter to Davies, 12 March 1970, *PP/CRI/E/1/18/9.*

28. Crick, letter to Davies, ibid.

29. Crick FHC. 1970. The scale of pattern formation, *Symp Soc Exp Biol* **25:** 429–438.

30. Munro M, Crick F. The time needed to set up a gradient: Detailed calculations, *Symp Soc Exp Biol* **25:** 439–453.

31. Lawrence PA, Crick FHC, Munro M. 1972. A gradient of positional information in an insect, *Rhodnius, J Cell Sci* **11:** 815–853.

32. Crick FHC, Lawrence PA. 1975. Compartments and polyclones in insect development, *Science* **189:** 340–347; Morata G, Lawrence PA. 1975. Control of compartment development by the *engrailed* gene in *Drosophila*, in *Ciba Symposium on Cell Development*, London, 1974, pp. 161–182. Elsevier, Amsterdam.

33. See UCSD Biologists visualize protein gradient responsible for dividing embryo into nervous system and different types of epidermis, *Science Daily*, 17 January 2002 (the work of Ethan Bier's group).

34. Watson JD, Crick FHC. 1953. Genetical implications of the structure of deoxyribonucleic acid, *Nature* **171:** 966; *DHn*, p. 246.

35. Warning M, Britten RJ. 1966. Nucleotide sequence repetition: A rapidly reassociating fraction of mouse DNA, *Science* **154:** 791–794; Britten RJ, Kohne DE. 1968. Repeated sequences in DNA. Hundreds of thousands of copies of DNA sequences have been incorporated into the genomes of higher organisms, *Science* **161:** 529–540.

36. Britten RJ, Davidson EH. 1969. Gene regulation for higher cells, *Science* **165:** 349–357; Britten RJ, Kohne DE. ibid. (The GGTTA repeat occurs in the end of a chromosome known as the telomere.)

37. Monod J, Jacob F. 1961. General conclusions: Teleonomic mechanisms in cellular metabolism, growth, and differentiation, *Cold Spring Harbor Symp Quant Biol* **26:** 393.

38. 1971. Report of the Port Cros Meeting, Circles, spacers and satellites on the Riviera, *Nature New Biol* **231:** 68. See *PP/CRI/E/1/19/6 Box 43.*

39. Crick, op cit (Note 18), pp. 766–769.

40. Molecular and Developmental Biology, Summer School held at La Fondazione Majorana, Erice, Sicily. See *PP/CRI/E/1/19/8 Box 44*; Crick, letter to Lawrence, undated, author's archive.

41. *Chromosome*, p. 25, footnote.

42. *Chromosome*, p. 25.

43. *Chromosome*, p. 27.

44. Crick, letter to the Editor of *Nature*, 28 September 1971, *PP/CRI/D/2/27.*

45. Crick, letter to Lawrence, 11 September 1971, author's archive.

46. *Chromosome*, p. 25.

47. Ibid. Italics added by the author.

48. Roger Kornberg, telephone interview with the author, 10 February 2005.

49. The best example of such loops is tRNA, the subject of intense study in the 1970s in Cambridge.

50. R. Kornberg, op cit (Note 48).

51. Klug, interview with the author, March 2004, Cambridge.

52. R. Kornberg, interview with the author, 4 October 2006, Pittsburgh.

53. van der Westhuyzen DR, von Holt C. 1971. A new procedure for the isolation and fractionation of histones. *FEBS Lett* **14:** 333–337.

54. Kornberg RD. 1974. Chromatin structure: A repeating unit of histones and DNA, *Science* **184:** 870.

55. Klug, op cit (Note 51).

56. Olins AL, Olins DE. 1974. Spheroid chromatin units (v bodies), *Science* **183:** 330–332; see also Olins AL, Olins DE. 2003. Chromatin history: Our view from the bridge, *Nat Rev Mol Cell Biol* **4:** 809–814.

57. Klug, op cit (Note 51).

58. Kornberg, op cit (Note 52).

59. Ibid.

60. Ibid.

61. See DuPraw E. 1970. *DNA and chromosomes.* Holt & Winston, New York.

62. Bak AL, Zeuthen J, Crick FH. 1977. Higher-order structure of human mitotic chromosomes, *Proc Natl Acad Sci* **74:** 1595–1599.

63. Crick, personal communication (2004).

64. Crick FHC, Klug A. 1975. Kinky helix, *Nature* **255:** 530–533.

65. Crick FHC. 1976. Linking numbers and nucleosomes, *Proc Natl Acad Sci* **73:** 2639–2643; Fuller FB. 1971. The writhing number of a space curve, *Proc Natl Acad Sci* **68:** 815–819.

66. Cook PR. 1995. A chrommeric model for nuclear and chromosome structure, *J Cell Sci* **108:** 2927–2935; Dorigo B, Schalch T, Kulangara A, Duda S, Schroeder RR, Richmond TJ. 2004. Nucleosome arrays reveal two-start organization of the chromatin fiber, *Science* **306:** 1571–1573.

67. Crick, personal communication.

68. *Life Itself*, p. 79.

69. *Life Itself*, p. 45.

70. Crick FHC, Orgel LE. 1973. Directed panspermia, *Icarus* **19:** 343.

71. *Life Itself*, p. 133.

72. *Life Itself*, p. 49.

73. *Life Itself*, p. 72. Crick was by no means alone in using this figure of speech 40 years ago.

74. *Life Itself*, p. 70.

75. *Life Itself*, p. 71.

76. *Life Itself*, p. 165.

77. *Life Itself*, p.153.

78. Montefiore H. 1990. *Reclaiming the high ground: A Christian response to secularism*, p. 91. Macmillan, New York.

79. *Life Itself*, p. 88.

80. *Life Itself*, p. 83.

81. 1966. Report on the meeting of the British Biophysical Society, *Nature* **212:** 1397; Crick FHC. 1972. Origin of the genetic code, *Nature* **213:** 119.

82. Woese C. 1967. *The genetic code; the molecular basis for genetic expression*, p. 190. Harper & Row, New York.

83. *Life Itself*, p. 82.

84. See Raymond Gesteland, Thomas Cech, and John Atkins (eds.), 1999. *The RNA world: The nature of modern RNA suggests a prebiotic RNA*. Cold Spring Harbor Laboratory Press, Cold Spring Harbor, New York.

85. Foreword, *Life Itself*, p. xiv.

86. Orgel LE. 1994. The origins of life on the Earth, *Sci Am* **267:** 52–61.

87. *WMP,* p. 136.

## Chapter 18

1. Crick, letter to Klug, 30 September 1976, *UCSD, MSS660:4/47.*

2. *Sunday Times*, 27 March 1977.

3. Ibid.

4. *The Times*, 29 March 1977.

5. Op cit (Note 2).

6. Ibid.

7. Richard Le Page, e-mail to the author, 6 April 2006, author's archive.

8. Crick, interview with the author, 9 January 1991, author's archive.

9. Crick, to Miss B.M.J. Brumfitt, 19 December 1963, *FD21/14.*

10. Crick, to Brumfitt, 8 March 1963, *FD21/14.*

11. *DTAH,* p. 17.

12. Crick, interview with the author, 9 January 1991, author's archive.

13. Ibid.

14. Ibid.

15. *WMP,* p. 145.

16. Crick, letter to Watson, 10 November, 1976, *CSHLA,W:9/36.*

17. Perutz, letter to Sir John Gray, 22 February 1977, *FD21/14.*

18. Crick, letter to Klug, 17 September 1976, *UCSD, MSS660:4/47.*

19. Crick, op cit (Note 1).

20. Crick, letter to André Lwoff, 5 August 1977, *PP/CRI/D/1/1/12.*

21. Richard Stevens, personal communication, 3 March 2005, Pittsburgh.

22. Brenner, letter to Crick, 2 November 1976, *PP/CRI/D/2/6.*

23. Crick, letter to Brenner, 16 November 1976, *PP/CRI/D/2/6.*

24. www.salk.edu/about/history. Salk was subsequently to meet Picasso's former mistress, Françoise Gilot, and in 1970 they married.

25. The architects Robert Venturi and Denise Scott Brown were impressed by the visual and symbolic power of this outdoor mecca. They wrote: "Its common space poised between a vast continent—symbolized by the bosk of trees—and a vast ocean—defined by an infinite horizon—is perceptually, physically and poignantly American. In framing the sea and the land in its composition, it marks the end of the Western frontier and the beginning of a new frontier." Quoted in Herbert Muschamp, "Critic's Notebook."—Art and science politely disagree on an architectural jewel's fate. *New York Times*, 16 November 1992.

26. Steele J. 2002. *Salk Institute: Louis I. Kahn*, p. 20. Phaidon Press, London and New York.

27. Frampton K. 1995. Studies in tectonic culture: *The poetics of construction in nineteenth and twentieth century architecture*. The MIT Press, Cambridge, quoted in www.galinsky.com/buildings.salk/index.htm. See also Frampton K. 1992. Salk Institute Controversy. *The Architectural Rev* **192**: 1142.

28. Behind the historic center with its restored laboratories and further up the hillside, there are now new buildings of a more modern style.

29. Crick, letter to Jonas Salk, 20 August 1963, *PP/CRI/C/1/3*.

30. Crick, letter to Weaver, 8 May 1964, *PP/CRI/C/1/2*.

31. Crick, personal communication, 2003.

32. Crick, op cit (Note 30).

33. Jacob Bronowski, letter to Salk, July 1962, *PP/CRI/C/1/7*.

34. Crick, letter to Salk, 19 January 1967, *PP/CRI/C/2/1/1*.

35. Crick, Contemporary frontiers, 20 September 1978, *PP/CRI/H/6/9*.

36. Ibid, p. 6.

37. Ibid, p. 7.

38. Ibid, p. 9.

39. Ibid, p. 11.

## Chapter 19

1. Crick, "Crick Takes," *Omni*, September 1994, p. 52.

2. Hubel D, Wiesel T. 2005. *Brain and visual perception: The story of a 25-year collaboration*, p. 50. Oxford University Press, Oxford.

3. Ibid.

4. *WMP*, p. 149.

5. David Hubel, letter to Crick, 13 November 2003, *UCSD, MSS600:92/H-2002/3*.

6. Crick, letter to Hubel, 17 November 2003, *UCSD, MSS600:92/H-2002/3;* op cit (Note 5).

7. Also known by the earlier Brodmann nomenclature as Area 17.

8. *AH*, p. 124.

9. Crick, letter to Hubel, 6 October 1977, *PP/CRI/D/1/5/7*.

10. Crick, letter to Brenner, 2 September 1977, *CSHLA,B*.

11. Crick FH. 1979. Thinking about the brain, *Sci Am* **241**: 219.

12. Ibid, p. 219.

13. Ibid.

14. Ibid, p. 221.

15. Ibid, p. 224.

16. Crick, *AH*, p. 33.

17. Ibid.

18. Crick, op cit, (Note 11), pp. 221–222.

19. Ibid, p. 223.

20. *AH*, p. 76.

21. Crick FHC, Marr DC, Poggio T. 1981. An information-processing approach to understanding the visual cortex, *Neuroscience Research Program, Colloquium.* pp. 505–533.

22. See Marr, 1982. *Vision: A computational investigation into the human representation and processing of visual information*, p. 335. Freeman, San Francisco. See also *AH*, p. 76.

23. Crick, Marr, and Poggio, op cit (Note 21), p. 505.

24. Crick et al., op cit (Note 21), p. 15.

25. Ibid, p. 336–337.

26. Ibid, p. 349.

27. Ibid, p. 37.

28. Ibid, p. 355.

29. Ibid, p. 339.

30. Crick, letter to Tonaso Poggio, 7 May 1979, *PP/CRI/D/2/32.*

31. Ibid.

32. Crick, letter to Poggio, 9 November 1979, *PP/CRI/D/2/32.*

33. Ramachandran VS. 1990. Vision: A computational investigation into the human representation and processing of visual information. In *AI and the eye* (ed. Andrew Blake and Tom Troscianko), p. 24. Wiley, Chichester, New York.

34. *WMP*, p. 156.

35. Crick, letter to Hubel, 17 November 2001, *UCSD, MSS600:92/H.*

36. Crick, letter to Poggio, 9 November 1979, *PP/CRI/D/2/32.*

37. Ramachandran, op cit (Note 33), p. 24.

38. This is a reference to the "von-Neumann architecture of the digital computer." See *Psychological and biological models,* D.A. Norman, 1987. Reflections on cognition and parallel distributed processing, in *Parallel distributed processing, Vol. 2,* (ed. J.L. McClelland and D.E. Rumelhart), p. 534. MIT Press, Cambridge, Massachusetts.

39. *AH*, p. 186.

40. Jay McClelland, interview with the author, 17 September 2005, Pittsburgh, author's archive. [Actually Crick's height was due chiefly to his very long legs, not his torso.]

41. Ibid.

42. Crick F, Asanuma C. 1986. Certain aspects of the anatomy and physiology of the cerebral cortex. In *Parallel distributed processing, Vol. 2, Psychological and biological models* (ed. J.L. McClelland and D.E. Rumelhart), p. 335. MIT Press, Cambridge, Massachusetts.

43. Crick F. 1989. The recent excitement about neural networks, *Nature* **337:** 129–132.

44. Ibid, p. 130.

45. Ibid, pp. 130–131.

46. Ibid.

47. Ibid, p. 132.

48. Crick, letter to Poggio, 15 May 1980, *UCSD, MSS600:5.*

49. Crick, transcript of public lecture on the function of dreaming, p. 5, author's archive.

50. Crick F, Mitchison G. 1983. The function of dream sleep, *Nature* **304**: 111.

51. Ibid, p. 112.

52. Hopfield JJ, Feinstein DI, Palmer RG. 1983. "Unlearning" has a stabilizing effect in collective memories, *Nature* **304**: 158–159.

53. Rex Chapman, letter to Crick, 28 July 1983, *UCSD, MSS600:3. Dhyāna* is a kind of meditation important in the Hindu and Buddhist religions. Of the eight stages in meditation described by the Buddha, "the first four are connected to the physical realm and the last four only with the mental realm . . ." *Wikipedia, The Free Encyclopedia*, http://en.wikipedia.org./wiki/Dhyāna.

54. Robin Clarke, letter to Crick, 9 August 1983; *UCSD, MSS600:38.*

55. Dick Swaab, letter to Crick, 26 July 1983, *UCSD, MSS600:38.*

56. Graeme Mitchison, e-mail to the author, 12 August 2005, author's archive.

57. Ivan Osorio and Robert Daroff, to the editors of *Nature*, Comments on dream sleep, undated *UCSD: MSS600:38.*

58. Osorio I, Daroff RB, Richey ET, Simon JB. 1980. Absence of REM and altered NREM sleep in patients with spinocerebellar degeneration and slow saccades, *Ann. Neurol.* **7**: 277–280.

59. Osorio and Daroff, unpublished letter to the editor of *Nature*, 29 September 1983, *UCSD, MSS600:38/comments on dream sleep.*

60. Alan Fine, letter to the editor of *Nature*, undated. *UCSD: MSS600:38.*

61. Jouvet M. 1999. *The paradox of sleep: The story of dreaming* (trans. Laurence Garey), p. 19. MIT Press, Cambridge, Massachusetts.

62. Ibid, p. 20.

63. Crick, letter to Mitchison, 28 October 1997, *UCSD: MSS600:8A-1.*

64. Flanagan O. 2000. *Dreaming souls: Sleep, dreams, and the evolution of the conscious mind*, p. 91. Oxford University Press, Oxford and New York.

65. See Stickgold R, Hobson JA, Fosse R, Fossel M. 2001. Sleep, learning, and dreams: Off-line memory reprocessing (Review), *Science* **294**: 1052–1057.

66. Crick, op cit (Note 11), p. 223.

67. Crick F, Jones E. 1993. Backwardness of human neuroanatomy, *Nature* **361**: 109–110.

### Chapter 20

1. Crick F, Koch C. 1990. Towards a neurobiological theory of consciousness. *Sem Neurosci* **2**: 263.

2. von der Malsburg C. "Correlation theory of brain function," Internal Report 81-2 Dept. of Neurobiology, Max Planck Institute for Biophysical Chemistry, Göttingen, 1981. von der Malsburg's concern was how cells representing different parts of the same object, when they fire, are related. He explored the possibility that the timing sequence of the firing might achieve this.

3. Crick F. 1982. Do dendritic spines twitch? *Trends Neurosci* **5**: 44; citing 1970's *Exci-*

*tatory synaptic mechanisms* (ed. P. Anderson and J.K.S. Jansen), Oslo Universitats-forlag, Oslo, in which the pioneer computational neuroscientist Wilfrid Rall and others discussed the role of spines in altering the weighting of dendritic synapses.

4. Ibid, p. 46.

5. See Idan Segav, John Rinzel, and Gordon M. Shepherd, eds. 1995. *The theoretical foundations of dendritic function, selected papers of Wilfred Rall with commentaries.* MIT Press, Cambridge, Massachusetts. See also Matus A. 2000. Actin-based plasticity in dendritic spines, *Science* **290:** 754–758.

6. *AH,* p. 99.

7. *Quest,* p. xvi.

8. Koch, interview with the author, 15 June 2006; *Quest,* p. 5, footnote 6.

9. *Quest,* p. xvii.

10. Crick F. 1989. Neural Edelmanism. *Trends Neurosci* **12:** 240–248. (Review of Edelman G. 1987. *Neural Darwinism.* Basic Books, New York).

11. Among others who collaborated with Crick on specific projects were the X-ray crystallographer Alexander Rich, the chemist Leslie Orgel, and the mathematician Graeme Mitchison.

12. See correspondence between Crick and Hugh Christopher Longuet-Higgins at the Royal Society of London.

13. Crick F. 1984. Function of the thalamic reticular complex: The searchlight hypothesis. *Proc Natl Acad Sci* **81:** 4586.

14. Ibid, p. 4588.

15. Ibid.

16. Ibid, p. 4590.

17. Koch C, Ullman S. 1985. Shifts in selective visual attention: Towards the underlying neural circuitry. *Hum Neurobiol* **4:** 219–227.

18. *AH,* p. v.

19. *Quest,* p. xvii.

20. Koch C. 1999. *Biophysics of computation: Information processing in single neurons.* Oxford University Press, New York.

21. Crick and Koch, op cit (Note 1), p. 263.

22. Ibid.

23. Ibid, p. 264.

24. Ibid, p. 265.

25. *Quest,* p. 42.

26. Horgan J. 1996. *The end of science: Facing the limits of knowledge in the twilight of the scientific age,* p.162. Little, Brown & Co., London.

27. Koch C. 1993. When looking is not seeing: Towards a neurobiological view of awareness. *Engineering & Sci* **56:** 12.

28. Later research into synchronization has not yielded decisive results. Note Wolf Singer's caution in 2004: ". . . there is converging evidence from different experimental approaches that cortical networks can handle temporal patterns with high precision and that precise timing relations among the discharges of distributed neurons are computationally relevant. Thus, the major constraints for the use of synchrony as a

relation defining code are met. But does the brain exploit this option?" Synchrony, oscillations, and relational codes, In *The visual neurosciences* (ed. Leo Chalupa and John Werner), p. 1670. MIT Press, Cambridge, Massachusetts and London.

29. *AH,* p. 3.

30. *AH,* p. 261.

31. *AH,* p. 262.

32. *AH,* p. 266.

33. Jackendoff R. 1987. *Consciousness and the computational mind,* p. 14. Cambridge, Massachusetts.

34. Crick and Koch, op cit (Note 1), p. 264 (*emphasis added*).

35. Wittgenstein L. 1997. *Philosophical investigations,* §§286. Quoted In P.M.S. Hacker, 1997, *Wittgenstein on human nature,* p. 49. Phoenix, London.

36. Crick and Koch, op cit (Note 1), p. 264.

37. *AH,* p. 250.

38. *AH,* pp. 250–251.

39. *AH,* p. 252.

40. AOL customer from New York City, "You can spend your time better," Google: Customer Book Reviews: Astonishing Hypothesis.

41. Eugene Goheen, from San Diego, Yahoo: Customer Book Reviews: Astonishing Hypothesis.

42. A doctor from Deltona, FL, Yahoo: Customer Book Reviews: Astonishing Hypothesis.

43. Searle JR. 1995. The mystery of consciousness: Part II, *NY Rev Books* **42:** 18. http://www.nybooks.com/articles/1741, p. 10.

44. *AH,* p. 11.

45. Ibid.

46. Searle, op cit (Note 43), p. 9.

47. Ibid, p. 10.

48. Jackendoff, op cit (Note 33), pp. 11–12.

49. *AH,* p. xii.

50. *AH,* p. 246.

51. Crick F, Koch C. 2003. A framework for consciousness, *Nat Neurosci* **6:** 119–126.

52. *Quest,* p. 46.

53. Crick F, Koch C. 2001. Consciousness and neuroscience, In *Philosophy and the neurosciences, A reader* (ed. Bechtel et al.), p. 261. Blackwell Publishers Ltd., Oxford.

54. Crick and Koch, op cit (Note 51).

55. Ibid.

56. Ibid.

57. Crick quoted in Horgan, op cit (Note 26), p. 162.

58. Crick, personal communication.

59. Crick F, Koch C. 1995. Are we aware of neural activity in primary visual cortex? *Nature* **375:** 121–123.

60. *AH,* p. 259.

61. *AH,* p. 261.

62. *AH,* p. 262.

63. Crick and Koch, op cit (Note 59), p. 123.

64. Chalmers D. 1995. *The conscious mind: In search of a fundamental theory.* Oxford University Press, New York.

65. Crick and Koch, op cit (Note 53), p. 274.

66. *Quest,* p. 318.

67. *Quest,* p. 327.

### Chapter 21

1. May Brenner, letter to Odile Crick, 1 August 2004, *UCSD, MSS660/11/25.*

2. Murray K. 2004. *Memorial,* p.35.

3. Koch, interview with the author, 7 March 2006, author's archive.

4. Crick, video to the Congress of the International Telecommunications Union, Seminar on Fifty Years after the Discovery of the DNA Structure, Bilbao, Spain, 6 February 2003.

5. Murray, interview with the author, 5 October 2004.

6. Klug, letter to Odile Crick, 4 August 2004, *UCSD, MSS660/11/34.*

7. Crick F, Koch C. 2005. What is the function of the claustrum? *Philos Trans R Soc Lond B Biol Sci* **360:** 1271–1279.

8. Ibid, pp. 1276 and 1277.

9. Crick FHC. 1965. Introduction: Recent research in molecular biology. *Br Med Bull* **21:** 186.

10. Sir Ernst Chain, 1973. Future contributions from biochemistry to the advancement of medical research, attached to letter to the Secretary of the Medical Research Council, dated 20 November. Wellcome Library, Chain Papers, E 93. See also *PDH,* p. 451.

11. Burnet FM. 1966. Men or molecules? *Lancet* **1:** 37.

12. Transcript of the film *Double Helix,* shown in 2003.

13. Odile Crick to Kreisel, 10 February 1971. Stanford University Libraries, Archives, (SC136) Georg Kreisel Papers: 3/9.

14. Odile Crick, telephone interview with the author, 6 June 2006, Fifth draft of the millenium report from Kedington, author's archive.

15. Joan Wolfenden, letter to Crick, 3 May 1994, *UCSD, MSS660/10/46.*

16. Koch, *Memorial,* p. 32.

17. Alison Auld, interview with the author, 16 July 2002.

18. Michael Crick, interview with the author, 11 July 2007.

19. Auld, interview with the author, 12 March 2004.

20. Jones L. 1968. *How to undress in front of your husband.* Tandem, London.

21. Pauline Finbow, letter to the author, 25 July 2007, author's archive.

22. Finbow, interview with the author, 12 March 2004, author's archive.

23. Crick, letter to Finbow, 17 May 2003, author's archive.

24. Few and Kester, letter to Crick, 17 April 1967, *UCSD, MSS660,3/32.*

25. Crick F. Sailing with Jacques, In *Origins of molecular biology. A tribute to Jacques Monod* (ed. André Lwoff and Agnes Ullman), p. 227. Academic Press, New York. Monod told Crick that an account of this trip appeared mysteriously in *Le Figaro*. Monod to Crick, 26 August 1969, *PP/CRI/D/2/26*.

26. Kendrew's quip is quoted in Jeffries' letter to Crick, 16 August 1991, *UCSD, MSS660:11/12*.

27. Foulkes, diary entry, 8 March 1972, author's archive.

28. http://vintagecars.about.com/od/historygreatmoments/a/lotus_elan.htm.

29. Gutfreund, e-mail to the author, 6 March 2007, author's archive.

30. Some of the menus can be found in 1988's *But the crackling is superb* (ed. Nicholas and Gina Kurti). Adam Hilger, Bristol and Philadelphia.

31. Odile Crick, letter to Gutfreund, cited in Note 29, author's archive.

32. Odile Crick, letter to Crick, 6 February 1967, *UCSD, MSS660:2/15*.

33. Crick, "refusals," example: 6 February 1998, *UCSD, MSS600:11/D*

34. Provost, Sir Ivor Evans, letter to Crick, 14 July 1960, *UCSD, MSS660,10/19*.

35. Handwritten draft of letter responding to Birthday Honours List Invitation, 14 May 1963, *UCSD, MSS660/12/17*.

36. List of people who declined a British honour, *Wikipedia* (taken from *The Sunday Times*, December 2003).

37. Hans Noll, letter to Crick, 17 June 1971. "The most authoritative *Encyclopedia Britannica* has you clearly down as Crick, Sir Francis. I suspect that you got knighted in a secret ceremony at Buckingham Palace (probably in the Chapel)." *PP/CRI/D/1/2/9*.

38. Georg Melchers, personal communication, undated (1972). John Mosley, BEA (airways) left a message addressed to Lord Crick for Crick to receive on his arrival, *UCSD, MSS660/8/22.iii.1957*.

39. Crick, personal communication.

40. Perutz, letter to Crick, 27 November 1991, *UCSD, MSS660/13/1*.

41. Once resident in the United States, Crick satisfied his hunger for the theatre on his annual summer visits to England. Peter Lawrence advised him of the agenda of shows in London and obtained the tickets for the many required performances. Odile and Francis did once take out a season's subscription to the opera in San Diego, but they did not repeat the experience. Odile, personal communication.

42. Jeffries, op cit (Note 26).

43. Crick, personal communication.

44. *WMP,* p. 78.

45. *WMP,* p. 76.

46. Crick, "The way we were" (adapted from a talk given on 1 March 1993 at Cold Spring Harbor Laboratory); see also Crick F. 1979. How to live with a golden heix. In *From Gene to protein: Information transfer in normal and abnormal cells.* (ed. TB Russell et al.), pp. 1–13. Academic Press, New York.

47. Crick, letter to Kendrew, 14 April 1967, Bodleian Library, *Kendrew Papers*.

48. Crick, op cit (Note 46), p. 2.

49. *WMP,* p. 75.

50. Maddox B. 2002. *Rosalind Franklin. The dark lady of DNA*. Harper Collins, London.

51. Hodgkin, note to David Sayre, 6 January 1975; copy enclosed with letter from Maddox to Crick, 10 July 2002, and Crick to Maddox, 11 July 2002, author's archive.

52. Pauline Harrison (née Cowan), letter to the author, 16 December 2005, author's archive.

53. Author's question to and response from Crick in 2003.

54. Crick, letter to Brenda Maddox, 9 January 2002, some general remarks, p. 2, author's archive.

55. *Chambers 20th Century Dictionary*, 1983, p. 322. Chambers, Edinburgh.

56. Crick, personal communication, undated (2003).

57. Crick FHC, Watson JD. 1954. The complementary structure of deoxyribonucleic acid. *Proc R Soc A* **223:** 80–96. Also Crick, personal communication, undated. See also Chapter 10.

58. Bernard Davis, Review of *What mad pursuit, Chemical and Engineering News,* 5 June 1989.

59. Charles Weissmann, e-mail to the author, 2 May 2008, author's archive.

60. Odile Crick, telephone conversation with the author, undated (February 2006).

61. Auld, interview with author, 12 March 2004.

62. John Maddox, personal communication.

63. Crick, as reported by Richard Murphy, President and CEO of the Salk Institute, *Memorial*, p. 5.

64. www.crick-jacobs.salk.edu.

65. Koch, *Memorial*, p. 33.

66. The original intention of the family was to bury Darwin in the cemetery at Down, but Francis Galton was successful in initiating a movement to have the location changed to Westminster Abbey. See Adrian Desmond and James Moore, 1991. *Darwin*, p. 664 ff. Michael Joseph, London.

67. Alcor, letter to Crick, April 2004, *PP/CRI/uncatalogued Batch 3 File 9/32.* On the letter is written: "Ignore per Crick's instructions."

68. Tomaso Poggio, *Memorial*, p. 26.

69. Jacqueline Crick, *Memorial,* p. 25.

70. Crick, 2003. Welcome to participants at the Golden Jubilee of the Discovery of the Structure of DNA. Cold Spring Harbor (Video).

# Illustration Credits

**Chapter 2**

2.1 Redrawn, with permission, from Michael Crick.

2.2 Adapted, with permission, from *Collins 100 miles around London* (© 2007 Harper Collins).

**Chapter 4**

4.1 (*A*) Adapted from Cowie, JS. 1949. *Mines, minelayers and minelaying;* reprinted, with permission, from Oxford University Press. (*B*) Reprinted from Science Museum South Kensington, *Naval mining and degaussing. Catalogue of an exhibition of British and German material used in 1939–1945* (London: H.M. Stationary Office, 1946), with permission, from the Science & Society Picture Library.

**Chapter 5**

5.1 Reprinted, with permission, from Wellcome Library, London.

5.2 Reprinted, with permission, from Darlington, CD. 1937. *Recent advances in cytology, 2e.* P. Blakiston's Son & Co., Philadelphia.

5.3 Reprinted from Schrader F. 1944. *Mitosis: The movement of chromosomes in cell division,* with permission, from Columbia University Press.

**Chapter 6**

6.1 Courtesy of Mike Fuller; reprinted, with permission, from the MRC Laboratory of Molecular Biology.

6.2 (*A*) Reprinted, with permission, from Pauling L, Corey RB, Branson HR. 1951. *Proc Natl Acad Sci* **37**: 206. (*B*) Reprinted from Bragg L, Kendrew J, Perutz M. 1950. *Proc Roy Soc London A* **203**: 336, with permission, from The Royal Society.

6.3 Reprinted, with permission, from Crick F. 1954. *Science Progress* **166**: 207.

6.4 Reprinted, with permission, from Bragg L. 1913. *Proc Roy Soc A* **88**: 428–438.

6.5 Adapted from Dickerson R. 2005. *Present at the flood*, with permission, from Sinauer Associates.

6.6 Reprinted, with permission, from Crick F. 1953. PhD Thesis, Cambridge.

6.7 Reprinted, with permission, from Kendrew J, Perutz M. 1949. *Haemoglobin*. Interscience Publishers, NY.

6.8 (*A,B*) Reprinted, with permission, from Kendrew J, Perutz M. 1949. *Haemoglobin*. Interscience Publishers, NY.

497

6.9     (*A,B*) Reprinted, with permission, from Kendrew J, Perutz M. 1949. *Haemoglobin.* Interscience Publishers, NY.

6.10    Adapted from Paige D. 1964. Rev. 1969 by Shelley HA. *Heffer's map of cambridge.* W. Heffer & Sons, Cambridge, with permission from Wiley-Blackwell.

6.11    Reprinted from Cochran W, Crick F. 1952. *Nature* **169:** 234–235, with permission, from Macmillan Publishers.

6.12    (*A,B*) Adapted from Dickerson R. 2005. *Present at the flood*, with permission, from Sinauer Associates.

### Chapter 7

7.1     Redrawn from Crick F, Watson J. 1954. *Proc Roy Soc Lond A* **223:** 276, with permission, from the Royal Society.

7.2     Courtesy of James D. Watson. Reprinted, with permission, from Watson, J. 1968. *The double helix.* Weidenfeld & Nicolson, London.

### Chapter 8

8.2     Reprinted from Crick F. 1953. *Acta Crystallographica* **6:** 695, with permission, from Wiley-Blackwell.

8.3     Reprinted from Dickerson R. 2005. *Present at the flood.* Sinauer Assoc., with permission, from David Goodsell.

8.4     Reprinted, with permission, from King's College London.

### Chapter 9

9.2     (*A*) Reprinted from Broomhead J. 1948. *Acta Crystallographica* **1:** 324–329, with permission, from Wiley-Blackwell. (*B*) Reprinted, with permission, from Olby R. 1974. *Path to the double helix.* University of Washington Press.

9.3     Courtesy of James D. Watson. Reprinted, with permission, from Watson, J. 1968. *The double helix.* Weidenfeld & Nicolson, London.

9.4     Reprinted from Watson J, Crick F. 1953. *Nature* **171:** 964–967, with permission, from Macmillan Publishers Ltd.

        Figures in letter from Crick to son Michael. Courtesy of the Crick Family.

### Chapter 10

10.1    Reprinted, with permission, from Wellcome Library, London.

10.2    Reprinted, with permission, from Wellcome Library, London.

10.3    Reprinted from Watson J and Crick F. 1953. *Nature* **171:** 737, with permission, from Macmillan Publishers Ltd.

### Chapter 11

11.1    Reprinted from Magdoff F, Crick F, Luzzati V. *Acta Crystallographica* **9:** 156–162, with permission from Wiley-Blackwell.

11.2    (*A,B*) Reprinted from Crick F. 1954. *J Chem Phys* **22:** 347–348, with permission, from the American Institute of Physics.

## Chapter 12

12.1   (*Left*) Redrawn, with permission, from Rodley G, Scobie R, Bates R, Lewitt R. 1976. *PNAS* **73:** 277. (*Right*) Redrawn from Watson J, Crick F. 1953. *Nature* **171:** 737–738, with permission, from Macmillan Publishers.

12.2   Reprinted from Gamow G. 1954. *Nature* **173:** 318, with permission, from Macmillan Publishers.

12.3   Reprinted, with permission, from James D. Watson and Wellcome Library, London.

12.4   Reprinted, with permission, from Crick F, Griffith J, Orgel L. 1957. *PNAS* **43:** 417–418.

Table 12.1 Reprinted from Crick F. 1958. *Symposia of the society for experimental biology* **12:** 140, with permission, from The Company of Biologists.

## Chapter 13

13.1   Reprinted from Franklin R, Klug A, Holmes K. 1957. *Ciba foundation symposium on the nature of viruses*, Wolstenholme and Millar, eds. Little Brown & Co.

13.2   Reprinted, with permission, from Wellcome Library, London.

## Chapter 14

14.1   Reprinted from Hoagland M. 1996. *Trends Biochem Sci* **21:** 77–80, with permission, from Elsevier.

14.2   (*A*) Redrawn from Brock T. 1990. *The emergence of bacterial genetics*, with permission, from Cold Spring Harbor Laboratory Press, NY. (*B*) Redrawn from Watanabe K. 2006. *Genetic code: Introduction*, from Encyclopedia of Life Sciences, John Wiley & Sons.

14.3   Reprinted from Olby R. 1972. Francis Crick, DNA, and the central dogma, in the twentieth century sciences: Studies in the biography of ideas. Holton, G. ed. Norton Publishing.

14.4   Reprinted, with permission, from Mark Bretscher and the Crick Family.

## Chapter 15

15.2   Reprinted, with permission, from Wellcome Library, London.

15.3   Reprinted from Crick F. 1967. *Proc Roy Soc B* **167:** 339, with permission, from The Royal Society.

15.4   Reprinted from Commoner B. 1964. *Nature* **203:** 486–491, with permission, from Macmillan Publishers Ltd.

15.5   Reprinted from Crick F. 1970. *Nature* **227:** 561–563, with permission, from Macmillan Publishers Ltd.

## Chapter 16

16.1   Reprinted, with permission, from UCSD Mandeville Special Collections Library.

## Chapter 17

17.1   Redrawn from Wikimedia Commons.

17.2   Adapted from Crick F. 1979. *Science* **204:** 265, with permission, from AAAS.

17.3   Reprinted from Crick F. 1971. *Nature* **234:** 26, with permission, from Macmillan Publishers.

17.4    Reprinted, with permission, from the *Annual review of biochemistry*, Volume 46, © 1977 by Annual Reviews.

## Chapter 19

19.1    Reprinted from Hubel D, Wiesel T. 2005. *Brain and visual perception. The story of a 250 year collaboration*, with permission, from Oxford University Press Inc.

19.2    Reprinted from Nicholls J, Martin R, Wallace B. 1992. *Neuron to brain. A cellular and molecular approach to the function of the nervous system, 3e*, with permission, from Sinauer Associates.

19.3    Reprinted from Hubel D, Wiesel T. 1961. *Journal of Physiology* **155**: 385–398, © Blackwell Publishing.

19.4    Reprinted, with permission, from Wiesel T, Gilbert C. 1989. *Neural mechanisms of visual perception: Proceedings of the retina research foundation symposia, 2.* Portfolio Publishing Co.

19.5    Reprinted from Harmon L, Julesz B. 1973. *Science* **180**: 1194–1197, with permission, from AAAS.

19.6    Reprinted from McClelland J, Rumelhart D, PDP Research Group. *Parallel distributed processing V.2: Explorations in the microstructure of cognition: Psychological and biological models*, p. 320 © 1986 Massachusetts Institute of Technology, with permission, from The MIT Press.

## Chapter 20

20.1    Reprinted from Crick F. 1982. *Trends in Neurosciences* **5**: 44, with permission, from Elsevier.

20.2    Redrawn from Wikipedia.

20.3    Reprinted from Wikipedia.

## Plate Section

1    Reprinted, with permission, from UCSD Mandeville Special Collections Library.

2    Reprinted, with permission, from the Northampton Public Library.

3    Courtesy of the Crick Family.

4    Courtesy of Ralph Followell.

5    Courtesy of Ralph Followell.

6    Courtesy of Doreen Scholfield-Fost.

7    Courtesy of Doreen Scholfield-Fost.

8    Courtesy of Doreen Scholfield-Fost.

9    Courtesy of the Crick Family.

10    Courtesy of the Crick Family.

11    Reprinted, with permission, from UCSD Mandeville Special Collections Library.

12    (*Left*) Courtesy of Mike Gunn. (*Top left*) Reprinted, with permission, from UCSD Mandeville Special Collections Library. (*Bottom left, near right,* and *far right*) Courtesy of Queen's University Belfast and Mark McCartney.

13    Courtesy of the Crick Family.

14    Courtesy of the Crick Family.

15    Reprinted, with permission, from UCSD Mandeville Special Collections Library.

16 Courtesy of Strangeways Research Laboratory.

17 Courtesy of Strangeways Research Laboratory.

18 Reprinted, with permission, from UCSD Mandeville Special Collections Library.

19 Courtesy of the Crick Family.

20 Courtesy of MCDOA, www.mcdoa.org.uk.

21 Courtesy of the Crick Family.

22 Reprinted, with permission, from UCSD Mandeville Special Collections Library.

23 Courtesy of the Crick Family.

24 Reprinted, with permission, from the Cambridgeshire Collection, Cambridge Central Library.

25 Reprinted, with permission, from UCSD Mandeville Special Collections Library.

26 Courtesy of Maurice S. Fox.

27 Reproduced, with permission, from Astbury, W. 1947. *Symposia of the society for experimental biology 1.* p.72.

28 Courtesy of Seweryn Chomet. Reprinted, with permission, from Chomet, S. ed. *D.N.A: Genesis of a discovery.* Newman-Hemisphere, London.

29 Courtesy of MRC Laboratory of Molecular Biology, Cambridge.

30 Reprinted, with permission, from Olby R. 1974. *The path to the double helix.* Macmillan, London.

31 Reprinted from Maddox B. 2002. *Rosalind Franklin. The dark lady of DNA*, with permission, from Victor Luzzati.

32 © American Philosophical Society, reprinted with permission.

33 Courtesy of Seweryn Chomet. Reprinted, with permission, from Chomet, S. ed. *D.N.A: Genesis of a discovery.* Newman-Hemisphere, London.

34 (*Left*) Reprinted from Franklin R, Gosling, R. 1953. *Nature* **171:** 740–741, with permission, from Macmillan Publishers Ltd. (*Right*) Reprinted from Franklin R, Gosling R. 1953. *Nature* **172:** 156–157, with permission, from Macmillan Publishers Ltd.

35 Reprinted, with permission, from the Ava Helen and Linus Pauling Papers, Special Collections, Oregon State University.

36 © Science Museum/Science & Society Picture Library, London, reprinted with permission.

37 Courtesy of Cold Spring Harbor Laboratory Archives.

38 Courtesy of the Perutz Family.

39 Reprinted from Hughes J. 2008. *Nature* **452:** 30, with permission, from Macmillan Publishers Ltd.

40 Courtesy of Professor Freddie Gutfreund.

41 © Nigel Unwin, courtesy of MRC Laboratory of Molecular Biology, Cambridge.

42 Reprinted, with permission, from Wellcome Library, London.

43 Reprinted, with permission, from Wellcome Library, London.

44 © Bettmann/Corbis.

45 Erhan Guner © Pressens Bild/Scanpix/Sipa Press.

46 Erhan Guner © Pressens Bild/Scanpix/Sipa Press.

47    (*Left, right*) Courtesy of Cold Spring Harbor Laboratory Archives.

48    Reprinted, with permission, from UCSD Mandeville Special Collections Library.

51    Courtesy of Professor Freddie Gutfreund.

52    Reprinted, from Gross C. 2005. *Nature* **434:** 1069–1070, with permission from Macmillan Publishers Ltd.

53    Courtesy of Professor Christof Koch.

56    Courtesy of Pauline Finbow.

57    Courtesy of Kent Schnoeker.

58    Courtesy of Pauline Finbow.

60    Reprinted, with permission, from Wellcome Library, London.

# Biographical Index

**Andrade, Edward Neville da Costa,** b. London (1887–1971). Andrade trained in physics at University College London; he carried out his doctoral research there and in Heidelberg. After serving in World War I, followed by a position at the Ordnance College, Woolwich, he became University College's Quain Professor of Physics in 1928. Crick was his research student from 1937 to 1939.

**Annan, Noel, Baron,** b. London (1916–2000). Annan studied history at King's College Cambridge and had a distinguished career in military intelligence during World War II. In 1956 he became Provost of his Cambridge College, and in 1966 Provost of University College London. He was the first chairman of the Trustee's education committee at Churchill College Cambridge.

**Barnett, Leslie,** b. London (1922–2002). Barnett came to work as a technician at the MRC Unit shortly before the move from the Cavendish Laboratory to the Hut. Her role was to help with computing for the crystallographers. On Sydney Brenner's arrival, however, help was needed to set up the phage research and then to prepare for the arrival of Seymour Benzer and other American visitors that fall. Barnett showed her versatility when she transferred to Brenner's program and made a major contribution to the laboratory work, becoming a co-author of their two major papers on the results. In 1966 she was appointed Senior Tutor at the new graduate College, Clare Hall. She proved very popular in this work. On Brenner's retirement in 1986 Barnett left the LMB to work for him in the new MRC Molecular Genetics Unit in Addenbrooke's hospital. She trained Crick in the experimental work on phage and assisted him.

**Bernal, John Desmond,** b. Menagh, Tipperary, Ireland (1901–1971). Bernal trained in physics at Cambridge, and under Sir William Bragg at the Royal Institution, London, he made important contributions to methods in X-ray crystallography. From 1927 to 1937 he taught at Cambridge. He discovered how to obtain X-ray diffraction patterns of a crystalline protein in 1934. Leaving Cambridge in 1937, he moved as Professor to Birkbeck College London, where Rosalind Franklin carried out her excellent work on the structure of tobacco mosaic virus. Bernal was considered by Crick to be the "grandfather" of molecular biology.

**Bragg, Sir (William) Lawrence,** b. Adelaide, South Australia (1890–1971). Bragg came to England at age 18 with a first-class degree in mathematics. Then as a scholar at Trinity College Cambridge he won first-class honors in the Natural Science Tripos in 1912. Before that year was out, he published the results of his first research into X-ray diffraction. Developing this research in collaboration with his father led to the joint award of the Nobel Prize in Physics in 1915, the only one ever given jointly to father and son. After service in World War I he became professor of physics at Manchester University (1919–1937), and then he followed Ernest Rutherford as Cavendish Professor and Director of the Cavendish Laboratory (1938–1953). Backed by so respected a figure as Bragg, Max Perutz was able to build his MRC Unit in the Cavendish and provide a place in it for both Crick and Sydney Brenner. Under Bragg's leadership, the Cavendish Laboratory became the world leader in protein structure research.

**Brenner, Sydney,** b. Germiston, South Africa (1927– ). Brenner matriculated from high school a shade under 15 years of age and entered the University of the Witwatersrand, Johannesburg, where he completed his medical degree in 1951. He also found time to study histology and begin research in this area. Keen to continue research, he won a scholarship to Oxford where he studied phage genetics for his doctorate. He met Crick and James Watson in Cambridge in April 1953, saw their model for DNA, and in due time came back to England in December 1956 to begin his career with Crick at the MRC Unit. Brenner succeeded Max Perutz as Director of the MRC LMB in 1979, retiring in 1986. In 2001 he joined Crick at the Salk Institute as Distinguished Professor. He received the Nobel Prize in Physiology or Medicine in 2002.

**Bronowski, Jacob,** b. Łódź, Poland (1908–1974). Bronowski read mathematics at Jesus College Cambridge and was with Operation Research during World War II. He directed the National Coal Utilization Research Board (1950–1959). In 1964 he became associate Director of the Salk Institute (1964–1974). Bronowski was famous as the presenter of the TV series *The Ascent of Man*.

**Brundrett, Sir Frederick,** b. Ebbw Vale, Monmouthshire, Wales (1894–1974). Brundrett was a Welsh mathematician (Sydney Sussex College, Cambridge) and a Naval Volunteer Reserve in World War I. He worked on the development of shortwave radio for the Navy (1919–1937). Then, after transfer to the Royal Naval Scientific Service headquarters in London and a series of promotions, he became Chief of the Royal Naval Scientific Service in 1946. CBE, 1946; KBE, 1950; KCB, 1956. He appointed Crick Temporary Experimental Officer in December 1939.

**Chargaff, Erwin,** b. Czernowitz, Austria, now Ukraine (1905–2002). Chargaff trained in chemistry at the University of Vienna. He spent 1928–1930 as a postdoctoral fellow at Yale University and then returned to Europe, working in Berlin and Paris. In 1935, he emigrated to the United States to join the chemistry department of Columbia University in New York. There, influenced by the work of Oswald Avery on the Transforming Principle, he revolutionized our understanding of the chemistry of the nucleic acids.

**Collingwood, Sir Edward,** b. Alnwick, Northumberland, UK (1900–1970). The Collingwoods were a naval and military family. Collingwood joined the Royal Navy at the age of 15 but was invalided out of the service following a fall. At this point he turned to mathematics and in 1918 entered Trinity College, Cambridge. He received his doctorate in 1929 and became Steward of Trinity College in 1930. World War II claimed him for research in the Navy as a Captain;

in 1943, put in charge of the Minesweeping division; and in 1945, as Chief Scientist in the Mine Design Department where Crick worked. 16 years older than Crick and coming from an historic and wealthy family, Collingwood knew how to handle and advise him.

**Crick, Doreen** (née Dodd), b. Nuneaton, Warwickshire, U.K. (1913– ). Doreen was a student of English literature at University College London. She worked as a clerk in London. She married Francis Crick in 1940, and they divorced in 1947. For their son Michael, see below. She subsequently married James Potter, a Canadian soldier.

**Crick, Gabrielle,** b. Cambridge, U.K. (1951– ). Elder daughter of Francis and Odile Crick. She was very supportive of her mother at the time of her father's death and visited old haunts in France with her in 2006 on what was to be Odile's last visit to France.

**Crick, Jacqueline,** b. Cambridge, U.K. (1954– ). Younger daughter of Francis and Odile Crick. She married Christopher John Nichols, who works in the construction business in Sussex. They have two adopted children, Mark and Nicholas.

**Crick, Michael Francis Compton,** b. Teddington, Middlesex, U.K. (1940– ). Only child of Francis and Doreen (née Dodd). He married Barbara Davis of Seattle in 1971. He won a First in Physics at University College London, and went on to graduate work in physiology for the M.Sc., followed by doctoral research at Harvard and MIT. There, working under Dr. Jerome Lettvin, he applied the emerging field of parallel processing to the design of games and gave up physiology to become one of the pioneers of computer games. He is known today for his news puzzles Cricklers and Enigma and the computer game WordZap. His four children are Alexander, Kindra, Camberley, and Francis.

**Crick, Odile** (née Speed), b. King's Lynn, Norfolk, U.K. (1920–2007). Odile was educated at King's Lynn, Paris, and Vienna; studied art in London; and worked for the Admiralty during World War II, where she met Crick. Married in 1949, they lived in Cambridge until 1976 and then moved to La Jolla, California.

**Delbrück, Max,** b. Berlin, Germany (1906–1981). Delbrück studied physics at Göttingen and obtained his doctorate in 1930. After acting as assistant to Lise Meitner in 1937, he left for the United States. In Berlin he had interacted with the Russian geneticist Nikolay Timofeeff-Ressovsky and had been one of the three authors of an important paper in which X-ray-induced mutations were used to explore the nature and size of the gene. This so-called "Three-Man-Paper" had been the key source used by Erwin Schrödinger in his book, *What is Life?* (1944), a work that played an important part in turning Crick to biology. Delbrück was one of James D. Watson's mentors, and with Salvador Luria, the founder of the Phage Group, he played an important role in shaping Watson's career.

**Fell, Dame Honor,** b. Fowthorpe, North Yorkshire, U.K. (1900–1986). Fell was Director of the Strangeways Laboratory in Cambridge from 1929 to 1970. She developed a method of tissue culture that permitted growth of differentiating cells culled from embryos. She accepted Crick into her laboratory when, still a physicist, he sought training in biology.

**Fisher, Sir Ronald,** b. London (1890–1962). Fisher won a scholarship to Gonville and Caius College, Cambridge in 1909. An enthusiast for eugenics and Mendelian heredity, he used his formidable skill in statistics to harmonize the conflicting schools of the Mendelians and the biometricians. His book, *The Genetical Theory of Natural Selection* (1930), was one of the key sources for the modern theory of evolution. As professor of Genetics in Cambridge (1943–1957) and Vice President of Gonville and Caius College, he and Crick met frequently and became friends.

**Foakes, Susan** (née Barnes). Foakes was the personal secretary to the two heads of the Cell Biology Division at the MRC, Francis Crick and Sydney Brenner, in 1971–1974.

**Fost, Harold,** b. Mill Hill, London (1916–2007). Fost was a close friend and contemporary of Crick at Mill Hill School. He later became an accountant in Jersey, Channel Islands.

**Franklin, Rosalind,** b. Bayswater, London (1920–1958). Graduated in chemistry from Newnham College, Cambridge in 1941, Franklin worked on the structure of coals at the British Coal Utilization Research Association (1942–1947) and at the Laboratoire Central des Services Chimiques de l'Etat in Paris (1947–1950). Following her research on DNA at King's College London (1951–1953) she moved to Bernal's department at Birkbeck College, London, where she built up an outstanding research program on the structure of viruses. Crick first met her in 1952. Ovarian cancer caused her early death at 37 years of age. Had she lived, she would surely have moved with her Birkbeck colleagues, Aaron Klug, John Finch, and Kenneth Holmes to the LMB in 1962.

**Gamow, George,** b. Odessa, Russian Empire (now in the Ukraine) (1904–1968). Gamow was educated at the Universities of Odessa and Leningrad, and his research into the atomic nucleus took him to Göttingen, Copenhagen, and Cambridge. These studies established a mechanism for nuclear decay, and thus for both the evolution of the elements and the Big Bang theory of the origin of the universe. He escaped from Russia in 1933, moving a year later to the United States. In 1953, Watson and Crick's structure for DNA stimulated him to discuss the nature of the genetic code. He first met Crick in the winter of 1953. Gamow was also a very successful popular science writer.

**Gutfreund, Herbert,** b. Vienna (1921– ). Known to his friends as Freddie, Gutfreund studied chemistry at the University of Vienna. His first research position in Cambridge was in the Sub-department of Crystallography, from whence he moved to the Low Temperature Laboratory and became an expert on enzyme kinetics. A close friend of the Cricks and the Orgels, Gutfreund lunched regularly with Crick and Watson at The Eagle. Owing to a part-time position at Yale, he was able to meet with Crick in the United States in 1953 and 1954. He is now Professor Emeritus at Bristol University. He married Mary Kathelen in 1958; the Cricks and the Orgels were present at the church service.

**Hammond, Edie,** Hammond was resident at 9, St. Martin's Square, Chichester, Hants, U.K. She met Crick toward the end of the war when he lodged at her mother's house. As her letters to him show, she was in love with him.

**Harker, David,** b. Mill Valley, California (1906–1991). Harker trained at Caltech under Linus Pauling. He went on to make his name as a theoretician in X-ray crystallography, beginning with his theory of Patterson sections and projections, known as Harker sections, and with John Kasper, the Harker–Kasper inequalities of "direct methods." He founded and directed the Protein Structure Project at Brooklyn Polytechnic in 1950, where Crick spent the academic year 1953–1954.

**Hill, Archibald Vivian,** b. Bristol, U.K. (1886–1977). Hill was the father of biophysics in Britain. He won a scholarship to Trinity College Cambridge and studied mathematics, becoming the Third Wrangler for the Mathematical Tripos in 1907. In 1910 he became a Fellow of the College. He received the Nobel Prize in Physiology or Medicine for his work on muscle in 1922. Hill advised Crick on seeking a laboratory in which to study biology.

**Hoagland, Mahlon,** b. Boston, Massachusetts (1921– ). After a medical training at Harvard, Hoagland began research at Massachusetts General Hospital, subsequently joining Paul Zamecnik's group in the Huntington Laboratories. He isolated a form of RNA subsequently called transfer RNA. He spent the year 1957–1958 working with Crick in the MRC Unit. From 1978 to 1995, Hoagland directed the Worcester Foundation, of which his father had been a co-founder.

**Hodgkin, Dorothy** (née Crowfoot), b. Cairo, Egypt (1910–1994). Hodgkin, who was educated in England, trained in chemistry at Somerville College Oxford. She moved to Cambridge University for her doctoral research in protein crystallography with J.D. Bernal (1932–1934). Returning to Oxford she accepted a fellowship at Somerville and tutored students in chemistry. Hodgkin's research led to the solution of the structures of penicillin (1945), vitamin $B_{12}$ (1954), and insulin (1969), and she received the Nobel Prize in Chemistry in 1964 and the Order of Merit from Queen Elizabeth in 1965.

**Hughes, Arthur,** Hughes was the microscopist at the Strangeways Laboratory who developed phase-contrast cinematography for the study of cell division. He helped Crick in his research on the viscosity of the cytoplasm. Later he wrote a history of cytology.

**Ingram, Vernon,** b. Breslau, Prussia (now Poland) (1924–2006). Anti-Semitism caused Ingram's family to leave Breslau for England in 1938. In London he won his degree in chemistry at night school while working at a pharmaceutical factory by day. As a postdoc, he carried out research at the Rockefeller Institute and Yale University. There Freddie Gutfreund put him in touch with Max Perutz at the MRC Unit in Cambridge. He demonstrated that sickle-cell anemia is associated with the change in a single amino acid in hemoglobin. In 1958 Ingram worked at MIT, where he remained for the rest of his career.

**Jacob, François,** b. Nancy, France (1920– ). Jacob was educated in Paris; his medical studies were interrupted by World War II. He served as a medical officer and was wounded in Africa and later in Normandy. His injuries preventing him from practicing as a surgeon, he found a new career in microbiological research. Joining the staff of the Pasteur Institute in 1950, his collaboration with Elie Wollman laid the foundation for the discovery of the mechanism controlling gene expression, for which Jacques Monod and François Jacob won the Nobel Prize in 1965. He met Crick when he attended the London symposium of 1957 and heard his lecture on protein synthesis. Their famous meeting in 1960 gave birth to the concept of messenger RNA.

**Keilin, David,** b. Moscow (1887–1963). Keilin studied at Magdalene College Cambridge. In 1915 he became Cambridge University's Quick Professor and in 1931 Director of the Molteno Institute. Keilin rediscovered cytochrome and played a key role in the foundation of the MRC Unit for the Study of the Structure of Biological Systems, also hosting Crick's biochemical work at his Institute.

**Kendrew, Sir John,** b. Oxford (1917–1997). Kendrew graduated in chemistry at Trinity College Cambridge (1939) and worked on radar and operational research in World War II. He joined the MRC Unit when it was formed in 1947 and became a Fellow of Peterhouse. The first two years were spent on doctoral research, completed in 1949. After eight further years and a pioneering effort, he became the first to achieve an approximate three-dimensional model of a protein. The Nobel Prize in Chemistry in 1962 was shared between Kendrew, for his structure for myoglobin, and Max Perutz for his structure for hemoglobin. Kendrew, meanwhile, was taking on other responsibilities. He was one of the founders of the *Journal of Molecular Biology*, serving as Editor-in-Chief from 1957 to 1987. He was the European Molecular Biology Organization's first Secretary-General. He advised the British Government on science and was a major figure in the foundation of the European Molecular Biology Laboratory in Heidelberg. Crick, on a number of occasions, relied on the always diplomatic Kendrew for his skill in dealing with awkward situations.

**Khorana, (Har) Gobind,** b. Raipur (Punjab) now East Pakistan (1922– ) Khorana came from a rural village, he studied chemistry of the Punjab University in Lahore, and in 1945 moved to Liverpool, UK, where he received his doctorate in 1948. He revisited the UK again in 1950 to work with Lord Todd and G.W. Kenner. Next he moved to Vancouver and in 1960 to Wisconsin. Crick visited him on a number of occasions there in the course of Khorana's exciting researches on the genetic code. He is now emeritus professor at MIT.

**Klug, Sir Aaron,** b. Želva, Lithuania (1926– ). Klug, who was raised in Durban, South Africa, is a graduate of the University of the Witwatersrand, Johannesburg. He began the study of crystallography at Capetown and continued it at Cambridge, where he received his doctorate. Postdoctoral research took him to Birkbeck College London, where he worked with Rosalind Franklin on the structure of viruses. After Klug moved to Cambridge in 1962, he and his group worked on the self-assembly of tobacco mosaic virus and the structure of tRNA. In 1982 he won the Nobel Prize in Chemistry for his work on self-assembly and his invention of crystallographic electron microscopy. He directed the LMB from 1986 to 1996 and was President of the Royal Society from 1995 to 2000.

**Koch, Christof.** b. Kansas City, Missouri (1956– ). Koch, who was brought up in Morocco and educated by Jesuits, received his doctorate for research into nonlinear information processing from the Max Planck Institute in Tübingen, Germany in 1982. Moving to the Artificial Intelligence Laboratory at MIT in 1982, and to Caltech in 1986, he is now Lois and Victor Troendle Professor of Cognitive and Behavioral Biology there. Koch collaborated with Crick in his latter years on the neural basis of consciousness, publishing their first joint paper in 1990.

**Kornberg, Arthur,** b. Brooklyn, New York (1918–2007). Kornberg, with a medical degree from the University of Rochester, began his research career at the National Institutes of Health (1942–1952). Then he chaired the microbiology department at Washington University, St. Louis (1953–1959). He moved to Stanford University in 1959, chairing the biochemistry department for a decade and continuing to work in the lab until the end of his life. He first met Crick at a conference in Baltimore in 1956. Kornberg's work on DNA polymerase played a major part in bringing biochemistry and molecular biology into fruitful relation. He won the 1959 Nobel Prize in Physiology or Medicine.

**Kornberg, Roger,** b. St. Louis, Missouri (1947– ). Roger Kornberg is the eldest son of Arthur Kornberg. He graduated from Harvard University in 1967 and moved to Stanford for doctoral research (Ph.D., 1972). His postdoctoral research at the MRC LMB, Cambridge, led him to discover the nucleosome, the histone particle forming the beads-on-a-string of chromosomal DNA in 1974. In 1976 Kornberg returned to Harvard and two years later he moved to Stanford, where the research of his group led to the discovery of the structural details of the transcription of DNA into RNA, and the central processing unit of gene transcription called the mediator. In 2006 these achievements were rewarded with the Nobel Prize in Chemistry.

**Kreisel, Georg,** b. Graz, Austria (1923– ). Kreisel's parents sent him to England before the *Anschluss*, and he graduated in mathematics at Trinity College Cambridge in 1947. During the war he worked for the Admiralty. He met Crick at Havant and they became lifelong friends. Kreisel taught at the Universities of Reading, Princeton, Paris, and Los Angeles and finally settled at Stanford (1958–1959, 1962–1985), now Professor Emeritus.

**Luzzati, Vittorio,** b. Gênes, Italy (1923– ) Luzzati's family emigrated to Buenos Aires, Argentina, before World War II. There he trained at the Colegio Nacionale Mariano Moreno, but returned to Europe in 1947. Settling in France, he worked for the CNRS in Paris and Strasbourg and finally at the Centre de Génétique Moléculaire at Gif-sur-Yvette.

**Magdoff, Beatrice** (née Fairchild). Magdoff trained with Lindo Patterson and then worked as a postdoc for David Harker's Protein Structure Project at the Brooklyn Polytechnic. Subsequently at Columbia University, New York, she made important contributions to the X-ray crystallography of sickle-cell hemoglobin.

**Marr, David,** b. Woodford, Essex, UK (1945–1980). Marr studied mathematics at Trinity College, Cambridge and, for his doctorate, developed a functional model of the cerebellum (1969). In 1976 he moved to the Artificial Intelligence Laboratory of MIT. Crick knew him in Cambridge and they renewed their acquaintance in the United States in 1979. His well known book, *Vision,* was published posthumously in 1982, in which he described his computational approach to the subject and his conception of levels of analysis.

**Massey, Sir Harrie,** b. Melbourne, Australia (1908–1983). Massey, who was trained at the University of Melbourne, obtained his doctorate in physics at Cambridge University in 1932. He was appointed a lecturer in physics at Queens University, Belfast (1933–1938), Goldsmid Professor of Mathematics at University College London (1938–1950), and Quain Professor of Physics (1950–1972). Crick worked under his direction in the Mine Design Department of the Royal Navy until Massey was sent to the United States to work on the Manhattan Project (the development of the first atomic bomb).

**McClelland, James (Jay) L.** (1948– ). McClelland, trained in cognitive psychology at the University of Pennsylvania, received his doctorate in 1975. After a decade teaching at the University of California at San Diego, he moved to Carnegie Mellon University, Pittsburgh. There he directed the Center for the Neural Basis of Cognition until moving back to California in 2006 to join the psychology department of Stanford University, where he is the founder and Director of the Center for Mind, Brain and Computation. The two-volume work *Parallel Distributed Processing* (1986) that he edited with David Rumelhart has been particularly influential.

**Medawar, Sir Peter,** b. Rio de Janeiro, Brazil (1915–1987). Trained in zoology at Magdalen College Oxford, Medawar held the Chair of Zoology at Birmingham University (1947–1951) and that of University College London (1951–1962), after which he directed the National Institute for Medical Research. In 1960 he was awarded the Nobel Prize in Physiology or Medicine with Sir Frank Macfarlane Burnet for his work on tissue graft rejection and acquired immune tolerance, work that had begun in World War II on the treatment of burn victims. A stroke in 1969 curtailed his stellar career, but he continued writing and research, adding to his impressive contributions on philosophical and speculative scientific topics for the general reader. He proved an admirable foil to Crick's more extreme views on eugenics.

**Mellanby, Sir Edward,** b. West Hartlepool, U.K. (1884–1955). Mellanby studied physiology at Emmanuel College Cambridge and medicine at St. Thomas's Hospital in London. Research that he carried out into the cause of rickets led him to the conclusion that a dietary factor formed by exposure to sunlight was missing in cases in which rickets occurred. This important discovery established Mellanby's reputation, and after 13 years as professor of pharmacology at the University of Sheffield, he was appointed secretary of the MRC in 1933. In 1947 he was able to arrange funding for Crick to retrain in biophysics.

**Mitchison, Graeme,** (1944– ). Mitchison, who was trained as a mathematician, joined the LMB in 1969 and worked on pattern formation. His work with Crick in La Jolla in 1981 led to their theory of the function of dreaming as "reverse learning."

**Monod, Jacques,** b. Paris (1910–1976). Monod was a French biologist who worked at the Sorbonne and the Institut Pasteur. From the study of growth, he turned to induced enzyme synthesis in bacteria and with colleagues discovered the mechanism of enzyme induction. For this work he shared the Nobel Prize in Medicine or Physiology in 1965 with André Lwoff and François Jacob. He was a good friend of Crick's, an experienced yachtsman, and an atheist.

**Montefiore, Hugh,** b. London (1920–2005). Montefiore was Fellow and Dean of Gonville and Caius College, Cambridge (1954–1963), Vicar of Great St. Mary's (University) Church, and Bishop of Birmingham 1978–1987. He was a controversial figure in the Anglican church, and he played a vital role in the fight to build a chapel at Churchill College, Cambridge.

**Mott, Sir Nevill Francis,** b. Leeds, U.K. (1905–1996). Mott, who trained as a physicist at Cambridge with Ernest Rutherford, was professor of physics at Bristol University (1933–1954); he then succeeded Bragg to the Cavendish Chair in

Cambridge (1954–1971). He served as Master of Gonville and Caius College (1959–1966). For Mott's work on semiconductors, he, with Philip W. Anderson and John H. Van Vleck, was awarded the Nobel Prize in Physics in 1977.

**Nirenberg, Marshall,** b. New York (1927– ). Nirenberg was trained in zoology at the University of Florida, Gainesville, and in biochemistry at the University of Michigan, Ann Arbor. He joined the National Institutes of Health, Bethesda, as a postdoctoral fellow in 1957. Two years later he and Heinrich Matthaei began their study of protein synthesis that in 1961 led to the first step in breaking the genetic code. Nirenberg went on to make major contributions to solving the rest of the code. For this work he was awarded the Nobel Prize in Physiology or Medicine in 1968.

**Orgel, Leslie,** b. London (1927–2007). Orgel studied chemistry at Oxford University, completing his doctorate in 1951 with work on bonding in transition metals. In 1955 he moved to the theoretical chemistry department at Cambridge University. He first met Crick in April 1953, when Orgel and three friends went from Oxford to see the DNA model at the Cavendish Laboratory. As he came to know Crick, Orgel became fascinated by the mystery of the origin of life. In 1964 he accepted the invitation to move to the Salk Institute, where he directed the Chemical Evolution Laboratory.

**Patterson, (Arthur) Lindo,** b. Nelson, New Zealand (1902–1966). After graduating from McGill University, Patterson learned X-ray crystallography at the Royal Institution under Sir William Bragg in London (1924–1926). In Berlin-Dahlem, Germany, he studied fiber X-ray analysis. Patterson, who is famous for his introduction of the Patterson Function (1934), first met Crick in 1954. Patterson analysis has often been used when attempting to solve the structure of biochemically important compounds. Crick was well aware of the limitations of the method and warned his boss, Max Perutz, about these limitations in connection with his work on hemoglobin.

**Pauling, Linus,** b. Portland, Oregon (1901–1994). Pauling was Caltech's famous physical chemist who in the 1930s broke down the distinction between ionic and covalent bonds and introduced the idea of resonance between single and double bonds. These concepts laid the foundations for his research into the structure of proteins for which he won the Nobel Prize in Chemistry in 1954. To Crick, it seemed in the 1940s and 1950s that Pauling was showing the way forward in applying chemistry to biology. The two men first met in Cambridge in 1952.

**Perutz, Max,** b. Vienna (1914–2002). Perutz, a chemist and molecular biologist, was Director of the MRC Unit for the Study of the Structure of Biological Systems that grew into the MRC Laboratory of Molecular Biology. He was trained at the University of Vienna and came to Cambridge for doctoral research. He met J.D. Bernal there and began his long career using X-ray crystallography to search for the structure of the protein, hemoglobin. His success came in 1959, and he shared the Nobel Prize in Chemistry with John Kendrew three years later. Behind the scenes, Perutz loyally defended his research student, Francis Crick, and, as director of the MRC Unit in Cambridge, he continued to look after Crick's career with the Medical Research Council.

**Poggio, Tomaso,** b. Genoa, Italy (1947– ). Poggio, who trained in physics in Italy, moved to Tübingen in 1971 to work at the Max Planck Institute for Biological Cybernetics. There he first met Crick in 1977 on one of his visits to Tübingen. He moved to the Artificial Intelligence Laboratory at MIT in 1981 and is now Eugene McDermott Professor in Brain Sciences and Human Behavior and Director of the Center for Biological and Computational Learning at MIT. Like Crick, Poggio believes that no theories "developed in a vacuum have any chance of success in brain science."

**Pontecorvo, Guido,** b. Pisa, Italy (1907–1999). Pontecorvo was trained at the University of Pisa; his early research was in animal genetics. When anti-Semitism became strong, he was dismissed from his teaching position in Florence. Emigrating to Edinburgh, he turned to the genetics of the fruit fly (*Drosophila*). There, with H.J. Muller and other émigrés, "Ponte" revolutionized Edinburgh's genetical research program. In 1945 he was appointed Glasgow's first lecturer in genetics, and from 1955 to 1968 he occupied the chair of genetics, building an international reputation for his pioneer work in microbial genetics, working on the mold *Aspergillus*.

**Randall, Sir John,** b. Newton-le-Willows, St. Helens, Lancashire, U.K. (1905–1984). Randall was a physicist who worked on shortwave radio and invented the cavity magnetron, a tool of great value to radar in World War II. After the war, he became Professor of Physics at St Andrews University in Scotland. In 1947 he moved to the physics chair at King's College London. There the MRC supported his bid to establish a research unit in biophysics. Maurice Wilkins and Rosalind Franklin worked in this Unit.

**Rich, Alexander,** b. Hartford, Connecticut (1925– ). Rich, who was trained at Harvard, worked at the National Institute of Mental Health (NIMH) during World War II. In 1949 he joined Pauling's group at Caltech, where he began work on the structure of nucleic acids and collaborated with James Watson on the structure of RNA. After a further four years at the NIMH, in 1958 he was appointed Associate Professor of Biophysics at MIT, where he has remained. In 1955 Rich spent six months with Crick in Cambridge working on the structure of polyglycine and collagen. Returned to Bethesda, he and David Davies succeeded in hybridizing two RNA polymers in 1956. Rich's group was later to find themselves in competition with Aaron Klug's group at the LMB in Cambridge over the structure of tRNA.

**Sanger, Frederick,** b. Rendcombe, Gloucestershire, U.K. (1918– ). Sanger, who was trained in biochemistry at Cambridge, took advantage of the newly developed separation techniques of partition chromatography to attack the structure of proteins. In 1958 he was awarded the Nobel Prize in Chemistry for his discovery of the amino acid sequence of the polypeptide chains that constitute the protein insulin. Crick often visited Sanger's laboratory in Cambridge to learn about the progress of his work and to discuss it with him. In 1962, Sanger was persuaded to join the former MRC Unit when it became the Laboratory of Molecular Biology on the Addenbrooke's site. In 1980 he won his second Nobel Prize in Chemistry for his development of sequence analysis in the nucleic acids.

**Schrödinger, Erwin,** b. Erdberg, Vienna (1887–1961). Schrödinger was educated at the University of Vienna 1906–1910. His research in physics plus a period of military service in World War I occupied him until 1918, when he obtained his Habilitation. In 1927, by which time he had held several physics chairs, he succeeded Max Planck to the chair at Berlin. This stellar rise was brought to an end by the rise of Hitler, and after several moves Schrödinger settled in Dublin as Director of the new School for Theoretical Physics (1940–1955). For his celebrated wave equation, he was awarded the Nobel Prize in Physics in 1933. His lectures in Dublin on the theme "What is Life?," published as a book in 1944, influenced Crick when he was thinking of moving to biology.

**Speed, Alfred Valentine,** b. King's Lynn, Norfolk, U.K. (1899–1974). Odile's father. He succeeded his father, Alfred Speed, in the family business of Speed and Sons, goldsmith and watchmaker.

**Speed, Marie-Therese** (née Jaeger), b. Paris (1894–1958). Odile's mother. She married Alfred Valentine Speed in 1919.

**Speed, Philippe,** b. King's Lynn, Norfolk, U.K. (1925– ). Odile's younger brother. He served in the rank of officer in the Indian army after the war, and then trained in horology before entering his father's business.

**Stanley, Wendell M.,** b. Ridgeville, Indiana (1904–1971). Stanley was an American virologist who trained as a chemist in the United States and Germany. Stanley moved to the Rockefeller Institute in 1931, where he was the first to crystallize a virus, and to the University of California at Berkeley in 1948, where he was appointed Professor of Biochemistry and Director of the Virus Laboratory. Crick lectured to his group on DNA in 1954.

**Streisinger, George,** b. Budapest, Hungary (1927–1984). Streisinger's family emigrated to the United States in 1937. He graduated from the Bronx High School of Science in 1944 and entered Cornell University. He moved to the University of Illinois genetics department, directed by Salvador Luria, for doctoral research. After four years at Caltech, Streisinger moved to Cold Spring Harbor Laboratory, where, apart from the academic year 1957–1958 spent in Cambridge with Brenner, he remained until 1960. Then he went to the University of Oregon in Eugene.

**Todd, Alexander, Baron,** b. Cathcart, Glasgow (1907–1997). Todd studied chemistry at Glasgow University and undertook doctoral research at the universities of Frankfurt-am-Rhine and Oxford. After several moves—the University of Edinburgh, The Lister Institute, London, and the University of Manchester—he was appointed to the chair of organic chemistry at Cambridge. Todd, a world authority on nucleosides and nucleotides, came to know Crick in 1953.

**Watson, James,** b. Chicago, Illinois (1928– ). Watson entered the University of Chicago High School (Robert Hutchins' Four Year College) in 1941, and at 15 he was accepted under the early admissions program into the University. Upon graduation in 1947, he entered the genetics program at the University of Indiana under Salvador Luria, obtaining his doctorate in 1950. Postdoctoral work took him to Copenhagen and in October 1951, to Cambridge, where he met Crick. Their discovery of the structure of DNA in 1953 gave proof of Watson's abilities, leading to his appointment to Harvard's Department of Biology in

1955. In 1968 he accepted the post of Director of the Cold Spring Harbor Laboratory, although he did not resign from Harvard until 1976. He became Chancellor Emeritus of the Laboratory in 2007.

**Wilkins, Frank Foster,** b. Northampton, U.K. (1850–1913). Wilkins was the founder of Wilkins and Darking Ltd, and Francis Crick's maternal grandfather.

**Wilkins, Maurice,** b. Pongaroa, New Zealand (1916–2004). Wilkins' family moved to Birmingham, U.K., in 1923. Wilkins, while a scholar at St. John's College, Cambridge, studied physics, followed by research at Birmingham University under John Randall. After war work in Birmingham and in Berkeley, California, he rejoined Randall in the physics department of St Andrews University, Scotland. In 1946 he followed Randall to the new MRC Biophysics Unit at King's College London where he succeeded Randall as Director in 1972. Wilkins and Crick met for the first time in 1947 and became lifelong friends.

# Index